WETLAND ENVIRONMENTS: A GLOBAL PERSPECTIVE

COMPANION WEBSITE

This book has a companion website:
www.wiley.com/go/aber/wetland
with Figures and Tables from the book

Wetland environments:
A global perspective

James Sandusky Aber
Emporia State University, Kansas

Firooza Pavri
University of Southern Maine

Susan Ward Aber
San José State University, California

A John Wiley & Sons, Ltd., Publication

This edition first published 2012 © 2012 by James Sandusky Aber, Firooza Pavri and Susan Ward Aber.

Blackwell Publishing was acquired by John Wiley & Sons in February 2007. Blackwell's publishing program has been merged with Wiley's global Scientific, Technical and Medical business to form Wiley-Blackwell.

Registered office:
John Wiley & Sons, Ltd, The Atrium, Southern Gate, Chichester, West Sussex, PO19 8SQ, UK

Editorial offices:
9600 Garsington Road, Oxford, OX4 2DQ, UK
The Atrium, Southern Gate, Chichester, West Sussex, PO19 8SQ, UK
111 River Street, Hoboken, NJ 07030-5774, USA

For details of our global editorial offices, for customer services and for information about how to apply for permission to reuse the copyright material in this book please see our website at www.wiley.com/wiley-blackwell.

The right of the author to be identified as the author of this work has been asserted in accordance with the UK Copyright, Designs and Patents Act 1988.

All rights reserved. No part of this publication may be reproduced, stored in a retrieval system, or transmitted, in any form or by any means, electronic, mechanical, photocopying, recording or otherwise, except as permitted by the UK Copyright, Designs and Patents Act 1988, without the prior permission of the publisher.

Designations used by companies to distinguish their products are often claimed as trademarks. All brand names and product names used in this book are trade names, service marks, trademarks or registered trademarks of their respective owners. The publisher is not associated with any product or vendor mentioned in this book. This publication is designed to provide accurate and authoritative information in regard to the subject matter covered. It is sold on the understanding that the publisher is not engaged in rendering professional services. If professional advice or other expert assistance is required, the services of a competent professional should be sought.

Library of Congress Cataloging-in-Publication Data
Aber, James S.
  Wetland environments : a global perspective / James Sandusky Aber, Firooza Pavri, Susan Ward Aber.
     p. cm.
  Includes bibliographical references and index.
  ISBN 978-1-4051-9841-7 (cloth) – ISBN 978-1-4051-9842-4 (pbk.) 1. Wetlands. 2. Wetland ecology. I. Pavri, Firooza. II. Aber, Susan Ward. III. Title.
  QH87.3.A24 2012
  551.41'7–dc23
                                        2012010853

A catalogue record for this book is available from the British Library.

Wiley also publishes its books in a variety of electronic formats. Some content that appears in print may not be available in electronic books.

Cover image: Panorama of Russell Lakes State Wildlife Area with the Sangre de Cristo Range in the far background; south-central Colorado, United States. Saline lakes and marshes occupy hollows between low, mesquite-covered dunes. Blimp aerial photo by James Aber and Susan Aber.
Cover design by: Design Deluxe

Set in 9.5pt ITC Garamond Std by Toppan Best-set Premedia Limited, Hong Kong

DEDICATION

In memory of Kiira Aaviksoo (1955–2011), who pursued her passion for wetlands, stimulated a generation of students, and inspired our interest in Estonian bogs.

Contents

Preface: Why wetlands? xi
Acknowledgements xiv

Part I
1 Wetland overview 1
  1.1 Introduction 1
  1.2 How much and where 4
  1.3 Wetland trends 8
  1.4 Wetland preservation and protection 10
  1.5 Wetland science 11
  1.6 Book approach and outline 13
  1.7 Summary 13
2 Wetland criteria 15
  2.1 Definitions 15
  2.2 Water 16
  2.3 Soil 18
  2.4 Vegetation 20
  2.5 Wetland classification 22
  2.6 Peatland 24
  2.7 Anthropogenic dimensions of wetlands 27
  2.8 Summary 28
3 Methods in wetland research 30
  3.1 Introduction 30
  3.2 Remote sensing 31
    3.2.1 Image resolution and interpretability 32
    3.2.2 Wetland image interpretation 35
    3.2.3 Macro-level systems 37
    3.2.4 Meso-level systems 41
    3.2.5 Micro-level systems 42
  3.3 Maps and geographic information systems 46
  3.4 Physical ground-based methods 48
    3.4.1 Surface methods 48
    3.4.2 Subsurface methods 51
  3.5 Flora, fauna and ecological monitoring and survey methods 53
  3.6 Social science methods and techniques 56
  3.7 Summary 57

Part II
4 Wetland hydrology 58
  4.1 Introduction 58
  4.2 Surface and ground water 61
  4.3 Floods and flooding 64
  4.4 Hydrologic functions of streams and wetlands 65
  4.5 Hydrochemistry 68
  4.6 Summary 70
5 Wetland soil 72
  5.1 Introduction 72
  5.2 Brief history and soil classification 73
  5.3 Hydric soil criteria 78
  5.4 Mineral and organic hydric soils 79
  5.5 Submerged wetland substrates 81
  5.6 Summary 82
6 Wetland vegetation 86
  6.1 Plant adaptations 86
    6.1.1 Structural adaptations 86
    6.1.2 Biochemical adaptations 88
  6.2 Ecological categories 89
    6.2.1 Shoreline plants 90
    6.2.2 Emergent plants 92
    6.2.3 Floating plants 94

		6.2.4	Submerged plants	96
		6.2.5	Plant zonation	97
	6.3	Indicator categories		99
	6.4	Plant hardiness zones		101
	6.5	Invasive plant species		102
	6.6	Summary		105
7	Wetland wildlife			107
	7.1	Introduction		107
	7.2	Wetland invertebrates		108
		7.2.1	Insects	108
		7.2.2	Mosquitos	110
		7.2.3	Corals	111
	7.3	Wetland vertebrates		113
		7.3.1	Amphibians	113
		7.3.2	Reptiles	115
		7.3.3	Birds	117
		7.3.4	Mammals	125
	7.4	Invasive animal species		130
	7.5	Summary		132

Part III

8	Wetland change			135
	8.1	Introduction		135
	8.2	Hydroseral succession		136
	8.3	Sea-level change and crustal movements		139
		8.3.1	Glacial eustasy	139
		8.3.2	Glacial isostasy	141
		8.3.3	Complicated responses	141
		8.3.4	Modern sea-level rise	143
	8.4	Climate change		147
		8.4.1	Climate basics	147
		8.4.2	Climate and wetlands	150
	8.5	Fire		152
	8.6	Summary		154
9	Wetlands through time			156
	9.1	Introduction		156
	9.2	Coal		157
		9.2.1	Paleozoic coal	157
		9.2.2	Cretaceous–Tertiary coal and lignite	158
	9.3	Amber		160
	9.4	Pleistocene and Holocene wetlands		163
		9.4.1	Nordic region	164
		9.4.2	North America	167
		9.4.3	Tropics and Antarctica	171
		9.4.4	Holocene climate and early man	171
	9.5	Summary		173
10	Environmental cycles and feedback			175
	10.1	Biogeochemical cycles		175
		10.1.1	Wetland elements	175
		10.1.2	Nitrogen	175
		10.1.3	Phosphorus, potassium and sulfur	177
	10.2	Carbon cycle		179
		10.2.1	Carbon reservoirs	179
		10.2.2	Carbon balance	179
		10.2.3	Carbon gases and climatic feedback	180
	10.3	Fossil fuels		181
		10.3.1	Fossil-fuel consumption	181
		10.3.2	Coal mining and acid rain	183
		10.3.3	Estonian oil shale	184
	10.4	Human experiment		185
	10.5	Summary		187

Part IV

11	Wetland services, resources and valuation			190
	11.1	Human use of wetland ecosystems		190
	11.2	Ecosystem services		191
		11.2.1	Habitats	191
		11.2.2	Wetlands and biogeochemical cycles	194
		11.2.3	Storm surge and coastal flood protection	195
	11.3	Hydrological services		196
		11.3.1	Flood abatement	196
		11.3.2	Water quality	196
		11.3.3	Water storage and diversion	197
	11.4	Economic services		199
		11.4.1	Extractive industries	199
		11.4.2	Pearl production	204
		11.4.3	Services industries	207
	11.5	Wetland valuations		211
		11.5.1	Why value wetlands?	212
		11.5.2	Property regimes and externalities in wetland use and valuations	212

		11.5.3	How to value wetlands?	215
	11.6	Summary		216
12	Conservation and management: Wetland planning and practices			218
	12.1	The conservation movement		218
	12.2	Wetland resource management		220
	12.3	Wetland management plans		221
	12.4	Wetland management practices		223
		12.4.1	Terrestrial and hydrologic-based strategies	223
		12.4.2	Biological and chemical strategies	226
		12.4.3	Socio-economic strategies	227
	12.5	Summary		229
13	Wetland restoration, enhancement and creation			231
	13.1	Introduction		231
	13.2	Terminology		232
	13.3	Wetland restoration, enhancement and creation design principles		235
	13.4	Restoration and enhancement considerations		238
	13.5	Approaches to wetland restoration and enhancement		240
		13.5.1	Active approaches	240
		13.5.2	Passive and hybrid approaches	242
	13.6	Artificial treatment wetlands		244
	13.7	Contaminated mine-water treatment		246
	13.8	Summary		249
14	Wetlands governance and public policy			251
	14.1	Wetlands governance and policy		251
	14.2	International wetland policy		251
	14.3	Wetland policy in the developed world		252
		14.3.1	United States	253
		14.3.2	Canada	255
		14.3.3	Western Europe	255
		14.3.4	Central Europe	257
		14.3.5	Commonwealth of Independent States	260

		14.3.6	Australia, New Zealand and Antarctica	262
	14.4	National wetland policy in the developing world		264
	14.5	Shared wetlands		264
	14.6	Summary		266

Part V

15	Low-latitude wetland case studies			268
	15.1	Introduction		268
	15.2	Sundarbans of South Asia		269
	15.3	Okavango Delta of southern Africa		274
	15.4	Pantanal of South America		276
	15.5	Gulf of Mexico, United States		279
		15.5.1	Florida Everglades	281
		15.5.2	Mississippi River delta	285
		15.5.3	Padre Island and Laguna Madre	291
	15.6	Summary		297
16	Middle-latitude wetland case studies			299
	16.1	Introduction		299
	16.2	Great Plains of North America		299
		16.2.1	Upper Arkansas River valley, Colorado and Kansas	300
		16.2.2	Biocontrol of saltcedar along the upper Arkansas River valley	305
		16.2.3	Cheyenne Bottoms, Kansas	309
		16.2.4	Nebraska Sand Hills	314
		16.2.5	Missouri Coteau, southern Saskatchewan	318
	16.3	Coastal wetlands of Maine and Massachusetts, United States		323
		16.3.1	Wells Reserve, southeastern Maine	325
		16.3.2	Plum Island Ecosystem, northeastern Massachusetts	326
	16.4	Estonia, eastern Baltic region		328
	16.5	Summary		333
17	High-latitude and high-altitude wetland case studies			336
	17.1	Introduction		336
	17.2	Andes Mountains, Venezuela		336

17.3	Southern Colorado, United States	340	18.3 Key opportunities in wetland conservation 362
	17.3.1 Culebra Range	341	18.4 Future directions 363
	17.3.2 San Luis Valley	345	
17.4	The Arctic	350	Glossary of wetland types and terms 364
	17.4.1 Arctic Coastal Plain, Alaska	351	References 372
	17.4.2 Yukon Delta, Alaska	353	Index 401
	17.4.3 Lena River delta, Russia	354	Color Plates are between pages 210 and 211
17.5	Summary	357	
18	Sustainability for wetlands	358	
18.1	Introduction	358	
18.2	Key risks to wetlands	359	

COMPANION WEBSITE

This book has a companion website:
www.wiley.com/go/aber/wetland
with Figures and Tables from the book

Preface

Why wetlands?

This question is often asked of the authors. Swamp, marsh, tundra, bog – these are places that are rarely visited and as such, perceptions prevail over observations so that people cannot visualize how wetlands might figure into their everyday lives. Wetlands simply do not appeal to most people from a practical or aesthetic point of view. Among the large urban and rural human populations of today, few are familiar with wetlands and fewer still have an active interest in understanding, enjoying, or protecting wetlands. In fact, for most people wetlands are wastelands – in other words, places to be converted, drained, filled, or exploited for industrial and economic uses.

For us, however, wetlands are intrinsically beautiful environments where one may see the natural and essential values in the interaction of water, soil, vegetation, wildlife, and humans (Fig. 1). Furthermore, individual wetlands are small pieces within the Earth's complex environment, a system that sustains us as well as all other life. At the transition from drylands to deep-water bodies, wetlands provide key links for the flux and temporary storage of energy and materials as well as crucial habitats for many plant and animal species. In addition to their modern environmental roles, wetlands also preserve in their sediments and fossils proxy records of past environments and climatic conditions, which help us to know how the present came to be and how the future might be. We are drawn to wetlands to observe and study their critical past and present environmental roles as well as to describe and enjoy their unique beauty.

The wetland-as-wasteland point of view may have been acceptable in the past, when human population was small and environmental effects were less understood and perceived to be insignificant. But this is no longer the case. Human population has surged from around one billion two centuries ago to seven billion today and will continue to grow rapidly in the near future. Population has more than doubled in the past half century, during which time global food and fresh-water consumption have more than tripled and use of fossil fuels has increased fourfold. Humans now co-opt at least one-third to as much as half of global photosynthesis (Foley 2010). In short, our exploitation of planetary resources has surpassed our quickly expanding population.

All aspects of the Earth's environmental system are impacted by this human assault on the planet, which includes vast conversions of land use, massive extraction of mineral resources and fossil fuels, heavy use of surface and ground water, and uncontrolled exploitation of many other non-renewable land and marine resources. The natural flows of energy and materials within the environment have been altered or disrupted as a result of modern human development.

Rockström et al. (2009a and 2009b) identified key processes for maintaining a sustainable global environment (Table 1). The boundaries set for these factors represent "tipping points" beyond which uncertain or irreversible consequences may take place. Three processes have

Figure 1. Aerial overview of the Rachel Carson National Wildlife Refuge along the Atlantic coast of southeastern Maine, United States (see Color Plate 1). The salt marsh, pools and tidal channels intervene between the beach front (right) and mainland (left), both of which have dense residential development. The human presence here has strong influence on the wetland water supply, vegetation and wildlife. View toward north; blimp airphoto by J.S. Aber, S.W. Aber and V. Valentine.

already exceeded their boundaries – biodiversity loss, nitrogen pollution, and atmospheric CO_2 increase – and other factors are approaching their limits (Foley 2010).

Of these critical processes, fresh-water use is most directly related to the subject of this book, namely wetlands. Several others are tied directly or indirectly to wetland environments as well – biodiversity loss, carbon cycle, nitrogen and phosphorus cycles, land use, ocean acidification, and chemical pollution. Wetlands are not only unique as individual environments, they also form critical connections between drylands and deep-water bodies with complex interactions and feedback relationships involving the atmosphere, hydrosphere, biosphere and lithosphere.

Why wetlands? We hope this book will foster a greater awareness and appreciation of wetlands, promote a culture of conservation and wise management, and spread the knowledge that wetlands are important, indeed crucial, elements of the global environment. Our attempts to understand, manage and enhance wetlands in the twenty-first century are parts of the larger effort to maintain a sustainable Earth for all people.

Table 1. Critical environmental processes, their boundaries, potential consequences, and possible solutions.
* process that has exceeded its boundary value; based primarily on Foley (2010).

Process	Measurable rate or quantity			Consequences	Solutions
	Pre-industrial	Current	Boundary		
Biodiversity loss*	Extinction rate (species per million per year)			Land and marine ecosystems fail	Slow land clearing and development; pay for ecosystem services
	0.1 to 1.0	>100	10		
Carbon cycle*	Atmospheric CO_2 concentration (ppm)			Ice sheets, glaciers, sea ice and permafrost melt; global rise in sea level; regional climatic shifts	Switch to low-carbon fuels and renewable energy; pay for carbon emissions
	280	387	350		
Nitrogen cycle*	Removal from atmosphere (million tons per year)			Fresh-water and marine dead zones expand	Reduce fertilizer use; process animal waste; switch to hybrid vehicles
	0	133	39		
Phosphorus cycle	Removal from atmosphere (million tons per year)			Marine food chains disrupted	Reduce fertilizer use; process animal and human waste better
	1	10	12		
Land use	Percentage converted to cropland			Ecosystems fail; CO_2 escapes	Limit urban sprawl; improve farm efficiency; pay for ecosystem services
	Negligible	11.7	15		
Ocean acidification	Aragonite saturation in surface water (Omega units)			Coral reefs and microorganisms die; carbon sink reduced	Switch to low-carbon fuels and renewable energy; reduce fertilizer runoff
	3.44	2.90	2.75		
Fresh-water use	Rate of human consumption (km^3 per year)			Aquatic ecosystems fail; water supplies disappear	Improve irrigation efficiency; install low-flow appliances
	415	2600	4000		
Ozone depletion	Stratospheric ozone concentration (Dobson units)			Radiation harms humans, plants and animals	Phase out hydrochlorofluorocarbons; test effects of new chemicals
	290	283	276		
Aerosol loading	Particulate concentration in atmosphere (values to be detemined)			Regional climatic shifts; variable and unpredicable	Reduce atmospheric emissions from power plants and vehicles
Chemical pollution	Amount emitted to or concentrated in environment (values to be detemined)			Toxic contamination; variable and unpredictable	Reduce all chemical emissions to the air, water, and ground

Acknowledgements

Many colleagues, friends and students have encouraged, supported, and assisted our work on wetland environments in various ways over many years, which resulted in this book. We thank the following in particular: Inge Aarseth, Kiira Aaviksoo, David Ackerman, Susan Adamowicz, Karl Anundsen, Andrzej Ber, Max Bezada, Lawrence A. Brown, Marc Carullo, Dan Charman, Abraham Dailey, Gayla Corley, David Croot, Michele Dionne, Debra Eberts, David Edds, Tom Eddy, Elder family, Jack Estes, Mark Fenton, Darek Gałązka, Marco Giardino, Maria Górska-Zabielska, Lixiao Huang, Juraj Janočko, Volli Kalm, Hemant Karkhanis, Edgar Karofeld, Barbara Kosmowska-Ceranowicz, Peder Laursen, Brooks Leffler, Dave Leiker, Linda Lillard, Kam Lulla, Holger Lykke-Andersen, Maya Mahajan, Gina Manders, Viktor Masing, Robert Nairn, Atle Nesje, Kate O'Brian, Lida Owens, Robert Penner, Matt Nowak, Tinaz Pavri, Johannes Ries, Tom, Darius and Ethan Rotnem, Hanna Ruszczyńska-Szenajch, David and Mary Sauchyn, Hans and Ingeborg Schlichtmann, Jean Schulenberg, Greg and Lynette Sievert, Rich Sleezer, Eva Stupnicka, Marsh Sundberg, Michele Tranes, Cheryl Unruh, Vinton Valentine, Steven Veatch, Elena Volkova, Ryszard Zabielski, Brenda Zabriskie, and Mark Zwetsloot.

We especially thank our colleagues and students who have contributed their excellent wetland photographs for this book: Jeremy W. Aber, Sara Acosta, Roy Beckemeyer, Ken and Marilyn Buchele, Peter Frenzel, Brian Graves, Nick Hubbard, William Jacobson, Paul and Jill Johnston, Scott Jones, Preben Jensen, Edgar Karofeld, Victor Krynicki, Margaret Martin, Irene Marzolff, Johannes Ries, Shawn Salley, Mel Storm, Elena Volkova and Brenda Zabriskie. We also thank Abraham Dailey for helping produce the maps used in the book. U.S. postage and duck stamps came from the collection of Jean Vancura. Special thanks to Lawrence A. Brown for his steadfast support and for providing the opportunity to learn from him.

Financial support and logistical assistance were provided by institutional grants from Emporia State University (USA), the University of Southern Maine (USA), the University of Tartu (Estonia) and the Technical University of Košice (Slovakia) as well as by NASA EPSCoR and JOVE awards. Additional financial support for JSA was given by the National Scholarship Programme of the Slovak Republic and the U.S. National Research Council's Estonian Twinning Program.

Special thanks to our parents who encouraged us to follow our scientific ambitions: Sarajane S. and R. Kenneth Aber, Gool and Bomi Pavri, Marian M. and Henry A. Ward.

1 Wetland overview

1.1 Introduction

Wetland. The name summons immediate images or experiences to most people – from the endless sand beach of Padre Island, Texas (Fig. 1-1), to wildlife in the Okavango Delta of Botswana (Fig. 1-2), to the deadly Great Grimpen Mire, as described in the Sherlock Holmes tale, *The Hound of the Baskervilles* by Sir Arthur Conan Doyle:

> Rank weeds and lush, slimy water plants send an odour of decay and a heavy miasmatic vapor into our faces, while a false step plunged us more than once thigh-deep into the dark, quivering mire, which shook for yards in soft undulations around our feet.

Whether real or fictional, wetlands have conspicuous roles in the physical, biological and cultural geography of the world. Wetlands are places where the ground is generally saturated or flooded for extended periods during the growing season such that distinctive soils form and specialized vegetation grows under conditions in which oxygen is depleted or absent. Such environments include marshes, fens, bogs, and swamps (see chapter 2). Wetlands occur at the confluence of unique terrestrial, hydrological and climatic conditions that give rise to some of the most biodiverse regions of the world. They also play a vital role in the cycling and storage of key nutrients, materials and energy through the Earth's system. Wetland components include water, soil, vegetation, and wildlife. Since the first human hunter-gatherers camped by springs and shores, people have utilized, modified, exploited or impacted wetlands in various ways. Moreover, the early establishment of human settlements and subsequent expansion were based on irrigated agriculture along major river floodplain valleys – Nile (Fig. 1-3), Tigris–Euphrates, Niger, Indus, Mekong, etc.

Wetlands continue to be essential for modern human society; they represent the primary sources of fresh water for people in most places around the world. Wetlands minimize flooding and storm damage, nourish fisheries (Fig. 1-4), produce fur-bearing animals, sustain irrigated agriculture, support herding and grazing (Fig. 1-5), recharge aquifers, provide shipping waterways (Fig. 1-6), supply hydropower, grow timber, yield fossil fuels (Fig. 1-7), are incubators for gemstones, and provide many other resources. These functions are clearly evident, as they influence the daily lives of people living in and deriving economic benefits from wetlands.

In spite of local recognition of wetland functions and values, however, the larger regional and global significance of wetlands is more difficult for many people to fathom. What is economically beneficial in upper portions of drainage basins – irrigation, timber harvesting, hydroelectric power, recreation and other human uses – is often deleterious for downstream inhabitants of wetlands and coastal

Figure 1-1. Padre Island National Seashore, southern Texas, United States. View northward showing Padre Island (left) and the Gulf of Mexico (right). More than 70 miles (110 km) of island and beach are protected. Note person standing at bottom for scale. Kite aerial photo by J.S. Aber and S.W. Aber.

Figure 1-2. Hippopotamus (*Hippopotamus amphibius*) displays its formidable jaws and teeth in a marsh of the Okavango Delta, Botswana, southern Africa. Photo courtesy of M. Storm.

regions. Upstream manipulations and exploitation of wetland water resources have resulted in serious degradation or dramatic changes lower in drainage basins (Fig. 1-8). In contrast, some exploitations of wetlands, for example pearl farming, actually add marine life and provide protected areas that are free from dynamite and cyanide fishing. It is safe to say, though, that few, if any, major wetland systems of the world have not been altered or changed in substantial or subtle ways by human activities.

Wetlands are situated at the transitions between dry uplands and deep-water lake and marine environments (Fig. 1-9). Wetlands, thus, may be viewed as the links that bind together all other habitats at the Earth's surface, and they play key roles in the overall environmental system through transfer and storage of materials and energy. Numerous feedback relationships exist between wetlands and their surroundings.

WETLAND OVERVIEW 3

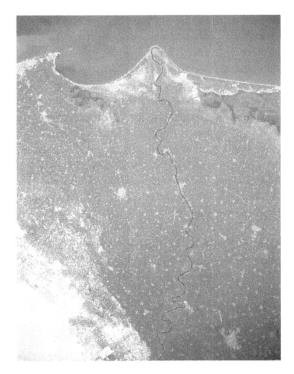

Figure 1-3. Near-vertical view of the River Nile and Mediterranean coast in the vicinity of Alexandria, Egypt. The Nile supported one of the earliest agricultural civilizations. Linhof large-format film camera, March 1990, STS36-151-101; image adapted from NASA Johnson Space Center.

Figure 1-4. Fishing nets and boats on the shore of Võrtsjärv, a large freshwater lake surrounded by marsh in southern Estonia. Photo by J.S. Aber.

Wetlands are, for example, significant sinks for carbon stored in their organic soil and sediment. They are likewise important sources for carbon dioxide (CO_2) and methane (CH_4), both greenhouse gases, released from the stored organic mass. Thus, wetlands are critical components of

Figure 1-5. Cattle grazing in the páramo (alpine) grassland-shrub vegetation in the Andes Mountains of Venezuela. The páramo zone is almost constantly in the clouds, rain or fog, as shown in the background. Photo by J.S. Aber.

Figure 1-6. Barges on the River Rhine at Andernach, Germany. Two loaded barges are moving upstream. The control house and living quarters are located at the stern of the barge. Photo by J.S. Aber.

Figure 1-7. Traditional hand cutting of peat in Ireland. After drying, the peat is used for home heating and cooking. Photo by J.S. Aber.

Figure 1-8. Impact of water diversions in the High Plains of the central United States. A. Dry channel of the Arkansas River at Ingalls, southwestern Kansas. Upstream reservoirs and extraction of water for irrigation have dried up the river, even in the spring of a wet year when this picture was taken. B. Center-pivot irrigation system. The sprayers are set too high for the winter wheat crop, so that considerable water is lost to wind drift and evaporation. C. Center-pivot irrigation system with the sprayers set just above the winter wheat crop to minimize evaporative loss. Photos by J.S. Aber.

the carbon cycle, which has significant implications for global climate.

The economic benefits and ecological functions of wetlands are numerous and varied, as noted above. For many people, nonetheless, wildlife is the most obvious and perhaps important aspect of wetlands. In some American and African wetlands, for example, millions upon millions of waterfowl and shorebirds visit briefly or remain seasonally during annual migrations. Such dramatic concentrations of wildlife have attracted hunters since prehistoric times, and hunting continues to be a major use of wetlands (Fig. 1-10). Wetlands are, in fact, among the most productive ecosystems in the world (Niering 1985).

Wetlands also harbor some of the greatest biodiversity found on the planet. Many aquatic animals are endemic to isolated wetlands, such as the hundreds of fish species found in the lakes of eastern Africa (Dugan 2005). In other cases, wetlands represent the last refuges of animals forced out of other habitats by human development – the Bengal tiger in the Sundarbans of India and Bangladesh and the jaguar in wetlands of South and Central America. Endangered species, such as the whooping crane (*Grus americana*), are often foremost in the public eye as symbols for the need to preserve wetland habitats (Fig. 1-11).

Costanza et al. (1997) attempted to estimate the economic value of ecosystem services for major biomes of the world. They identified 17 ecosystem services (Table 1-1), many of which are connected to or interact with wetland environments, particularly those involving water, soil, gases, nutrients and climate. The value of these services was determined using market and non-market means, such as the willingness-to-pay method (see chapter 11). They found that wetlands in general and estuaries, seagrass/algae beds, coral reefs, tidal marsh/mangroves, and swamps/floodplains in particular have the highest values for providing a broad array of ecosystem services. Wetland values are considerably greater, in fact, than tropical forest or other biomes.

1.2 How much and where

The total coverage of existing wetlands is estimated to range from at least 7 million km^2 to about 10 million km^2, or 5–8 percent of the land

Figure 1-9. Presque Isle is a sandy spit that extends from the mainland into Lake Erie in northwestern Pennsylvania, United States (see Color Plate 1-9). The transition from sandy shore, to shallow water, to deep lake is depicted in this panoramic view looking toward the northeast. Kite aerial photo by J.S. Aber and S.W. Aber.

Figure 1-10. Hunters in a camouflaged "duck boat" return from a venture in the marsh at Cheyenne Bottoms, central Kansas, United States. Photo by J.S. Aber.

surface of the world, depending upon the definition for what is included (Mitsch and Gosselink 2007). Bog, swamp, marsh, fen, muskeg, and similar habitats are represented in this total. The broader definition of Ramsar (see chapter 2) includes lakes, rivers, and coastal marine water bodies (up to 6 m deep), which pushes the wetland coverage to more than 12 million km^2. Peatland (mire) includes those types of wetlands that accumulate peat at least one foot (30 cm) in thickness (Fig. 1-12), which may happen in swamp, bog, muskeg, and fen environments. Peatlands cover approximately 4 million km^2 worldwide (Dugan 2005).

Figure 1-11. U.S. postage stamp issued in 1957 depicts a family of whooping cranes. Original stamp printed in blue, brown, and green colors. From the collection of J. Vancura.

Wetlands of diverse types are found in all land and coastal regions of the world; however, the distribution of wetlands is certainly not uniform (Fig. 1-13). The greatest concentration of wetlands is found in boreal and subboreal

Table 1-1. Ecosystem services and functions used for estimating the value of major biomes of the world. Based on Costanza et al. (1997, Table 1).

Ecosystem services	Ecosystem functions	Examples
Gas regulation	Atmospheric chemical composition	Oxygen, ozone, sulfur oxides, UV protection
Climate regulation	Global, regional and local weather and climate	Greenhouse gases, cloud formation
Disturbance regulation	Absorbing and damping ecosystem responses	Storm protection, flood control, drought recovery
Water regulation	Hydrological flows	Irrigation, transportation, industrial applications
Water supply	Storage and retention	Soil moisture, aquifers, streams and reservoirs
Erosion and sedimentation	Retention of soil and sediment	Prevention of soil loss, siltation in lakes
Soil formation	Soil-forming processes	Rock weathering, organic matter accumulation
Nutrient cycling	Storage, processing and transfer of nutrients	Nitrogen fixation, K and P cycles
Waste treatment	Nutrient recovery, removal of harmful substances	Pollution control detoxification
Pollination	Movement of floral gametes	Pollinators for plant reproduction
Biological control	Regulation of populations	Predator control of prey
Refugia	Habitat for resident and migratory populations	Nurseries, regional habitats, migratory routes
Food production	Gross primary production for food	Crops, fish, game, fruits and nuts, livestock
Raw materials	Gross primary production for materials	Timber, fiber, fuel, fodder, minerals and ores
Genetic resources	Unique biological materials and products	Medicines, plant and animal varieties, ornamental species
Recreation	Recreational opportunities	Ecotourism, sport fishing and hunting, bird watching
Cultural	Non-commercial uses	Artistic, aesthetic, spiritual, religious, or scientific values

Figure 1-12. Layer of postglacial peat (*) ~1m thick resting on glacial till in western Poland. Photo by J.S. Aber.

regions of high to middle northern latitudes, namely from about 50–70 degrees north (Matthews and Fung 1987; Matthews 1993; Mitsch and Gosselink 2007). Most of the wetlands north of 60° latitude are affected by permafrost (U.S. Department of Agriculture (USDA) 1996). This includes large areas in Alaska, Canada, Scandinavia, and Russia. A second concentration in wetland distribution is found in the tropics (±15° latitude) with a peak abundance just south of the Equator. Central Africa, northern South America, northern Australia, Indochina, and Indonesia possess substantial tropical wetland regions. Significant temperate wetlands are situated in middle latitudes (30–50°) in eastern China, the eastern United States, and central Europe (USDA 1996).

This global pattern of wetland distribution occurs at the confluence of local terrestrial and hydrological conditions and general climatic circulation. Influential climatic conditions include heavy precipitation and evapotranspiration in the tropics and moderate precipitation with limited evapotranspiration at middle to high latitudes. The results in both cases are surplus surface waters. Parts of the subtropical zone (~15–30° latitude), in contrast, are characterized by scarce precipitation and high rates of evapotranspiration, which lead to well-known deserts – Sahara, Kalahari, Gobi, southwestern United States, western Australia, etc. Still, major wetlands such as the marshes of southern Iraq

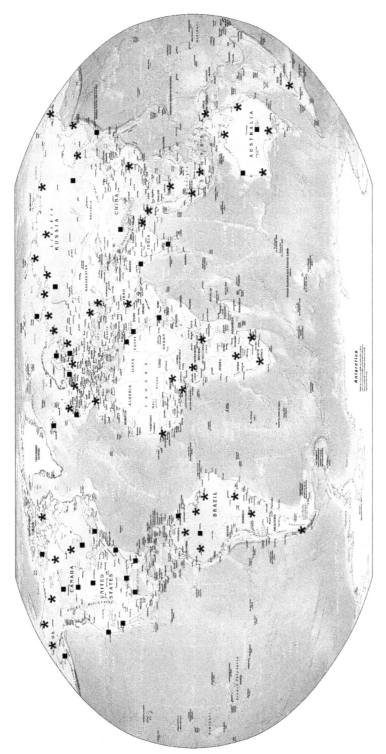

Figure 1-13. General distribution of wetlands around the world. Major wetlands (asterisk) and abundant wetlands (square). Based on Mitsch and Gosselink (2007) and other sources. Map adapted from *CIA World Factbook*, version of June 2009.

are found in the subtropical zone, where the configuration of terrestrial, hydrological and local climatic conditions gives rise to wetland habitats. Many of these subtropical wetlands are affected by high salinity, especially in the Middle East, central Asia, and Australia (U.S. Department of Agriculture (USDA) 1996).

The general climatic pattern and distribution of wetlands are influenced substantially by the positions of continents, flow of ocean currents, prevailing winds, mountains, and other major geographic features. For example, the largest mangrove swamp in the world, the Sundarbans covering 6000 km^2 (Dugan 2005), is located in the subtropics of Bangladesh and eastern India, where monsoons and runoff from the Himalaya combine to create a vast floodplain and delta complex (see chapter 15).

1.3 Wetland trends

Humans have modified and exploited wetlands in many ways, most of which have led to loss of wetland habitats and their conversion to other land uses and covers. Draining wetlands is observed globally and for various purposes – agriculture, forestry, grazing, peat mining, etc.

Water supply to wetlands may be reduced by levees, canals and dams as well as by extraction of ground water. Filling is another common means of converting wetlands for building construction, urban expansion, and industrial development (Fig. 1-14). Acid rain, shoreline erosion, and pollution (Fig. 1-15) are further factors for wetland loss. Such human development began with the advent of agriculture in the Neolithic, particularly once early civilizations arose, and the pace of wetland loss has

Figure 1-14. Overview of Port Bienville, an industrial park built on the Pearl River delta in southwestern Mississippi, United States. A dredged canal provides a connection via the Pearl River to the Intracoastal Waterway along the Gulf of Mexico coast. Kite aerial photo by S.W. Aber, J.S. Aber, and M. Giardino.

Figure 1-15. Huge chat piles are a legacy of lead-and-zinc mining at Picher, Oklahoma. Considered among the most seriously degraded sites in the United States, the landscape is essentially destroyed; toxic pollution of ground and surface water extends many kilometers downstream with severe impacts on human health and wetland habitats. View toward the southwest; blimp airphoto by J.S. Aber and S.W. Aber.

accelerated since the Industrial Revolution in the eighteenth century.

Dugan (2005) emphasized the distinction between "hydraulic" and "aquatic" civilizations in terms of how they utilized wetland resources. Hydraulic civilizations usually developed in upstream or inland settings in which water resources were seasonal or limited. Storage and distribution of water were controlled via engineering structures such as dams, levees, reservoirs and canals for irrigating farm land, all of which degrade or eliminate wetlands. Aquatic civilizations, in contrast, were situated in downstream or coastal settings where water was generally abundant. These civilizations utilized the annual flood cycle to farm deltas and alluvial plains, which had lesser impacts on wetlands. During the European period of exploration and colonization, beginning in the late fifteenth century, the hydraulic approach ruled at home and was exported throughout the world.

It is generally agreed that worldwide at least half of all pre-development wetlands have been lost to human activities (Mitsch and Gosselink 2007). This global loss of wetland habitats may be attributed primarily to the hydraulic emphasis of the past five centuries (Dugan 2005). The same holds true for the coterminous United States since the birth of the country. The 20 northeastern states are representative of this trend (Table 1-2). In general, the relatively rocky New England states (Maine, New Hampshire, Vermont, Massachusetts) had lower losses, whereas the Midwestern corn belt (Ohio, Indiana, Illinois, Iowa, Missouri) had the greatest conversions of wetland along with the Atlantic states of Connecticut and Maryland. The western Great Lakes and Appalachian states experienced intermediate reductions in wetlands. The distribution of wetland losses from state to state reflects primarily the extent of agriculture and amount of urban development.

Elsewhere around the world, similar wetland losses range from over 90 percent in New Zealand (Dugan 2005) to minimal impacts in remote and little-developed regions (Table 1-3). However, oil-and-gas and diamond exploration and extraction threaten once pristine wetlands in the circumarctic region, and human

Table 1-2. Wetland losses in the 20 northeastern states of the United States from c. 1780 to 1980. Areas given in hectares; based on Dahl (1990).

State	1780	1980	% loss
Maine	2584	2080	19
New Hampshire	88	80	9
Vermont	136	88	35
Massachusetts	328	236	28
Rhode Island	42	26	37
Connecticut	268	70	74
New York	1025	410	60
Pennsylvania	450	200	56
New Jersey	600	366	39
Delaware	192	90	54
Maryland	660	176	73
West Virginia	54	40	24
Ohio	2000	194	90
Michigan	4480	2234	50
Indiana	2240	300	87
Wisconsin	3920	2132	46
Illinois	3285	502	85
Minnesota	6030	3480	42
Iowa	1600	170	89
Missouri	1938	258	87

Table 1-3. Estimated wetland losses for selected regions of the world. Adapted from Mitsch and Gosselink (2007, Table 3.2).

Region		% loss
United States		53
Canada	Atlantic tidal/salt marshes	65
	Lower Great Lakes/St. Lawrence	71
	Prairie pothole region	71
	Pacific coastal estuaries	80
Australia	Swan Coastal Plain	75
	Coastal New South Wales	75
	Victoria	33
	River Murray basin	35
New Zealand		>90
Philippine mangrove swamps		67
China		60
Europe		60

encroachment on wetlands continues apace throughout the developing tropical world, both inland and offshore. The 2010 BP Deepwater Horizon oil spill in the Gulf of Mexico demonstrates that further degradation of wetland habitat may take place even in places already heavily affected by intensive human exploitation.

1.4 Wetland preservation and protection

Recognition of the importance of wetlands emerged gradually during the twentieth century, and now wetland conservation is a cause with considerable public support around the world. Early efforts focused on wildlife. As long ago as 1916 the United States and United Kingdom agreed to what became the Migratory Bird Treaty Act (1918), which protected birds migrating between the U.S. and Canada (Fig. 1-16). Specifically this act made it illegal for people to take migratory birds, their eggs, feathers or nests (U.S. Fish and Wildlife Service 2010a). Similar bilateral treaties were established by the United States with Mexico (1936), Japan (1972) and the Soviet Union (1976).

Another early and quite successful program was the U.S. Migratory Bird Hunting and Conservation Stamp, commonly known as "duck stamps," which began in the 1930s as a means to raise money for preservation of duck and goose habitat (Fig. 1-17). As of 2008, sales of duck stamps had generated US$700 million, which was used to purchase more than two million hectares (>5 million acres) of wetland habitat for the National Wildlife Refuge system.

As wildlife protection efforts in North America spread to other parts of the world, an international consensus emerged for the preservation of wetlands. Negotiations between various countries and non-governmental agencies in the 1960s culminated with a treaty adopted in the Iranian city of Ramsar in 1971. This treaty, which came into force in 1975, dealt with conserving wetland habitats necessary for migratory waterbirds. The number of contracting parties (countries) has reached 160, representing all parts of the world, and nearly 1900 sites have been listed as wetlands of international importance covering more than 185 million hectares (Ramsar 2010a).

Wetlands are a high priority also for many science-based and non-governmental

Figure 1-17. U.S. Migratory Bird Hunting and Conservation Stamp. Above: First U.S. "duck stamp" issued in 1934. The original stamp was flat-plate printed in blue monotone and depicted two mallards. For valid use, the stamp had to be signed; original signature can be seen faintly across the top of this example. Below: U.S. postage stamp issued to recognize the 50th anniversary of the duck stamp program, in the same blue monotone. From the collection of J. Vancura.

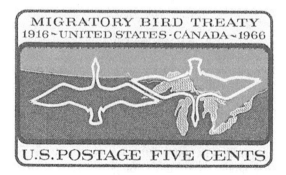

Figure 1-16. U.S. postage stamp marking the 50th anniversary of the convention on U.S.–Canada migratory birds. Original stamp printed in red, blue and black. From the collection of J. Vancura.

organizations (NGOs), such as Ducks Unlimited, Audubon, the Nature Conservancy in North America, the Wildfowl and Wetland Trust in the United Kingdom, and WetlandCare Australia, among others. As this list suggests, interested parties fall into two general categories – hunters and wildlife enthusiasts, again with wildlife conservation and sustainable management as the main themes. Such organizations have much in common; they strive in various ways to maintain, restore and protect native habitats for wildlife populations, so that future generations may enjoy the benefits of diverse wild animals thriving under natural conditions. In the case of migrating shorebirds and waterfowl, this means wetlands in summer and winter grounds as well as along the flyways during spring and autumn migrations.

These NGOs often work in close cooperation with local and national governmental agencies in order to complement or enhance efforts for wetland preservation. In the United States, the Environmental Protection Agency (EPA), Fish and Wildlife Service (FWS), Army Corps of Engineers (ACE), and Natural Resources Conservation Service (NRCS) are major agencies involved with wetland research, protection, and regulation. North of the border, Environment Canada is the lead national agency for various aspects of wetlands.

Wetlands International (WI) is the main global NGO concerned with restoring and sustaining wetland habitats, resources, and biodiversity. Headquartered in the Netherlands, WI deals mostly with wetlands in developing countries of South America, eastern Europe, southern and eastern Asia, and Africa. As a partner of Ramsar, WI has developed the Ramsar Site Information Service, which allows users to display map and statistical information about any Ramsar site online (see http://ramsar.wetlands.org/). Some WI projects highlight difficult situations; for example, efforts to protect wetlands in western Africa contradict attempts to control mosquito-borne malaria, which is epidemic in the region and a growing problem worldwide. West African malaria is a complex issue involving natural wetlands, rice agriculture, climate change, pharmacology,

Figure 1-18. Emergent wetland vegetation growing in shallow, muddy water of the Baía de Marajó, part of the Amazon Delta complex, near Belém, Brazil. Photo courtesy of K. Buchele.

economic policy, and many other aspects with no easy solutions (Gwadz 2001; Touré 2001).

As global recognition of and support for environmental issues has expanded during the past half century, so has ecotourism, supported by a growing middle class with interests ranging from whale watching to tropical wildflowers. Exotic adventures into Amazonia (Fig. 1-18) or the Okavango Delta (Fig. 1-19) have brought many more people into direct contact with natural environments and wetlands. Governments and NGOs in developing countries recognize that wetland preservation makes good economic sense in order to gain further financial support.

1.5 Wetland science

The scientific study of wetlands was traditionally considered to be part of biology, and this is still often the case. Terms such as "wetland ecology" or "mire ecology" reinforce this biological emphasis. However, wetlands are integrated systems based on water, soil, climate, vegetation and wildlife as utilized and modified by human activities. Focusing mainly on biology, thus, may overlook many other fundamental aspects of wetlands (Fig. 1-20). Mitsch and Gosselink (2007) identified four factors that are unique to wetland science.

Figure 1-19. Ecotourism camp in the Okavango Delta of Botswana. The tent structures are elevated on wooden posts to minimize surface impact, and the interior accommodations are quite comfortable. Photo courtesy of M. Storm.

- Wetlands have many special properties not adequately covered in biological specialties.
- Wetlands of disparate types do have some common properties.
- Wetland studies require an interdisciplinary approach that may involve several subdisciplines not commonly included in university academic programs.
- Strong scientific understanding of many facets is necessary for the development of policy, regulation, and management of wetlands.

Several wetland scientific societies and publications support this multidisciplinary approach. A major organization in North America is the Society of Wetland Scientists (SWS), which publishes the journal *Wetlands*. According to its own description, it is "an international journal concerned with all aspects of wetlands biology, ecology, hydrology, soils, biogeochemistry, management, laws and regulations" (SWS 2010). This description certainly highlights the many disciplines and subjects of wetland research.

The International Peat Society (IPS) was constituted in Canada and is now based in Finland. The International Mire Conservation Group (IMCG) is likewise based in Europe. Together IPS and IMCG publish *Mires and Peat*, an online journal (see http://www.mires-and-peat.net/).

Figure 1-20. S. DeGraaf prepares to place a soil-temperature logger into a water-filled hole ~½ m deep in a subalpine bog at ~3200 m elevation, Colorado, United States. Study site is part of a long-term climate investigation. Photo by J.S. Aber.

Recent articles spanned the globe from ecohydrology of mires in Tierra del Fuego, Argentina (Grootjans et al. 2010) to a carbon-fiber-composite Byelorussian peat corer (Franzén and Ljung 2009). The open-access nature of this journal illustrates the desire of some societies to make their publications freely available to everybody.

The scientific organizations noted above are large and international in character. Wetlands exist throughout the world, so many smaller scientific organizations deal with local or regional interests in more specialized ways. A good example is Suoseura, the Finnish Peatland Society. It serves as the Finnish National Committee of the IPS, organized the 12th International Peat Congress in 2004, and publishes the journal *Suo*. The society has a membership of approximately 450, clearly indicating that wetlands in general and peat in particular are major scientific issues for the small country of Finland.

1.6 Book approach and outline

In this book, we recognize that a complete study of wetland environments requires the assessment of the physical and biological attributes, properties and functions of these ecosystems, and the economic, political and social aspects that mediate their use globally. We adopt a systems approach, which emphasizes simultaneously examining component parts of a system in the context of the whole. Such an approach allows us to consider the interactions between the physical, biological and human elements of wetland ecosystems. Moreover, selected examples from across the world are used to illustrate wetland characteristics and interactions. Collectively, these provide a broad understanding of the global scope of wetlands, their contributions to natural processes and human societies.

Part I of the book provides a general overview and introduction to the study of wetlands. Chapter 2 considers the physical and social components of wetland systems, while chapter 3 discusses the methods used to study and monitor these systems. Part II focuses on the fundamental physical and biological aspects of wetlands including wetland hydrology (chapter 4), soils (chapter 5), vegetation (chapter 6) and wildlife (chapter 7). In Part III, we consider short- and long-term changes in wetland environments and their roles in environmental cycles and feedback. Autogenic and allogenic change and the influence of climate, fire, tectonic activity, sea-level fluctuation, and animal activity on wetlands are addressed in chapter 8, while chapter 9 provides a long-term record of wetland formation and development through geologic time. The important role of wetlands in biogeochemical cycles and climate regimes by acting as carbon reservoirs is addressed in chapter 10.

Part IV of the book focuses on the human use and governance of wetland environments. Topics addressed in this section of the book include wetland services, resources and methods for valuation (chapter 11), wetland conservation planning and management practices (chapter 12), wetland restoration, enhancement and creation (chapter 13), and finally global wetland governance and public policy (chapter 14). The final section of this book (Part V) provides regional case studies focusing on the unique social and physical characteristics of both large and internationally renowned as well as smaller wetland sites from the low latitudes (chapter 15), middle latitudes (chapter 16), and high altitudes and latitudes (chapter 17).

1.7 Summary

Wetlands include water, soil, vegetation and wildlife, as modified and exploited by human activities. Early civilizations arose in fertile river valleys, and wetlands continue to be essential for modern human society. Wetlands provide many resources for people who live in or derive economic benefits from them. In addition, wetlands serve important, but less tangible functions for water supplies and high levels of biological productivity and biodiversity. As major sources and sinks for carbon, wetlands play critical roles in the global carbon cycle with significant consequences for greenhouse gases and potential climate change.

In spite of these direct and indirect values, people have a difficult time balancing their own, local, economic gains with broader regional or global issues concerning wetland development. Humans modify wetlands for various purposes in many ways, most commonly by artificial draining, filling, and reducing inflow or extracting water. What may be good for upstream water users, however, often leads to undesirable effects for downstream wetland habitats, water resources, and the people who depend on them.

The total extent of existing wetlands is estimated to range from 7 to 12 million km^2, depending upon the definition for what is included. These wetlands are found primarily in two latitudinal zones – boreal and sub-boreal (50–70°N) and tropical (±15°), and many more wetlands of diverse types are found in all other parts of the world. Aquatic civilizations adapt to seasonal resources in wetlands, such as annual flood cycles, in order to practice agriculture and extract resources with lesser impacts on wetland environments. Hydraulic civilizations, in contrast, undertake intensive modifications of water supplies, which lead to degradation and loss of wetlands. The hydraulic approach has been dominant around the world for the past five centuries; as a consequence, the modern wetland total is no more than half of pre-development coverage.

Wetland conservation began about a century ago with initial emphasis on protecting habitats for migrating waterbirds. The United States undertook bilateral treaties with the United Kingdom (Canada), Mexico and other countries. Among the most successful early programs was the U.S. Migratory Bird Hunting and Conservation Stamp, commonly known as "duck stamps," which still continues today. Many federal agencies are involved in wetland research and management in the United States, and similar agencies exist in many other countries. Likewise several non-governmental wetland organizations are active in North America and other parts of the world. International efforts culminated with the Ramsar Convention on Wetlands in 1971; to date, the number of contracting parties has reached 160, and nearly 1900 sites covering more than 185 million hectares have been listed as wetlands of international importance.

Given the broad range of wetland functions, their feedback relationships with other environmental factors, and human impacts, it is not surprising that scientific investigations of wetlands have become important for many reasons. So-called wetland science involves biological, physical and cultural aspects of environments and human impacts on wetlands. Several scientific societies and publications are devoted to multidisciplinary wetland science and management. In addition, ecotourism to wetlands has emerged as a popular leisure activity, thereby exposing many more people to the beauty and values of wetlands.

2 Wetland criteria

2.1 Definitions

Many definitions for wetlands have been proposed and utilized over the years. In fact, more than 50 wetland definitions may be cited (Dugan 2005), ranging from the broadly inclusive definition of the Ramsar Convention to much more specific and restricted definitions. The approach taken by Ramsar is simply to name typical kinds of wetlands with common words familiar to the public.

> Wetlands are areas of marsh, fen, peatland or water, whether natural or artificial, permanent or temporary, with water that is static or flowing, fresh, brackish or salt, including areas of marine water the depth of which at low tide does not exceed six metres (Ramsar 1971).

Many terms derived from several languages have come into English usage to describe wetlands of various types in different geographic settings – bayou, billabong, bog, fen, mangrove, marsh, muskeg, playa, pocosin, sabkha, slough, swamp, and wad – to name just a few (see Glossary). Many of these names are now united under the general term mire, which refers to any wetland that accumulates peat. Equivalent words in other European languages include mose (Danish), tourbière (French), Moor (German), bagno (Polish), suo (Finnish), soo and raba (Estonian), pântano (Portuguese) and boloto (Russian). At least 19 words refer to different kinds of mires in dialects of northern Finland, including the terms aapa, palsa and tundra, which are used internationally in ecological research nowadays (Aapala and Aapala 1997).

Still, all these names and terms do not specify the basic characteristics of wetlands. A definition of what is a wetland often depends on who is asking the question and what development, exploitation, preservation, or study is proposed for a particular wetland site. The fact that wetlands may dry out from time to time complicates the attempt to define wetlands in a simple fashion. In fact, some wetlands may be dry more often than they are wet. With applications ranging from urban real estate to wildlife nature refuge, a great many points of view may be expressed for proper definitions, classification, and management techniques for wetland environments.

In the United States, one of the earliest modern definitions was developed by the U.S. Fish and Wildlife Service in the 1950s for categorizing waterfowl habitat.

> The term "wetlands" . . . refers to lowlands covered with shallow and sometimes temporary or intermittent waters. They are referred to by such names as marshes, swamps, bogs, wet meadows, potholes, sloughs, fens and river overflow lands. Shallow lakes and ponds, usually with emergent vegetation as a conspicuous feature, are included in the definition, but the permanent waters of

streams, reservoirs, and portions of lakes too deep for emergent vegetation are not included. Neither are water areas that are so temporary as to have little or no effect on the development of moist-soil vegetation (Shaw and Fredine 1956).

The definition of wetlands utilized by the U.S. Fish and Wildlife Service was elaborated subsequently by Cowardin et al. (1979), and this definition has been cited widely, adopted or modified by others.

> Wetlands are lands transitional between terrestrial and aquatic systems where the water table is usually at or near the surface or the land is covered by shallow water ... wetlands must have one or more of the following attributes: 1) at least periodically, the land supports predominantly hydrophytes, 2) the substrate is predominantly undrained hydric soil, and 3) the substrate is nonsoil and is saturated with water or covered by shallow water at some time during the growing season of each year.

The U.S. Army Corps of Engineers (USACE) and the Environmental Protection Agency (EPA) have agreed on the following definition, which they use for legal recognition of wetlands under the U.S. Clean Water Act.

> The term "wetlands" means those areas that are inundated or saturated by surface or ground water at a frequency and duration sufficient to support, and that under normal circumstances do support, a prevalence of vegetation typically adapted for life in saturated soil conditions. Wetlands generally include swamps, marshes, bogs, and similar areas (USACE 1987).

According to Charman (2002, p. 3), the distinguishing features of wetlands are: a) the presence of water at or near the land surface, b) unique soil conditions that are most often characterized by low oxygen content, and c) specialized biota, particularly plants, that are adapted to growing in these environments. Greb, DiMichele and Gastaldo (2006, p. 2) stated that wetlands are characterized by water at or near the soil surface for some part of the year, soils that are influenced by water saturation all or part of the year, and plants that are adapted to living in conditions of water saturation all or

Figure 2-1. The island of Vormsi merges into the shallow Väinimeri (Strait Sea) in the left foreground with the deeper Gulf of Finland in the far background. Water salinity here, at the eastern end of the Baltic Sea, is quite low. View northward from Förby, Vormsi, Estonia. Kite aerial photo by J.S. Aber and S.W. Aber.

part of the year. And finally, Mitsch and Gosselink (2007, p. 25) emphasized the presence of standing water for some period during the growing season, unique soil conditions, and organisms, especially vegetation, adapted to or tolerant of saturated soils.

From these several definitions, it is clear that three aspects – water, soil, and vegetation – are accepted as the basis for recognizing and describing wetland environments (Schot 1999). This triad is the modern approach for wetland definition under many circumstances that include vastly different environments. Furthermore, wetlands typically occupy transitional settings or intervening positions between dry, upland environments and deep-water habitats (Fig. 2-1). Each of these primary aspects is briefly described in the following sections and elaborated in subsequent chapters.

2.2 Water

Ground water (water table or zone of saturation) is at the surface or within the soil root zone during all or part of the growing season. Water may be obvious in shallow pools, pannes, puddles and channels, or it may lie just beneath the surface (Fig. 2-2). Water flow ranges from turbulent streams to essentially stagnant ponds, and water level may fluctuate; tidal flats are

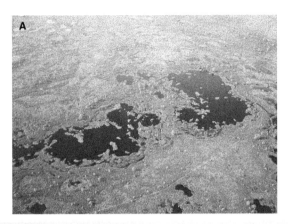

Figure 2-2. Water is the primary ingredient for wetlands. A. Nigula Bog, southwestern Estonia (see Color Plate 2-2A). Water fills numerous shallow pools of irregular size and shape. *Sphagnum* moss (reddish brown) surrounds each pool, and in between the pools, low hummocks are covered with heather and dwarf pines. A wooden walkway about half a meter wide is laid directly on the bog surface and runs across the bottom and right sides of the scene. Kite aerial photo (Aber et al. 2002). B. Salt marsh and tidal-channel complex of the Rowley River in northeastern Massachusetts, United States. Note linear drainage ditches in lower portion of scene. View eastward; Atlantic Ocean on the horizon. Blimp aerial photo by J.S. Aber, S.W. Aber and V. Valentine.

flooded daily, and many wetlands experience deeper floods from time to time. Standing water in pools and hollows is usually too deep to walk through but too shallow to swim in, although coral reefs are a popular exception to the swimming limitation. For quiet, shallow water, hip waders, canoes, pirogues, or kayaks are the best way to get around (Fig. 2-3).

Notice that water quality is not specified – salinity varies from fresh, to brackish, to marine, to hypersaline. Acidity and oxygen content may span the entire scales of natural pH and Eh values, and temperature may range from permafrost to hot springs (Fig. 2-4). Turbidity varies from crystal-clear tropical seas to glacier outwash rich with rock flour (Fig. 2-5). Quantity and quality of surface and ground water may be quite stable for some wetlands. In contrast, the water conditions often change dramatically in other wetlands as a consequence of tides, storms, seasons, long-term drought-and-flood cycles, and human activities.

From Siberian tundra to Saharan oasis, wetlands are present in all climatic and topographic settings around the world. Wetlands are relatively common in tropical and temperate

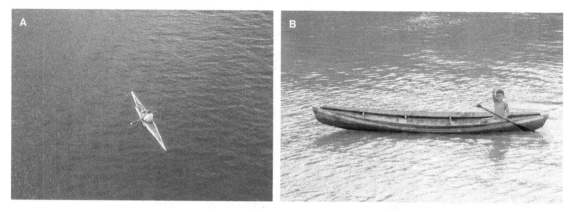

Figure 2-3. Traditional small boats suitable for wetlands and shallow water. A. Vertical view of Innuit kayak in coastal water of southwestern Greenland. Photo taken from a helicopter; courtesy of P. Jensen. B. Boy in small wooden canoe, Amazonia, Brazil. Photo courtesy of K. Buchele.

Figure 2-4. Panorama of the Blue Lake hot spring and marsh complex in the desert basin of western Utah, United States. The boardwalk and dock in lower center provide access for scuba divers to enter the deep pool. Fresh water in the foreground becomes highly saline in the background. View toward southeast. Photo taken from a model airplane; courtesy of B. Graves.

lowlands, because of abundant water supply. Alpine wetlands are also fairly common, as mountains typically receive more precipitation than do adjacent lowlands. Even deserts have wetlands supported by drainage from adjacent mountains, ground-water discharge, or infrequent storm runoff (Fig. 2-6). Much of the circum-arctic region turns to wetland during the brief period of summer melting. This rich diversity of wetland environments requires a flexible definition in terms of water conditions.

2.3 Soil

Soils are characterized by frequent, prolonged saturation and low oxygen content, which give rise to anaerobic chemical environments where

WETLAND CRITERIA

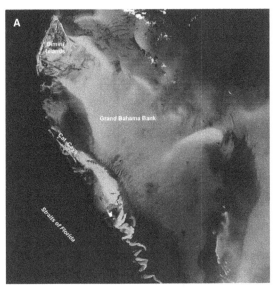

Figure 2-6. Marsh complex fed by ground water from springs and artesian wells in the San Luis Valley, a desert in south-central Colorado, United States. Russell Lakes State Wildlife Area; altitude ~2300 m, view toward northeast with Sangre de Cristo Mountains in the far background. Kite aerial photo by J.S. Aber and S.W. Aber.

Figure 2-5. A. Bimini Islands and Cat Cays, Bahamas (see Color Plate 2-5A). Shoals of carbonate sediment show distinctly as pale blue through the shallow, clear water of Grand Bahama Bank. False-color Landsat image in which vegetation on islands appears in red. Field of view ~75 km across; image courtesy of NASA Goddard Space Flight Center. B. The Tjórsa River carries highly turbid glacial melt water through a basalt gorge in south-central Iceland. Photo by J.S. Aber.

Figure 2-7. Salt accumulation along the margin of Dry Lake during a low-water phase, western Kansas, United States. Note small overturned boat in lower left corner for scale. Kite aerial photo by J.S. Aber and S.W. Aber.

reduced iron is present. Such soils are called hydric soils, and they range from organic-rich peat and muck to highly saline evaporite deposits (Fig. 2-7). The key condition in terms of soil development is water logging and extended saturation such that available oxygen is consumed and the chemical environment becomes reducing. When this condition develops, anaerobic bacteria thrive, and organic matter may accumulate to form peat (Fig. 2-8). In ancient rocks, peat is compressed and converted into lignite and coal, which represent former wetlands (Fig. 2-9).

An important indicator is the status of iron in the soil, which depends on the combination of acidity (pH) and oxidation potential (Eh). Low Eh and pH favor reduced iron (Fe^{2+}), whereas high Eh and pH lead to the oxidized state of iron (Fe^{3+}) (Davis and DeWiest 1966). This does not require any chemical instrumentation to determine; the status of iron is readily portrayed by its color – greenish gray for the reduced phase and rusty red-orange for the oxidized

Figure 2-8. Peat exposed in a cliff along the North Sea coast, northwestern Denmark. The peat accumulated in bogs during the Iron Age about 2000 years ago. Sand dunes later buried the peat, and recent coastal erosion has uncovered the peat layer. Scale pole marked in 20-cm intervals; photo by J.S. Aber.

Figure 2-9. Highvale coal mine in central Alberta, Canada. A thick seam of Cretaceous coal is extracted for generating electricity in a nearby power plant. Photo by J.S. Aber.

state. The latter is often observed in ancient soils, known as paleosols, preserved in the geologic record, for example red beds (Fig. 2-10). Modern wetland soils typically display both mottled greenish-gray and orange patches, lenses or layers, which demonstrate alternating reducing and oxidizing chemical environments.

2.4 Vegetation

Plants that grow in standing water or saturated soils, such as moss, sedges, reeds, cattail and horsetail, mangroves, cypress, rice and cranberries, are called hydrophytes. Emergent vegetation ranges from heavily forested swamps to

Figure 2-10. Sangre de Cristo Formation exposed near Cuchara in south-central Colorado, United States (see Color Plate 2-10). Several thousand meters of red sandstone, shale and conglomerate accumulated as alluvial fans during the Permian when the Ancestral Rocky Mountains were uplifted. The red color indicates oxidizing conditions in the depositional environment. These strata were tilted upward later, when the modern Rocky Mountains were deformed in the Eocene. Photo by J.S. Aber.

nearly bare playas and mudflats. The inclusion of rice, the world's most important crop, shows that wetlands may be cultural as well as natural features.

Hydrophytes have various special adaptations for dealing with submergence and lack of oxygen. Some wetland plants can tolerate substantial variations in soil moisture and water level, but others have strict water requirements for survival. On this basis, wetland vegetation is grouped into four general ecological categories, depending mainly on growth position in relation to water level (Whitley et al. 1999):

- Shoreline – Plants that grow in wet soil on raised hummocks or along the shorelines of streams, ponds, bogs, marshes, and lakes. These plants are situated at or above the level of standing water; some may be rooted in shallow water (Fig. 2-11).
- Emergent – Plants that are rooted in soil that is underwater most of the time. These plants grow up through the water, so that stems, leaves and flowers emerge in air above water level (Fig. 2-12).
- Floating – Plants whose leaves mainly float on the water surface. Much of the plant body

Figure 2-11. Buttonbush (*Cephalanthus occidentalis*), a shoreline wetland plant. It is a deciduous bush or small tree that may exceed 3 m (10 feet) in height, growing immediately beside lakes and streams, sinkholes and swamps. Most distinctive are the showy, white flower clusters that are shaped like spherical pincushions. Buttonbush is a member of the Quinine family, and its inner bark was once used as a quinine substitute. Lake Kahola, east-central Kansas, United States; photo by J.S. Aber.

Figure 2-12. Watercress (*Rorippa nasturium-aquaticum*), an emergent wetland plant (see Color Plate 2-12). A member of the mustard family (Cruciferae), it is a succulent, long-stemmed plant growing in tangled masses or low mounds up to ~30 cm (1 foot). The leaves have a strong peppery taste; watercress is highly valued for food flavoring and medicinal uses (Tilford 1997). Watercress absolutely requires clear, flowing water with temperatures <18 °C (65 °F), which means it favors spring-fed streams. Flint Hills of east-central Kansas, United States; photo by J.S. Aber.

Figure 2-13. Egyptian lotus (*Nymphaea caerulea*) has day-blooming flowers. The large round leaves are floating in a lily lagoon of the Okavango Delta, Botswana. Photo courtesy of M. Storm.

is underwater and may or may not be rooted in the substrate. Only small portions, namely flowers, rise above water level (Fig. 2-13).
- Submerged – Plants that are largely underwater with few floating or emergent leaves. Flowers may emerge (briefly) in some cases for pollination (Fig. 2-14).

Wetland vegetation is typically found in distinct spatial zones that are related mainly to water depth and salinity. As the groupings above suggest, many wetland plants have rigorous preferences for soil moisture and water depth or salinity. Some occupy primarily shoreline or emergent habitats, whereas others grow in floating or submerged situations (Fig. 2-15).

Thus, distinct zones of vegetation are developed across the transition from dry, upland positions into deep-water environments. Similar transitional zones mark the change from fresh to salt water along marine coastlines (Fig. 2-16) and around inland saline lakes and playas.

Figure 2-14. Coontail, also known as hornwort (*Ceratophyllum demersum*), a submerged wetland plant. It has stout stems up to 2 m long with featherlike leaves that grow almost entirely underwater. The long, ropelike stems drift in slow-moving water of springs and spring-fed streams. Coontail prefers cool, slightly alkaline water with a high calcium content (hard water). Seen here in a shallow spring-fed stream with watercress in the foreground. Turner Falls, southern Oklahoma, United States; photo by J.S. Aber.

2.5 Wetland classification

Given the great variety of wetland environments, many classification schemes have been proposed and utilized over the years. The U.S. Fish and Wildlife Service developed a hierarchy that consists of wetland systems, subsystems, and classes (Table 2-1). At the highest level are five wetland systems (Schot 1999):

- Marine – Open ocean, continental shelf, including beaches, rocky shores, lagoons, and shallow coral reefs (Fig. 2-17). Normal marine salinity to hypersaline water chemistry; minimal influence from rivers or estuaries. Where wave energy is low, mangroves, mudflats or sabkhas may be present.
- Estuarine – Deep-water tidal habitats with a range of fresh-brackish-marine water chemistry and daily tidal cycles. Salt and brackish marshes, intertidal mudflats, mangrove swamps, bays, sounds, and coastal

Figure 2-15. Zonation of fresh-water marsh vegetation: a) spatterdock *(Nuphar luteum)*, b) pickerelweed *(Pontederia cordata)* and narrow-leaved cattail *(Typha angustifolia)*, and c) duckweed *(Lemna minor)*. Conneaut Marsh, northwestern Pennsylvania, United States. View southward with the village of Geneva in the background. Blimp airphoto by J.S. Aber and S.W. Aber.

Figure 2-16. Salt marsh and swamp at the Wells National Estuarine Research Reserve (see Color Plate 2-16). Panorama looking toward the southwest along the Atlantic coast of southeastern Maine, United States. A bridge on Drakes Island Road (*) is part of a water-control structure that limits tidal flow between the marine lagoon in the background and marsh in the foreground. Notice the distinct vegetation zones, which reflect variations in water depth and salinity. Blimp airphoto by J.S. Aber, S.W. Aber and V. Valentine.

Table 2-1. Classification of wetlands and deep-water habitats utilized by the U.S. Fish and Wildlife Service. Based on Cowardin et al. (1979).

System	Subsystem	Class	System	Subsystem	Class
Marine	Subtidal	Rock bottom		Upper perennial	Rock bottom
		Unconsolidated bottom			Unconsolidated bottom
		Aquatic reef			Aquatic bed
		Reef			Rocky shore
	Intertidal	Aquatic bed			Unconsolidated shore
		Reef		Intermittent	Streambed
		Rocky shore	Lacustrine	Limnetic	Rock bottom
		Unconsolidated shore			Unconsolidated bottom
Estuarine	Subtidal	Rock bottom			Aquatic bed
		Unconsolidated bottom		Littoral	Rock bottom
		Aquatic bed			Unconsolidated bottom
		Reef			Aquatic bed
	Intertidal	Aquatic bed			Rocky shore
		Reef			Unconsolidated shore
		Streambed			Emergent wetland
		Rocky shore	Palustrine		Rock bottom
		Unconsolidated shore			Unconsolidated bottom
		Emergent wetland			Aquatic bed
		Scrub-shrub wetland			Unconsolidated shore
		Forested wetland			Moss-lichen wetland
Riverine	Tidal	Rock bottom			Emergent wetland
		Unconsolidated bottom			Scrub-shrub wetland
		Aquatic bed			Forested wetland
		Rocky shore			
		Unconsolidated shore			
		Emergent wetland			
	Lower perennial	Rock bottom			
		Unconsolidated bottom			
		Aquatic bed			
		Rocky shore			
		Unconsolidated shore			
		Emergent wetland			

Figure 2-17. Both rocky cliffs and sandy beach are well-known tourist attractions at Canon Beach on the Pacific coast of Oregon, United States. Kite aerial photo by J.S. Aber and S.W. Aber.

Figure 2-18. Aerial overview of fjords along the southwestern coast of Greenland. These deep glacier-carved valleys descend hundreds of meters below sea level and extend from the inland ice sheet (visible in background) to the edge of the continental shelf. Photo by J.S. Aber.

rivers. Fjords and drowned coasts, where supply of river sediment is insufficient to infill estuary basins (Fig. 2-18).

- Riverine – Freshwater, perennial streams comprising the deep-water habitat contained within a channel. This restrictive system excludes floodplains adjacent to the channel as well as habitats with more than 0.5‰ salinity (Fig. 2-19).
- Lacustrine – Inland water bodies that are situated in topographic depressions, lack emergent trees and shrubs, have less than 30% vegetation cover, and occupy at least 20 acres (8 ha). Includes lakes, larger ponds, sloughs, lochs, and bayous (Fig. 2-20).
- Palustrine – All non-tidal wetlands that are substantially covered with emergent vegetation – trees, shrubs, moss, etc. Most bogs, swamps, floodplains and marshes fall into this system (Fig. 2-21), which also includes small bodies of open water (<8 ha), as well as playas, mudflats and salt pans that may be devoid of vegetation much of the time. Water chemistry is normally fresh but may range to brackish and saline in semiarid and arid climates.

All of these systems may interface with each other or with dryland and deep-ocean habitats, so that an environmental continuum exists across the Earth's surface. Flows of materials and energy occur within wetlands and between wetlands and other habitats; a holistic approach is necessary to investigate, understand, develop, and manage wetlands successfully.

A significant portion of this book deals with the palustrine system of wetland environments as well as other water bodies not more than 6 m (20 feet) deep, according to the broad Ramsar definition. The palustrine system includes several classes that are recognized on the basis of substrate conditions or dominant vegetation cover (Schot 1999), including rock bottom, unconsolidated bottom, aquatic bed, unconsolidated shore, moss-lichen wetland, emergent wetland, scrub/shrub wetland, and forested wetland.

2.6 Peatland

The classification scheme outlined above deals mainly with the surficial aspects of wetlands. As most wetlands exist in topographic depressions, they tend to accumulate sediment through time. Furthermore, wetland soils are typically anaerobic, which often leads to the preservation of organic remains, namely peat. It is the living vegetation combined with accumulation of preserved plant debris that distinguishes peatlands

Figure 2-19. Niobrara River in north-central Nebraska, United States. The river channel has a sand bed, in which bars and banks are visible through the clear water. Photo by J.S. Aber.

Figure 2-20. Lake Tahoe occupies a tectonic basin formed between faults in the Sierra Nevada Mountains, United States. The lake is over 500 m deep and covers nearly 500 km² in area. Seen from Emerald Bay, California, looking northeast toward Nevada on the opposite side; photo by J.S. Aber.

Figure 2-21. Linnusaare (bird island) Bog, part of the Endla mire complex in east-central Estonia. A sparse pine forest grows on peat hummocks. Pine is the only tree that can tolerate the acid water and lack of nutrients that are typical in mature, raised bogs. This bog is given the highest level of protection in order that a natural environment may continue to exist without human disturbance. Photo courtesy of J.W. Aber.

from mineral wetlands (Charman 2002). The term peatland (American) or mire (European) denotes those wetlands in which substantial peat accumulation – at least one foot (30 cm) – has taken place, which is typical in bogs, fens, and swamps. The peatland substrate is in reality an organic structure built by biological activity. This gives peatland depth – a temporal dimension that contains a record of changing environments and past climatic conditions.

In fact, bogs have been called "monuments of nature" (Masing 1997). Natural monuments contain in their flora, fauna, rocks, sediments, fossils, and landforms unique records of environmental conditions and physical processes that have shaped the Earth. Bogs may be considered as repositories of environmental and climatic data through the surrogate of plant remains, namely peat, which contains both macrofossils (Fig. 2-22) and microfossils (pollen). Indeed much of what we know about climate and environment of the Holocene Epoch (past 10,000 years) has been gleaned from intensive investigations of bogs.

Peat is intrinsic to many wetlands around the world. Peat is partly decomposed plant remains

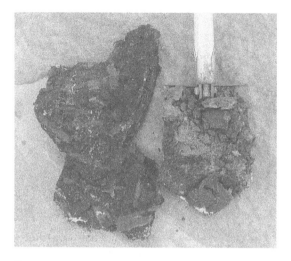

Figure 2-22. Peat samples from the Danish Stone Age (early Holocene) along the northwestern coast of Denmark. Plant macrofossils are well preserved and clearly evident. Photo by J.S. Aber.

Figure 2-23. Moss, lichens and small shrubs cover a knob of basalt in south-central Iceland. These plants contribute to the formation of peat in blanket bogs and depend for nutrients entirely on rain, snow, and atmospheric deposits. Knob is ~3 m across; photo by J.S. Aber.

that consist of more than 65 percent organic matter (dry weight). Moss, grass, herbs, shrubs and trees may contribute to the buildup of organic remains, including stems, leaves, flowers, seeds, nuts, cones, roots, bark and wood. Through time, the accumulation of peat creates the substrate, influences ground-water conditions, and shapes surface morphology of the wetland. Several factors are considered important for classification of peatland types (Charman 2002):

- Floristic – Plant composition of vegetation communities, which may be used as proxies for environmental factors.
- Physiognomy – Structure of the dominant plants, particularly used in Scandinavian and Russian schemes.
- Morphology – Three-dimensional shape of the peat deposit and geomorphology of the wetland surface.
- Hydrology – Source of the supply and flow regime for surface and ground water.
- Stratigraphy – Vertical layering, composition, and nature of underlying peat deposits.
- Chemistry – Chemical characteristics of surface water, particularly acidity and content of nutrients.
- Peat characteristics – Botanical composition, nutrient content and structure, or human applications – fuel, horticulture, etc.

From these diverse criteria, water supply and nutrient status are considered the most fundamental elements for classifying peatlands. A gradient exists from nutrient-rich, alkaline water (fens) to nutrient-poor, acidic water (bogs), and so peatlands are generally divided in two broad categories (Charman 2002):

- Ombrotrophic – Raised or blanket bogs that receive all water and nutrients from direct precipitation (Fig. 2-23). Neither ground water nor runoff from surrounding land reaches the surface of the bog. Rain and snow provide the water source and nutrients. The latter also are derived from whatever is carried or blown in – dust and ash, leaves, bird droppings and feathers, spider webs, and animal fur (Fig. 2-24). Water chemistry tends to be acidic, and nutrients for plant growth are in extremely short supply. Only a few plants can survive such harsh conditions; namely *Sphagnum* (peat moss) and pine.
- Minerotrophic – Fens and swamps located in depressions that receive surface runoff and/or ground-water discharge from adjacent or underlying mineral-soil sources (Fig. 2-25). Nutrients are more abundant and

water ranges from slightly acidic to alkaline – conditions that are suitable for a wide range of plants and which give rise to greater floristic diversity compared with bogs. The terms oligotrophic and eutrophic refer respectively to nutrient-poor and nutrient-rich fens.

2.7 Anthropogenic dimensions of wetlands

Human settlement across wetland ecosystems dates back millennia. Our early ancestors used the diverse resources provided to sustain themselves, while later riverine civilizations from Egypt to Mesopotamia and the Indus flourished on their contributions. Today, few wetland habitats exist without some form of anthropogenic influence or modification. Numerous extractive and service industries are based on the resources and functions provided by wetlands. These ecosystems support a wide diversity of flora and fauna, they contribute to biogeochemical cycling of nutrients, sequester carbon and other greenhouse gases regulating climate, act as natural filters to purify water, and protect against destruction from floods and storm events.

Humans extract a wide array of economic goods and services from these ecosystems. In some regions of the world, wetlands provide for basic needs of food, fuel, water and shelter. In other areas, technology has allowed humans to mimic wetland functions and take advantage of the ecosystem services they offer by constructing artificial wetlands where none existed, or restoring and rehabilitating other sites that may have been degraded. In still other regions, wetlands provide intrinsic aesthetic value to communities dependent on tourism and recreation.

Given the multiple uses and services of wetlands, managing this resource requires an integrated systems approach. Such an approach recognizes the complex and latticed nature of

Figure 2-24. High air humidity combined with smoke from autumn burning of agricultural wastes form a haze over Männikjärve Bog (foreground) and Endla Lake (left background) in eastern Estonia. Ash settles and contributes to nutrients and influences surface-water chemistry in the bog environment. Kite airphoto by J.S. Aber and S.W. Aber.

Figure 2-25. Nigula Bog and Salupeaksi mineral island, southwestern Estonia. A. Overview of the raised bog with the mineral island on right side. B. Close-up view of tree-covered mineral island developed on a drumlin (see Color Plate 2-25B). Note distinct vegetation zones: (a) *Sphagnum* moss, (b) pine, (c) birch (partly bare), and (d) ash, elm, maple and other deciduous hardwoods, some of which display fall colors. Kite aerial photo; Aber et al. (2002).

wetland ecologies and human systems while acknowledging that sustainable management strategies ought to balance the needs of both. It also recognizes the essential role played by humans in arbitrating wetland resource use and allocation.

Socio-cultural, economic and political conditions and constraints mediate the use of wetlands and their management globally. Understanding these underlying conditions and building appropriate and responsive institutional frameworks are vital to ensuring the sustainability of wetland ecosystems (Ostrom 1990; Bromley 1991). An institutional analysis of wetland systems focuses attention on building and supporting clearly established formal resource rights regimes that may promote the sustainable use of wetlands. Such resource regimes detail wetland resource rights, access and rules of use, and elaborate on monitoring and enforcement methods to guard against resource misuse. Globally, natural resources are governed by essentially four broadly defined property rights regimes (Hanna, Folke, and Maler 1995, p. 17):

- Private property systems identify individuals or corporate entities as owners that control access to resources and enjoy the benefits of such rights by excluding non-owners. Even though individual rights are paramount, the use of resources under such systems is subject to constraints set by society.
- State or public property systems provide rights of management and use to federal, state and local governments. As steward, the state manages a resource by generally attempting to balance ecosystem and societal needs.
- Common property systems include resources that are held collectively by a group of people. Similar to private property, members of such groups may exclude use by non-members and reserve rights of access.
- Open access systems are those lacking established resource rights or rules governing their use and observe virtually no monitoring of misuse. The overuse and destruction of resources may often occur under such circumstances, particularly when resource demands are high. Individual users lack any incentives to adopt sustainable resource-use practices as there are few assurances that others would not act in their own self interest.

Yet, merely enforcing appropriate property regimes may not be sufficient to use wetlands sustainably. Countries across the globe implement wetland conservation and management policies. Regulatory approaches emphasize laws, enforcement mechanisms and penalties to promote wise use and temper and balance development objectives with ecosystem health. In other cases, countries implement incentives and disincentives to encourage or discourage wetland management activities.

Alongside conservation policies, a balanced valuation of wetland resources outlines the economic advantages accrued, but also includes quantifiable ecological, social, and cultural benefits provided by wetlands. Wetland management from an integrated systems perspective could ensure that the multifaceted nature of wetland ecosystems and their complex interactions are accounted for. Such plans are designed to reflect locally specific conditions, incorporate the views and knowledge of a broad range of stakeholders, and are mindful of the essential role these resources play in providing livelihoods for dependent communities (Ramsar Convention Secretariat 2007a).

2.8 Summary

Among more than 50 definitions for wetlands, the broadly inclusive statement of Ramsar is considered most applicable around the world. It embraces all types of marsh, fen, peatland or water bodies, natural or artificial, permanent or temporary, in which water depth (at low tide) does not exceed 6 m (20 feet). Water, soil, and vegetation form the triad for modern wetland science. Water is either at the surface or just beneath the surface, so the plant root zone is saturated during all or part of the growing season. Hydric soil is waterlogged, such that anaerobic bacteria thrive and reducing chemical conditions develop. Plants are adapted for growing in standing water or saturated soils, which include shoreline, emergent, floating and submerged types of vegetation. Wetlands intervene between dry, upland and deep-water habitats. Distinct zones are evident in vegetation and soils based mainly on water depth and salinity.

Wetland classification is a hierarchical scheme of systems, subsystems, and classes. At the highest level are the five wetland systems: marine, estuarine, riverine, lacustrine, and palustrine. Within each system, classes are based primarily on substrate conditions or dominant vegetation cover. The primary emphasis of this book is the palustrine system of wetland environments, as well as other water bodies not more than 6 m deep.

Peatland is one of the most widespread groups of palustrine wetlands around the world. Peat consists of partly decayed plant remains that accumulate in bogs, fens and swamps to thickness of one foot (30 cm) or more. Peat contains in its macro- and microfossils a record of past environmental conditions, and this record is crucially important for understanding climates and environments of the past several millennia. The nature of water supply and nutrients are key criteria for classifying peatlands. Ombrotrophic peatlands include raised and blanket bogs that are characterized by acidic water and scarce nutrients, whereas fens and swamps fall into the minerotrophic category in which water is more alkaline and nutrients more abundant.

The anthropogenic dimension of wetland ecosystems accounts for the important role wetlands play in providing services and resources to humans, while also recognizing the important influence of human systems in managing these habitats and allocating resources obtained from them. Even though the importance of wetland conservation and protection is widely acknowledged, countries face daunting challenges as demands for land and resources continue to increase and the vagaries of climate change heighten vulnerabilities and complicate management strategies.

Methods in wetland research

3.1 Introduction

Wetland environments comprise water, soil, vegetation, and wildlife. Thus, all manner of physical and biological scientific methods may be applied for investigations, which range from microscopic laboratory analysis to space-borne satellite observations of continental regions. When human influence is added, then various cultural and socioeconomic approaches may be integrated into wetland studies. In other words, potential methods for wetland research are quite diverse and include nearly all scientific means used to study the Earth's environments and human interactions with those environments.

The dean of Estonian mire research, Viktor Masing, reviewed historical issues of wetland mapping, research and classification, particularly in Germany, Russia and Scandinavia (Masing 1998). The earliest studies were limited necessarily to ground-based observations, which were, and still are, logistically difficult to accomplish in many bog, marsh and swamp situations. By the 1930s, aerial photography had already come into use for geobotanical mapping purposes, and three basic levels or scales were recognized for mire classification (Galkina 1946):

- Micro landscape – individual hummocks, ridges, hollows, pools and other forms within bogs.
- Meso landscape – bog complex, mesotope, or *Grossmoor* made up of many individual components.
- Macro landscape – mire system consisting of multiple bogs, lakes, streams, intervening uplands, and related features.

From this beginning, Masing (1984) elaborated a multi-level research approach (Table 3-1). His classification scheme was developed specifically for mires in the boreal and subboreal regions of the northern hemisphere. Nonetheless, its basic structure could be adapted for other types of wetlands in different regions. In addition to the spatial dimensions of this scheme, the temporal element must be added (Charman 2002). All wetlands are dynamic through time, either because of internal evolution (autogenic change) or forcing from external factors (allogenic change). Time scales vary from diurnal to millennial.

Here we emphasize those methods that reveal the spatial relationships of organic and inorganic constituents and temporal variability that characterize wetland environments, namely geographic information systems (GIS) and remote sensing (RS). GIS/RS must be combined with conventional field observations and sampling in order to provide ground truth for validating, analyzing, and interpreting the remotely sensed information. It is this multi-level approach that yields comprehensive and convincing results.

Wetland Environments: A Global Perspective, First Edition. James Sandusky Aber, Firooza Pavri, and Susan Ward Aber.
© 2012 James Sandusky Aber, Firooza Pavri, and Susan Ward Aber. Published 2012 by Blackwell Publishing Ltd.

Table 3-1. Levels of boreal mire research. Based primarily on Masing (1998); spatial resolution added by the authors.

Level	Map scale	Resolution	Subjects	Main methods	Discipline or application
Clonal and population	>1:10	<1 mm	Plant clones and tussocks, population types, reproduction (life) strategies	Direct measurements, point quadrats, small permanent plots, cultivation and replacement experiments, study of moss increment; study of degeneration, decomposition, competition	Primary productivity of moss layer and single species, population ecology
Ceonotic	1:10 to 1:100	1 mm to 1 cm	Plant communities, micro-associations, synusia, etc. Syntaxonomical units		Synecology, primary productivity of plant communities
Microstructural	1:100 to 1:1000	1 to 10 cm	Pattern of mire surface, microforms (hummocks, hollows) and compound microforms	Measurement of hydrological parameters, large-scale plans, step method, line-intercept method	Modeling of mire surface features, primary productivity and decomposition, succession studies
Microtope or coenocomplex	1:1000 to 1:10,000	10 cm to 1 m	Homogeneous microtopes, heterogeneous sites, site types, coenocomplexes	Hydrological parameters, aerial photography, large-scale mapping, dendrometry	Large-scale mapping, habitat conservation, site evaluation for forestry, berries, cultivation, etc.
Mesotope or mesostructural	1:10,000 to 1:100,000	1 to 10 m	Bog complexes and other mire mesotopes; landscape units; mire (complex) types	Hydrological parameters, satellite imagery, transect methods, stratigraphy along transects	Mire hydrology, bog mesotope modeling, landscape ecology, vegetation and habitat mapping, nature reserve management; land evaluation for peat, forestry or agriculture
Macrotope or macrostructural	1:100,000 to 1:1,000,000	10 to 100 m	Mire systems, peatland basins; landscape classification units	Hydrography and geomorphology of mire landscape, stratigraphy; geographical mapping based on topography, satellite images, aerial photographs, etc.	
Regional	<1:1,000,000	>100 m	Mire regions, mire provinces, mire zones; regional units		Physical geography, biogeography, landscape ecology

3.2 Remote sensing

Aerial photography and satellite imagery are examples of remote sensing; in other words, collecting information about objects from a distance. Information is conveyed by the electromagnetic spectrum in visible and invisible radiation that is reflected or emitted from objects on the ground, transmitted through the atmosphere, and collected by various types of cameras or detectors. For most situations involving wetlands, natural sunlight illuminates the scene. Reflected visible, near-infrared, and mid-infrared wavelengths are utilized to create images. In some cases, emitted thermal infrared radiation may be exploited for surface temperature measurements. For a full discussion of remote sensing methods, the reader is advised to consult recent textbooks on the subject (e.g. Jensen 2007; Lillesand, Kiefer and Chipman 2008).

Figure 3-1. The tranquil beauty of the bog is apparent in this early morning, foggy view at Nigula, southwestern Estonia. Photo courtesy of E. Karofeld.

As Tiner (1997) emphasized, aerial photography continues to be essential for wetland studies. With the launch of the first Earth Resources Technology Satellite in 1972 (later renamed Landsat I), applications were immediately recognized for wetlands and related water resources (e.g. Carter, McGinness and Anderson 1976; Higer, Cordes and Coker 1976; Krinsley 1976). From the frog's-eye view on the ground (Fig. 3-1) to the vantage from outer space, we now have the ability to observe wetlands from many heights and resolutions (Fig. 3-2).

3.2.1 Image resolution and interpretability

The term "resolution" is used in widely different senses in various disciplines. In the general context of remote sensing, it refers to the smallest object or feature that can be identified by an experienced image interpreter. For digital imagery, the smallest object is a single cell or picture element (pixel) in the scene. Each pixel corresponds to an actual spot on the ground, and the linear size of this spot is known as ground sample distance (GSD), which is an appropriate measure for image scale (Comer et al. 1998). However, the identity of a single pixel can rarely be determined by itself. For positive identification, usually a group of four to nine pixels is necessary, which leads to a general rule of thumb (Hall 1997):

> Positive recognition of objects in remotely sensed images requires a GSD at least three times smaller than the object size.

The spatial resolution (GSD) of an aerial photograph or satellite image is only one factor that comes into play for image interpretation. Other basic aspects of resolution are temporal and spectral characteristics for different kinds of remote sensing. Additional factors include the following features (based on Avery and Berlin 1992; Teng et al. 1997; Jensen 2007):

- Tone or color – Gray tone (b/w) and color are how we recognize and distinguish objects. The tone or color of an object helps to separate it from other features in the scene, especially for features with high contrast.
- Shape – Natural shapes tend to follow the lay of the land. Cultural (human) shapes, on the other hand, are often geometric in nature with straight lines, sharp angles, and regular forms. In single images only plan shape can

Figure 3-2. Platforms for manned photography of the Earth's surface. A. Discovery space shuttle as seen from the International Space Station above the southwestern coast of Morocco; ISS026-E-032253, 7 March 2011. Modified from NASA Human Space Flight; accessed online <http://spaceflight.nasa.gov/gallery/images/station/crew-26/html/iss026e032253.html> January 2012. B. Challenger II light sport aircraft for low-height aerial photography. Overhead wings and large open windows allow good views for the photographer. Photograph courtesy of W.S. Lowe.

be appreciated fully; in stereoscopic images, the full three dimensions become apparent.
- Size and height – Absolute size and height are important clues that depend on scale of the photograph. Always check photo scale (GSD) for a guide to object size.
- Shadow – Shadows may be useful clues for identifying objects, as seen from above. Trees, buildings, animals, bridges and towers are examples of features that cast distinctive shadows. Shadows on the landscape also help provide depth perception.
- Pattern – The spatial arrangement of discrete objects may create a distinctive pattern. This is most apparent for cultural features, for example city street grids, airport runways, agricultural fields, etc.
- Texture – This refers to grouped objects that are too small or too close together to create distinctive patterns. Examples include tree crowns in a forest canopy, individual plants in an agricultural field, ripples on a water surface, etc. The difference between texture and pattern is determined largely by photo scale.

Table 3-2. The civilian U.S. National Imagery Interpretability Rating Scale levels (0–9) and examples of interpretation tasks. Levels based on Leachtenauer, Daniel and Vogl (1997); scale/resolution categories correspond approximately with Table 3-1.

Rating level	Scale/resolution	Interpretation tasks
0	–	Interpretability precluded by obscuration, degradation, or poor resolution
1	MACRO	Distinguish between major land-use classes – urban, agricultural, forest, etc. Detect medium-sized port facility Distinguish between runways and taxiways at a large airport Identify regional drainage patterns – dendritic, trellis, etc.
2		Identify large (800 m diameter) center-pivot irrigated fields Detect large buildings – hospital, factory, etc. Identify road pattern – highway interchange, road network, etc. Detect ice-breaker tracks in sea ice; detect wake of large (100 m) ship
3		Detect individual houses in residential neighborhoods Detect trains or strings of railroad cars on tracks Identify inland waterways navigable by barges Distinguish between natural forest and orchards
4	MESO	Identify farm buildings as barns, silos, or residences Count tracks along railroad right-of-way or in train yards Detect basketball court, tennis court, volleyball court in urban area Detect jeep trails through grassland
5		Recognize individual train cars by type – box, flat, tank, etc. Identify large tents within camping areas Distinguish between deciduous and coniferous forest during leaf-off season Detect large animals (elephant, rhinoceros, bison) in grassland
6		Distinguish between row crops and small grains Identify passenger vehicles as sedans, station wagons, etc. Identify individual utility poles in residential neighborhoods Detect foot trails in barren areas
7	MICRO	Identify individual mature plants in a field of known row crops Recognize individual railroad ties Detect individual steps on stairways Detect stumps and rocks in forest clearings or wet meadows
8		Count individual baby farm animals – pigs, sheep, etc. Identify a survey benchmark set in a paved surface Recognize individual pine seedlings or individual water lilies on a pond Detect individual bricks in a building
9		Identify individual seed heads on grain crop – wheat, oats, barley, etc. Recognize individual barbs on a barbed-wire fence Detect individual spikes in railroad ties Identify individual leaves on a tree

- Context – The association and site location of objects are often important for aiding interpretation. Note land cover and land use as clues to help identify related features in the scene, and refer to existing maps or census data for ancillary information.

Each of these basic features is seldom recognized on its own. Rather it is the combination of visual elements that allows interpretation of objects depicted in aerial photographs. With these factors in mind, a perceptual measure of image quality or interpretability has been developed – the U.S. National Imagery Interpretability Rating Scale (NIIRS). An image rating depends on the most difficult interpretation task that can be accomplished (Table 3-2). Levels 1–3 represent the largest and most generalized features, which correspond to regional and macro levels (>10 m) of mire classification (see Table 3-1). The middle three interpretability levels (4–6) are at the meso/micro levels (>10 cm to 10 m), and the highest interpretability levels (7–9) represent the micro and sub-micro classes (<10 cm).

3.2.2 Wetland image interpretation

Water, vegetation and soil are the principal features of interest for interpretation of wetland imagery. Water is the most widespread feature on the surface of the Earth, and water bodies exist in many forms – seas, lakes, rivers, ponds, estuaries, bayous, lagoons, etc. In aerial photography, water color is most obvious, and the color of water bodies is a good indication of suspended sediment. Clean water reflects blue light weakly, but reflectance drops off sharply for green and red light and is essentially zero for infrared radiation. Thus, clean water typically looks dark blue or sky colored in visible imagery (Fig. 3-3). Suspended sediment influences the color of water, which depends on the sediment composition and turbidity (Fig. 3-4).

Aerial photography of water bodies often displays sun glint and glitter (Fig. 3-5). Although usually considered undesirable, under some conditions sun glint is advantageous for identifying small water bodies that otherwise would be difficult to distinguish. Conversely, sun glint

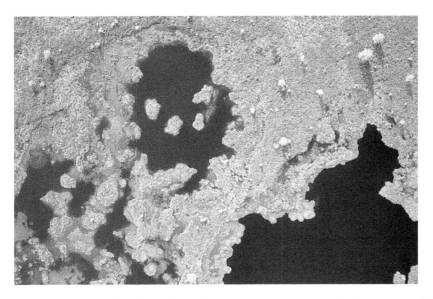

Figure 3-3. Close-up vertical view of pools in the Nigula Bog, Estonia. Open water bodies are essentially black, as the clear water reflects almost no sunlight. Dwarf pine trees cast shadows on hummocks between pools. Kite airphoto by S.W. Aber and J.S. Aber.

Figure 3-4. The man-made pond in the foreground trapped recent runoff and has a high content of yellowish-brown suspended sediment (see Color Plate 3-4). The next pond downstream did not receive this sediment and has a clean, dark blue appearance. Central Kansas, United States. Kite aerial photo by J.S. Aber and S.W. Aber.

Figure 3-6. Sun glint highlights tiny islands in the shallow Väinameri (Strait Sea) south of Vormsi, Estonia. Kite aerial photo by J.S. Aber and S.W. Aber.

Figure 3-5. Sun glint from smooth water (*), and sun glitter from ripple and wave surfaces elsewhere in scene. Southern tip of South Padre Island, Texas, United States. Kite airphoto by J.S. Aber and S.W. Aber.

may aid recognition of small, emergent islands within larger water bodies (Fig. 3-6).

Many wetlands are largely, if not entirely, covered with emergent vegetation. Photosynthetically active "green" leaves absorb blue and red light, weakly reflect green light, and strongly reflect near-infrared radiation (Fig. 3-7). Green plants are the only objects at the Earth's surface with this spectral signature. For this reason, color-infrared photographs are recommended for wetland applications (Tiner 1997), because emergent vegetation appears in pink and red colors (Fig. 3-8). The unique spectral signature of active green plants has given rise to several vegetation indices, such as the normalized difference vegetation index (NDVI), which are based on the ratio of red to near-infrared bands (Tucker 1979; Murtha et al. 1997).

Although color-infrared imagery is ideal for separating emergent vegetation from other objects in wetlands, the authors have found that conventional color-visible imagery is often better for recognizing fine distinctions in vegetation. Subtle variations in plant height, texture, pattern, and visible color may be quite revealing (Fig. 3-9). In temperate climates, deciduous vegetation undergoes marked phenological changes from season to season. These changes may be used to advantage for plant identification and are sometimes most striking in the autumn (Fig. 3-10).

Soil, when visible from above, is the third main feature of interest for wetland image interpretation. Like water, the most obvious aspect of soil is its color. Yellowish-brown and reddish-brown colors indicate ferric iron (rust) in oxidized soils (Fig. 3-11). In contrast, bluish- or greenish-gray colors are typical of reducing conditions (Fig. 3-12), and dark brown to black colors suggest high contents of organic matter. In saline soils, surface evaporation may deposit a crust of salt that appears quite bright (see Fig. 2-7). However, soil color must be treated with some caution, as moisture content has a strong

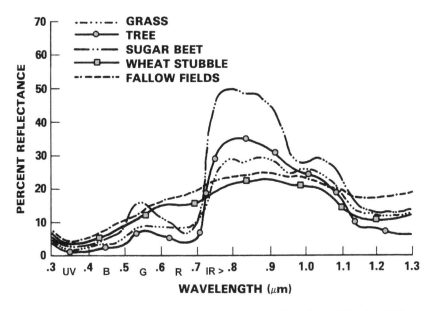

Figure 3-7. General spectral reflectance curves for selected vegetation. Note blue (B) and red (R) absorption, weak green (G) reflection, and strong near-infrared (IR) reflection of active plants in contrast to stubble or fallow soil. Adapted from Short (1982, Fig. 3-5B).

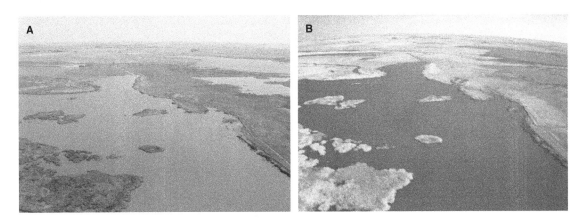

Figure 3-8. Color-visible (A) and color-infrared (B) digital images of marsh at the Nature Conservancy, Cheyenne Bottoms, central Kansas, United States (see Color Plate 3-8). Active vegetation appears in bright red-pink colors in the latter. Kite aerial photographs from Aber et al. (2009, Fig. 5).

impact on visible color. For further discussion of soil color, see section 3.4.1 below.

3.2.3 Macro-level systems

Several satellite systems provide low spatial resolution imagery in visible and infrared portions of the spectrum. These include the Advanced Very High Resolution Radiometer (AVHRR) carried onboard several NOAA weather satellites as well as the Moderate Resolution Imaging Spectroradiometer (MODIS) instrument on NASA's Terra and Aqua satellites (Fig. 3-13). AVHRR typically provides data at 1 km spatial resolution; MODIS normally operates at 250 m resolution. Of particular importance for

Figure 3-9. Close-up view of marsh and shore at Luck Lake, Saskatchewan, Canada (see Color Plate 3-9). Distinctive vegetation zones are revealed by differences in plant texture, pattern, and color. The maroon plant is red samphire (*Salacornia rubra*), which grows on saline mudflats. Kite aerial photo by J.S. Aber and S.W. Aber.

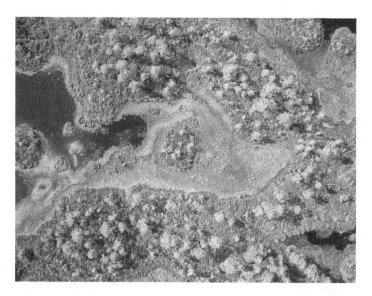

Figure 3-10. Vertical kite aerial photograph in visible light showing water pools and vegetated hummocks in the central portion of Männikjärve Bog, Estonia (see Color Plate 3-10). *Sphagnum* moss species display distinctive green, gold, and red autumn colors along with pale green dwarf pine trees on hummocks. These dramatic colors are not displayed so clearly at other times of the year. Field of view ~60 m across. Based on Aber et al. (2002, Fig. 2).

METHODS IN WETLAND RESEARCH   39

Figure 3-11. Red Hills in Barber County, southern Kansas, United States (see Color Plate 3-11). Alluvial sediment accumulated in an ancient desert floodplain environment and is now eroding in a badlands topography. Bright red color indicates oxidizing conditions of deposition. Kite airphoto by S.W. Aber.

Figure 3-12. Exposed mudflat on the shore of Luck Lake, Saskatchewan, Canada (see Color Plate 3-12). Color analysis of this photograph gives light gray to pale blue-green colors for the mud, indicating reducing conditions in the sediment. Kite aerial photo by J.S. Aber and S.W. Aber.

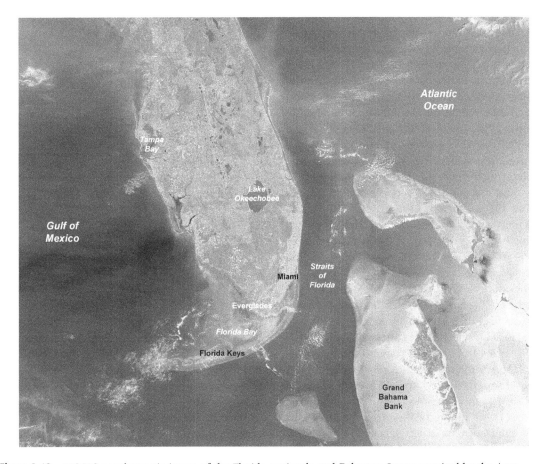

Figure 3-13. MODIS panchromatic image of the Florida peninsula and Bahamas. Image acquired by the Aqua satellite on 16 May 2001; adapted from NASA Visible Earth (http://visibleearth.nasa.gov).

Table 3-3. Spectra, bands, wavelengths (μm) and spatial resolutions for the Thematic Mapper of Landsats 4 and 5 and the Enhanced Thematic Mapper Plus of Landsat 7. Bands are arranged according to wavelength.

Spectrum	TM (4 & 5)	Wavelength	Resolution	ETM+ (7)	Wavelength	Resolution
Blue	Band 1	0.45–0.52	30 m	Band 1	0.45–0.52	30 m
Green	Band 2	0.52–0.60	30 m	Band 2	0.52–0.60	30 m
Red	Band 3	0.63–0.69	30 m	Band 3	0.63–0.69	30 m
Near-infrared	Band 4	0.76–0.90	30 m	Band 4	0.77–0.90	30 m
Panchromatic	–	–	–	Band 8	0.52–0.90	15 m
Mid-infrared	Band 5	1.55–1.75	30 m	Band 5	1.55–1.75	30 m
Mid-infrared	Band 7	2.08–2.35	30 m	Band 7	2.09–2.35	30 m
Thermal infrared	Band 6	10.4–12.5	120*m	Band 6	10.4–12.5	60 m

* data collected at 120 m spatial resolution, but resampled and distributed at 60 m. Based on U.S. Geological Survey (http://eros.usgs.gov/#/Find_Data/Products_and_Data_Available/band).

wetlands, these instruments detect red and near-infrared wavelengths for vegetation analysis over broad regions. Thus, the state of the world's vegetation can be monitored for seasonal and interannual changes.

At the moderate-resolution level, Landsat is the longest running of all Earth resources satellite systems. The success of Landsat spawned many similar earth-resources satellites in the late twentieth century by the United States and several other countries (Lauer, Morain and Salomonson 1997; Tatem, Goetz and Hay 2008). Late in 2008, the U.S. Geological Survey made all archive Landsat datasets freely available to the public, which opened tremendous potential for long-term analysis of land-cover conditions.

The 1970s Landsats (1, 2, 3) carried a Multi-Spectral Scanner (MSS) which detected two visible bands (green, red) and two near-infrared bands at 80 m resolution. Landsats 4 and 5, launched in the early 1980s, had the MSS as well as the more advanced Thematic Mapper (TM), which has seven bands and improved spatial resolution (Table 3-3). In the 1990s, Landsat 6 failed to achieve orbit, but Landsat 7, launched in 1999, carries an Enhanced Thematic Mapper Plus (ETM+) instrument. In spite of some technical issues, Landsats 5 and 7 are still functioning with Landsat data now spanning four decades, and the Landsat Data Continuity Mission is scheduled for launch in December 2012 (National Aeronautical and Space Administration (NASA) 2011a).

Landsat imagery has proven of great value for wetland studies and is widely utilized for water-resource investigations (e.g. USGS 1995). We employ Landsat imagery extensively in this book to illustrate spatial and temporal characteristics of selected wetlands. Single-band images are presented in gray tones. Composite images employ three bands that are color coded to produce false-color pictures; most examples presented herein are TM bands 3, 4, 5 and 2, 5, 7 (Fig. 3-14). Other band combinations are possible as well as multi-temporal composites (Pavri and Aber 2004).

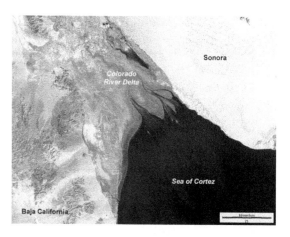

Figure 3-14. False-color composite Landsat TM image of the Colorado River delta in the Sea of Cortez, Mexico (see Color Plate 3-14). Bands 3 (red), 4 (near-infrared) and 5 (mid-infrared) color coded as blue, green and red. Irrigated crops are bright yellow-green; dry mud/salt flats are bright cyan; sand dunes are near white. Landsat 5, March 2004. Image from NASA; processing by J.S. Aber.

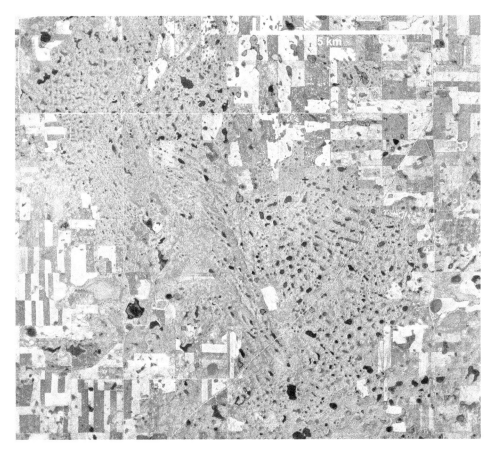

Figure 3-15. Conventional panchromatic (visible, gray tone), vertical, aerial photograph of the Crestwynd vicinity, southern Saskatchewan, Canada. Part of the prairie pothole region, this area contains countless small lakes and marshes of variable sizes and irregular shapes. Photograph A21639-7 (1970); original photo scale 1:80,000. Reprocessed from the collection of the National Air Photo Library, Natural Resources Canada.

3.2.4 Meso-level systems

Conventional aerial photographs have been the workhorse for wetland mapping and investigations of all sorts since the mid-twentieth century. These are most typically panchromatic (visible, gray tones), color, or color-infrared vertical pictures taken from several thousand meters above the surface at original (film) scales of 1:12,000 to 1:80,000 and spatial resolutions of 1–4 m (Fig. 3-15; Tiner 1997). In many countries, useful aerial photographs date back 60 years or more and represent important historical images for documenting wetland conditions (e.g. Aaviksoo, Kadarik and Masing 1997). In some countries, however, aerial photography is classified for security reasons and is not available to the public or scientific researchers.

A popular trend during the past decade is conversion of conventional aerial photographs into digital orthophoto quadrangles (DOQ). These images have been resampled to remove geometric distortions and to fit accurately onto a ground coordinate grid with high pixel resolution – usually 1 or 2 m. In the United States, DOQs are produced in the NAD83 geodetic datum and UTM grid projection. Each DOQ corresponds to 3.75 or 7.5 minutes of latitude and longitude referenced to standard 7½-minute map sheets (U.S. Geological Survey 2001). DOQs may be utilized for GIS in combination with other kinds of cartographic data

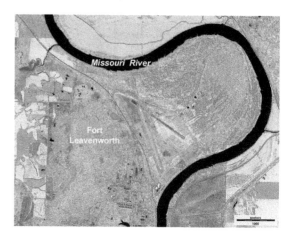

Figure 3-17. Ikonos false-color composite of Fort Leavenworth, northeastern Kansas, United States (see Color Plate 3-17). Green, near-infrared and red bands color coded as blue, green and red; active vegetation appears dark green to yellow-green colors. Dataset acquired August 2000; compare with Figure 3-16. Image from NASA; processing by J.S. Aber.

Figure 3-16. Digital orthophotographs of Fort Leavenworth, northeastern Kansas, United States. A. Four DOQs were concated together to create this mosaic. The tile boundaries are most obvious in the Missouri River because of different lighting conditions. B. Close-up view of the central fort complex. The star-shaped building in upper right corner is the former disciplinary barracks (military prison). Note the level of detail visible in this scene. Image processing by J.S. Aber.

(Fig. 3-16). A parallel trend is replacement of conventional aerial cameras employing large-format film by digital aerial cameras with similar spatial resolution.

Beginning in 1999 with Ikonos, high-resolution commercial satellites have provided imagery with GSD of 1-4 m and, most recently, as small as half a meter (Tatem, Goetz and Hay 2008). This spatial resolution rivals conventional airphotos and DOQs. These satellites also provide multispectral capability, typically in blue, green, red, and near-infrared bands, which is highly desirable for wetland imagery (Fig. 3-17). The main drawback for this type of data is its high commercial cost, which amounts to several thousand US dollars per scene, compared with essentially free DOQ, Landsat, and other data types. Google Earth provides high-resolution imagery for many parts of the world, which is quite valuable for general geography; however, metadata are not given, which limits the usefulness of such images for detailed interpretation purposes.

3.2.5 Micro-level systems

To reach the micro level of wetland research, GSD <10 cm is necessary. This can be achieved through small-format aerial photography (SFAP). Manned or unmanned platforms include ultralights, paragliders, balloons, blimps, model airplanes, kites, and other aircraft that typically fly at heights of just a few hundred meters (Aber, Marzolff and Ries 2010). Many types of film or digital cameras may be utilized to acquire photographs in visible and near-infrared portions of the spectrum. We have used kites and a small helium blimp extensively to document wetlands in many settings (Fig. 3-18 and Fig. 3-19).

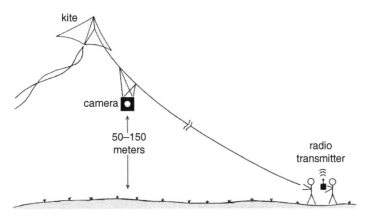

Figure 3-18. Cartoon showing the typical arrangement for kite aerial photography. The camera rig is attached to the kite line, and a radio transmitter on the ground controls operation of the camera rig. The kite line normally is anchored to a secure point on the ground. Not to scale; adapted from Aber, Zupancic and Aber (2003, Fig. 1).

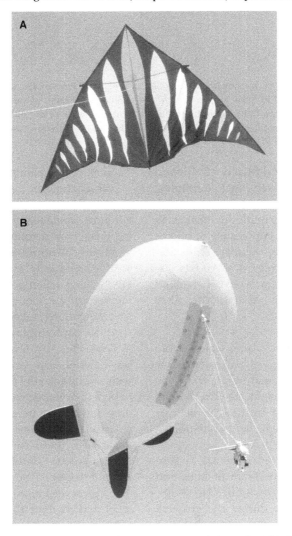

Figure 3-19. Platforms for SFAP. A. Large delta kite has a wing span of about 6 m. This kite is particularly well suited for high altitudes because of its large surface area. Photo by J.S. Aber. B. Helium blimp ready for flight with radio-controlled camera rig attached to keel. Blimp is 4 m (13 feet) long and contains about 7 m^3 of helium; tether line extends to lower right. Photograph courtesy of N. Hubbard.

Figure 3-20. Close-up vertical view of elephant seals on the beach at Point Piedras, California, United States. These juvenile seals are from about 2 to 2.5 m long, and most are sleeping on a bank of seaweed. People were not allowed to approach the seals on the ground, but the seals were not aware of the photographic activity overhead. The spatial detail depicted in such images is extraordinary; individual pebbles are clearly visible on the beach. Kite aerial photo by J.S. Aber and S.W. Aber.

Tiner (1997) noted the difficulty of wetland interpretation using conventional airphotos, because of the highly variable water, topographic, and vegetation conditions that may apply. He emphasized the importance of photographic scale for setting spatial limits on wetland mapping. In this regard, SFAP has distinct advantages for certain types of wetland investigations (Aber and Aber 2001; Aber et al. 2002).

- High-resolution (2–5 cm), large-scale, visible and near-infrared imagery are suitable for detailed mapping and analysis.
- Equipment is light in weight, small in volume, and easily transported by foot, vehicle or small boat under difficult field conditions – peat, mud, water.
- Minimal impact on sensitive habitat, vegetation and soil. Silent operation of kites and blimps does not disturb wildlife (Fig. 3-20).
- Repeated photography during the growing season and year to year documents changing environmental conditions (see Fig. 4-3).
- Lowest cost by far relative to other manned or unmanned types of remote sensing to achieve comparable high spatial and temporal resolutions.

For most satellite systems, time of day for acquiring imagery is determined by the satellite orbit, and nearly all airborne and space-based systems are designed only for vertical (nadir) imagery. SFAP allows much greater flexibility for acquiring vertical and oblique vantages (Fig. 3-21) in all orientations relative to the sun position, shadows, and ground targets (Fig. 3-22).

Over the past few decades, such remotely sensed data have provided wetland managers with important information about the effectiveness of their management efforts. In some cases, they have also persuaded policy makers to endorse more stringent management strategies. Regardless, these data give the scientific community a better understanding of the functioning and health of these ecosystems and indicate that a great deal more needs to be done to document and monitor these habitats. Such studies not only improve our understanding of these unique landscapes and their functioning, but perhaps also provide convincing evidence for their future conservation.

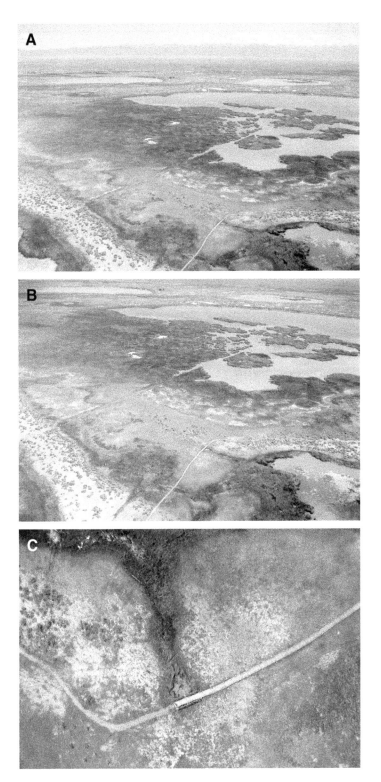

Figure 3-21. Russell Lakes State Wildlife Area, San Luis Valley, southern Colorado, United States. A. High-oblique view looking toward the northeast with the Sangre de Cristo Mountains visible on the horizon. B. Low-oblique vantage with the camera tilted toward the side, but the horizon is not visible. Notice footpath in lower portion of scene. C. Close-up vertical shot looking straight down – nadir view. The boardwalk is about 4 feet (1.2 m) wide; the GSD is ~1 inch (2.5 cm); interpretability level = 7 (individual boards in the walkway). Blimp airphotos by J.S. Aber and S.W. Aber.

Figure 3-22. Eroding cliff on the North Sea coast at Bovbjerg, Jylland, Denmark. A. View toward the north with good illumination of the scene; note shadow of lighthouse on right side. B. View of same cliff looking toward the south. Sun glint from North Sea and heavy shadows on cliff face result in poor depiction of details. Kite aerial photographs; after Aber, Marzolff and Ries (2010, Fig. 4-12).

3.3 Maps and geographic information systems

A map is a graphic representation or scale model of spatial concepts. It is a means for conveying geographic information, namely those physical, biological, and cultural phenomena related to wetlands. Maps are a universal medium for communication, easily understood and appreciated by most people, regardless of language or culture. Incorporated in a map is the understanding that it is a "snapshot" of an idea, a single picture, a selection of concepts from a constantly changing database of geographic information (Merriam 1996).

Geographic information systems (GIS), also known as geographic information science, is a multidisciplinary technology that emerged in the 1970s and 80s and became a routine working method in the 1990s. GIS represents a major shift in the cartographic paradigm. In traditional (paper) cartography, the map was both the database and the display of geographic information. For GIS, the database, analysis, and display are physically and conceptually separate aspects of handling geographic data. Geographic information systems comprise computer hardware, software, digital data, people, organizations, and institutions for collecting, storing, analyzing, and displaying georeferenced information about the Earth (Nyerges 1993).

GIS combines traditional surveying, geodesy, and cartography with remote sensing, database systems, global positioning system (GPS), and sophisticated statistical, spatial, and temporal data analysis (e.g. Burrough 1988; Chang 2008). It has become a primary method for wetland research. One example of this approach is documenting playas of the southern High Plains of the central United States (Fig. 3-23). Previous estimates for the total number of playas in this region varied greatly; Steiert and Meinzer (1995) cited a value of 25,000 with 2000 in Kansas. Other estimates ranged as high as 60,000 total for the region with 10,000 in Kansas (Evans 2010). A GIS analysis of high-resolution aerial imagery, digital raster graphics, and soil data has revealed more than 22,000 playas in Kansas alone (Fig. 3-24), more than twice the previous high estimate (Kansas Geospatial Community Commons 2010).

Having emphasized the advantage of GIS for wetland research, it is wise to note some cautions. Maps and GIS databases are never perfect pictures of the so-called real world. Field measurements are subject to errors of accuracy and precision. Aerial photographs and satellite

METHODS IN WETLAND RESEARCH   47

Figure 3-23. Playa basins on the nearly flat, featureless High Plains in west-central Kansas, United States. Ephemeral lake in a playa depression during a wet period (see Color Plate 3-23). Water fills a shallow basin in a fallow field with winter wheat fields in the background. Kite aerial photograph; after Aber and Aber (2009, Fig. 17).

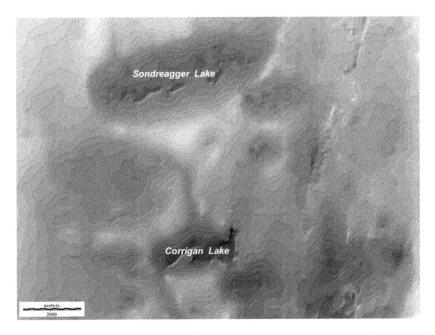

Figure 3-24. Highly exaggerated, shaded-relief digital elevation model. Numerous closed depressions (playas) are evident. Derived from 1-arc-second (~30m) National Elevation Dataset from the U.S. Geological Survey (http://seamless.usgs.gov/). Image processing by J.S. Aber with Idrisi Taiga software.

images represent only certain portions of the light spectrum, as filtered through the atmosphere and detected by cameras. No map can depict all physical, biological, and cultural features for even the smallest area. A map may display only a few selected features, which are portrayed usually in highly symbolic styles according to some kind of classification scheme. In these ways, all maps are estimations, generalizations, and interpretations of true geographic conditions.

Furthermore, all maps and GIS datasets are made according to certain basic assumptions, for example sea-level datum, which are not always true or verifiable. Finally any map is the product of human endeavor and, as such, may be subject to unwitting errors, misrepresentation, bias, or outright fraud. In spite of these limitations, maps and now GIS have proven to be remarkably adaptable and useful through several millennia of human civilization. Maps of all kinds are fundamentally important for wetland research.

3.4 Physical ground-based methods

3.4.1 Surface methods

Ground-based means of study include all types of traditional biological and physical methods for sampling, measuring, and describing the water, soil, vegetation, and wildlife of wetlands. Detailed descriptions of these methods are beyond the scope of this book; the reader is advised to consult specific procedures for each discipline. A few of the more basic ground methods are presented here, beginning with measuring the quantity of water flowing into or out of a wetland system.

Surface water flow through well-defined channels is measured by stream gauging. Measuring discharge at a gauging station depends on knowing the cross-sectional area of the channel and the velocity of water flow at different depths and positions across the channel. Once these have been determined by preliminary field measurements, the discharge may be calculated from stream stage (height). Gauging stations

Figure 3-25. Temporary gauging station (box in center) erected on a glacial melt-water stream that flows into the lake in the background (A). View of recording instrument that measures water height in the pipe below the box (B). Photos by J.S. Aber at Boundary Glacier, Alberta, Canada.

range from temporary, seasonal setups (Fig. 3-25) to permanent, automatic stations that transmit real-time data for online display (Fig. 3-26). Direct water gain (precipitation) and loss of water via evapotranspiration may be determined

Figure 3-26. Permanent gauging station on the upper Missouri River at Culbertson, Montana, United States. A. Cable (∧) supports a manned carriage for measurements of depth and velocity across the channel. The carriage comes from the far side (*). B. Closeup view of manned carriage that rides across the cable. U.S. Geological Survey gauging station 06185500 in operation since 1941; current and historical data online (http://nwis.waterdata.usgs.gov/nwis).

Figure 3-27. Portable, automatic, solar-powered weather station operating at an artificial wetland site for passive treatment of contaminated ground water. All conventional meteorological factors are measured for site and regional weather monitoring and analysis. Photo by J.S. Aber at Commerce, Oklahoma, United States.

by meteorological means (Fig. 3-27). Measuring underground water movement is much more complicated and requires monitoring wells and some knowledge of the aquifer characteristics (see below).

Color is an important attribute of wetland soils, as noted above. Color vision is among the most important human senses, and a great deal has been written about human perception of color both from the aesthetic and technical points of view. A full discussion of color is well beyond the intent of this book, so only a few basic aspects are presented here. For a good discussion of color in nature, see Lynch and Livingston (1995). Several means exist to define color quantitatively. One approach is based on the proportions of primary colors (blue, green, red) and their intensities, which applies to digital photography, color monitors, and human vision (e.g. Drury 1987).

Another well-established means for defining color is the Munsell Color system, which is based on three attributes of color that are often illustrated in a wheel diagram (Fig. 3-28). It is widely used in the natural sciences for describing soil and rock colors (Fig. 3-29).

- Hue – actual spectral color such as red, yellow, green, blue, etc. Hue is designated by a number and letter.
- Value – lightness or brightness of the color. Value ranges from zero for pure black to 10 for pure white.
- Chroma – intensity or saturation of the color. Chroma begins with zero for neutral (gray) and increases with no set upper limit.

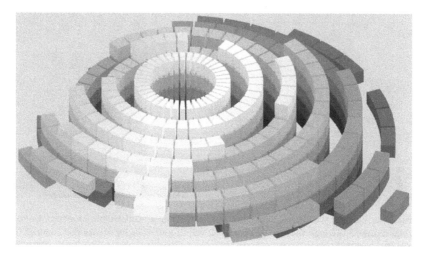

Figure. 3-28. Schematic illustration of the Munsell Color system (see Color Plate 3-28). Hue is the spectral color (circumference); chroma is the intensity of color (radius); value is the brightness (vertical axis). Modified from original illustration by SharkD; obtained from Wikimedia Commons <http://commons.wikimedia.org/>.

Figure 3-29. Distinctive colors are displayed in this wetland soil (see Color Plate 3-29). Dark gray/brown indicates a high content of organic matter, and orange mottles show oxidized iron. Taken from Vasilas, Hurt and Noble (2010, Fig. 7).

Figure 3-30. Collecting a peat/sediment core from Skansen Bog on Askøy Island, near Bergen, Norway. A. Peat drill. The cutting and sampling part is 1 m long (to left) with extension poles added in 1-m sections to right. B. Pulling up the drill. Note the risk of sinking into the soft turf. C. Inge Aarseth examines the resulting core sections laid out for inspection. Such peat cores may be collected up to 10–12 m deep. Photos by J.S. Aber.

3.4.2 Subsurface methods

Learning about the subsurface characteristics of wetlands usually involves test drilling and geophysical prospecting. Wetland soils and underlying sediments are generally unconsolidated materials that are relatively easy to penetrate with manual or powered drilling equipment. For the study of peat deposits, manual drilling equipment is usually sufficient to recover continuous core samples (Fig. 3-30). When doing any type of sampling or measuring work in a bog, fen or swamp, the soft turf and watery surface may not fully support the weight of people and equipment. It may be necessary to lay down a temporary wooden platform, or in some cases permanent platforms are constructed for research purposes (Fig. 3-31). The recovered samples may be subjected to various types of laboratory analysis for clay mineralogy, pollen content, radiocarbon age, etc.

Radiocarbon (^{14}C) is undoubtedly the most important radioactive isotope for dating organic materials of the late Pleistocene and Holocene. Wood, charcoal, peat, shell, bone, and other organic remains are suitable sample materials (Fig. 3-32). Several analytical techniques are in use for radiocarbon dating – proportional gas counting, liquid scintillation, and negative-ion accelerator mass spectrometry (AMS). Each type of sample must be prepared in special ways, which is best done by a dedicated radiometric-dating laboratory. In any case, all radiocarbon dates are given with a standard deviation, for example 10,000 ± 200 years BP. This indicates the dated age likely lies between 9,800 and 10,200 years ago; BP means before present. Radiocarbon dates may be calibrated by comparison with the tree-ring record established by dendrochronology.

Palynology is the study of pollen. Pollen is well preserved as a record of ancient land vegetation and hence is useful for reconstructing past environmental and climatic conditions. Pollen is the resistant, dust-sized (0.01–0.1 mm) male reproductive apparatus of plants. Pollen grains can usually be identified on the basis of size, shape and ornamentation to the generic and sometimes specific level of classification. Most common trees and grasses are wind pollinated, and their pollen grains are widely scattered and preserved in lake and mire sediments. The science of palynology began in Scandinavia, where a classic late glacial/Holocene pollen sequence has been established (Fig. 3-33). The major periods correspond to climatic conditions that controlled vegetation. Additional factors include soil development, depletion of nutrients,

Figure 3-31. Platforms for mire research. A. Temporary wooden pallet laid around a peat coring site on a watery subalpine fen in Colorado, United States. Photo by J.S. Aber. B. Aerial view of raised wooden boardwalk (*) in Männikjärve Bog, Estonia. Note foot trail leading to the boardwalk from the lower left corner; the boardwalk is ~10 m long. This platform is reserved for scientific research and is not open to the public. Kite airphoto by J.S. Aber and S.W. Aber. C. Flux chambers with tubes and battery-driven pumps to collect samples of gas emitted from the bog at the research boardwalk shown in B. Image courtesy of P. Frenzel (Frenzel and Karofeld 2000).

rate of plant migration, competition between plants through time, and early human impact on vegetation (Iversen 1973).

Power drilling equipment or geophysical instruments are necessary in the deeper subsurface. Direct-push drilling, such as Geoprobe, is a good way to collect samples, prepare monitoring wells, and conduct well logging tens of meters deep in unconsolidated sediments typical of most wetlands (Fig. 3-34). Geophysical techniques with potential for use in wetlands include the transient electromagnetic (TEM) method and ground-penetrating radar (GPR). Both are suitable for shallow subsurface investigations in unconsolidated sediment.

TEM is based on the response of subsurface materials – solids and fluids – to an electromagnetic impulse (Jørgensen, Sandersen and Auken 2003). TEM is done with a transmitter loop carrying a steady current (3 A) and a receiving coil in the center of the loop. The ground response is measured when the transmitted signal is turned off for brief intervals (9 μs to 9 ms). The sampling may be done in a discrete grid pattern or along a continuous profile. Depth of penetration varies from 100 to 300 m, and spatial resolution spreads out with greater depth. TEM has proven valuable for investigating buried valleys and their aquifers; for example in Denmark (Jørgensen et al. 2005).

GPR has become relatively common for shallow-subsurface prospecting and mapping purposes. Short pulses of high-energy microwaves are transmitted from the surface into the

GPR is useful for determining deeper and larger geologic structures. Shorter wavelengths (higher frequency, 300 to 1000 MHz) are better for identifying shallow, small objects such as buried pipelines or individual boulders. However, microwave interactions are influenced strongly by water content of soil and sediment; the best subsurface conditions are either dry or frozen materials (e.g. Lønne and Lauritsen 1996).

3.5 Flora, fauna and ecological monitoring and survey methods

Using well-tested spatially distributed point, transect or grid sampling techniques to assess ecological conditions across a study site may provide valid geographic generalizations about the same. Following a well-defined sampling process ensures a reliable data gathering effort. The sampling process starts with identifying the target geographic area, constructing a sampling frame, selecting an appropriate sample design, identifying data to be collected and proceeding with data collection (Burt, Barber and Rigby 2009). For wetland environments, spatial probability samples often employ stratified or cluster designs to adequately capture the complex ecological conditions present at any given site.

Like ecological surveys, periodic and systematically conducted species-specific surveys are one way to provide regional population estimates for wetland plants, birds, mammals, amphibians, reptiles and other animals. Such surveys are undertaken in numerous countries across the world and can provide reliable local or regional presence, absence or abundance indices for target species over time. The results may help prioritize conservation and management goals by providing population trends and spatial distribution patterns. Moreover, periodically conducted surveys could offer a proxy for the general health of the larger ecosystem, while also providing early warning signals for disturbance or degradation within an ecosystem. Beyond population estimations, a variety of species-specific data also could be gathered through such surveys. These may

Figure 3-32. Sample materials suitable for radiometric dating. A. Carefully removing a peat sample from the coring tool. The age of basal peat from a bog establishes the time when peat accumulation began. B. Fossil shells (*Litorina* and *Cardium*) in shallow marine deposits of latest Pleistocene age, Skærbæk, northern Jylland, Denmark. Dates on seashells must be corrected for the marine carbon-reservoir effect. Photos by J.S. Aber.

ground. The microwaves propagate through the ground, interact with subsurface materials in complex ways, and may be scattered or reflected. Microwave frequency (wavelength) determines the depth of penetration as well as spatial resolution of the resulting data. Longer wavelengths (lower frequency, 25 to 200 MHz) result in deeper penetration, but lower resolution. Such

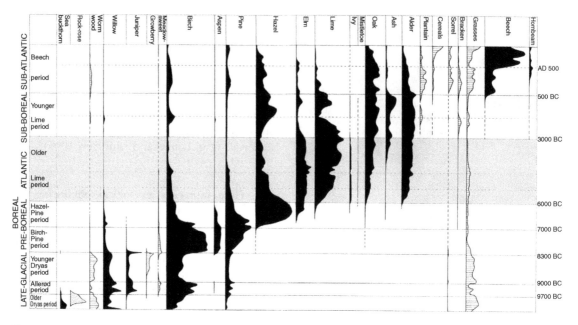

Figure 3-33. Composite pollen diagram based on multiple sample sites in eastern Denmark, in which pollen production of each species is taken into account. Black = trees and shrubs, dotted-gray = dwarf shrubs, lined = herbs and grass. The periods shown on left side are developed throughout much of northern Europe and reflect climate change, soil development and plant migration. The Atlantic (gray) represents a climax forest dominated by elm, linden (lime), oak, and other hardwood trees. The impact of early agriculture becomes marked in the Sub-boreal and Sub-Atlantic periods. Dates are uncorrected radiocarbon ages. Adapted from Iversen (1973, Fig. 34). Copyright: Geological Survey of Denmark and Greenland (GEUS).

include information on a given species' habitat, distribution, range and health, reproductive and breeding patterns.

To produce reliable results, monitoring methods should employ scientifically tested survey sampling designs and follow set field recording protocols (Fig. 3-35). New technologies such as GPS, GIS and remote sensing also may be used to aid in capturing, analyzing and publishing these data. Monitoring methods that are easy to implement and quick to conduct, include stakeholders, and are repeatable, scale appropriate and sensitive to local geographic contexts increase data reliability.

In addition to these elements, field survey methods documenting fauna must be mindful of seasonal or episodic patterns of occurrence with birds, fish, amphibians and other migrating species. To increase chances of recording species, careful attention must be given to the activity patterns of target animals. Field visits should be coordinated during peak activity time, keeping in mind that some species may be more active at night or during early morning and evening hours. All the same, as Tyre et al. (2003) suggested, several factors would influence the ability to observe a species at any given time. These include the spatial extent of the home range of a species, prevailing climatic conditions, the sample and survey design, and the expertise of the surveyor to identify and record species. Scientists conducting faunal surveys especially guard against false absences or the failure to detect a species, which can skew the results of a study. Increasing the number of site visits could increase the chances of spotting target species.

One noteworthy example of a monitoring and survey method is the North American Breeding Bird Survey (BBS). Initiated in 1966, this annual breeding-season bird survey uses government and non-government experts and local

Figure 3-34. Constructing a ground-water monitoring well in floodplain alluvium with a Geoprobe direct-push drilling rig. Campus of Emporia State University, Kansas, United States. Photo by J.S. Aber.

volunteers from the United States and Canada to engage in a systematic bird survey based on a set methodology along thousands of kilometers of roads within the two countries (U.S. Geological Survey (USGS) 2007). Volunteers methodically record every bird vocalization or sighting at several predetermined stops along an assigned 39.2 km route. At each stop, a volunteer spends a few minutes recording birds within a 400 m radius. As the North American BBS mission statement suggests, this survey is to provide "scientifically credible measures of the status and trends of North American bird populations at continental and regional scales to inform biologically sound conservation and management actions" (USGS 2007, p. 2).

The British Trust for Ornithology (BTO), a non-profit, citizen-science organization in the United Kingdom has been conducting a similar BBS across the country since 1996 (British Trust for Ornithology 2011a). Civilian volunteers are trained to collect data and monitor changes of some 100 bird species on randomly sampled 1 km^2 sites across the United Kingdom. The BTO survey includes three field visits over the breeding season with an average visit time of

Figure 3-35. Experimental prairie plots designed to test how different treatments (burning, mowing) may influence vegetation and wildlife. Each plot measures 30 × 30 m; small pond in upper right corner of view. Plots are sampled periodically on the ground and documented by aerial photography. Kite aerial photo by S.W. Aber and J.S. Aber; Kansas, United States.

approximately 90 minutes each (British Trust for Ornithology 2011b). In addition to bird sightings, data on habitat conditions are also recorded and monitored at survey sites. These data are shared to promote appropriate conservation and management strategies.

3.6 Social science methods and techniques

Wetlands provide fundamental and life sustaining resources to humans, and their use by humans has been well documented in historical and contemporary studies. Recognizing and documenting the cultural and socio-economic value of these environments is vital to their sustainable management. Established social science methods including the analysis of census data, resource-use surveys, in-depth interviews, oral histories and other economic valuations such as resource assessments are used to evaluate human use and value of wetlands.

It is essential to recognize that most wetland landscapes across the globe have been settled and modified by humans over millennia. Recent information on the scale and scope of human habitation in wetland ecosystems may be obtained through census counts. A census provides a full count of individuals and often includes their demographic and socio-economic characteristics. Such systematic counts provide necessary information on not just the number of individuals that live within wetland ecosystems but also how these ecosystems might be used. For instance, district and village level censuses conducted every 10 years across India provide information on economic activities by trade (such as fishing, aquaculture, etc.), in addition to the population's socio-demographic characteristics of age, sex, and education. When analyzed over time, these data provide valuable insights to ongoing socio-economic changes across wetland regions.

In the United States, the Bureau of the Census provides valuable decadal socio-demographic and income data on residents of the country at various scales of aggregation from the individual household to the census track, township, county and state levels. Similarly, the Economic Census conducted every five years provides data by economic sector. Other data relevant to wetlands are collected and disseminated by various governmental agencies such as the U.S. Army Corps of Engineers, Bureau of Land Management, Department of Agriculture, Environmental Protection Agency, Geological Survey, Fish and Wildlife Service, National Marine Fisheries Service, and the Natural Resources Conservation Service.

Survey instruments may be used to collect a variety of information from individuals who live in or use resources from wetlands. Sampling frameworks provide statistically robust ways to collect such data from a smaller subset of a larger population. Similar to ecological surveys, such resource utilization surveys may be administered to target populations using random, systematic, stratified, or cluster sampling techniques. Surveys allow for more targeted and in-depth data collection specific to wetland activities and resources and, as such, could yield valuable information on a number of issues (Secretariat of the Convention on Biological Diversity 2006). These include:

- Land tenure
- Traditional management practices
- Ecosystem services and benefits
- Resource use strategies
- Harvesting techniques
- Site-specific social or cultural traditions that influence resource use

In addition to surveys, in-depth interviews with key informants, local experts, indigenous groups, stakeholders, wetland managers and the like may provide more detailed perspectives on wetland-related issues under scrutiny. Oral traditions passed down through generations often disseminate practical knowledge regarding fishing, trapping, harvesting, agricultural techniques and traditional medicine and, as such, comprise the cultural wealth of many traditional societies (Papayannis and Pritchard 2008). Oral histories from key informants documenting such wetland-related traditions and practices could serve to maintain the knowledge base and inform culturally sensitive management strategies. Social science methods including interviews, participant observation, and recording oral histories may collectively help in documenting such data.

Lastly, resource assessments may be used to evaluate the potential economic value of wetland resources from a particular site. Data may be collected on the abundance and stock of a particular resource, its economic value and importance to local economies, extraction practices used for harvesting, and information about dependent local communities (Secretariat of the Convention on Biological Diversity 2006). Such assessments could provide essential information for maintaining a sustainable resource base.

3.7 Summary

All manner of physical, biological, and social science methods may be applied for wetland investigations, which range from microscopic laboratory analysis, to economic modeling, to space-borne satellite observations of continental regions. Based on spatial resolution of surficial environments, methods for investigating wetlands may be divided roughly into three levels – macro (>10 m), meso (10 m to 10 cm), and micro (<10 cm), which may be accomplished by various means of space-borne, aerial, and ground-based cameras, instruments, sampling, and observations. Emphasis usually falls on the primary aspects of water, soil and vegetation as well as wildlife and human interactions with wetlands. Remote sensing and geographic information systems play key roles for acquiring, assembling, displaying, analyzing, and interpreting diverse types of data.

The temporal dimension ranges from diurnal cycles to wetland development over millennia. For longer time frames, the vertical dimension (depth) gives stratigraphic information about changing conditions, as documented by layers of sediments, fossils, and other factors. Subsurface prospecting and logging include various types of manual and powered drilling equipment as well as shallow-subsurface geophysical techniques. The combination of multi-spatial and multi-temporal methods applied to wetlands yields comprehensive and convincing results.

Ecological and socio-cultural inventorying through biological and social science methods are widely used in collecting wetland-related data. Ecological, species-specific and resource-use surveys collected through sampling may provide valuable information for wetland conservation and management. Cultural histories and traditions could also be documented through in-depth interviews and oral histories with key informants.

Wetland hydrology 4

4.1 Introduction

The quantity and quality of water are crucial factors for wetlands as well as all other environments (Niering 1985). Hydrology is the science of water, particularly its flow and storage, physical and chemical properties at the surface and in the subsurface. Water is constantly in movement at the Earth's surface via the water or hydrologic cycle, which is powered by solar energy (Fig. 4-1). Wetlands represent a storage point within the hydrologic cycle, and the relationship of water gains, losses and storage is known as the water budget. In simple terms, these three components must balance for a given wetland (Charman 2002), so that:

Gains – Losses – Storage = zero

Wetlands may gain water from direct precipitation, surface inflow, and ground-water discharge. Water is lost via evaporation, transpiration, surface outflow, and ground-water recharge. The amounts lost by evaporation from soil and water bodies and transpiration by plants are usually difficult to measure separately, so the term evapotranspiration is often used for the combined effects of both mechanisms. Wetlands represent a delicate balance between these factors. The balance may be relatively stable through time, so that a wetland does not appear to change much (Fig. 4-2), or may fluctuate seasonally or erratically from year to year (Fig. 4-3). The former is common in relatively humid climates, whereas the latter is typical of semiarid and arid environments.

Storage capacity of a wetland refers to the volume of water held in pools, soil, and the shallow subsurface. Storage capacity is determined by many factors – bedrock, sediment, soil, landform, water level, and vegetation. For peatland, Charman (2002, p. 42) likened a mire to a "giant bubble of water held together by a mass of living and dead plant material." The storage capacity may be relatively large or small in relation to the area of the wetland. Likewise, storage capacity may be relatively large or small compared with the annual water gains and losses. These factors, working together, influence the hydroperiod, which is the seasonal pattern of water level (Welsch et al. 1995).

Marine coastal and estuarine wetlands exhibit daily and monthly fluctuations associated with tides (Fig. 4-4). Given the widespread occurrence of wetlands along the continent–ocean interface, tides play an important and influencing factor in determining coastal wetland compositions and distributions. Hydrological conditions shaped through tidal action along with the configuration of the coastline, slope of the land, and influx of fresh water determine the two principal types of coastal wetlands – saline or brackish – found across the world. Tides caused by the gravitational pull of the moon and sun flood coastal regions diurnally, submerge coastal soils and plants, and deposit, erode and redistribute sediment in somewhat

Wetland Environments: A Global Perspective, First Edition. James Sandusky Aber, Firooza Pavri, and Susan Ward Aber.
© 2012 James Sandusky Aber, Firooza Pavri, and Susan Ward Aber. Published 2012 by Blackwell Publishing Ltd.

WETLAND HYDROLOGY 59

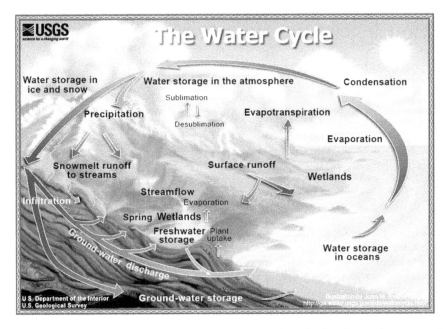

Figure 4-1. The general water cycle showing wetland storage points for inland and coastal settings. Adapted from an illustration by Evans (2011).

Figure 4-2. The Fegge peninsula extends into the Limfjord estuary, northern Jylland, Denmark. Tidal effect is minimal; water level and chemistry rarely vary for the wet meadow, marsh, shore, and shallow-water environments. Cattle in lower left corner for scale; kite airphoto by S.W. Aber and J.S. Aber.

predictable patterns (Mitsch and Gosselink 2000).

Coastal wetland regions with ocean frontage and the unrelenting onslaught of tidal action followed by salt-water submergence provide conditions conducive to the establishment of salt marshes dominated by grasses in the upper latitudes and mangroves that include trees with special adaptations in tropical regions (van den Bergh, Barendregt and Gilbert 2004). Coastal areas that observe the mixing of fresh and salt water along estuarine zones, tidal creeks and inlets provide an environment ideal for the development of brackish marshes. In both salt and brackish marshes, salinity levels fall away from the open ocean and in proximity to freshwater source inputs.

On the other hand, inland wetlands tend to exhibit seasonal hydroperiod patterns based on precipitation, spring thaw, and storm events (Fig. 4-5). In headwater settings, wetlands typically have limited storage capacity and respond rapidly to local storms or snow melt, especially in areas with silty or clayey soils and steep slopes that generate fast runoff. Sandy soils and more gentle slopes allow more infiltration, so runoff is slower. Hydroperiods for headwater wetlands may be quite variable, depending on local weather events.

Large riparian wetlands in downstream locations tend to respond more slowly to individual storm events. Such wetland hydroperiods display average conditions over the larger watershed

Figure 4-3. Overview of Dry Lake, an ephemeral lake at the terminal point of an enclosed basin on the High Plains in west-central Kansas, United States (see Color Plate 4-3). A. May 2007, a wet year with the lake full of water; note small overturned row boat in lower left corner for scale. B. May 2008 displays a wet mudflat surrounded by salty soil. Similar views toward the southwest; kite airphotos by S.W. Aber and J.S. Aber.

Figure 4-4. Tidal effects on Rowley River and salt-marsh complex in northeastern Massachusetts, United States. A. Morning high tide. Rowley River and smaller tributaries are full. B. Afternoon low tide. Note exposed mudflats. The river is much narrower and some smaller channels are nearly drained. Tidal ebb and flow generate noticeable currents in the channels. Blimp airphotos by S.W. Aber, J.S. Aber, and V. Valentine.

and tend to reflect seasonal patterns in precipitation and snow melt. Wooded swamps are typically associated with larger river systems and, thus, also follow seasonal water supply from the entire drainage basin. These hydroperiods are more predictable.

Water enters a wetland with certain physical, chemical, and biological properties. Physical properties are temperature and turbidity, which is the content of suspended fine sediment (mud = silt + clay) particles. Chemical properties include salinity, acidity (pH), oxidation potential (Eh), hardness, and other dissolved solids or compounds. Biological properties refer primarily to microorganisms, namely bacteria and viruses, as well as other plants and animals (Fig. 4-6). As the water moves through a wetland, all of these properties may be altered by physical, chemical, and biological processes within the wetland environment. Residence time is a measure of how long it takes a parcel of water to move through the wetland, typically weeks, months or years for many wetlands. Longer residence times allow substantial modifications in water properties to occur. Human structures – ditches, canals, drains – that reduce residence

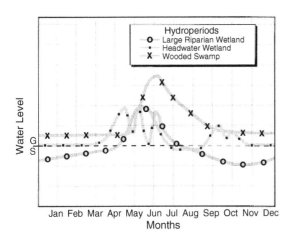

Figure 4-5. Idealized hydroperiods for wetlands in three settings in temperate climates, such as the northeastern United States. GS indicates ground surface level. Adapted from Welsch et al. (1995).

Figure 4-7. Drainage ditch in the headwaters of the Middle Loup River in the Sand Hills of western Nebraska, United States. Water level is lowered in the wet meadows and marshes to increase hay production, and water flow is accelerated downstream. Photo by J.S. Aber.

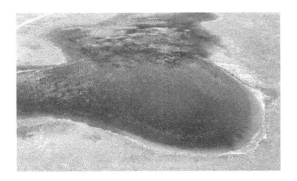

Figure 4-6. Marsh-pond complex in the Sand Hills near Lakeside in western Nebraska, United States (see Color Plate 4-6). Bright maroon and golden-orange colors in this pond are presumably caused by carotenoid pigments of invertebrates, such as brine shrimp, brine flies and rotifers, in the hyperalkaline water typical of the western Sand Hills region (Bleed and Ginsberg 1990). Kite airphoto by S.W. Aber and J.S. Aber.

Figure 4-8. Spring at the head of Trinchero Creek emerges from a fault in hard sandstone at scene center. The spring is surrounded by mounds of brook crest (*Cardamine cordifolia*). Culebra Range, Sangre de Cristo Mountains, southern Colorado, United States. Photo by J.S. Aber.

time also decrease the amount of water processing within the wetland (Fig. 4-7).

4.2 Surface and ground water

The relationship of wetlands to surface water inflow and outflow is often obvious and may be readily measured in many cases (see Figs. 3-25 and 3-26). Wetlands also interact with ground water in ways that are less obvious and more difficult to monitor. Aquifers are underground formations that both store and transmit water. The practical manifestations of aquifers are springs (Fig. 4-8), where ground water comes to the surface naturally, and man-made wells that produce water. In most wells, water rises to the level of the water table below ground and must be pumped to the surface (Fig. 4-9). In some cases, the aquifer has sufficient hydrostatic

Figure 4-10. Water flows to the surface under its own pressure in this artesian well in the Flint Hills, east-central Kansas, United States. Photo by J.S. Aber.

Figure 4-9. High-capacity water well next to the Oldman River in south-central Alberta, Canada. Electric pump lifts water for irrigation on adjacent uplands. Ground water is recharged from the river in the background. Photo by J.S. Aber.

pressure to lift water to the surface in so-called artesian wells (Fig. 4-10). Springs are important water sources for many wetlands (see Fig. 2-4). Strictly speaking, wells represent artificial water production. Once created, however, artesian wells may become permanent sources to maintain wetlands (Fig. 4-11).

The rate at which ground water moves through an aquifer or wetland sediment and soil depends on the nature of the material and the water pressure. This relationship is expressed as Darcy's Law (Davis and DeWiest 1966):

$$Q = k \, (dh/dl)$$

where Q is the flow or discharge (volume per time), k is the hydraulic conductivity or permeability for the material, and dh/dl is the hydraulic gradient (change in height over length) of the water surface, which is a measure of pressure.

Hydraulic conductivity varies greatly depending on the porosity of mineral and organic materials; the sizes, shapes, and connections between pores as well as water adhesion on particles all influence permeability. In general, coarse-grained mineral soils have relatively high permeability, whereas fine-grained organic soils have much lower permeability. Hydraulic conductivity values ($k = \text{cm/s} \times 10^{-5}$) range from >5000 for sand to <0.05 for clay and dense peat (Mitsch and Gosselink 2007). In the case of peat, pore size depends on the amount of decomposition of the plant material, the degree of compaction, and the type of parent plant. *Sphagnum* peat generally has the lowest permeability; peats of higher plants, such as *Carex* and *Phragmites*, are more permeable (Charman 2002).

Ground water moves both horizontally and vertically in response to the hydraulic gradient

Figure 4-11. Artesian well flows into a marsh at Russell Lakes State Wildlife Area in the San Luis Valley, south-central Colorado, United States. Photo by J.S. Aber.

at rates determined by hydraulic conductivity and hydrostatic pressure. Because of these movements, ground water has a close relationship with surface water bodies – streams, lakes, and wetlands; they are hydrologically connected. In the case of gaining streams, discharging ground water moves into the stream channel, and this baseflow maintains stream flow between surface runoff events. Conversely, water in a losing stream soaks into the ground, thus recharging the aquifer and reducing downstream surface flow.

Wetlands have similar relationships to ground water. In a recharging wetland, water moves from the wetland into the ground-water system. Thus, wetlands may be important sources of recharge for aquifers that feed into surrounding springs or supply well water. A discharging wetland is the opposite; water migrates from the aquifer into the wetland (Welsch et al. 1995). The state of recharge or discharge may fluctuate with the hydroperiod. During times of high water level in the wetland, ground-water recharge may take place. When water level in the wetland falls during the dry season or a drought period, the situation may reverse and become discharging.

On a regional basis, many wetlands may exist at different elevations and may be connected via various surface- and ground-water flows. Consider a series of linked wetlands at the end of a drought phase when water storage has been greatly depleted. As a wet climatic phase begins, wetlands at higher elevations must fill their storage capacity before they can begin to release surface water or recharge aquifers that feed into lower wetlands. Thus, filling of lower wetlands in the series may be delayed in a cascade manner (Welsch et al. 1995). The case of the Pantanal wetlands in South America illustrates this process nicely. The northern wetlands are replenished by rain-fed streams and rivers including the Paraguay River flowing from the highland regions of Brazil during the wet season of November through March. This water slowly makes its way into the southern Pantanal, nourishing the vast wetlands as it flows southward (see section 15.4).

Human diversions or extractions of either surface flow or ground water have the potential to impact integrated surface–ground-water systems both locally and downstream. A common scenario is overpumping of ground water, such that local or regional water tables are drawn down to a deep level at which springs, streams, lakes, and wetlands are diminished or dry out completely (Fig. 4-12). In coastal marine settings, overpumping of ground water may lead to ground subsidence and intrusion of salt water, such that fresh-water marshes are converted into open marine bays. Impoundment and diversion of surface water for irrigation, cities, or industry may reduce downstream flows substantially with severe impacts on wetlands. A famous case is the Colorado River, which no longer reaches its delta in the Sea of Cortez (see Color Plate 3-14); all the fresh water is extracted along the way. The Colorado River delta is considered among the world's most severely affected ecosystems (Avila-Serrano, Téllez-Duarte and Flessa 2009).

Figure 4-12. Finney Wildlife Area in southwestern Kansas, United States. Originally designed for a much larger lake, upstream ground-water pumping has reduced spring flows, and diversion of surface water has rendered this artificial "lake" into a wetland wildlife area. A. During rare wet years, a tiny puddle of water is held behind this dam (see Color Plate 4-12A). Asterisk indicates position of the outlet tower. Kite airphoto by J.S. Aber. B. No surface water is present during dry years, which are common. View from dam and outlet tower; photo by J.S. Aber.

4.3 Floods and flooding

From large rivers to small creeks, streams flow into, through or out of nearly all types of palustrine wetlands. Thus, stream flow is intimately connected with wetland hydrology. Because most streams and rivers flood, many wetlands are subject to flooding. In fact, flooding is a universal phenomenon which may occur in all climatic regimes and any size or type of drainage basin. The U.S. Geological Survey operates gauging stations and collects water-resource data on large rivers and smaller streams at approximately 1½ million sites throughout the country (see Fig. 3-26); many of these are automated with real-time and historical data available online. Flooding is an entirely natural event that takes place on most streams every few years.

Flooding becomes a hazard when high water leads to human casualties, damage to structures, and impairment of human land use.

Figure 4-13. Rural results of 1997 southern Polish flooding. Buildings were destroyed, and fine-grained alluvial topsoil was stripped away down to a base of cobble gravel. All inundated crops were declared a total loss, unfit for human or animal consumption due to contamination by raw sewage. Trees in fruit orchards were killed by prolonged submergence; many cattle and other livestock died. Village of Pilce next to the Nysa River, southwestern Poland. Photo by J.S. Aber.

Figure 4-14. Urban consequences of 1997 southern Polish flooding. The Nysa River inundated low-lying sectors in the city of Kłodzko. There was heavy damage to buildings, streets, vehicles, buried pipes, and other infrastructure. Lower floors of buildings were submerged and were temporarily abandoned for human occupation. The high water mark can be seen by the whitish stain on building walls. Photo by J.S. Aber.

Development of drainage basins for shipping, agriculture, water resources, recreation, hydropower, and urban growth often leads to increased frequency and larger magnitude of flooding. Flood water usually contains much fine sediment along with raw sewage, oil and grease, insoluble compounds, animal carcasses, insect eggs, weed seeds, floating debris and many other dangerous, undesirable, or toxic materials that may be introduced into wetlands. Erosion and deposition of sediment may cause permanent changes in channel and basin geomorphology.

Both rural and urban areas may experience substantial damage, as flooding in southern Poland demonstrated in July 1997. Prolonged, heavy rains were generated by a stalled weather system that persisted for several weeks over the region. Major flooding took place on the Odra and Vistula rivers and on many of their tributaries. Flooding on many streams and rivers was the greatest of the twentieth century, and flood recurrence intervals on various streams are estimated in the 200- to 500-year range. More than 200 bridges were washed away. Homes, vehicles, farms, and factories were completely destroyed in several locations (Figs. 4-13 and 4-14). Fifty-five people died during floods, and many more suicides took place in the following weeks. The Polish floods of 1997 were an immediate economic and personal catastrophe for thousands of people in the affected regions and led to permanent changes in floodplain soils and drainage. On the other hand, the damaged areas were reconstructed with modern building techniques and appropriate hydrologic designs for flood control and management of drainage basins. In the long run, safer living conditions and stronger economic activity will result.

4.4 Hydrologic functions of streams and wetlands

Rivers and streams serve two basic hydrologic functions in the landscape – to remove excess surface water and to carry away sediment. Wetlands also serve two basic hydrologic functions – to store excess surface water and to retain sediment. Any human modifications of stream channels, wetland drainage or watershed conditions inevitably lead to changes in the hydrologic and sedimentologic regimes. All types of agricultural, transportation, industrial, and urban development involve modifications of or changes

Figure 4-15. Impact of bridges on flooding. A. Traditional double-arch stone bridge on Rock Creek in south-central Kansas. As water level rises, flow width under the bridge decreases, which exacerbates flooding around the bridge. B. Modern bridge design on the Cottonwood River at Emporia, Kansas. As water level rises, flow width under the bridge increases to accommodate greater discharge. Photos by J.S. Aber.

in soils, land slopes, drainage, surface cover, and other attributes that affect water infiltration, runoff, flow rate, residence time, and other hydrologic factors. A simple example is building bridges, something people do everywhere (Fig. 4-15). In fact, just about every kind of human activity results in some sort of hydrologic impact.

Such changes may benefit certain segments of society while proving deleterious for other economic or recreational interests. Thus, implementing stream and river control is a complex societal issue in the modern world. As noted before, human activities have led to substantial decreases in wetlands worldwide. Nonetheless, floodplain wetlands often play a significant role for lessening the impact of downstream flooding. Wetlands function essentially as sponges, soaking up excess water, recharging aquifers, and then releasing surface water slowly. Thus, the pulse of flood water is spread over a longer time period, so the downstream flood peak is lowered. Furthermore, flood water stored temporarily in wetlands and then released gradually is cleansed of much sediment and debris.

In the summer of 1993, the upper Mississippi drainage basin experienced major floods, affecting the upper Mississippi, lower Missouri, Kansas, Des Moines, Illinois, and other rivers. Millions of acres of farmland and urban areas were inundated for weeks, and property damage exceeded $10 billion (Parrett, Melcher and James 1993). Persistent rains fell throughout much of the region during the spring and summer. Most of the region received 150 percent of normal precipitation, and some spots had more than 200 percent of normal rainfall (Melcher and Parrett 1993).

Peak discharges were all-time records on many rivers, and other rivers recorded the greatest discharges since the time their flows had been regulated by reservoirs and canals. Prior to the 1993 floods, it is estimated that floodwater storage capacity of the Mississippi River had been reduced by up to 80 percent, because of loss of forested wetlands and confinement by levees (Gosselink et al. 1981). The inability to absorb excess water undoubtedly exacerbated flooding, which was unusual because of its widespread and long-lasting nature (Larsen 1996). At Saint Louis, the Mississippi River exceeded flood stage on 26 June, reached its peak discharge of 1,080,000 cfs on 1 August (Parrett, Melcher and James 1993), and remained in flood until mid-August (Fig. 4-16). The magnitude of peak floods would probably have been much higher without substantial regulation of stream flow. On the other hand, the long duration of flooding was probably caused in part by stream regulation.

Management of flood-control reservoirs is to reduce downstream peak discharge by spreading the flow over longer time intervals. Many dikes, levees, and small dams failed during these floods. However, large structures of the U.S. Army Corps of Engineers functioned properly without serious failures, although some

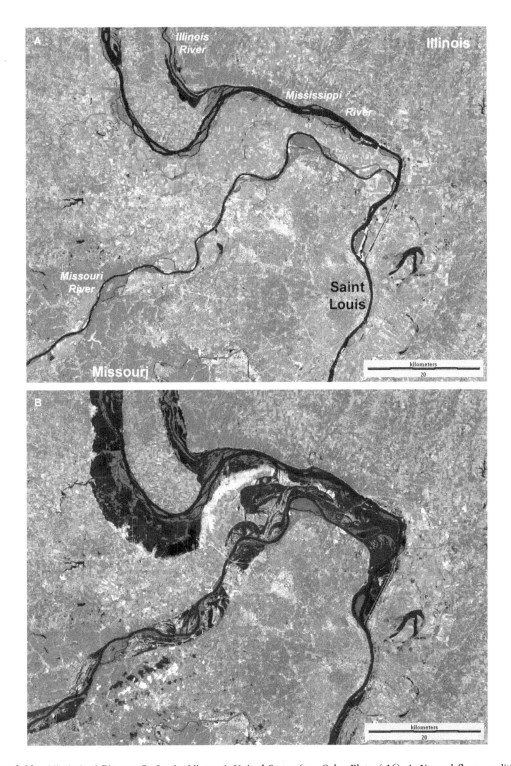

Figure 4-16. Mississippi River at St. Louis, Missouri, United States (see Color Plate 4-16). A. Normal flow conditions in August 1991. B. Waning flood conditions in August 1993. Landsat TM bands 2, 5 and 7 color coded as blue, green and red; active vegetation appears in green colors. Image from NASA; processing by J.S. Aber.

Figure 4-17. Tuttle Creek Lake on the Big Blue River at Manhattan, Kansas was forced to open its spillway gates (background) to release water as the reservoir threatened to overflow the dam in July 1993. The resulting downstream flow excavated deeply into bedrock, creating a series of plunge pools (foreground). Note people for scale in the spillway channel. Photo by J.S. Aber.

Figure 4-19. Tourists visit Wieliczka salt mine, near Kraków, Poland. The underground cathedral was carved from within a chamber of the salt mine; the chandeliers and other decorations also are made of crystalline salt. This salt mine has been worked since Medieval times and is still active; it is designated as a World Heritage site. Photo by J.S. Aber.

Figure 4-18. Little Arkansas River at Halstead in south-central Kansas. The levee in the left background was under construction, but not finished, in 1993 when flooding devastated the city of Halstead on the other side. The levee was subsequently completed and has protected the city since. However, downstream flooding of unprotected cities has increased as a consequence. Photo by J.S. Aber.

reservoirs had to release water to avoid being overwhelmed (Fig. 4-17). Based on the outcome of the 1993 flooding, engineers and planners are adopting new strategies for flood management, based on less artificial control (dams and levees) and more natural river behavior, namely floodplains and wetlands. In spite of this effort, many levees have been constructed or rebuilt (Fig. 4-18), which may lead to more damage in the future (Welsch et al. 1995).

4.5 Hydrochemistry

Water chemistry is a rich and complex subject involving both organic and inorganic aspects that are affected by diverse natural processes and human activities (e.g. McMurry, Castellion and Ballantine 2007). Here we review several factors of special importance for understanding wetland hydrochemistry. Water contains dissolved solids in the form of ions – positively charged cations and negatively charged anions. Salt, the mineral halite (NaCl), is a simple example (Fig. 4-19). As a dissolved solid, salt exists as Na^+ cations and Cl^- anions. Water solutions typically contain many dozens of dissolved solids, such that overall charges balance for electrical neutrality. In addition, soluble organic compounds and gases may dissolve into water. Common gases in water include oxygen (O_2), carbon dioxide (CO_2), methane (CH_4), hydrogen sulfide (H_2S), and helium (Fig. 4-20).

Water molecules naturally dissociate into H^+ and OH^- ions:

$$H_2O = HOH = H^+ + OH^-$$

Pure water at 25 °C contains 10^{-7} moles per liter of H^+ ions. This is the basis of the pH scale; neutral pH = 7 (inverse log of H^+ concentration). For acidic solutions with more H^+, the pH

Figure 4-20. Hot springs at Uunartoq, southwestern Greenland. Helium, nitrogen, argon and other gases bubble out of the water (Persoz, Larsen and Singer 1972). The helium is presumably a byproduct of deep-seated radioactive decay. Photo courtesy of P. Jensen.

value is lower. For example vinegar has a pH = 3. Fresh water in contact with the atmosphere absorbs CO_2, which creates carbonic acid and lowers pH to 5.6. In basic solutions with less H^+, the pH value is higher; for example, standard sea water pH = 8.3. Note that this scale is logarithmic; each whole value represents a 10-fold increase or decrease from the next higher or lower value.

Many chemical reactions depend upon pH; for example the behavior of calcium carbonate, which is among the most common dissolved solids of both fresh and sea water. The following reactions are related to its stability:

$CaCO_3 = Ca^{2+} + CO_3^{2-}$ (carbonate)

$CaCO_3 + H^+ = Ca^{2+} + HCO_3^-$ (bicarbonate)

$CO_2 + H_2O = H^+ + HCO_3^- = H_2CO_3$ (carbonic acid)

In most waters, thus, the stability of calcium carbonate depends mainly on how much CO_2 is dissolved, which affects acidity (pH). More CO_2 drives the reaction toward solution; less CO_2 favors precipitation of solid $CaCO_3$. The reaction also depends on temperature, which affects the solubility of CO_2 and other gases, and the content of other dissolved solids such as sodium and potassium (Davis and DeWiest 1966). Nonetheless, pH is the primary controlling factor for calcium carbonate in most surficial and shallow natural waters.

Mires, especially bogs with *Sphagnum* moss and peat, are strongly acidic, typically in the pH range 3.5 to 4.5 and reaching as low as 3. Cation exchange is a principal source for hydrogen ions, moss releases organic acids, and sulfide oxidation may form sulfuric acid (Charman 2002). *Sphagnum* is particularly effective at obtaining cations from solution, which are necessary nutrients for survival in ombrotrophic situations, and releasing hydrogen ions. This cation exchange is facilitated by uronic acids that are held in cell walls and make up 10–30 percent of the dry mass of *Sphagnum*, and phenolic compounds also aid cation exchange (Australian Bryophytes 2008). The spreading rope rush (*Empodisma minus*) in New Zealand raised mires is another plant that has similar high cation exchange capacities in its root layer (Charman 2002).

In general warm, tropical sea water of the continental shelf environment favors precipitation of calcium carbonate, which may form the mineral calcite or aragonite depending on subtle differences in other chemical constituents. For example, corals build aragonite skeletons, oysters construct multilayered calcite and aragonite shells, and brachiopod shells are calcite (Blatt, Middleton and Murray 1972). Ground water in contact with limestone and dolostone is often highly charged with dissolved calcium and magnesium carbonate, so-called hard water. Water hardness is generally reported as ppm (or mg/L) $CaCO_3$ equivalent (40 ppm Ca^{2+} = 100 ppm $CaCO_3$). When hard water emerges from springs, dissolved CO_2 is lost and lime precipitates. In some cases, impressive deposits of tufa or travertine may accumulate (Fig. 4-21).

In addition to pH, the oxidation potential of water is extremely important for many chemical reactions. Oxidation or redox potential (Eh) refers to changes in electrical valence of ions in solution and is measured as a voltage required to force a change in valence. Iron is among the most common and important elements for wetland environments. Iron exists in two valence states – ferrous (reduced), which is relatively soluble, and ferric (oxidized), which is highly insoluble. The standard potential is:

Figure 4-21. Turner Falls, south-central Oklahoma, United States. The falls are developed on massive tufa, which forms the cliffs on either side. The tufa precipitated from hard spring water derived from ground water in thick limestone and dolostone bedrock of the Arbuckle Mountains. Photo by J.S. Aber.

Fe^{2+} (ferrous) = Fe^{3+} (ferric) + e^-

(Eh = 0.77 volts)

Furthermore, iron is usually present along with other dissolved solids and gases in natural waters, which gives rise to more complex reactions. Consider, once again, the case of dissolved carbon dioxide:

$2Fe^{2+} + 4HCO_3^- + H_2O + \frac{1}{2}O_2 = 2Fe(OH)_3 + 4CO_2$

It is clear that both oxygen (Eh) and carbon dioxide (pH) control this reaction toward ferrous (left) or ferric (right) states. This situation is illustrated with an Eh-pH diagram, also known as Pourbaix diagram, which plots Eh on the vertical axis and pH on the horizontal axis (Fig. 4-22). Wetlands are places where reducing conditions often develop, particularly within saturated soils, so the status of iron is a critical indicator (see chapter 5). Eh and pH may be measured quickly in the field using portable instruments in order to establish the chemical environment.

Ground water often contains considerable dissolved iron in the ferrous state. When such ground water emerges in springs, it comes into contact with air, is rapidly oxidized, and iron precipitates as insoluble ferric deposits, namely rust (Fig. 4-23). Bog iron ore forms in a similar way, when iron-bearing water is oxidized by algae, bacteria or the atmosphere. It consists mainly of goethite or limonite ($HFeO_2$) in spongy, porous, impure deposits often mixed with plant debris and other sediments.

4.6 Summary

Water quantity and quality are essential ingredients for all environments and life on Earth. Hydrology is the science of water, and wetlands represent a storage point within the hydrologic cycle. Wetlands may gain water from direct precipitation, surface inflow, and ground-water discharge; water is lost by evapotranspiration, ground-water recharge, and surface outflow. Water storage capacity, hydroperiod, and residence time vary greatly for different kinds of wetlands; some are quite stable through time, others change regularly or erratically.

Many wetlands are hydrologically connected with underlying ground water, which moves in response to the hydraulic gradient at rates determined by hydraulic conductivity and hydrostatic pressure. Some wetlands are sustained by ground-water discharge; in other cases, wetlands may recharge aquifers. Human diversions or extractions of either surface flow or ground water have the potential to impact integrated surface–ground-water systems. Overpumping ground water and diverting flow from surface streams are among the common means by which people adversely affect wetlands both locally and downstream.

Flooding is a universal phenomenon that affects many wetlands in all climatic regimes and any size or type of drainage basin. Flooding is an entirely natural event that takes place on most streams every few years. However, flooding becomes a hazard when high water leads to human casualties, damage to structures, and impairment of human land use. Most types of human land use and development lead to increased frequency and magnitude of flooding, because the flood-moderating influence of natural wetlands is reduced.

Water chemistry involves both organic and inorganic aspects that are affected by diverse natural processes and human activities. Natural

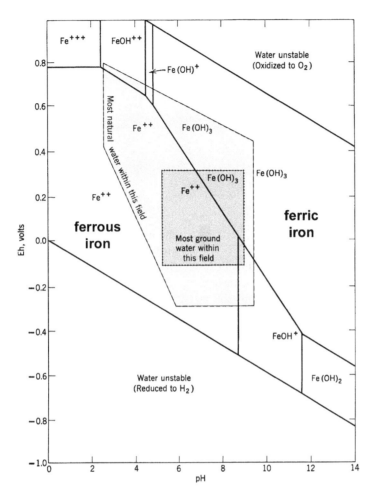

Figure 4-22. Eh-pH diagram for iron. The main boundary between ferrous and ferric iron runs diagonally across the middle of the field in which most natural waters occur. Slight changes in either Eh or pH may affect the valence state of iron. Adapted from Davis, S.N. and DeWiest, R.J.M. 1966. *Hydrogeology*. John Wiley & Sons, New York (Fig. 3.11).

Figure 4-23. Iron-cemented rødsten (redstone) formed where ground water emerged from a gravel bed exposed in the cliff (see Color Plate 4-23). Photo by J.S. Aber; Ristinge Klint, Denmark.

waters contain dissolved solids, soluble organic compounds, and dissolved gases in solutions ranging from highly acidic to highly alkaline and from well oxygenated to lacking in free oxygen. In wetland environments, acidity (pH) and oxidation potential (Eh) are two key factors that control many chemical reactions that may take place.

Wetland soil 5

5.1 Introduction

"Soil" is a common English word, both a noun and verb, which is often synonymous with dirt, earth or ground. To a highway engineer, soil is any unconsolidated material that can be moved by a bulldozer without the need for explosives. For a farmer, soil is the medium in which crops are rooted and draw their water and nutrients for growth. For our purposes, soil is the complex interface between the solid continental crust below and the atmosphere, hydrosphere and biosphere above. This interface contains organic and inorganic components along with air, water and living organisms, which are often distinctly layered in so-called soil horizons. In most situations, soil represents the upper few to several meters of the surface that is densely penetrated by terrestrial plant roots. This common view of soil is subject to wide variations, however. In some soils, surficial plants are virtually absent, whereas in other cases plant roots may grow many tens of meters deep.

Soils may form in two basic geomorphic situations. The first is by weathering, decomposition, and alteration of the underlying parent material; in other words, in-place soil development. The primary means of weathering are oxidation and hydration of minerals, often aided by microorganisms, as well as freezing–thawing, wetting–drying, animal burrowing, root growth, and other mechanical actions. The classic example is granite or other crystalline rocks of the crust that break down when exposed at the surface to air, water, and living organisms (Fig. 5-1). Soils also may form by the accumulation of unconsolidated sediment deposited by wind, water, or ice. For example, fine-grained deposits on floodplains are the basis for much alluvial soil (Fig. 5-2). Wind-blown dust, known as loess, is an important component of many upland soils (Fig. 5-3), and till deposited by former glaciers covers vast regions in northern Eurasia and North America (Fig. 5-4). These deposits often contain organic matter as part of the sedimentary accumulation (Fig. 5-5).

From permafrozen tundra to hyper-arid sabkha environments, soil traits vary tremendously around the world. Wetland soils are known as hydric soils, which were "formed under conditions of saturation, flooding or ponding long enough during the growing season to develop anaerobic conditions in the upper part" (Natural Resources Conservation Service 2010a). Thus, it is specifically the lack of oxygen that influences development of wetland soils. Three factors are key attributes of wetland soils; when a dominant portion of the soil exhibits all three factors it is classified as a hydric soil (Welsch et al. 1995).

- Saturation – water saturates the soil for prolonged or extended periods (Fig. 5-6), particularly during the growing season, which limits diffusion of air into the soil and allows organic matter to accumulate at the surface.

Wetland Environments: A Global Perspective, First Edition. James Sandusky Aber, Firooza Pavri, and Susan Ward Aber.
© 2012 James Sandusky Aber, Firooza Pavri, and Susan Ward Aber. Published 2012 by Blackwell Publishing Ltd.

5.2 Brief history and soil classification

In the United States, soil science and survey began in the late nineteenth century, primarily for agricultural purposes, and wet soils were generally mapped in two categories (quoted from Whitney 1909, pp. 116–117).

- Muck and peat – These soils are composed largely of organic matter in various conditions of decay, the muck representing an advanced stage of change in peat areas. Of relatively limited extent and poorly drained, these soils are highly valued for their adaptation to special crops, such as celery, onions, peppermint and cabbage.
- Swamp, tidal swamp and marsh – Under these heads are grouped areas covered with water the greater part of the year and unfit for agriculture except where drained and protected from tidal or fluvial overflow. When reclaimed, much of this land will become quite productive.

The need for reclamation, namely drainage, was a recurring theme for productive use of wetland soils. Furthermore, once drained, wetland soils were thought to be highly valuable for crops and other agriculture (Bonsteel 1912a & b). Management of wetlands for non-agricultural purposes hardly existed, and the accepted practice was to drain all such areas wherever economically feasible and to put them into agricultural production. Even though the chemical properties and texture of wetland soils were not well suited to crop production, drained wetlands with inputs of chemicals allowed vast parts of the temperate latitudes and tropics to be converted to agriculture (Guthrie 1985). The costs associated with the constant maintenance of proper drainage, fertility, and erosion control of wetland soils may be quite high. Even so, rice is probably the most ubiquitous crop grown on floodplain and coastal wetland soils across Asia (Fig. 5-7). Meanwhile, sugarcane, soybean, corn, peanut, cotton, and other forage crops are common elsewhere including across North and South America (Guthrie 1985).

Figure 5-1. A. Weathering of granite along the Baltic shore, island of Bornholm, Denmark. B. The fragmental debris, called grus, has accumulated in layered deposits nearby. Scale pole marked in 20-cm intervals. Photos by J.S. Aber.

- Reduction – oxygen is depleted leading to anaerobic conditions for microbes and chemical reactions. Oxidation of organic matter slows or is halted.
- Redoximorphic features – mobilization and removal of ferrous (reduced) iron and deposition of ferric (oxidized) iron as mottles, streaks, and patches.

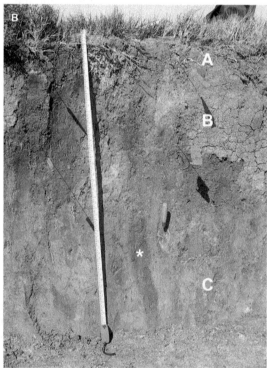

Figure 5-2. Alluvial soil formed by periodic overbank flooding in a closed depression, Cheyenne Bottoms, central Kansas, United States. The soil consists mainly of loam, a mixture of fine sand, silt and clay. A. Long profile is ~5 m across. Note numerous vertical structures in lower portion of profile. B. Profile detail; tape measure is ~1.3 m. Organic-rich material from the A and B horizons has migrated downward in probable prairie-dog burrows (*) into the C horizon. Photos by J.S. Aber.

Figure 5-3. Olpe Soil is a compound upland soil (see Color Plate 5-3). The lower portion (C) is a paleosol developed in highly weathered (leached and oxidized), older (Neogene), alluvial gravel. The upper part (A & B) is composed of younger (Holocene) loess. Scale pole marked in feet (30-cm intervals). Flint Hills, Kansas, United States. Photo by J.S. Aber.

Figure 5-4. Till exposed in a coastal cliff at Galway, western Ireland. Also known as boulder-clay, till is an unsorted, unstratified mixture of anything and everything over which the glacier moved prior to deposition. Pocket knife for scale; photo by J.S. Aber.

Figure 5-5. Hardwood tree trunks exposed by recent flood erosion in Holocene alluvial gravel of southwestern Poland. Photo by J.S. Aber.

Figure 5-6. Standing water saturates the soil throughout the year in this Florida cypress swamp. Photo courtesy of P. Johnston.

By the 1930s, however, doubts began to appear about converting wetlands to other land uses. Kenney and McAtee (1938) noted two issues. Firstly they questioned the economic advantages of converting wetlands into croplands, and secondly they pointed out the ecological consequences of widespread wetland drainage, particularly the detrimental impacts on populations of waterfowl and fur-bearing animals. Nevertheless, the continued and rising demand for food production across the developing world has meant an unrelenting push for wetlands and forested areas to be brought under agricultural production.

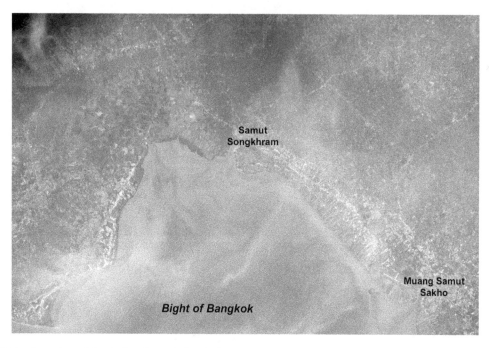

Figure 5-7. Overview of the Bight of Bangkok, Thailand showing extensive rice agriculture in the low-lying coastal plain and suspended sediment in the bight. International space station, photo ID: ISS023-E-32634. View toward west, 5 May 2010. Adapted from NASA Gateway to astronaut photography of Earth <http://earth.jsc.nasa.gov/>.

As soil science matured, the need for a systematic soil classification system grew as well. During the mid-twentieth century a series of soil classifications was developed, which culminated in the work *Soil Taxonomy* (Soil Survey Staff 1975). The nomenclature was derived mostly from Greek and Latin words or parts of words that are added entirely or in part as formative elements at each successive classification level (Buol et al. 1997). Of the ten original soil orders, the only one that is clearly associated with wet soils is the Histosols (Table 5-1). The root element "histos" was used to indicate that these soils are composed largely of plant tissue (at least 20–30% in upper parts). They are essentially the peat, bog, half bog, and muck soils of earlier classifications.

Hydric soils are more common in the next level of classification, the suborder. The formative element "Aqu," from the Latin *aqua* meaning water, is added to soils that have aquic moisture regimes. This implies reducing conditions that are virtually free of dissolved oxygen, because the soil is saturated by ground water or by water of the capillary fringe (Fig. 5-8). These soils are generally identified in the field as soils whose dominant Munsell colors have chromas of 2 or less and also have brighter colored mottles (Soil Survey Staff 1975, p. 109). In terms of taxonomic nomenclature, then, a wet (aquic moisture regime) Alfisol would belong to the suborder Aqualf (Aqu for wet and alf for Alfisol). The same procedure is used for the other orders as appropriate forming six wet-aquic suborders: Aqualfs (wet Alfisols), Aquents (wet Entisols), Aquepts (wet Inceptisols), Aquolls (wet Mollisols), Aquox (wet Oxisols), Aquods (wet Spodosols), and Aquults (wet Ultisols).

The U.S. Fish and Wildlife Service requested help in the 1970s to develop a practical and useful classification of wet soils for the National Wetlands Inventory (Mausbach 1994). The intent

Table 5-1. Original ten United States soil orders and a description of their formative elements. Based on Soil Survey Staff (1975, Table 8, p. 87).

Soil order	Formative element	Derivation	Pronunciation
Alfisols	Alf	Meaningless	Pedalfer
Aridisols	Id	Latin – aridus, dry	Arid
Entisols	Ent	Meaningless	Recent
Histosols	Ist	Greek – histos, tissue	Histology
Inceptisols	Ept	Latin – inceptum, beginning	Inception
Mollisols	Oll	Latin – mollis, soft	Mollify
Oxisols	Ox	French – oxide	Oxide
Spodosols	Od	Greek – spodos, wood ash	Odd
Ultisols	Ult	Latin – ultimus, last	Ultimate
Vertisols	Ert	Latin – verto, turn	Invert

Figure 5-8. Shallow ground water, ~20 cm deep, soaks upward through clayey silt to evaporate at the surface of this saline mud flat in western Kansas, United States. Note dark masses of organic-rich sediment just below the surface. This soil displays a strong rotten-egg odor. Comb is ~5 inches (12.5 cm) long; photo by J.S. Aber.

was to create a hydric soil classification that correlated wet soils with hydrophytic vegetation. The definition of what hydric soils are, how long they take to form, and how long they must be saturated to be called hydric soils has been refined considerably. Such issues are now under the purview of the National Technical Committee for Hydric Soils, which was originally charged with finalizing a definition of hydric soils and creating a list of hydric soils in the early 1980s and continues to serve five major functions with regard to wet-hydric soils (National Technical Committee for Hydric Soils 2010):

- Provide technical leadership.
- Update and distribute a national list of hydric soils.
- Refine and maintain "Field Indicators of Hydric Soils."
- Communicate and respond to public comments regarding hydric soil definitions.
- Conduct research regarding data necessary to better define hydric soils.

The best and most up-to-date source for information about hydric soils and hydric-soil definitions in the United States is the USDA/NRCS hydric soils website, which includes definitions of terms, criteria, technical notes, lists of field indicators, and state and national lists of hydric soils (see http://soils.usda.gov/use/hydric/).

The Canadian system of soil taxonomy follows that of the United States closely, although with some distinct differences in hierarchical structure and field criteria; a greater emphasis is placed on the uppermost soil horizon (Soil Classification Working Group 1998). The Organic order includes soils that develop in poorly drained settings that are saturated for prolonged periods. Organic soils are separated into four great groups: Fibrisol, Mesisol, Humisol, and Folisol. Fibrisols are composed of relatively undecomposed plant tissues (peat), whereas Humisols represent advanced decomposition (muck). Mesisols are intermediate; all three great groups are hydric soils. The Folisols, in contrast, form in upland forest settings rather than wetlands.

At present, various other national soil classification systems differ considerably from the U.S. and Canadian approach, and knowledge of soils is far from complete around the world, "therefore it is not possible to develop an international system of classification for the whole population of known and unknown soils" (Soil Classification Working Group 1998).

5.3 Hydric soil criteria

Wetlands are areas of predominantly hydric soils that may support a prevalence of water-loving plants, known as hydrophytic vegetation, which are adapted for growing in anaerobic conditions. Anaerobiosis refers to the combined chemical and biological processes operating within soils for which oxygen is depleted. The field criteria for recognizing hydric soils are mostly consequences of anaerobiosis. These criteria are uniquely associated with wet soils and allow in-field identification of hydric soils (Vasilas, Hurt and Noble 2010). Four key elements are involved – iron, manganese, sulfur, and carbon.

Iron and manganese behave in a similar way chemically, although iron is more abundant and commonly recognized in soils than is manganese. Under anaerobic conditions, soil microbes reduce iron (Fe^{3+} to Fe^{2+}) and manganese (Mn^{4+} to Mn^{2+}). Ferric iron is highly insoluble, but ferrous iron is quite mobile and may migrate easily in solution. Transitions from ferric to ferrous and back to ferric may take place in different locations within a soil or at different times, depending mainly on water level. In this manner, iron may be relocated such that some zones become depleted in iron and other zones are enriched. Ferric iron tends to precipitate as soft, rusty masses, particularly along root channels, cracks, and other pores where air is able to penetrate. Such redoximorphic features are likely to develop in soils that alternate between aerobic and anaerobic chemistry.

Microbes likewise reduce sulfate (SO_4^{2-}) to hydrogen-sulfide (H_2S) gas, which has a strong "rotten-egg" odor. This olfactory indicator is a sure sign of a hydric soil (Fig. 5-9). However, the rotten-egg odor is present only in the wettest

Figure 5-9. Hydrogen-sulfide odor is common in salt marshes and other extremely wet environments. View over Atlantic coast and salt marsh at Moody Division, Rachel Carson National Wildlife Refuge, Maine, United States. Blimp aerial photo by S.W. Aber, J.S. Aber, and V. Valentine.

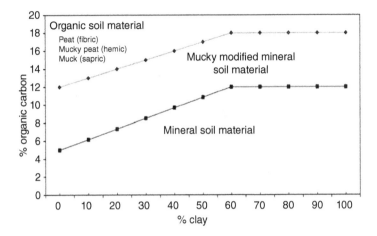

Figure 5-10. Classification of organic and mineral hydric soils according to contents of clay and organic carbon. Adapted from Vasilas, Hurt and Noble (2010, Fig. 53).

soils that contain sulfur-bearing compounds, so it is not found in all hydric soils. In aerobic soils, most carbon compounds are oxidized by soil microbes. However, this process slows considerably under anaerobic conditions, particularly in colder climates, and partly decomposed organic matter may accumulate as peat, muck, or organic-rich mineral layers. In the field, organic content may be estimated by rubbing a sample between the fingers and examining with a hand lens (Vasilas, Hurt and Noble 2010). The organic material is further classified as sapric, hemic, or fibric based on the fibers visible with a hand lens. These terms correspond respectively with muck, mucky peat, and peat (Fig. 5-10).

Color is among the best indicators for the status of iron and organic matter in hydric soils, as noted in section 3.4.1. Soil color is observed under moist conditions. Wet soils should be allowed to dry out until they do not glisten, and dry soils should be moistened until color becomes stable (Vasilas, Hurt and Noble 2010). It is important to examine color in the field soon after collecting samples, as ferrous iron may oxidize rapidly when exposed to air (Fig. 5-11).

5.4 Mineral and organic hydric soils

Soils provide the nutrients that plants need for growth. In principle, organic-rich soils have more of the essential nutrients than do mineral soils, but in practice the nutrients are bound in organic compounds and are not readily available for plant growth. Thus peat and muck soils tend to be oligotrophic, meaning that they are deficient in nutrients. Furthermore, organic-rich soils tend to be acidic. The combination of scarce nutrients and low pH limits the types of plants that can live in such soils. *Sphagnum* moss, for example, thrives under oligotrophic and acidic soil conditions (Fig. 5-12), but relatively few other plants can survive. This situation explains why conversion of organic-rich hydric soils into cropland often fails (see Fig. 14-7). Nutrients are insufficient to support crop plants, so fertilizer must be added in large quantities to sustain crop production.

Mineral hydric soils generally have more nutrients available for plant growth and less acidic conditions compared with organic-rich soils. Inflow of surface or ground water may also deliver nutrients leading to mesotrophic or eutrophic conditions. The result is greater floristic diversity and higher primary productivity (Fig. 5-13). Of course, many variations or exceptions exist in terms of hydric soil conditions. In semi-arid or arid environments, for example, high soil salinity may be a limiting factor for plant growth. Where limestone or dolostone is present, acids may be neutralized, so that soil and water pH become intermediate to basic.

Figure 5-11. Hydric soil profile from a playa mud flat at Dry Lake, west-central Kansas, United States (see Color Plate 5-11). A. Freshly dug pit, ~30 cm deep, showing distinct horizons from the surface downward. Note the redoximorphic iron accumulation (*) and dark gray color at depth. B. Clump of greenish-gray (5 GY 6/1), clayey silt with ferric iron accumulation (*). Photos courtesy of B. Zabriskie.

Figure 5-12. Close-up view of *Sphagnum warnstorfia* moss (see Color Plate 5-12). This moss forms red tussocks at the edge of oligotrophic bogs. Karelia, Russia; photo courtesy of E. Volkova.

Figure 5-13. Tropical rain forest in northern Brazil near Manaus. The forest is inundated for months at a time during the rainy season; flood waters bring nutrients to support the rich species diversity and high productivity of this wetland environment. Photo courtesy of K. Buchele.

Organic matter has quite high cation exchange capacity. What this means practically is that organic-rich hydric soils act like sponges to soak up excess nutrients and other pollutants (Welsch et al. 1995). Positively charged cations, such as ammonium (NH_4^+) and potassium (K^+), are absorbed and held loosely by organic molecules. The same principle is used for water softening, in which Na^+ in the softener filter is exchanged for Ca^{2+} in hard water. The stored cations may be retained for long periods, or taken up by other chemical or microbial activity, or become buried in the wetland sediment. In this way, organic-rich hydric soils filter

Figure 5-14. Blanket bog covers much of Dartmoor in southwestern England. Traces of former peat cutting are evident on the hillside and three tors (bedrock outcrops) are visible in the background. Modified from original photograph by Nilfanion; obtained from Wikimedia Commons <http://commons.wikimedia.org/>.

water and remove potentially harmful substances. This ability is exploited in artificial wetlands designed to treat contaminated water (see chapter 13).

It should be stressed at this point that prolonged saturation by water and the resulting lack of oxygen are key factors for the development of hydric soils in which organic carbon may accumulate. Hydric soils are, thus, sources and sinks for greenhouse gases, namely carbon dioxide and methane. Although the summer growing season determines the annual carbon gas exchanges for boreal mires, winter fluxes may also be important as long as the soil is not frozen (Leppälä, Laine and Tuittila 2011). Likewise, hydric soils, especially peat, are major reservoirs for carbon at the Earth's surface (see section 10.2). Raised and blanket bogs cover substantial regions and represent organically constructed deposits that build up through time and, thereby, transform the landscape (Fig. 5-14).

5.5 Submerged wetland substrates

The substrates of many wetlands are submerged more or less continuously under a few to several meters of water. In customary usage, the term "soil" would not apply to the floors or shores of shallow lakes, streams or seas below the limit of low tide or continuous submergence. According to the Ramsar wetland definition, however, the continental shelf out to six meters deep is included (see section 2.1). Thus, rocky and sedimentary substrates must be considered in fresh and marine environments (see Table 2-1). Types of substrate materials range from soft mud to crystalline bedrock and fall into several discrete categories:

- Well-consolidated bedrock, such as granite, slate, sandstone, limestone, etc. Bedrock is typically exposed where erosion takes place in high-energy environments with rapid water flow, strong wave action, and high turbulence. Mountain streams and coastal headlands are typical situations (Figs. 5-15 and 5-16).
- Unconsolidated terrigenous (clastic) sediment consisting of gravel, sand, silt and clay derived from erosion of land areas (Fig. 5-17). Rock fragments, quartz and feldspar grains, and clay minerals are typical components and are deposited according to sediment size, water flow, turbulence, and distance from sources.
- Unconsolidated bioclastic sediment comprised of whole or fragmented hard parts of carbonate or siliceous invertebrate organisms. Some are microscopic – radiolarians, diatoms, foraminifera, and others are macroscopic – echinoderm spines, gastropod

Figure 5-15. High-energy stream flows across a bouldery bed flanked by shoreline hydrophytic vegetation. Baker Creek, Culebra Range of the Sangre de Cristo Mountains, Colorado, United States. Photo by J.S. Aber.

Figure 5-16. High-energy coast at Point Pinos, Pacific Grove, California, United States. A. Overview of waves pounding against granite headland. B. Vertical shot of waves swirling around granite knobs and loose boulders. Field of view ~40 m across. Kite airphotos by S.W. Aber and J.S. Aber.

shells, brachiopods, etc. As with clastic sediment, deposition is according to sediment size and water energy (Fig. 5-18; see Color Plate 2-5A).

- Chemical precipitates include various soluble salts such as halite, gypsum and mirabilite (Fig. 5-19), as well as less soluble carbonates – calcite, aragonite and dolomite. The precipitation of these minerals depends on water chemistry as well as temperature, Eh and pH, and is often mediated by micro-organisms.
- Consolidated, rocky structures constructed by organic activity of corals (Fig. 5-20), algae (see Fig. 9-1), and other reef-building organisms. These are associated mostly with shallow, tropical or subtropical marine environments in which terrigenous sediment is scarce to absent (Fig. 5-21).

5.6 Summary

Soil is the complex interface between the solid continental crust below and the atmosphere, hydrosphere and biosphere above. Wetland soils, known as hydric soils, were "formed under conditions of saturation, flooding or ponding long enough during the growing season to develop anaerobic conditions in the upper part" (Natural Resources Conservation Service 2010a). Three factors are key attributes of wetland soils – saturation, reduction, and redoximorphic features; when a dominant portion of the soil exhibits all three factors it is classified as a hydric soil.

Figure 5-17. Sand shoal visible through shallow water in a bay on Presque Isle, Lake Erie, Pennsylvania, United States. Bare sand is visible at scene center; submerged vegetation covers the bay floor in lower right portion. Note vehicle to left for scale; kite aerial photo by S.W. Aber and J.S. Aber.

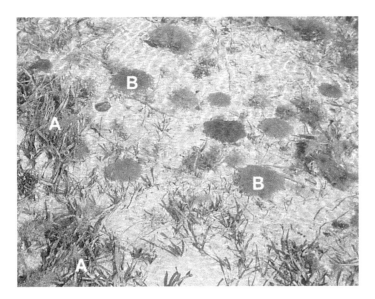

Figure 5-18. Looking through shallow water (<1 m deep) in Florida Bay near Big Pine Key, United States. Sandy, unconsolidated, bioclastic (carbonate) sediment with clumps of *Thalassia* sea grass (A) and upside-down jellyfish (B). Photo by J.S. Aber.

Figure 5-19. Accumulation of salt on hydric soils. A. Sodium-sulfate salt on mudflat adjacent to Frederick Lake, southern Saskatchewan, Canada. B. Salt crust including sodium and potassium chlorides at Dry Lake, western Kansas, United States. Comb is ~5 inches (12.5 cm) long; photos by J.S. Aber.

Figure 5-20. Vertical section in Key Largo Limestone, a fossil coral reef that makes up the northern portion of the Florida Keys, United States. This reef built up about 130,000 years ago when sea level was 5–8 m higher than today. The head coral in growth position is about half a meter across. Photo by J.S. Aber.

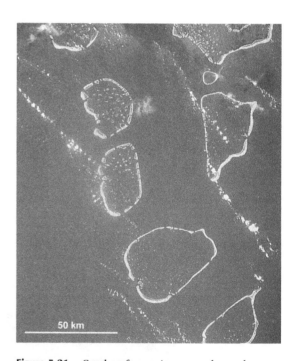

Figure 5-21. Coral reefs growing on sunken volcanoes formed the Maldive Islands in the Indian Ocean. The Maldives archipelago contains more than 1000 islands, 80% of which rise less than one meter above sea level. Adapted from NASA space-shuttle photograph, June 1983, STS07-19-904.

In modern soil classification, the order Histosols represents wetland soils; the root word "histo" refers to plant tissues, as in peat and muck soils. At the suborder level of classification, hydric soils are identified by the prefix "aqu," as in Aquoll, Aqualf, etc. Field criteria for recognizing hydric soils are mostly consequences of anaerobiosis, which is the combined chemical and biological processes operating within soils without oxygen. Four key elements to consider are iron, manganese, sulfur, and carbon. Iron and manganese may be reduced or oxidized and removed or accumulated within different portions of the soil. Sulfates may be reduced to hydrogen-sulfide gas, which gives off a distinctive rotten-egg odor. In the absence of oxygen, organic matter tends to accumulate as peat, muck, or organic-rich mineral layers. Color is among the best indicators for the status of iron and organic matter in hydric soils.

Organic-rich soils (peat and muck) are typically oligotrophic and acidic, which limit plant diversity and productivity. Mineral hydric soils, in contrast, are mesotrophic to eutrophic and less acidic to basic. These conditions support greater floristic diversity and primary productivity in most cases. Organic matter has quite high cation exchange capacity and, so, organic-rich hydric soils filter water and remove excess nutrients and other potentially harmful substances.

The term "soil" would not normally apply to the floors or shores of shallow lakes, streams or seas below the limit of low tide or continuous submergence. Yet, these types of wetlands are included in the Ramsar definition (to a depth of 6m). Thus, all types of substrates must be considered in fresh and marine environments, including well-consolidated bedrock, unconsolidated clastic and bioclastic sediments, chemical precipitates, and reefs built by organic activity.

Wetland vegetation 6

6.1 Plant adaptations

Vegetation is one of the key attributes of wetlands. Aquatic and wetland plants, known as hydrophytes, must cope with a lack of oxygen in the soil as well as potential flooding and drying, lack of nutrients, low pH, high salinity, and other limiting factors that may be present in certain kinds of wetlands. The primary problem is providing oxygen to roots submerged in water or growing in anaerobic soil or sediment (Dugan 2005). Many plants have responded with special adaptations that allow them not only to survive, but to thrive, under these stressful conditions. Some of the more important plant adaptations are discussed here, and other hydrophyte characteristics are mentioned below.

6.1.1 Structural adaptations

Among the most important features are structures designed to move oxygen from the leaves and stems above water to the roots in the saturated zone. Extremely porous tissue, called aerenchyma (pronounced air-ENK-a-ma), has large air-filled spaces in which gases may diffuse rapidly. Aerenchyma stores oxygen and allows it to move from leaves to roots and rhizomes. Aerenchyma also allows the reverse transport of methane from the anaerobic zone to the air. Many common wetland plants have aerenchyma, such as cattail (Fig. 6-1), water-hemlock (*Cicuta maculata*), green ash (*Fraxinus pennsylvanica*) and northern white-cedar (*Thuja occidentalis*). Cordgrass (*Spartina* sp.) delivers so much oxygen to its roots, in fact, that iron and manganese in adjacent soil are often oxidized into rusty streaks (Welsch et al. 1995). During the night, when photosynthesis stops but respiration continues, oxygen stored in aerenchyma is gradually consumed (Lahring 2003).

Another way to deliver oxygen to the roots is via pneumatophores, as in some mangroves (Fig. 6-2), or woody "knees" in swamp cypress (Fig. 6-3). These structures grow upward from lateral roots near the surface and project above ground into air. The exact function of pneumatophores has been debated, but it seems clear they are involved with root respiration (Dugan 2005).

Other trees, such as green ash and red mangrove (*Rhizophora mangle*), have enlarged pores on the bark called lenticels, which allow oxygen diffusion to submerged roots. Certain trees grow shallow lateral roots above the wetter soil level, and other trees develop adventitious roots, which are extra roots that sprout from the stem above the wetter soil level (Welsch et al. 1995). In mires, the slow accumulation of peat and upward growth of the substrate gradually raise the surface above the water table so that more oxygen is available.

Large trees growing in soft, water-saturated soil or water have a further structural challenge, which is simply to remain standing. Stilt or prop roots and trunk buttresses serve an architectural

Wetland Environments: A Global Perspective, First Edition. James Sandusky Aber, Firooza Pavri, and Susan Ward Aber.
© 2012 James Sandusky Aber, Firooza Pavri, and Susan Ward Aber. Published 2012 by Blackwell Publishing Ltd.

Figure 6-1. Cattails are among the most common and easily recognized wetland plants. Narrow-leaved cattail (*Typha angustifolia*) at Stillwater National Wildlife Refuge, Nevada, United States. Photo by J.S. Aber.

Figure 6-2. Black mangrove (*Avicennia germinans*) grows in the saline tidal zone, inland from the open sea. Breathing roots, called pneumatophores, grow up from soil for air exchange. A single tree may generate up to 10,000 pneumatophores (Newfoundland Harbor Marine Institute 2010). Photo by J.S. Aber; Florida Keys, United States.

Figure 6-3. Knees of swamp cypress (*Taxodium distichum*). The exposed knees stand 20–40 cm tall and allow gas exchange with the atmosphere. Widely planted as an ornamental tree around ponds and streams. Photo by J.S. Aber; Emporia, Kansas, United States.

Figure 6-4. Structural support for wetland trees. Red mangrove (*Rhizophora mangle*) lives in shallow marine water of bays and estuaries. The tree is supported by curving stilt (or prop) roots that are covered with lenticels for air exchange. Florida Keys, United States. Photo by J.S. Aber.

function similar to the flying buttresses of Medieval cathedrals, namely to hold the plant upright (Fig. 6-4). Stem hypertrophy (trunk buttress) is formed by lower-density wood that may also facilitate oxygen exchange. Submerged plants likewise have enlarged air spaces in their leaves, stems and roots. The plants are buoyant because of the gas-filled chambers and are partly supported by water; they have no need for a rigid or self-supporting structure as do emergent plants. When these plants are removed from water, they typically collapse (Fig. 6-5).

Chloroplasts are the organelles in which photosynthesis takes place. Strong radiation may damage the chloroplasts in plants exposed to direct sunlight and, thus, chloroplasts are embedded deeply in emergent leaf tissue. For submerged plants, however, light is diffuse and

Figure 6-5. Looking downward into slow-flowing water filled with submerged water milfoil (*Myriophyllum* sp.). Field of view ~2 m across. Photo by J.S. Aber; Wascana Creek, Regina, Saskatchewan, Canada.

Figure 6-6. Willow (*Salix* sp.) trees behind cattails. Willow converts the harmful byproducts of anaerobic glycolysis into pyruvic and glycolic acids (Dugan 2005). Photo by J.S. Aber; shore of Lake Kahola, Kansas, United States.

chloroplasts are found near the surface of leaves and stems. The amount of sunlight that penetrates determines how deeply plants could grow. This depends on water turbidity (suspended fine sediment) and light wavelength (Lahring 2003). In clear water, blue-green light penetrates deepest; for medium turbidity yellow light reaches deepest, and in highly turbid water only orange-red light is able to penetrate deeply. The depth at which one percent of surface light reaches is usually the lowest limit that may support photosynthesis (Mitchell 1974).

6.1.2 Biochemical adaptations

Even with all the structural features noted above, some hydrophytes have insufficient oxygen for normal plant respiration. This is particularly likely to happen during periods of prolonged flooding, saturation or submergence. Some plants are able to switch to respiration without oxygen. Anaerobic glycolysis converts food into energy, but at a much lower rate than normal respiration and with the added problem of toxic byproducts such as alcohol and acetaldehyde (Niering 1985; Welsch et al. 1995; Dugan 2005). Some plants are able to excrete these compounds through finely divided roots, and others are able to immobilize or convert the toxic compounds into organic acids that are used for growth (Fig. 6-6). Still, most hydrophytes can survive only for short periods in this manner.

In saline, tidal, or marine environments, salt is another factor that limits most hydrophytes. Plants that tolerate high salinity are known as halophytes. They have few competitors and may flourish in such environments. Certain mangroves (*Rhizophora* sp.) have specialized root cells that block sodium while allowing desirable nutrients (potassium) to pass through (see Fig. 6-4). Other mangroves, such as *Avicennia marina*, and cordgrass take up salt and secrete excess salt from their leaves. Saltcedar (Fig. 6-7), for example, may grow in soil with salinity up to 50‰ (sea water is ~35‰). Excess salt is collected in special glands in the leaves and then excreted onto the leaf surface; when these leaves fall to the ground, they contribute to soil salinity (Zouhar 2003). Red samphire (*Salicornia rubra*) is another hydrophyte that takes salt into its body (Fig. 6-8).

The first step in photosynthesis for most kinds of vegetation is the production of phosphoglyceric acid ($C_3H_7O_7P$), which is a three-carbon compound. Hence, these species are known as C_3 plants. Many hydrophytes, however, produce oxaloacetic acid ($C_4H_4O_5$), a C_4 compound. C_4 plants are more efficient for both rate of carbon fixation and water use. Increased water efficiency means that C_4 plants need less water. This reduces the rate at which potentially

Figure 6-7. Saltcedar (*Tamarix* sp.) growing beside an irrigation canal near Fallon, Nevada, United States. Saltcedar is a large bush or small tree, up to 4 m tall, with attractive pink flowers, often used as an ornamental garden tree. Saltcedar has become an invasive plant that grows in dense thickets along streams, rivers and wetlands in the western United States. Photo by J.S. Aber.

Figure 6-8. Red samphire, also known as saltwort (*Salicornia rubra*; see Color Plate 6-8). Succulent stems turn red in summer and may form bright maroon carpets on saline mud flats (see Color Plate 3-9). Cattle relish the high salt content; as human food, it may be eaten in salads and stews or used for pickling (Lahring 2003). Comb is ~5 inches (12.5 cm) long; photo by J.S. Aber at Frederick Lake, Saskatchewan, Canada.

harmful compounds, which are often abundant in anaerobic environments, are drawn into the plant (Dugan 2005). Oligotrophic wetlands, particularly raised and blanket bogs, are severely lacking in nutrients for plant growth. Few plants, notably *Sphagnum* (see Color Plate 5-12), can tolerate this situation. Some wetland plants have become carnivores in order to supplement their diets (Figs. 6-9 and 10).

In the general scheme of life, those species that reproduce the most effectively are also the most successful. Some plants reproduce sexually, usually via pollination, and others reproduce vegetatively by sending out runners or growing from plant fragments. Certain hydrophytes can do both. Papyrus (*Cyperus papyrus*), for example, reproduces in both ways and is among the fastest-growing plants in the world, which explain its phenomenal success in Africa (Dugan 2005). Another such plant is mosquito fern (*Azolla* sp.), a small aquatic fern that has worldwide distribution (Fig. 6-11). The plant usually reproduces asexually by fragmentation of the fronds as frequently as every two days (Watanabe 1982), and it may also reproduce sexually under special, poorly defined conditions.

6.2 Ecological categories

Wetland vegetation may be grouped into four general ecological categories, depending mainly on growth position in relation to water level (Fig. 6-12; Whitley et al. 1999). Of course, water levels and chemistry tend to vary in wetlands on seasonal and interannual time periods according to climatic conditions, human management, and ephemeral events. Whereas some wetland plants may tolerate substantial variations in soil moisture and water level, others have strict water requirements for survival. Thus, the following groups and illustrative

Figure 6-9. Block of United States postage stamps depicting carnivorous wetland plants. These plants trap insects, spiders, and other small animals in various ways to supplement their intake of nutrients. Original stamps printed in full color; issued in 2001.

Figure 6-10. Hooded pitcher plant (*Sarracenia minor*), Okefenokee Swamp, Georgia, United States. The white dots on the hood are presumably designed to attract insects into the hood, where they fall down into the long funnel that is lined with downward pointing hairs and become trapped in narcotic liquid at the bottom. Photo courtesy of M. Martin.

Figure 6-11. Close-up, vertical view of *Azolla cristata* on marsh margin (see Color Plate 6-11). Green-colored algae is below water level; maroon-colored *Azolla* is above water. *Azolla* spreads rapidly by vegetative reproduction and may form extensive mats (CAIP 2008). Field of view ~25 cm across; taken from Aber et al. (2010, Fig. 3).

plants represent only a rough guide to typical wetland plant habitats.

6.2.1 Shoreline plants

Shoreline plants grow in wet soil on raised hummocks or along the shores of streams, ponds, bogs, marshes, and lakes. These plants usually grow at or above the level of standing water, although some may be rooted in shallow water. Because these plants often grow on land, not all people immediately recognize them as hydrophytes. This riparian zone is marginal to upland plants and is subject to fluctuating water levels either intermittently or seasonally. During dry

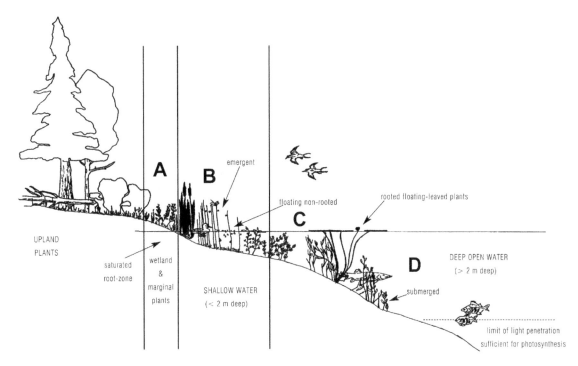

Figure 6-12. General ecological zones for wetland plants: shoreline (A), emergent (B), floating (C) and submerged (D). Shown schematically, not to scale. Adapted from Lahring (2003, p. 12).

periods, salt may accumulate so that high salinity becomes a limiting factor, in which case plant cover may be sparse.

Horsetails (*Equisetum* sp.) – Erect, segmented tubes without true leaves, up to 1.2 m tall (Fig. 6-13). Stems contain silica crystals that feel rough, like sandpaper. Young shoots may be eaten as asparagus, but mature plants are toxic, especially for livestock. Used for centuries for scouring, sanding and polishing; common names, pewterwort and scouring-rush, derived from use for scrubbing pots and pans.

Buttonbush (*Cephalanthus occidentalis*) – Deciduous bush or small tree that may exceed 3 m in height, growing immediately beside lakes and streams, sinkholes and swamps. Most distinctive are the showy, white flower clusters, which are shaped like spherical pincushions. The fragrant flowers draw numerous butterflies, bees, and other insects (Fig. 6-14). The seeds are eaten by waterfowl and other birds; shrub provides good cover for nesting.

Mangroves – Mangroves comprise a diverse group of salt-tolerant trees that live in tropical

Figure 6-13. Dense stand of horsetail (*Equisetum hyemale*) in a wet seep. Stalks stand ~2 feet (60 cm) tall. Photo by J.S. Aber, Sangre de Cristo Mountains, southern Colorado, United States.

Figure 6-14. Red admiral butterfly (*Vanessa atalanta*) feeds on the dazzling white flower cluster of the buttonbush (*Cephalanthus occidentalis*). See Color Plate 6-14. Photo by J.S. Aber and S.W. Aber, Lake Kahola, Kansas, United States.

Figure 6-15. Mangrove on the island of Carabane, which is part of the delta of the Casamance River, Senegal. The stilt roots support the mangrove in shallow water or on emergent shore. Modified from original photograph by Ji-Elle; obtained from Wikimedia Commons <http://commons.wikimedia.org/>.

and subtropical coastal marine environments worldwide (Fig. 6-15). Their habitats depend primarily on water salinity and depth. Some are primarily shoreline trees; others fall in the emergent category. Three species of mangroves are common in the Florida Everglades and Keys region (Stevenson 1969):

- Red mangrove (*Rhizophora mangle*). Lives in shallow marine water of bays and estuaries (see Fig. 6-4). The tree is supported by curving stilt (or prop) roots. Trees may form dense thickets that protect the coastline from storm erosion and provide cover for wildlife. The seed pod (radicle) floats to become rooted elsewhere.
- Black mangrove (*Avicennia germinans*). Grows on land in the saline tidal zone, inland from the open sea (see Fig. 6-2). Numerous breathing roots, called pneumatophores, grow up from soil for air exchange.
- White mangrove (*Laguncularia racemosa*). Inhabits the higher, landward, less salty zone. It lacks prop roots and has only a few pneumatophores.

Peat moss (*Sphagnum* sp.) – *Sphagnum* moss includes many species that inhabit bogs and accumulate peat (see Color Plate 5-12). For example, about 40 species are found in bogs of Finland (Laine and Vasander 1996). Through its biochemistry, *Sphagnum* creates highly acidic conditions that exclude most other plants, and moss is not eaten by animals. These factors allow *Sphagnum* to expand and accumulate peaty soil in bogs, which are particularly common in middle to high latitudes in North America and Eurasia.

Brook cress (*Cardamine cordifolia*) – Small four-petaled flowers cluster on the ends of leafy stems, up to 2 feet (60 cm) tall, in bushy clumps beside spring-fed streams and in wet meadows in montane and subalpine habitats (Dahms 1999). Also known as bitter cress, this plant is a member of the mustard family (Cruciferae); its flavor is stronger than watercress (see Fig. 4-8).

6.2.2 Emergent plants

Next are emergent plants that are rooted in soil that is underwater most of the time. These plants grow up through the water, so that stems, leaves and flowers emerge in air above water

Figure 6-17. Cluster of blunt spike rush (*Eleocharis obtusa*) growing in a shallow stream channel. The plants are about one foot (30 cm) tall. Photo by J.S. Aber near Tishomingo, Oklahoma, United States.

Figure 6-16. Water willow (*Justicia americana*). A. Carpet of water willow emerging from shallow water along a lake shore. Plants stand 1–2 feet (30-60 cm) above water level. B. Close-up view of bee on water willow flower. Photos by J.S. Aber and S.W. Aber; Lake Kahola, Kansas, United States.

level. In this zone, the water level is more dependable, but still subject to intermittent or seasonal fluctuations. Because water level and chemistry are more stable, this zone is often densely covered with hydrophytes. This category is recognized by most people as wetland plants without question, simply because the plants grow in water and emerge above the surface where they are easily visible.

Water willow (*Justicia americana*) – Erect stems with opposed, willow-shaped leaves, for which the plant is named (Fig. 6-16). Small, orchidlike, white flowers bloom on long-stemmed spikes and attract many insects. Roots are usually submerged in shallow water along stream or pond margins. Greatest value of water willow is for stabilizing streambeds and shorelines.

Spike rushes (*Eleocharis* sp.) – Grasslike plants that grow in clumps from 10 cm to 1.5 m tall, depending on species (Fig. 6-17). Characterized by leafless stems, each of which has a small fruiting spike at the top. Spike rushes are quite common in and diagnostic of wetland environments in temperate regions around the world. They provide shelter for fish, amphibians and insects, and are a food resource for many wetland birds and mammals.

Arrowheads (*Sagittaria* sp.) – Several common species, most of which are characterized by large, arrow-shaped or sagittate leaves, standing up to 1 m above water (Fig. 6-18). Leaves grow in dense clusters, and some species grow large, starchy tubers at the ends of roots. These bulbs are commonly known as duck potatoes and are prized by both wildlife and humans. Nearly all parts of the plant are valuable as food for waterfowl, songbirds, muskrat, porcupine and beaver. Arrowheads have been introduced in wildlife refuges to improve food resources, and they are cultivated in China and Japan for human food.

Bulrushes (*Scirpus* sp.) – Bulrushes, among the most beneficial emergent wetland plants, are actually members of the sedge family. They are found in all types of fresh and alkali wetland settings – marshes, river banks, and

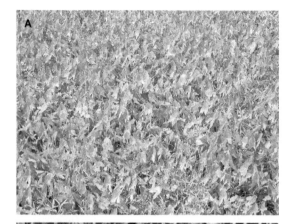

Figure 6-18. Midwestern arrowhead (*Sagittaria brevirostra*). A. Arrowheads form a dense carpet on a pool at Squaw Creek National Wildlife Refuge, Missouri River bottomland, northwestern Missouri, United States. B. Close-up view of a single arrowhead leaf, about one foot (30 cm) tall. Kahola Creek, Kansas, United States. Photos by J.S. Aber.

lake shorelines (Fig. 6-19). They may form dense thickets along the margins of water bodies. The seeds are particularly valuable for ducks; bulrush provides nesting habitat, and it binds wet soils quite effectively. Also known as tule or club rush.

Cattails (*Typha* sp.) – Among the most common wetland plants worldwide. Long, blade-like leaves and stiff flower stalks. The mature seed head looks like a brown sausage (Fig. 6-20). Two varieties are common in North America: common cattail (*T. latifolia*), up to 2.5 m tall, and narrow-leafed cattail (*T. angustfolia*), up to 1.5 m tall, as well as a hybrid (*Typha x glauca*). Normally found in shallow water of pond margins and marshes, but may thrive in almost any saturated soils from roadside ditches to sinkholes. Cattail is commonly used for wetland reconstruction and water treatment in artificial wetlands (Anderson 2006).

Pickerel weed (*Pontederia cordata*) – A perennial plant that reaches four feet (1.2 m) in height and grows in dense colonies along shallow shorelines (Fig. 6-21). It tolerates low nutrients and temporary flooding. Its natural range is from Nova Scotia to Texas in eastern North America, and it is widely used in water gardens because of its showy violet flowers. Seeds are eaten by waterfowl; geese and muskrats eat the vegetation; fish and some birds and mammals use pickerel weed for cover (Miller 2002).

6.2.3 Floating plants

This category includes plants whose leaves mainly float on the water surface. Much of the plant body is underwater. Some are rooted in the substrate, generally <2 m deep, and others are completely free floating near the surface in water of any depth. Only small portions, namely flowers, rise above water level. Again, this category is easily recognized by most people as hydrophytes.

Duckweeds (family Lemnaceae) – Numerous genera and species make up this family, which comprises the smallest and simplest of all flowering plants (Fig. 6-22). They are also among the most common plants worldwide. The plant consists of a photosynthetic floating body, called a

Figure 6-19. Bulrushes (*Scirpus* sp.). A. Great bulrush (*S. validus*) growing in proximity with B. Alkali bulrush (*S. maritimus*). Photos by J.S. Aber; Stillwater National Wildlife Refuge, northwestern Nevada, United States.

frond, and some species have tiny roots dangling below. Duckweeds are incredibly productive and are utilized as human food and feed for domestic animals. They are major food resources for birds, mammals, and fish in wetland environments.

Pondweeds (*Potamogeton* sp.) – The pondweed genus is large and varied. All pondweeds have rooted, submerged portions, and some have floating leaves as well (Fig. 6-23). Leaf form shows considerable variations from large and round to small, needlelike shapes. Underwater and floating leaves may be quite different in form on the same plant. Identification of pondweed species is difficult and requires a botanical key. Pondweeds are among the most important food plants for waterfowl in North America.

American white waterlily (*Nymphaea odorata*) – A popular water-garden plant, also known as fragrant waterlily, this species is native to eastern North America, but now is found from Alaska to Florida (USDA Plants 2010a). The waterlily casts shade that cools the water body, and it provides shelter and food for small aquatic animals (Fig. 6-24). In eastern North America, native people used the plant for medicines and food. Where introduced, however, it has a tendency to become invasive and is listed as a noxious weed in some western states (Washington Ecology 2010).

Giant waterlily (*Victoria amazonica*) – This spectacular wetland plant, known as the queen of waterlilies, is native to shallow waters of the Amazon basin in South America (Fig. 6-25). Waterlily leaves may exceed 2 m in diameter, and the lip is notched in two places to drain rainwater. The huge leaf may support a person (up to 135 kg), when the load is uniformly distributed (Davit and Cebrian 2007). The top surface is covered with minute stomata that allow gas exchange via aerenchyma with the roots. It was first discovered in Bolivia in 1801 and created quite a sensation when introduced into England in the mid-nineteenth century.

Figure 6-21. Pickerel weed (*Pontederia cordata*). The spade-shaped leaves and delicate violet flowers are quite attractive, and pickerel weed is a popular ornamental plant (see Color Plate 6-21). It may form dense cover in shallow marsh, pond and lake environments. Photo by J.S. Aber; Conneaut Marsh, northwestern Pennsylvania, United States.

Figure 6-20. Cattail (*Typha* sp.). The mature seed head (spadix) is about to disperse a multitude of tiny seed-bearing fruits (achenes) that are carried by numerous long hairs in the wind. Cattail has a tendency to become invasive, especially the hybrid variety (*T. glauca*). Photo by J.S. Aber; Flint Hills, east-central Kansas, United States.

6.2.4 Submerged plants

The submerged category has mostly underwater plants with few floating or emergent leaves. Flowers may emerge briefly in some cases for pollination, and some plants may have floating or aerial leaves that are quite different from submerged leaves, as with many pondweeds. Because these plants are largely submerged, they are rarely seen by the general public and are not readily photographed in nature. The depth of such plants is limited by light penetration, which depends primarily on suspended sediment (see section 6.1.1).

Northern water milfoil (*Myriophyllum sibiricum*) – *Myriophyllum* means "many" or "countless" (Greek: *myrios*) leaves (Latin: *phyllus*), which refers to its many, finely divided leaves (Fig. 6-26). The genus is widely spread across North America, Greenland, and Eurasia. Northern water milfoil is an important food and shelter plant for fish and invertebrates from Alaska to Texas. However, Eurasian water milfoil (*Myriophyllum spicatum*) has been introduced in North America, where it is more competitive than native species (Lahring 2003).

Bladderworts (*Utricularia* sp.) – Several species of largely underwater plants with some floating leaves (Fig. 6-27). The name comes from small bladders for catching aquatic invertebrates. The lip of the bladder has hair triggers; when touched the bladder springs open and the prey is sucked inside and trapped. Enzymes then dissolve the body, and nutrients are absorbed into the plant. In this manner, the

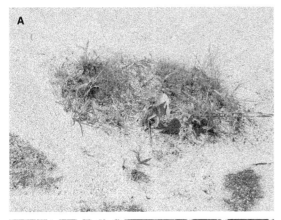

Figure 6-23. Tangled mass of pondweed with a small feather on top, floating in a slow-moving stream. Photo by J.S. Aber; Wascana Creek, Regina, Saskatchewan, Canada.

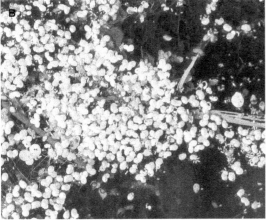

Figure 6-22. Lesser duckweed (*Lemna minor*). A. Duckweed covers the water surface with a small arrowhead emerging in the center. B. Close-up view of duckweed fronds, which resemble split green peas. Photos by J.S. Aber; drainage ditch in Middle Loup River valley, north-central Nebraska, United States.

Figure 6-24. American white waterlily (*Nymphaea odorata*), Okefenokee Swamp, Georgia, United States. Note the distinctive notch in the floating leaf and white emergent flower. Photo courtesy of M. Martin.

plant supplements its diet to survive in nutrient-poor waters. Bladderworts have a wide circumpolar distribution extending as far south as California, Texas and Florida in the United States (Lahring 2003).

Coontail (*Ceratophyllum demersum*) – Also known as hornwort, this plant resembles the tail of a raccoon, hence its common name. It grows on long stalks that may reach 2 m or more in length, but it is not rooted (Fig. 6-28). It may grow in relatively deep water (down to 7 m) and has a circumpolar distribution (Lahring 2003). In North America it extends from Alaska to Florida. The seeds and leaves provide food for waterfowl, and muskrats also eat the leaves. Dense stands of coontail are excellent shelter for fish and invertebrates.

6.2.5 Plant zonation

Wetland vegetation is typically found in distinct zones that are related mainly to water depth and salinity (see Fig. 6-12). As the groupings above suggest, many wetland plants have strict preferences for soil moisture and water depth. Some occupy primarily shoreline or emergent habitats, whereas others grow in floating or submerged situations. Thus, distinct zones of

Figure 6-25. Giant waterlily (*Victoria amazonica*). A. White flower blooms at night and turns a pale pink color when pollinated. B. Bottom side is protected with numerous flesh-piercing spines. The plant is quite popular in tropical and subtropical water gardens around the world. Photos from Brazil, courtesy of K. Buchele.

Figure 6-26. Northern water milfoil (*Myriophyllum sibiricum*) seen through shallow, clear water at Blue Lake, near Cuchara, Colorado, United States. Photo by J.S. Aber and S.W. Aber.

Figure 6-27. Common bladderwort (*Utricularia vulgaris*) seen early in the spring before water surface becomes covered with vegetation. Photo by J.S. Aber; Mined Land Wildlife Area, southeastern Kansas, United States.

vegetation are developed across the transition from dry, upland positions into deep-water environments.

In some cases, wetland vegetation zones are well defined and continuous along a shoreline or around a body of water (Fig. 6-29). In other cases, an irregular patchwork of vegetation zones reflects microtopography of the wetland environment (Fig. 6-30). Similar transitional zones mark the change from fresh-water to saltwater chemistry along marine coastlines and around inland saline lakes and playas (see Color Plate 2-16).

It should be noted that not all wetland plants fit easily into this zonation scheme. This is particularly true of plants that may occupy shorelines, wet meadows, stream banks, exposed mudflats, and other marginal zones (Larson 2006). As water level fluctuates, these plants may come into either shoreline or emergent categories. Some hydrophytes may grow either as submerged or emergent types, and the emergent morphology may be quite different from the submerged plant form. Such plants are called amphibious (Larson 2006). Typical examples include yellow water-crowfoot

(*Ranunculus flabellaris*), pepperwort (*Marsilea vestita*), and water smartweed (*Polygonum amphibium*).

Many plants found in wetlands also grow in dry, upland settings. These plants are opportunistic; they rapidly occupy disturbed sites, such as where receding water has exposed mudflats. Common examples are barnyardgrass (*Echinochloa muricata*), common plantain (*Plantago major*), and foxtail barley (*Hordeum jubatum*), which are just as likely to be found in wetlands as on uplands (Fig. 6-31).

6.3 Indicator categories

Given that some plants have specific wetland preferences, the occurrence of these plants may be a key factor for identifying wetland habitats. In the United States, hydrophytes are used for regulatory purposes to define wetlands. These plants are listed by the U.S. Fish and Wildlife Service, and macrophytes (vascular plants) are placed into five categories (Table 6-1). Mosses and algae are not included on the list. It should be noted that a plant's wetland status may change from region to region. For example, the slash pine (*Pinus elliottii*) is considered FACW in the southeast and FAC in the southern plains of the United States, but is listed as UPL in the Caribbean region.

Figure 6-28. Coontail (*Ceratophyllum demersum*), looks like a tangled pile of ropes in this shallow, spring-fed stream. In alkaline (hard) water, coontail becomes encrusted with lime, which makes the plants relatively stiff and brittle (Lahring 2003). Photo by J.S. Aber; Turner Falls, Oklahoma, United States.

Figure 6-29. Old Wives Lake, Saskatchewan, Canada (see Color Plate 6-29). Notice distinct and continuous vegetation zones marking high shorelines along the margin of the lake. Kite aerial photograph by J.S. Aber and S.W. Aber.

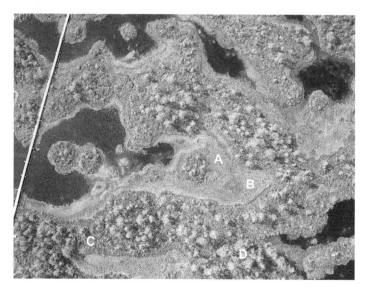

Figure 6-30. Männikjärve Bog, east-central Estonia (see Color Plate 6-30). Autumn vegetation zones follow irregular pools, hollows and hummocks in the raised bog. A. *Sphagnum cuspidatum* floating in water. B. *S. cuspidatum* around pool shore. C. *S. rubellum* above water. D. Pine trees on hummocks along with dwarf shrubs. Elevated boardwalk is ~2 feet (60 cm) wide. Photo by J.S. Aber.

Figure 6-31. Foxtail barley (*Hordeum jubatum*) is a long-awned foxtail found in saline marshes, sloughs, and along streams (Lahring 2003), but it may also occupy disturbed upland sites. Photo by J.S. Aber; Wascana Creek, Regina, Saskatchewan, Canada.

Table 6-1. Wetland plant indicators classified according to their occurrence by percentage in wetland habitats. The FACW, FAC, and FACU categories may have + and − values to represent species closer respectively to wetter and drier conditions. Based on Welsch et al. (1995).

Category	Symbol	Percent
Obligate wetland	OBL	>99%
Facultative wetland	FACW	67–99%
Facultative	FAC	34–66%
Facultative upland	FACU	1–33%
Upland	UPL	<1%

A wetland is indicated where an abundance of obligate and facultative wetland indicator species are present (Fig. 6-32). However, the absence of these species does not demonstrate that a site is *not* a wetland. Many other factors may cause indicator species not to occupy a particular wetland site. Common wetland indicator plants are listed in Table 6-2. The original National List of Plant Species that Occur in Wetlands was published by Reed (1988) and contained more than 6700 plant species. This list was subsequently revised and now has more

Figure 6-32. Skunk-cabbage (*Symplocarpus foetidus*) covers a subalpine wet meadow near Cucharas Pass at ~10,000 feet (3000 m) elevation in south-central Colorado. The flower stalk stands ~2 m tall. This species is an obligate hydrophyte, and its abundance here marks a wetland habitat. Photo by J.S. Aber and S.W. Aber.

Table 6-2. Selection of common North American wetland plants and their indicator categories. Based on Welsch et al. (1995).

Common name	Scientific name	Category
Black willow	*Salix nigra*	OBL
Green ash	*Fraximus pennsylvanica*	FACW
American sycamore	*Platanus occidentalis*	FACW
Eastern cottonwood	*Populus deltoides*	FAC+
Reed canary grass	*Phalaris arundinacea*	FACW+
Swamp white oak	*Quercus bicolor*	FACW
Black spruce	*Picea mariana*	FACW
Water tupelo	*Nyssa aquatica*	OBL
Small cranberry	*Vaccinium oxycoccos*	OBL
Buttonbush	*Cephalanthus occidentalis*	OBL
Swamp azalea	*Rhododendron viscosum*	OBL
Common cattail	*Typha latifolia*	OBL
Skunk-cabbage	*Symplocarpus foetidus*	OBL
Common reed	*Phragmites australis*	FACW

than 8000 plant species (U.S. Fish and Wildlife Service 1998).

For evaluating a site, the basic wetland criterion is that more than half of the dominant species are hydrophytic (OBL, FACW+, FACW, FACW-, FAC+ and FAC). The plant community is divided into strata or layers – trees, sapling/shrub, herb, and woody vines. Within each stratum, the dominant species are determined by measuring basal area (trees only), stem counts, frequency of occurrence or percentage cover. The 50/20 rule is applied, which states that for each stratum the dominant species are the most abundant plants that immediately exceed 50% of the total plus any additional species comprising 20% or more of total dominance (Table 6-3). In this example, red maple (*Acer rubrum*), swamp cypress (*Taxodium distichum*) and live oak (*Quercus virginiana*) meet this threshold. Red maple and swamp cypress are both wetland trees and make up more than half of the dominant tree species, so this site represents a wetland.

6.4 Plant hardiness zones

The ability of a particular plant to thrive in the wild or under human management depends on many factors – sunlight, water, nutrients, temperature, etc. Of these factors, the most critical is the minimum winter temperature that a given species can survive (Table 6-4). Additional variables taken into account are length of the frost-free growing season, summer rain, maximum temperature, snow cover, wind, and other factors (Lahring 2003). A plant may sometimes appear in a colder zone than it would normally occupy, but the plant may not flower. Its survival may depend on vegetative reproduction or reintroduction via waterfowl or human intervention. Plants may also survive in milder microclimates

Table 6-3. Example of the 50/20 rule for the tree stratum of vegetation.

Species	Total basal area	Relative contribution (%)	Cumulative dominant total	Yes/No	Category
Acer rubrum	80	80/200 = 40	40	Y	FAC, FACW
Taxodium distichum	60	60/200 = 30	70	Y	OBL
Quercus virginiana	40	40/200 = 20	90	Y	FACU
Pinus elliotti	10	10/200 = 5	95	N	UPL, FACW
Ulmus americana	5	5/200 = 2.5	97.5	N	FAC, FACW
Carpinus caroliniana	5	5/200 = 2.5	100	N	FAC
Totals	200	100			

Table 6-4. Plant hardiness zones of the U.S. Department of Agriculture based on average annual minimum low temperature. In Canada, zone 0 represents the harshest growing conditions.

Hardiness zone	Temperature	
	Celsius	Fahrenheit
1	<−46	<−50
2	−40 to −46	−40 to −50
3	−34 to −40	−30 to −40
4	−29 to −34	−20 to −30
5	−23 to −29	−10 to −20
6	−18 to −23	0 to −10
7	−12 to −18	10 to 0
8	−7 to −12	20 to 10
9	−1 to −7	30 to 20
10	4 to −1	40 to 30
11	>4	>40

within a colder zone, for example on a protected, sun-facing slope or within a large urban area.

In North America, the harshest growing conditions (zones 0–2) are found in the far north of Canada and Alaska; the mildest zones (10–11) are located in southern parts of California, Texas and Florida, as well as much of Mexico. Similar plant hardiness zone maps have been developed for South America, Europe, China, and Australia (BackyardGardener 2010).

It is most interesting to note how the plant hardiness zones have changed over the past half century in North America. The state of Kansas, in the center of the coterminous United States, is a good example. In the original USDA map of 1965, nearly all of Kansas was shown in zone 6 (0 to −10°F). This map was based on climatic data from the 1940s and 50s. The map was updated in 1990 using climatic data from 1974 to 1986; the revised map showed a significant shift toward colder zones (Fig. 6-33). On this version, Kansas is split between zone 5 (−10 to −20°F) in the north and zone 6 in the south. The Arbor Day Foundation subsequently revised the map in 2006 toward warmer zones; once again most of Kansas is in zone 6 (Arbor Day 2006). These changes reflect real climatic shifts, affecting primarily winter conditions. Minimum low temperatures decreased during the 1970 and 80s then rebounded in the 1990s and early 2000s.

6.5 Invasive plant species

An invasive species can be defined as "an alien species whose introduction does or is likely to cause economic or environmental harm or harm to human health" (National Invasive Species Information Center 2008). This definition applies to both plants and animals in all types of habitats. Introduced wetland plants may become invasive if they possess some competitive advantage compared with pre-existing (native) plants. Invasive plants tend to grow fast, disperse rapidly, and reproduce quickly, which may lead to replacement of the native flora and substantial changes in habitat conditions.

Such invasions may take place quite naturally in the course of plant migrations, but human intervention has accelerated the rate of plant movements over long distances. When people carry a plant species from one continent or island to another, for purposes of agriculture or

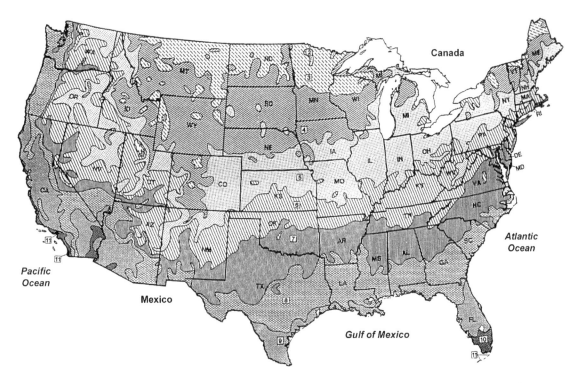

Figure 6-33. Portion of the *USDA Plant Hardiness Zone Map* for the coterminous United States, edition of 1990; see Table 6-4 for zone values. These zones are based on average minimum temperatures for the 1970s and 1980s. These zones were revised southward from the previous map of 1965, because of colder winter conditions.

gardening, the plant's natural herbivores are usually left behind, which may allow the plant to flourish unchecked. Wetlands have many such invasive plants. Selected invasive hydrophytes are described below for the United States (based on National Invasive Species Information Center 2010). Other invasive wetland plants and animals are discussed elsewhere in this book.

Purple loosestrife (*Lythrum salicaria*) – Purple loosestrife was introduced from Europe in the ballast of early ships and has now spread across substantial portions of southeastern Canada and the eastern United States (Lahring 2003). It is a shoreline plant that grows up to 2 m tall and displays showy purple spikes (Fig. 6-34). It produces seeds prolifically, is difficult to eradicate, spreads rapidly, and has almost no value for food or shelter by native wildlife. In marshes, purple loosestrife may form dense coverage that crowds out native plants and destroys wildlife habitat. It is considered "the worst of all plant pests in American wetlands" (Whitley et al. 1999).

Common reed (*Phragmites australis*) – This is a tall perennial grass that resembles bamboo and is common in temperate and tropical wetlands around the world. It is native to North America, but more invasive strains were introduced from Europe in the late 1800s. It forms dense colonies that spread by rhizomes or seeds along shorelines and emergent zones in fresh or brackish water. It does provide shelter and food for wildlife; people have used it for food and building materials, and it is excellent for wastewater treatment (Lahring 2003). Nonetheless, the common reed is taking over many wetlands, particularly in New England states (Fig. 6-35).

Eurasian water milfoil (*Myriophyllum spicatum*) – Native to Eurasia and northern Africa, this plant was introduced in the United States around 1900 and is commonly used in the aquarium trade. It readily outcompetes native

Figure 6-34. Purple loosestrife (*Lythrum salicaria*). Close-up view of flower stalk (see Color Plate 6-34). Photo by J.S. Aber; Conneaut Marsh, northwestern Pennsylvania, United States.

Figure 6-35. Common reed (*Phragmites australis*) at Batchelder's Landing in northeastern Massachusetts, United States. A small helium blimp is visible in upper left background. Photo by J.S. Aber.

Figure 6-36. Large Caterpillar tractor and disking implement used to control cattail infestation at the Cheyenne Bottoms State Wildlife Area, central Kansas, United States. Photo by J.S. Aber.

milfoils, although without flowers or fruits it is quite difficult to distinguish it from other milfoils (Wisconsin Department of Natural Resources 2008). Eurasian water milfoil reproduces vegetatively and may form dense mats on lakes and ponds, thus crowding out native species and impairing human use. It is transported easily via boats, trailers, and fishing gear.

Many means exist to control invasive wetland plants. The best one is to prevent the invasive species from reaching new territory. In many cases, invasive species were introduced by unwitting people purposefully or accidentally. Education about the dangers of invasive plants and mandatory inspections of boats and fishing gear are ways to minimize the potential for transporting invasive species. Once an invasive wetland plant has become established, however, other methods may be attempted.

- Physical methods include blocking sunlight by placing black mats or sheets on the water surface or bottom. Artificially raising or lowering water levels is a simple way to either drown or dry out the infested area. Mechanical removal or weeding of the offending plant might involve pulling, digging, cutting or raking (Fig. 6-36). For plants that

Figure 6-37. Controlled prairie burning at a small spring in the Flint Hills, Kansas, United States (see Color Plate 6-37). Such burning is typically done in latest winter or earliest spring to remove dead thatch of previous year. Nutrients in the thatch are returned to the soil to enrich spring growth. Photo by J.S. Aber.

reproduce vegetatively, removal must be complete. Controlled burning is another means to remove the above-ground plant bodies during the dry (dormant) season (Fig. 6-37). All of these methods are labor intensive and, in practice, generally lead only to partial or temporary results.

- Chemical control with herbicides was once the favored approach for dealing with aquatic "weeds" (e.g. Wilson and Boles 1967). However, wetland application of herbicides is now considered a last resort or is ruled out under many circumstances. Herbicides may kill beneficial plants as well as aquatic animals. The sudden death and decay may deplete oxygen and lead to fish kills. Most chemical treatments are temporary and must be repeated periodically to be effective (Whitley et al. 1999). Many herbicides have a copper base, and repeated use could lead to accumulation of copper in toxic amounts (Lahring 2003). Given the sensitive nature of wetland waters, it is best to avoid chemical control as much as possible, as the long-term consequences are poorly known.
- Biological controls include a variety of animals that may reduce or eliminate the undesirable plant. Muskrat (*Ondatra zibethicus*), for example, is well known for its ability to clear out cattails. Grass carp and triploid white amur are herbivorous fish that graze on filamentous algae, duckweed, and other submerged plants (Lahring 2003). However, these and other biological control agents often are non-native species, and their introduction must be considered with due caution, so they do not become invasive species as well. Government regulations restrict the use of certain biological control agents in some regions.

From this brief discussion, it should be clear that there is no simple way of eliminating invasive wetland plant species. Most of the methods produce only temporary or partial results. Physical methods are labor intensive and, thus, expensive. Chemical methods should be used sparingly in order to protect the environment, and biological controls also have limitations. Effective control of an invasive species generally requires a combination of methods carried out over a period of years or even decades.

6.6 Summary

Wetland plants, known as hydrophytes, must deal with a lack of oxygen in the soil as well as potential flooding and drying, lack of nutrients, high acidity and salinity, and other limiting factors. Many plants have responded with special adaptations that allow them not only to survive, but to thrive, under these stressful conditions. Structural adaptations allow oxygen to move from emergent plant leaves and stems into submerged roots and rhizomes (aerenchyma, pneumatophores, lenticels). For some hydrophytes, biochemical adaptations include the ability to switch to respiration without oxygen (anaerobic glycolysis) and tolerance for high salinity. Many hydrophytes conduct C_4 photosynthesis, which has some distinct advantages in aquatic environments, and some wetland plants may reproduce both sexually and by vegetative growth.

Wetland vegetation is typically arranged in distinct ecological zones – shoreline, emergent, floating, and submerged – depending mainly on growth position in relation to water level and salinity. Many wetlands experience fluctuations in water levels, so the shoreline and emergent categories are subject to change through time. Some wetland plants are able to grow in more than one ecological zone, and some plants found in wetlands also grow in dry, upland settings.

Hydrophytes are used for regulatory purposes in the United States to define wetlands. A wetland is indicated where an abundance of obligate and facultative wetland indicator species are present. However, the absence of these species does not prove that a site is *not* a wetland. For evaluating a site, the basic wetland criterion is that more than half of the dominant species are hydrophytic. The ability of a particular plant to thrive in the wild or under human management depends on many factors and is indicated by the plant hardiness zone, which is based primarily on average annual minimum temperature.

Introduced wetland plants may become invasive if they possess some competitive advantage compared with native plants. Invasive plants tend to grow fast, disperse rapidly, and reproduce quickly. Wetlands have many such invasive plants, which may cause substantial changes in habitat conditions. Many methods exist to control invasive wetland plants – prevention, physical means, chemical control, and biological agents. However, each of these methods has limitations, and no simple means exists for eliminating invasive wetland plant species. Effective control of an invasive species generally requires a combination of methods carried out over a period of years or decades.

7   Wetland wildlife

Chinese characters for Mandarin duck (*Aix galericulata*) courtesy of L. Huang.

7.1   Introduction

Wetlands are some of the most productive ecosystems in the world and are home to a rich variety of invertebrate and vertebrate animals ranging from delicate damselflies (suborder Zygoptera) to the Bengal tiger (*Panthera tigris*). As with hydrophytes, many of these animals have special adaptations for living exclusively in wetland environments; others may visit wetlands frequently or occasionally, but are not dependent on wetlands for their survival. Wildlife is the most visible and understandable aspect of wetlands for most people. We hunt, fish, and harvest all manner of wildlife – diverse fish and shellfish, ducks and geese, fur-bearing mammals, etc. Wetlands are particularly popular for bird watching (Fig. 7-1) as well as observing all other types of common and rare animals.

Protecting wildlife has been the primary inspiration for preserving and restoring wetland habitats worldwide, as demonstrated by Ramsar, Ducks Unlimited, the Nature Conservancy, and many other organizations as well as governmental agencies. Even though other wetland functions are equally important, it is wildlife that most attracts public interest and political support for conserving wetland environments. Typically it is animals near the top of the food chain that garner the most attention, namely birds, mammals and large reptiles (Fig. 7-2). Less attention is given to smaller and inconspicuous animals, although colorful butterflies are exceptions (see Color Plate 6-14). Nevertheless, smaller vertebrates and invertebrates, particularly insects, are present in great abundance and diversity within wetlands.

Wetland animals must deal with low oxygen levels and other constraints just as the plants must. Many animals, unlike plants, can move, which gives them more ways to cope with wetland environments. They have adapted several morphological and behavioral means to handle anaerobic conditions (Table 7-1). Many other animal adaptations are concerned with feeding, breeding, and raising young in order to exploit the rich potential and relative safety of wetlands (Dugan 2005). The method of feeding and means of locomotion are usually most instructive for understanding these adaptations. In other words, mouths and jaws indicate how and what an animal eats, and limbs are specialized for swimming, digging, climbing, running, flying, and other means of moving. In the following discussion, representative invertebrates and vertebrates are considered for selected groups of wetland animals.

Figure 7-1. The African fish eagle (*Haliaeetus vocifer*) ranges over most of sub-Saharan Africa, where it prefers fresh-water lakes and rivers. Okavango Delta region, Botswana. Photo courtesy of M. Storm.

Figure 7-2. African crocodile (*Crocodylus niloticus*), also known as the Nile crocodile, seen here in a marsh of the Okavango Delta region, Botswana. Hunted for high-quality leather, the crocodile was on the brink of extinction in the mid-twentieth century. Now protected by national laws and international trade regulation, the African crocodile has recovered. Photo courtesy of M. Storm.

7.2 Wetland invertebrates

Animals lacking backbones or internal skeletons may be soft bodied, such as worms and jellyfish (see Fig. 5-18), or possess a hard exoskeleton. The latter include myriad shellfish – oysters, clams, mussels, snails, crabs, shrimp, and crayfish. Corals build hard reef structures upon which the soft animals live. Invertebrates inhabit fresh, brackish, and marine wetlands. They may burrow into the soil or substrate, live on the soil or plants, swim freely, or fly. In other words, invertebrates exploit all types of niches within wetland environments.

Table 7-1. Morphological and behavioral means of animal adaptations for living in anaerobic wetland environments. Based mainly on Mitsch and Gosselink (2007).

Adaptation	Morphology or behavior
Special organs	Structures or regions for gas exchange; i.e. gills (fish and crustacea), parapodia (polychaetes), breathing tubes (insect larvae)
Improved oxygen conditions	Means to improve oxygen gradient across a membrane; moving to oxygen-rich environment or moving water across gills by ciliary action
Internal structural changes	Increased vascularization, better circulation system, stronger heart; to increase delivery of oxygen
Respiratory pigments	Enhanced respiratory pigments to improve oxygen-carrying capacity
Physiological functions	Shifts in metabolic pathways and heart pumping rate
Behavioral patterns	Decreased motor activity, closing shell at night, moving with tides or water level, etc.

7.2.1 Insects

The insect class, as a group, is the most successful of all animals; the number of known insect species is about equal to the total of all other species of plants and animals (Salsbury and White 2000). Thus, it is not surprising that insects comprise the largest and most diverse group of wetland invertebrates, and they are key parts of the food chains in most wetlands. Insects have numerous adaptations for life on, under, and above the water, and the ways in

Figure 7-3. Great diving beetle (*Dytiscus marginalis*). Note the feather-like legs for swimming; body is about one inch (25 mm) long. Modified from original photograph by H. Gröschl; obtained from Wikimedia Commons <http://commons.wikimedia.org/>.

Figure 7-4. Water strider (*Gerris remigis*) on a shallow clear stream. Distortions on water show where the legs have depressed the surface tension (*) and refracted sunlight creating the dark spots on gravel bed. Photo by J.S. Aber and S.W. Aber; Kahola Creek, Kansas, United States.

which they cope may change substantially as they undergo metamorphosis in life stages.

Insects breathe with a system of internal air tubes, known as tracheae that connect to every cell in the body (Salsbury and White 2000). Those that live under water must obtain oxygen in some way. In their larval stage, many aquatic insects, including mosquitos, have tiny abdominal tubes that act like snorkels so the larvae can breathe air directly (Dugan 2005). Other aquatic nymphs, such as mayflies and damselflies, breathe with tracheal gills on the abdomen. Some insect larvae, such as caddisflies (*Limnephilus* sp.), absorb oxygen directly from water through thin portions of their outer body casing, known as the cuticle. The great diving beetle (*Dytiscus marginalis*) breathes from bubbles of air trapped under its wings before each hunting dive.

Swimming insects have natatorial legs, which are flattened, laterally extended, and often have rows of tiny hairs so they resemble feathers (Fig. 7-3). The legs function like oars or paddles, so the insect may swim powerfully and gracefully under water. Another group of insects inhabits the surface tension. Whirligig beetles (family Gyrinidae) move rapidly in large groups half in and half out of the water. Each compound eye is divided in half to view above and below water at the same time. Water striders (family Gerridae), also known as pond skaters, have long slender legs. Fine unwettable hairs on the tarsal segments support the strider on the water surface (Fig. 7-4). Both of these groups scavenge for insects and other animals trapped on the surface tension.

Damselflies and dragonflies (suborder Anisoptera) are among the most spectacular of aerial wetland insects. The immature nymphs live under water for one to three years, where they undergo 10–15 instar stages in which they molt their skin and grow larger each time. They are ferocious predators using a hinged and hooked lower lip to capture insects, especially mosquito larvae, and other small animals (Beckemeyer and Huggins 1997, 1998). As adults, damselflies are delicate creatures and fold their wings upward together, when resting, but dragonfly wings remain fixed out to the sides (Fig. 7-5A & B).

Dragonflies are among the strongest insect flyers, reaching speeds up to 30 km/h, and they have huge eyes for tracking prey. They have extreme maneuverability, including hovering and even flying backward. The common green darner exhibits a distinctive color change that depends on temperature. At low temperature, the abdomen is a dark red-purple color, which changes to bright blue at higher temperature (Fig. 7-5C). This blue is thought to reflect most solar radiation in order to limit overheating during the day. Conversely the darker color absorbs solar energy to warm up the body

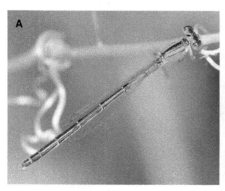  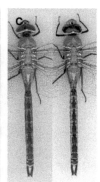

Figure 7-5. Damselfly and dragonfly comparison (see Color Plate 7-5). A. Eastern forktail female damselfly (*Ischnura verticalis*) resting with wings folded together over her back. B. Common green darner dragonfly (*Anax junius*) sitting on foliage with its wings extended to the sides. Wing span ~10 cm. C. Temperature-dependent color change of the common green darner: left at 7°C, right at 24°C. Photos A and C courtesy of R. Beckemeyer; photo B by J.S. Aber.

quickly to flight temperature in the morning (Beckemeyer, pers. com.).

7.2.2 Mosquitos

Among all wetland insects, mosquitos are the most hated and misunderstood. Almost every person alive has been bitten by mosquitos. Beyond the temporary nuisance of their bites, mosquitos transmit human diseases, particularly malaria, that are endemic in many parts of the world. Among some 3500 mosquito species, several species of the genus *Anopheles* are the vectors for malaria (Curtis 2007). More than 60 mosquito species may carry the West Nile virus in the United States, mostly in the genera *Aedes* (Fig. 7-6), *Anopheles* and *Culex* (Centers for Disease Control 2009). For this reason, individuals and health organizations have sought to eradicate all mosquitos, regardless of species, through various means, which often include the draining and destruction of natural wetlands.

All mosquitos require water to complete their life cycles – from a prairie marsh to a beer can in a roadside ditch. Some mosquitos lay their eggs on the water surface, and these eggs hatch within one or two days. In other cases, the eggs are laid in depressions and may remain dormant for many years until the site is flooded. Some eggs may hatch with the initial flooding event; others may require several wet–dry cycles before hatching (North Carolina State University 2004).

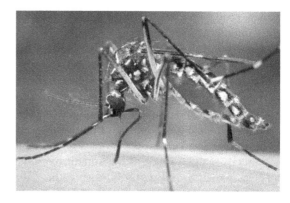

Figure 7-6. Close-up view of a female *Aedes aegypti* mosquito taking a blood meal from a human host (see Color Plate 7-6). This is the most common species of *Aedes* and a major vector for West Nile Virus as well as Dengue and Dengue hemorrhagic fever. Image by J. Ganthany (2006), courtesy of U.S. Centers for Disease Control and Prevention (CDC).

The authors experienced one such event in southern Saskatchewan, Canada in 1991. After several years of drought, a wet period caused a massive summer outbreak of mosquitos. All outdoor human events had to be cancelled, and cattle were stricken in the field.

Mosquitos breed in many potential sites in a typical American community (Purdue 2008). For urban, commercial, suburban, stream, woodland, and catch-basin environments, mosquitos breed and some may carry viruses that cause human diseases. Favored breeding sites include old tires, gutters and drains, potted plants, rain barrels, children's toys and wading pools, litter (cans and bottles), wheel ruts, tree holes, tall grass, stored boats, bird baths and pet bowls, wheel barrows, ornamental ponds, septic fields, and any other small depression that holds water. For natural ponds, in contrast, the mosquitos are typically not nuisance biters or disease carriers (Purdue 2008).

In the agricultural sector, any kind of irrigation greatly increases the potential for mosquito breeding in furrows, ditches, runoff basins, and other water-storage structures. Rice agriculture, which depends on flooding fields, is particularly susceptible to mosquito breeding. In sub-Saharan Africa, for example, the malaria epidemic is attributable largely to human activities – irrigation, deforestation, mining, hydroelectric power, and human population growth, which have drastically altered the natural environment and brought many more people into contact with mosquitos (American Association for the Advancement of Science 1991). Similarly, while the city of Mumbai, India experienced contained malaria outbreaks primarily during the monsoon season in the past, the recent boom in residential construction across the city has allowed for year-round mosquito breeding habitats at building sites close to dense human settlements. Widespread malaria outbreaks are now common throughout the year across the city, and in recent years outbreaks of dengue fever and chikungunya spread by the *Aedes aegypti* mosquito have also posed a troublesome menace.

In the authors' experience, mosquitos are generally more prevalent and problematic in urban, suburban and irrigated agricultural settings than in natural environments. Human activities and structures greatly increase the number of potential breeding sites for mosquitos. For control of mosquitos, four steps are recommended (North Carolina State University 2004):

- Identification of species – many are non-threatening and some are beneficial.
- Knowing about biology and behavior of these species.
- Eliminating breeding sites, particularly artificial sites close to human habitations.
- Using appropriate personal protection, such as clothing, bed netting, and chemical controls.

In some parts of the world, DDT (dichlorodiphenyltrichloroethane) continues to be used for mosquito control long after it was banned in the United States and most other countries (Curtis 2007). Although the environmental damage of DDT is well documented, it is difficult to control the illegal trade and use of DDT in many poorly developed countries. DDT has already been replaced by organophosphate and carbamate insecticides in many countries of South Asia and Central America, but these chemicals are more expensive and not as persistent. Pyrethroids also are viable for households; treating of bednets is among the most effective means to protect people in poorly developed regions (Curtis 2007).

From this discussion, it should be clear that out of thousands of mosquito species only a few genera are truly dangerous, most importantly *Anopheles* and *Aedes*; most mosquito species are either relatively harmless or even beneficial. The primary sources for mosquitos that affect human populations are, in fact, mostly man-made structures that provide mosquito-breeding habitats. The most effective controls are those that limit such breeding and larval habitats in close proximity to human populations, for example, screening water tanks and careful use of appropriate insecticides (Curtis 2007). Most mosquito species living in natural wetland environments are of minimal consequence from a human health and disease perspective.

7.2.3 Corals

Coral reefs are typically considered to be tropical and subtropical features of the world's shallow continental shelves (<100m deep). However cold-water coral reefs also exist at depths that may exceed 1000m (United Nations Environmental Program 2004). The deep-water habitats of cold-water coral reefs are beyond the scope of this book, so we focus herein on shallow coral-reef wetlands, as defined by Ramsar (see chapter 2). Such reefs are typically found at low latitudes (<30° N and S) in clear water with normal marine salinity (33–36‰) and temperature of 20–29 °C (United Nations Environmental Program 2004). About 800 reef-building coral species (order Scleractinia) live in symbiotic relationships with algae and may grow at rates up to 15 cm/year. More than 100 coral-reef complexes exist around the world; most are found in the Caribbean Sea, Red Sea, Indian Ocean, and equatorial Pacific (World Resources Institute 2011). Indonesia has the greatest area of coral reefs (>50,000 km^2), and the largest single reef complex is the Great Barrier Reef of Australia (Fig. 7-7). In the United States, coral reefs are abundant in islands of Polynesia and Micronesia – Hawaii, Guam, etc. The Florida Keys contain the most extensive coral reef in the coterminous United States (Fig. 7-8).

Reefs are organically constructed environments, built on the aragonite (CaCO$_3$) foundation secreted by coral polyps as their living platforms. Coral reef wetlands are among the most biologically diverse environments in the world, even though they occupy only about 0.1% of the world's oceans. According to the United Nations, one million species of plants and animals inhabit warm-water coral reefs including some 4000 fishes, which represent one-quarter of all marine fish species (United

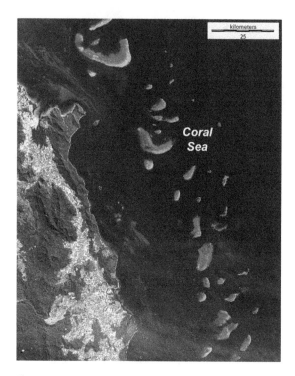

Figure 7-7. Portion of Great Barrier Reef and Coral Sea near Cairns off the coast of northeastern Australia (see Color Plate 7-7). Altogether, this reef covers approximately 30,000 km^2 (UNEP 2004). Landsat TM satellite image composite derived from TM bands 1, 3 and 5; TM dataset acquired 16 July 2009. Image from NASA; processing by J.S. Aber.

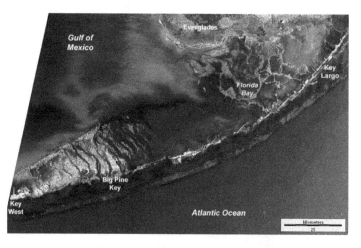

Figure 7-8. Florida Keys region south of the Everglades on mainland Florida, United States (see Color Plate 7-8). The modern coral reef forms a shallow, well-defined ridge (light blue color) along the edge of the continental shelf east and south of the keys. Landsat TM satellite image composite made from TM bands 1, 3 and 7 color coded as blue, green and red; TM dataset acquired 23 April 2008; processing by J.S. Aber.

Nations Environmental Program 2004). Reefs take on many different morphological forms – fringing, barrier, patch and atoll (see Fig. 5-21) – depending on local conditions of the continental shelf and marine environment.

Reefs typically build up to low-tide level, so they are subject to constant wave action. Storm surges and tsunamis bring the greatest water turbulence and potential for damage. During the Indian Ocean tsunami of December 2004, for example, some corals suffered severe, highly localized physical damage, particularly where already weakened by human activities, such as coral-rock mining. On the whole, however, most coral reefs experienced minimal damage and provided coastal protection in many places (Srinivas 2008).

Coral reefs are sensitive to environmental conditions, especially water temperature. A short-term increase of only 1–2 °C may cause bleaching of corals, and a long-term 3–4 °C increase leads to significant coral death (Ramsar 1999). The first major case of coral bleaching was observed in 1983, and since then several more episodes of bleaching have taken place around the world with economic consequences for tourism in the Maldives, Palau, and Caribbean islands (Global Coral Reef Monitoring Network 2010).

7.3 Wetland vertebrates

Vertebrate animals are members of the subphylum Vertebrata, which derives its name from the vertebral column, or backbone, that protects the spinal cord. Wetland vertebrates include representatives from five taxonomic classes. Of these, fishes are purely aquatic (Fig. 7-9), and amphibians are semi-aquatic. Reptiles, birds and mammals depend on water to varying degrees based on their individual adaptations and behaviors. Within these classes are some animals that are highly adapted for wetlands and could not survive in other environments (Fig. 7-10) as well as many species that frequent wetlands, but may live in other habitats just as well (Fig. 7-11). Selected wetland vertebrates are considered in the following sections, particularly for palustrine environments.

Figure 7-9. Longear sunfish (*Lepomis megalotis*) displays brilliant orange and turquoise colors (see Color Plate 7-9). This male is quite possibly guarding a nest. Native to eastern North America, the longear sunfish prefers shallow, clear water with slow to moderate current in streams with clean gravel beds (Ohio Department of Natural Resources 2010). Photo by J.S. Aber and S.W. Aber; Kahola Creek, eastern Kansas, United States.

Figure 7-10. American alligator (*Alligator mississippiensis*), a semi-aquatic reptile that lives near rivers, swamps, marshes, sloughs, and other permanent bodies of water. It eats just about any kind of animal – vertebrate or invertebrate. Eggs are laid in a large nest built of mud and vegetation; heat given off by rotting vegetation warms the eggs. Found in the southern United States from North Carolina to Texas. Seen here in Okefenokee Swamp; photo courtesy of M. Martin.

7.3.1 Amphibians

Herpetology is the study of amphibians and reptiles, many of which are closely associated with wetland habitats. These animals share nearly all the same body parts, functions and senses as humans – brain and well-developed

Figure 7-11. Ferruginous hawk (*Buteo regalis*), the largest hawk in North America, sits on a post beside Dry Lake, a salina in western Kansas, United States. It prefers open prairie habitats of the Great Plains region from Canada to northern Mexico. It eats a wide variety of prey from both wetland and upland settings; small mammals comprise the majority of its diet. Photo by J.S. Aber.

Figure 7-12. Adult American bullfrog (*Rana catesbeiana*) sits in a patch of mosquito fern (*Azolla* sp.) in a prairie marsh (see Color Plate 7-12). This frog is a game species; the largest native frog in North America, its body may reach 20 cm in length and weigh nearly 1 kg (Elliot, Gerhardt and Davidson 2009). Photo by J.S. Aber; central Kansas, United States.

nervous system, backbone, skin, eyes, ears, etc. However, in contrast, most lay eggs for reproduction, and they lack hair or feathers. They have lower rates of metabolism and require less food than do birds or mammals; body temperature is regulated by behavior, such as basking in the sun or seeking shade (Sievert and Sievert 2006). Amphibians depend on water; their soft, jellylike eggs must be laid in water. Most have two life stages, an aquatic larval (tadpole) phase followed by a semi-aquatic or terrestrial adult phase. This is true of all frogs, toads, and most salamanders, but some salamander species hatch from eggs directly into miniature versions of adults. Frogs and salamanders must maintain moist skin to avoid rapid dehydration, so they remain in wet environments as adults (Fig. 7-12); toads have dry skin, however.

Most amphibians have mucus glands in their skin that produce toxic or unpleasant-tasting secretions that presumably are a defense against predators. Human eyes and mouth are sensitive to these compounds, so it is important to wash after handling amphibians (Sievert and Sievert 2006). Further defenses include leaping into water, camouflage that changes to match the surroundings, retreating in crevices or burrows, sounding loud distress calls, and inflating to a

Figure 7-13. Plains gartersnake (*Thamnophis radix*) has caught a small American bullfrog (*Rana catesbeiana*) at the edge of a prairie marsh (see Color Plate 7-13). The snake has already swallowed the entire right hind leg of the frog, which has ceased to struggle. Photo by J.S. Aber and S.W. Aber; central Kansas, United States.

size too large for the predator to swallow (Elliot, Gerhardt and Davidson 2009). Adult salamanders, frogs and toads are carnivores eating mainly insects, spiders, worms, slugs, and other small invertebrates. Most tadpoles eat algae, and some may prey on other tadpoles. Amphibians are, in turn, prey for larger carnivores – fish, turtles, snakes (Fig. 7-13), birds and mammals (Briggler and Johnson 2008).

Around the world, amphibians face a crisis of extinctions, which is taking place even in well-protected nature refuges. Of nearly 6000 amphibian species, more than 40% have experienced recent declines, and some 30% are faced with possible extinction – a situation that is much worse than for birds and mammals (Elliot, Gerhardt and Davidson 2009). The reasons for this precipitous decline are many and varied; five main causes are cited: introduced (non-native) species, increased ultraviolet radiation, diseases and parasites, climatic change, and toxic pollution.

Non-native species and increased ultraviolet radiation appear to be marginal or insignificant factors in most cases. Disease caused by a chytrid fungus (*Batrachochytrium dendrobatidis*) is the leading explanation in the Americas, Africa, Europe, Australia and New Zealand (Elliot, Gerhardt and Davidson 2009). It is possible that global warming contributes to chytrid fungus outbreaks, and toxic pollution may weaken amphibian immune systems. The grotesquely deformed frogs discovered in Minnesota in 1995 and elsewhere since then were results of a parasitic trematode fluke, which thrives on agricultural fertilizer runoff. Still many uncertainties remain, and investigations continue into the causes and solutions for amphibian declines.

7.3.2 Reptiles

Reptiles, including alligators, turtles, lizards and snakes, are generally less dependent on aquatic or wetland habitats for their life cycles than are amphibians. They do not need to return to water or moist places to lay their eggs. Most reproduce via eggs that are laid on land, although some snakes bear living young without an external egg phase. The turtle shell and scales of other reptiles protect them from rapid dehydration, so they may venture far from water bodies. Nonetheless, many reptiles are adapted for life in or near water either through morphology or behavior and, so, spend most of their time in wetlands close to water bodies.

All turtles (order Testudines) have shells. Some shells are hard structures covered with

Figure 7-14. Aquatic turtles. A. Group of red-eared sliders (*Trachemys scripta*) sunning on a rock in a pond display their distinctive hard carapaces and scutes. B. The spiny softshell turtle (*Apalone spiniferus*) has a smooth carapace that lacks scutes. The elongated snout is used as a snorkel while under water, and the olive-brown shell is well camouflaged on sandy or muddy substrates. Quivira National Wildlife Refuge, Kansas, United States. Photos by J.S. Aber.

large scales (scutes), and others have soft, leathery shells that lack scutes (Fig. 7-14). The identification of many turtles is based on the size, shape, and color of scutes (Fig. 7-15). Many turtles are primarily aquatic; for example, of 17 turtle species found in Oklahoma, 15 inhabit rivers, lakes, marshes and other aquatic environments (Sievert and Sievert 2006). Turtles are for the most part omnivores that eat a variety of small animals, carrion, and plant materials. Among some species, however, the young are carnivores and adults become primarily herbivores, such as the painted turtle (*Chrysemys*

Figure 7-15. Northern painted turtle (*Chrysemys picta*), ventral (bottom) view (see Color Plate 7-15). The brilliant colors form an intricate pattern on the plastron and bottom edges of the carapace. This species thrives in shallow marsh pools and eats both plant and animal foods; adults prefer plants and juveniles are more carnivorous (Collins and Collins 2006). Photo by S.W. Aber; central Kansas, United States.

Figure 7-16. Freshly constructed crayfish burrow distinguished by a chimney-like structure of mud built above the ground surface. This is the favored habitat for Graham's crayfish snake (*Regina grahamii*). Comb is ~5 inches (12.5 cm) long. Photo by J.S. Aber; wet meadow in Cheyenne Bottoms, central Kansas, United States.

picta), eastern river cooter (*Pseudemys concinna*), and red-eared slider (*Trachemys scripta*).

Snakes (order Squamata) are carnivores that consume a variety of vertebrates and invertebrates. The jaw can open widely, and the left and right sides of the jaw move independently, which allow snakes to swallow whole prey animals of considerable size (see Fig. 7-13). Given the productivity of wetlands, prey is abundant, and snakes are common. Most snakes are opportunistic for their diets, but some species are quite particular about what they eat. Graham's crayfish snake (*Regina grahamii*), as the name suggests, feeds primarily on crayfish; it rests in the shade of crayfish burrows during hot summer days and retreats into these same burrows to hibernate in winter (Fig. 7-16).

Rattlesnakes are perhaps the best known of all venomous snakes; most species inhabit dry uplands, but some prefer wetlands, such as the Western massasauga (*Sistrurus catenatus tergeminus*). Despite the common fear of snakes, most snake species are relatively harmless and many are beneficial for humans (Sievert and Sievert 2006). As carnivores, snakes play important roles in controlling overpopulation among other small animals; they are themselves prey for still larger predators and, so, are integral parts in many wetland ecosystems (Collins and Collins 2006).

Alligators and crocodiles (order Crocodylia; see Figs. 7-2 and 7-10) are among the oldest living reptiles with a fossil record that goes back to the late Triassic, 230 million years ago (Elgin 2004). Many types of crocodilians evolved during the Age of Dinosaurs (Jurassic and Cretaceous) for terrestrial, fresh-water aquatic, and marine environments. But only three families survive today in the suborder Eusuchia – Crocodylidae, Alligatoridae and Gavialidae (gharials). Crocodiles and alligators are widespread in tropical and subtropical environments of the Old and New Worlds, although gharials

Figure 7-17. American crocodile (*Crocodylus acutus*) displaying the long, narrow snout and interlocking teeth that distinguish it from the alligator. Adapted from Everglades NPS <http://www.nps.gov/ever/parknews/evergladeswildlifeimages.htm>.

are native only to India and the Malay peninsula. Most crocodilian species cannot tolerate temperature below ~10 °C (San Diego Museum of Natural History 2010), although the American alligator (*Alligator mississippiensis*) may survive short-term freezing conditions if it is in water underneath ice (Britton 2009a).

Crocodiles and alligators inhabit primarily fresh-water lakes, rivers and swamps, as well as brackish lagoons, estuaries, and mangroves. Some species, such as the American crocodile (Fig. 7-17), tolerate marine water and may venture far offshore. The American crocodile is found also in hypersaline lakes, such as Lago Enriquillo (Dominican Republic), in which salinity may reach 100‰: three times normal marine salinity (Britton 2009b). Crocodilians are, in general, opportunistic eaters that take any available prey or carrion appropriate to the animal's size (juvenile or adult). The American alligator may scoop out so-called gator holes in soft soil during drought episodes. These depressions become miniature marshes in which many other wetland plant and animal species take refuge during the dry period (Niering 1985). Alligators and crocodiles occupy the top of the food chain in many wetlands; thus, they serve key roles for controlling animal populations and creating microhabitats with their nests and gator holes.

The main threats to crocodilians are habitat destruction and hunting for their valuable skins. The American alligator, for example, was heavily hunted in the first half of the twentieth century and was on the brink of extinction. Protective hunting laws in the 1960s and trade restrictions in the 1970s along with captive and wild sustainable harvesting have led to a dramatic recovery, so that now millions of American alligators survive in the wild (Britton 2009a). Similar conservation efforts are underway for other crocodilian species in many other countries, but enforcement is difficult, and poaching and habitat destruction continue to impact these animals in many parts of the world.

7.3.3 Birds

For many people, birds are the most obvious and attractive wildlife in wetlands – thousands of birds, millions of birds. The sheer number of birds and diversity of species demonstrate the biological richness and productivity of wetlands. At Tule Lake in California, for example, as many as four million birds may be present at one time during migration season on the Pacific flyway (Niering 1985). At the Laguna Atascosa National Wildlife Refuge in southern Texas, more than 400 resident and migrant bird species have been recorded (Fig. 7-18), the most of any national wildlife refuge in the United States (U.S. Fish and Wildlife Service 2009a). Nearly all bird species may visit wetlands occasionally to take advantage of food, protection from predators, or nesting (Fig. 7-19). Among these, some species are adapted specially for wetlands through their morphology or behavior.

Waterfowl and shorebirds comprise many species, genera, and families of birds that inhabit wetlands for most or part of their seasonal and life cycles. Ranging from the mouse-sized least sandpiper (*Calidris minutilla*) to the great blue heron (Fig. 7-20), these wetland birds have diverse bodies, behaviors, food and habitat requirements, life cycles, and migratory patterns. Waterfowl include those birds that typically swim or dive, such as ducks, geese, swans, coots, pelicans, grebes, teals, cormorants, and many other types. Shorebirds, on the other hand, walk on beaches, mudflats, sand banks and wet meadows or wade in shallow water. Herons, egrets, cranes, ibises, sandpipers, terns,

Figure 7-18. Laguna Atascosa National Wildlife Refuge, southern Texas, United States. This refuge is a key site on the central flyway; migrating birds avoid crossing over the Gulf of Mexico to the east and inland deserts to the west. A. Overview looking toward the northeast with Lagua Madre, a hypersaline lagoon, on the right and the mainland on the left (see Color Plate 7-18A). B. Detail view of tiny, crescentic island and shallow shoals within Laguna Madre. Such islands are important bird sanctuaries, as predators cannot reach them. Kite aerial photographs by S.W. Aber and J.S. Aber.

and gulls are among the many kinds of shorebirds. As with other animals, it is most instructive to consider the beaks and feet to understand how birds eat and how they move around on land or in water.

Shorebirds comprise more than 200 species that have long legs and highly specialized beaks – most are members of the Charadriidae (lapwing and plover) or Scolopacidae (sandpiper and snipe) families (Dugan 2005). They are normally seen wading in shallow water or walking on mudflats in which they probe for various insects, other small animals, and vegetation. Many species may live and feed together in these habitats, often in large numbers; each species is specialized for a particular method of probing for food in different depths of water or levels below the mudflat surface (Fig. 7-21).

The black-necked stilt (*Himantopus mexicanus*) favors shallow marshes and flooded fields and is found throughout the Great Plains of the United States. It has a long neck, long legs, and a straight beak that is needle sharp (Fig. 7-22). Different races of this species are found around the world in temperate and tropical environments (Fellows and Gress 2006). The American avocet (*Recurvirostra americana*) has quite a different beak that is slender, long, and curved

Figure 7-19. Least bittern (*Ixobrychus exilis*) nest. A. Nest suspended and hidden among emergent vegetation. B. Least bittern defending the nest. Photos courtesy of S. Acosta; Squaw Creek National Wildlife Refuge, Missouri, United States.

Figure 7-21. Selection of shorebirds seen at Quivira National Wildlife Refuge, Kansas, United States. Black-necked stilt (*Himantopus mexicanus*) lower center; American avocets (*Recurvirostra americana*) upper left; snowy egrets (*Egretta thula*) upper right. Photograph by J.S. Aber.

Figure 7-20. Great blue heron (*Ardea herodias*) skims along the shore of Padre Island, Texas, United States. Its wing span is nearly 6 feet (1.8 m); it flies with its neck folded, which distinguishes it from cranes. Photo by S.W. Aber.

Figure 7-22. Black-necked stilt (*Himantopus mexicanus*) wades in a shallow, saline marsh at Russell Lakes State Wildlife Area, San Luis Valley, Colorado, United States. Photo by J.S. Aber.

Figure 7-23. American avocet (*Recurvirostra americana*) pauses after stirring up the water in a shallow marsh at Quivira National Wildlife Refuge, Kansas, United States. The copper-colored head and striking black-and-white pattern on its back are quite distinctive. Photo by J.S. Aber.

Figure 7-24. Ibises possess long, strong, curved beaks for deep probing. A. White-faced ibis (*Plegadis chihi*) stalks through a shallow marsh. Photo by J.S. Aber and S.W. Aber, western Nebraska, United States. B. Scarlet ibis (*Eudocimus ruber*) seeks crabs and mollusks in mangrove thickets and coastal mudflats (see Color Plate 7-24B). Photo courtesy of M. Buchele, Brazil.

upward (Fig. 7-23). It uses the beak to sweep back and forth in muddy water to stir up aquatic insects and vegetation (Collins, Collins and Gress 1994). Ibises have long, strong, curved beaks to probe deeply for insects, crayfish, crabs and mollusks (Fig. 7-24).

Flamingos (family Phoenicopteridae) are particularly showy birds; six species are recognized (Table 7-2). They have long legs and long, flexible necks; frequently they stand on a single leg. Flamingos are filter feeders; they scoop up mud and strain out phytoplankton, zooplankton, and small invertebrates – mollusks, crustaceans, and brine shrimp. Flamingos favor salt lakes, where such food is abundant because fish are not present to eat the invertebrates (Dugan 2005). Color ranges from gray to light pink to bright red depending on age and diet (New World Encyclopedia 2008). The great blue heron (see Fig. 7-20) is a large predatory shorebird, feeding primarily on fish, amphibians and reptiles as well as small birds and mammals. It stands motionless in shallow water waiting for an unsuspecting prey animal to come within range, then the long, sharp beak is used to spear the prey.

Ducks and geese are the most commonly recognized waterfowl the world over, and some have been domesticated. Ducks and geese are particularly abundant in the prairie pothole region of the recently glaciated portion of the Great Plains in the United States and Canadian Prairie. Millions of undrained depressions provide all manner and sizes of wetland habitats that may support as many as 140 ducks per

Table 7-2. Species and geographic distribution of flamingos around the world (based on New World Encyclopedia 2008).

Species		Geography
Greater flamingo (*Phoenicopterus roseus*)	Old World	Parts of Africa, southern Europe, southern and southwestern Asia
Lesser flamingo (*Phoenicopterus minor*)		Eastern Africa to northwestern India
Caribbean flamingo (*Phoenicopterus ruber*)	New World	Everglades, Caribbean and Galápagos islands
Chilean flamingo (*Phoenicopterus chilensis*)		Temperate southern South America
James's flamingo (*Phoenicoparrus jamesi*)		High Andes in Peru, Chile, Bolivia, and Argentina
Andean flamingo (*Phoenicoparrus adinus*)		High Andes in Peru, Chile, Bolivia, and Argentina

square mile (2.6 km²; Niering 1985). Prairie marsh ducks fall in two general groups based on their food-collecting behavior – dabbling ducks and diving ducks. Dabblers favor shallow water in marshes; they tip up their tails and browse on aquatic vegetation and small animals that they can reach from the water surface. The mallard and cinnamon teal are typical examples of these relatively tame ducks (Fig. 7-25).

Diving ducks prefer the deeper water of ponds and lakes, in which they dive to the bottom in search of food. To facilitate underwater swimming, they have shorter legs set back farther on their bodies compared with dabblers (Niering 1985). Canvasback and redhead ducks represent the diving category (Fig. 7-26). Another group, known as wood ducks, inhabits forests, wooded ponds and mountain streams. They are represented by the wood duck (*Aix sponsa*) in North America and the Mandarin duck (*A. galericulata*) in eastern Asia. They build nests in tree cavities up to 15 m (50 feet) above the ground (Collins, Collins and Gress 1994). Both these ducks have spectacular plumage (Fig. 7-27), and the Mandarin duck has been

Figure 7-25. Dabbler ducks (see Color Plate 7-25). A. Mallard (*Anas platyrhynchos*) drake displays its distinctive color pattern with a green head and white collar. B. Cinnamon teal (*A. cyanoptera*) has a broad bill. Both are native to North America. Photos by J.S. Aber.

widely exported and bred in captivity. In fact, Mandarin ducks released from captive breeding have formed a wild-living population in Great Britain (Harris 2009).

Coots often feed with and are mistaken for ducks, but coots are not ducks. Coot feet are not webbed; rather, they have lobes on each toe that act like tiny paddles. Coots also seem to have trouble taking off and must run along the water surface for some distance to become airborne. The American coot (*Fulica americana*) has a completely gray (almost black) body and a white bill (Fig. 7-28).

Geese and swans are larger cousins of ducks. Both the lesser snow goose and white-fronted goose (Fig. 7-29) breed and summer in the Arctic tundra of Eurasia and North America,

Figure 7-26. Diving ducks (see Color Plate 7-26). A. Redhead (*Aythya americana*) drake. Redheads usually feed at night and rest during the day. B. Canvasback (*A. valisineria*) drake; note the long bill and large webbed feet. Both are native to North America. Photos by J.S. Aber.

Figure 7-27. The Mandarin duck (*Aix galericulata*) drake is among the most colorful and beautiful of all birds (see Color Plate 7-27). Native to eastern Asia, it breeds in eastern Siberia, China and Japan, and winters in southern China and Japan (Harris 2009). Photo by J.S. Aber.

Figure 7-28. An individual American coot (*Fulica americana*) slowly cruises along in search of aquatic vegetation to eat. Photo by J.S. Aber; Lake Kahola, Kansas, United States.

then migrate into the southern United States for winter (Niering 1985). The Canada goose is a familiar sight in all U.S. states and Canadian provinces, especially in spring and fall when flocks migrate in characteristic V-shaped formations. They vary significantly in size with large and small races or subspecies (Fig. 7-30; Niering 1985). They are highly tolerant of people, and their numbers have increased dramatically in recent years (Maccarone and Cope 2004).

The black swan (*Cygnus atratus*) inhabits lakes, rivers, and swamps in southern Australia and Tasmania (Fig. 7-31). It uses its exceptionally long neck to reach submerged aquatic vegetation and insects. It has been exported to New Zealand, Europe and North America; in Europe it has escaped into the wild (Jackson 2009). Not all geese are associated with wetlands, however. The Cereopsis goose (*Cereopsis novaehollandiae*) of Australia and the Hawaiian Nene goose (*Branta sandvicensis*), for example, are both primarily land-dwelling birds.

Fishing birds include pelicans, cormorants, the anhinga, several raptors, and penguins

Figure 7-29. A. Lesser snow goose (*Anser caerulescens*) displays the "blue goose" plumage. B. White-fronted goose (*A. albifrons*). Both these species are slightly smaller than domestic geese. They spend summers in the Arctic and head to California and Texas for winters. Photos by J.S. Aber.

Figure 7-30. Small Canada goose (*Branta canadensis*) displays its characteristic black-and-white neck while resting on a grassy lawn. Photo by J.S. Aber.

Figure 7-31. The black swan (*Cygnus atratus*) is native to Australia and Tasmania, but has been widely exported and bred for pets and zoos. Photo by J.S. Aber.

(Fig. 7-32). Their fishing techniques vary considerably. The anhinga (*Anhinga anhinga*) swims with only its head and neck above water, which gives rise to its nickname, snakebird. Like the great blue heron, it spears fish with its long, sharp beak. Returning to its nest, it tosses the fish into the air and swallows it head first (Niering 1985). Unlike most aquatic birds, however, its feathers are not waterproof, so it must spend considerable time drying in the sun (Fig. 7-33).

The American white pelican (Fig. 7-34) uses its extremely large bill like a dip net to scoop fish out of the water. Pelicans often practice group hunting, in which they form long lines and beat their wings to drive fish into a circle or toward a shallow shore where they capture

Figure 7-32. Snares crested penguin (*Eudyptes robustus*) is endemic to the Snares Island group, which is part of the New Zealand Subantarctic World Heritage Site (TerraNature 2007). Seen here at Station Cove, North East Island, Snares Islands. It spends much of the year at sea and feeds mainly on krill, squid, and fish. Modified from original photograph by T. Mattern; obtained from Wikimedia Commons <http://commons.wikimedia.org/>.

Figure 7-33. Anhinga (*Anhinga anhinga*) dries its wings at Loxahatchee National Wildlife Refuge, Florida, United States. Modified from original photograph by D. Schwen; obtained from Wikimedia Commons <http://commons.wikimedia.org/>.

the fish. Cormorants dive into water and swim using their wings and feet. They catch fish with their hooked beaks and can reach depths of 40 m (Dugan 2005). People have long used cormorants as traditional fishing aids, and cormorant fishing is still practiced in parts of Japan, China, and Macedonia. Like the anhinga, cormorant feathers are wettable, which allows them to swim deeply, so they must dry their feathers frequently.

Many raptors exploit the aquatic environment to catch fish, amphibians, rodents, snakes, crayfish, and other small animals. Several raptors are adapted for this lifestyle, including the northern harrier (*Circus cyaneus*), osprey (*Pandion haliaetus*), red-shouldered hawk (*Buteo lineatus*), and African fish eagle (see Fig.

Figure 7-34. American white pelicans (*Pelecanus erythrorhynchos*) wait quietly for fish to wander by in a lagoon next to the Missouri River in western North Dakota, United States. Photo by J.S. Aber.

Figure 7-35. Snail kite (*Rostrhamus sociabilis*) holding a snail shell. Photo adapted from Everglades NPS <http://www.nps.gov/ever/naturescience/Snail-Kite.htm>.

Figure 7-36. Bald eagle (*Haliaeetus leucocephalus*) dries its feathers on a utility pole, Kodiak, Alaska, United States. Modified from original photograph by M.L. Stephenson; obtained from Wikimedia Commons <http://commons.wikimedia.org/>.

7-1). Among the most specialized is the snail kite (*Rostrhamus sociabilis*). It is widespread in Central and South America, Mexico and Cuba, but is found only in vicinity of the Everglades and Lake Okeechobee, Florida in the United States (Tyson 2009). This kite feeds almost exclusively on apple snails (*Pomacea paludosa*), which it catches in shallow water and extracts from shells with its slender, hooked beak (Fig. 7-35). The brightly colored snails are heavily collected by shell hunters; the snails are further limited by loss of fresh-water habitats. Thus, the snail kite is listed as endangered in the United States, but is not considered threatened for the rest of its range.

Continuing loss of wetland habitat is the single greatest threat to wetland birds. Many are especially sensitive to any kind of human disturbance, such as nearby airboats and motorboats, low-flying airplanes, and walking too close to nesting sites. Raptors and other fishing birds are at the top of the food chain in many wetland ecosystems, and the fish they consume may have accumulated heavy metals and other toxic compounds derived from human pollution. Widespread use of pesticides has caused thinning of egg shells for many bird species, as documented by Rachael Carson in her influential book *Silent Spring* (1962). A combination of these factors brought many wetland bird species to the brink of extinction in the mid-twentieth century. Increased public awareness and conservation efforts since then have led to remarkable recovery for some species, most notably the bald eagle (Fig. 7-36).

7.3.4 Mammals

Mammals, like other animals, have adapted to diverse niches throughout the Earth's biosphere. Several familiar mammals are specialized for wetland habitats; for example, beaver, otter, muskrat, mink, moose (Fig. 7-37), hippopotamus (see Fig. 1-2), and seals (Fig. 7-38). These animals spend most, if not all, their life cycles

Figure 7-37. Moose (*Alces alces*), also known as Eurasian elk, have a circumpolar distribution. They inhabit boreal forests; in summer they eat large quantities of aquatic vegetation, and in winter they feed on various trees and shrubs (De Bord 2009). Modified from original photograph by the U.S. Fish and Wildlife Service; obtained from Wikimedia Commons <http://commons.wikimedia.org/>.

in or close to water bodies. Other mammals may visit wetlands on a regular or occasional basis to partake of the rich food resources (Fig. 7-39), dry-season water holes (Fig. 7-40), or as a last refuge from humans. Still other mammals are strictly dry, upland dwellers.

Other than humans, beavers are the most effective hydraulic engineers in the animal world. They are the largest rodents in North America and are able to transform a creek into a complex of pools, marshes and flooded meadows, in effect building their own wetland habitat. According to Niering (1985, p. 121), "no single wetland animal is more important than the beaver in determining the fate of wetlands." The primary technique is to construct a dam on a small stream in order to make a pond, in which the beavers build a lodge as their living quarters (Fig. 7-41). They use their ever-growing chisel-like incisor teeth to cut down trees and strip off the bark, which is their primary food. Aspen and poplar are their favorites, but they eat bark from many other trees as well. They continue to harvest trees in the vicinity of the dam and lodge until no more desirable trees are left, then they must relocate. Once constant maintenance of the dam ends, it begins to leak or could be breached by flooding. The pond partly drains and may turn into a marsh or so-called beaver meadow (Fig. 7-42). Meanwhile, the beavers move on and build a new dam, pond, and lodge, and the process of creating wetlands is repeated.

The American beaver (*Castor canadensis*) and Eurasian beaver (*C. fiber*) are generally similar in appearance and behavior, but they are not genetically compatible. Both were hunted extensively and intensively for their fur and castoreum (from the castor sacs) during the eighteenth and nineteenth centuries. As in the past, today beaver fur products are primarily coats and hats, and castoreum is valued for its taste and pheromonal or scent attraction. The Eurasian beaver had been reduced to only 1300 individuals by the beginning of the twentieth century (Holden 2010). Populations have been re-established in several countries, but their numbers remain small and populations are scattered. Beaver in North America suffered a similar fate and had been extirpated from much of their original territory by the start of the twentieth century (Anderson 2009). Beaver have since been reintroduced to many of their former habitats and are now found throughout North America, except for the northern tundra region and deserts in the southwestern United States and Mexico.

In many ways, beavers are crucial for creating and maintaining palustrine habitats that benefit many other plant and animal species. Numerous small dams serve to minimize downstream flooding and to recharge local ground water, and beaver meadows accumulate organic-rich soil. Nevertheless, the reintroduction and success of beavers in recent decades is not without

Figure 7-38. Hundreds of juvenile northern elephant seals (*Mirounga angustirostris*) relax on the beach at Point Piedras, California, United States (see also Fig. 3-20). Seen here in November, they await the return of adults who have been foraging at sea as far away as the Aleutian Islands (Warburton 2010). Kite aerial photo by S.W. Aber and J.S. Aber.

Figure 7-39. Polar bear (*Ursus maritimus*) on ice floe in Wager Bay, northern Canada. These bears are among the largest and most powerful predators on land; they hunt primarily ringed seals (*Pusa hispida*) and bearded seals (*Erignathus barbatus*). Modified from original photograph by A. Walk; obtained from Wikimedia Commons <http://commons.wikimedia.org/>.

consequences from a human perspective. Among other things, beavers cut down trees. That is what they do for a living, but not all property owners are happy to see their trees disappear overnight. In addition, beaver dams may lead to flooding of nearby properties, such as agricultural fields, roads, recreational sites, and so on (Figs. 7-43 and 7-44). Once beavers have decided to move in, little can be done short of trapping and relocating them far away.

Among the mammals that benefit from beaver activity are muskrat and mink. Muskrats (*Ondatra zibethicus*) are found just about everywhere that cattails grow – in marshes, swamps, ponds, ditches, etc. Native to North America, they have also been introduced into northern Eurasia, Japan and South America for their valuable fur (Newell 2010). Cattail is the primary food of muskrats, which eat the roots and rhizomes, but they may also eat other vegetation and small animals. They dig tunnels through the roots and use the stalks to build lodges, much like beavers do (Fig. 7-45). Muskrats are important for controlling overgrowth of cattails.

Mink are semi-aquatic animals with partly webbed toes; they can swim like otters and live on land like weasels (Collins, Collins and Gress 1994). They are generalist carnivores preying on small animals of all types depending on seasonal availability. Mink usually make their

Figure 7-40. Plains zebras (*Equus burchelli*) normally inhabit grasslands and wooded savannas, but during the dry season they congregate at a water hole in Okavango Delta of Botswana. Photo courtesy of M. Storm.

Figure 7-41. Recently constructed beaver dam on Cucharas Creek, south-central Colorado, United States. A. Overview of dam with spring snow melt flowing over the top; lodge in right background. B. Detail of freshly cut aspen to be used in the dam and for food. Aspen shoots ~5–10 cm in diameter. Photos by J.S. Aber.

homes inside muskrat or beaver lodges, but stay above the waterline (National Trappers Association 2005), and mink eat muskrat (Shier and Boyce 2009). Mink were once widespread in both North America (*Neovison vison*) and Europe (Fig. 7-46), but like beaver and muskrat were heavily hunted for their luxurious fur. Mink remain common in Canada and the United States, but their wild populations have declined seriously in Europe. As top predators, mink accumulate toxic contaminants, such as mercury and hydrocarbon compounds, which lead to problems with reproduction and health.

Besides beaver, muskrat and mink, many other aquatic mammals have been hunted for their fur, including several species of seals and otters. The sea mink (*Neovison macrodon*), for example, inhabited eastern coastal North America and was hunted to extinction in the nineteenth century for its rich pelt, which was

Figure 7-42. Panoramic view of a beaver meadow in a glacial kettle hole near Bear Lake in the San Isabel National Forest, Colorado, United States. Photo by J.S. Aber taken from a tree on the edge of the meadow.

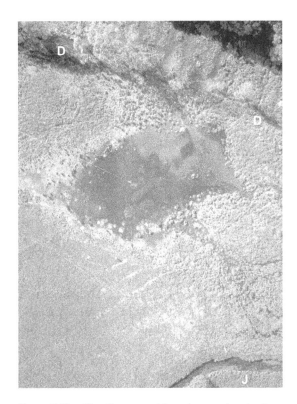

Figure 7-43. Flooding caused by a beaver dam in the Jossa Valley near Mernes (Spessart), Germany. On the inundated floodplain between the Jossa River (J) and the main irrigation ditch crossing the upper part of the image (D), a multi-channel drainage network and small lake have developed. Field of view ~150 m across. Hot-air blimp photograph courtesy of J.B. Ries, I. Marzolff and A. Fengler.

about twice the size of other mink (Maas 2010). Hunting pressure on wild populations of fur-bearing aquatic animals has lessened in recent decades for several reasons:

- Fur farms, in which animals are raised and bred for enhanced fur production.
- Limitations on quantities and seasons for harvesting wild fur-bearing animals.
- Synthetic fibers, so-called faux furs, have replaced natural furs for garments.
- Global demand for natural fur has declined due to environmental concerns.
- Reintroduction of wild populations into areas from which they had been extirpated.

As a result, many populations of fur-bearing wetland animals have stabilized, and some species are even expanding their ranges. The American beaver, for example, is now common across the Great Plains region (Fig. 7-47), where it once had been eliminated by fur trapping. Many wetland mammals remain in jeopardy, nonetheless, primarily because of human encroachment, hunting, and loss of habitat. The Florida panther (Fig. 7-48) is a well-known example. It once ranged across the southern United States from Texas to South Carolina, but nowadays is restricted to a few small pockets in southern Florida. It is listed as an endangered species and is on the brink of becoming extinct

Figure 7-44. Beaver dams in a roadside ditch, Neosho Wildlife Area, Kansas, United States. A. Elongate pools at scene center are impounded by a series of beaver dams next to the road. B. Recent beaver dam backs up a small pool that submerges trees next to road embankment. Kite aerial and ground photos by J.S. Aber and S.W. Aber.

Figure 7-45. Muskrat lodge in shallow marsh in the Sand Hills of north-central Nebraska, United States. Photo by J.S. Aber; Valentine National Wildlife Refuge.

Figure 7-46. Sketch of European mink (*Mustela lutreola*). Modified from original drawing by G. Mützel; obtained from Wikimedia Commons <http://commons.wikimedia.org/>.

with only 20–50 individuals remaining (Howard 2010).

7.4 Invasive animal species

Invasive wetland animal species come in all sizes and shapes (see http://www.invasivespeciesinfo.gov/). For example, the Asian tiger mosquito (*Aedes albopictus*) arrived in the continental United States from Japan, accidentally imported in tires in 1985 (U.S. Department of Agriculture 2010b). Old tires continue to be its preferred breeding habitat, usually in urban areas (Salsbury and White 2000). The American bullfrog (*Rana catesbeiana*) is native to eastern North America. Because it is highly desirable as a game animal, it has been introduced in the western United States and other parts of the world. It eats almost anything smaller than itself and outcompetes native frogs (Elliot, Gerhardt and Davidson 2009). Where introduced, the American bullfrog may become a pest that is quite difficult to eradicate (Sievert and Sievert 2006).

The nutria (*Myocastor coypus*) is a large semiaquatic rodent that was imported into the United States from South America in 1899 for its fur

Figure 7-47. Recently cut tree stumps and fresh wood chips demonstrate the return of beaver in the central Great Plains region. Photo by J.S. Aber; slough on the floodplain of the Cottonwood River valley, eastern Kansas, United States.

(Animal and Plant Health Inspection Service 2005). Thousands were released into the wild in the 1940s, when the nutria fur market collapsed. Nutria were also widely distributed for weed control by wildlife agencies and entrepreneurs. Nutria have a high rate of reproduction and have proliferated in the southeastern United States along the Gulf and Atlantic coasts as well as the Pacific coast of Oregon and Washington. Their preferred habitat is fresh-water marshes, where they cause considerable damage by eating vegetation and burrowing into banks and levees. This activity weakens the vegetative cover and breaks up soil, thus contributing to loss of coastal wetlands. Controlling nutria has proven difficult. Physical means, such as manipulating water levels and protecting structures and vegetation, are costly and not always effective. Trapping and killing is the only effective technique, but requires a sustained effort over many years. In Europe, attempts to eradicate nutria have been underway since the mid-twentieth century, and Great Britain succeeded in eliminating nutria in the 1980s (Colona et al. 2003).

Zebra mussels (*Dreissena polymorpha*) represent one of the greatest aquatic invasions

Figure 7-48. The Florida panther (*Puma concolor coryi*) is in grave danger of becoming extinct. Adapted from original photo by R. Cammauf; obtained from Everglades NPS <http://www.nps.gov/ever/parknews/evergladeswildlifeimages.htm>.

currently underway in North America. Zebra mussels have triangular shells with distinctive brown stripes, which give rise to the common name (Fig. 7-49). They originated in drainage basins of the Black and Caspian seas in central and eastern Europe; from there they followed shipping canals into non-native lakes and rivers of western Europe in the eighteenth and nineteenth centuries (Murphy 2010). Zebra mussels are filter feeders, straining planktonic algae, zooplankton, and other fine food particles from water, thus depleting food available for native fish and shellfish. Like barnacles, they encrust any available hard surface with their shells, thereby damaging or destroying vessels and water-handling structures. In Europe, they have seriously impaired water supplies, hampered shipping, and altered fresh-water ecosystems for more than a century.

They first entered North America in 1988 at Lake St. Clair, which connects Lake Erie and

Figure 7-50. Public awareness about zebra mussels is crucial in fighting this pest. Display on back of pickup truck of the Kansas Department of Wildlife and Parks. Photo by J.S. Aber.

Figure 7-49. Cluster of zebra mussels (*Dreissena polymorpha*) from the Deûle canalisée, Lambersart, northern France. Shells are up to 2 inches (5 cm) long. Modified from original photograph by F. Lamiot; obtained from Wikimedia Commons <http://commons.wikimedia.org/>.

Lake Huron, presumably from the fresh-water ballast of a transatlantic ship (U.S. Army Corps of Engineers 2002). Zebra mussels have many natural enemies in their native eastern European habitat, but few of these are present in North America. In effect, zebra mussels are unchecked by existing predators, parasites or benthic competitors in North America. They have spread rapidly and are now found throughout the Great Lakes, Mississippi, and most other eastern drainage basins. Zebra mussels are transported primarily in the free-swimming larval (veliger) stage in the bilge water of boats, fishing buckets, and any flowing water body – creeks, rivers, canals, etc. According to the U.S. Army Corps of Engineers (USACE), "the zebra mussel has become the most serious nonindigenous biofouling pest ever to be introduced into North American freshwater systems" (USACE 2002).

Potential means to control zebra mussels fall into three general categories – physical, biological and chemical (USACE 2002). The problem is to control both adult and veliger stages in situ and in transport. Physical controls to remove adult mussels include high-pressure water jetting, freezing, scraping and desiccation. Many possible biological controls are under investigation, but so far none is available (U.S. Geological Survey 2008). For chemical control, water-treatment plants may inject peroxide at the point of water intake to prevent zebra mussels from entering the plant. Molluscicides could be used to kill adult and veliger zebra mussels. However, zebra mussels are so tenacious that any effective chemical control would kill everything else in the water. Unfortunately, "once zebra mussels become established in a water body, they are impossible to eradicate with the technology currently available" (National Atlas 2009). At this point, the best strategy is to prevent, or at least slow, the spread of zebra mussels through public education and mandatory inspection programs (Figs. 7-50 and 51).

7.5 Summary

Wetlands are some of the most productive ecosystems in the world and are home to a diverse variety of invertebrate and vertebrate animals. Protecting this wildlife has been the primary inspiration for preserving and restoring wetland habitats worldwide. Aquatic animals must deal with low oxygen levels and other constraints, just as the plants must. In relation to wetlands, animals fall into three general categories. First are those that are wholly adapted for wetlands through their morphology and behavior; they spend most, if not all, of their life cycles in wetlands. Second are animals that frequent wetlands to take advantage of the rich resources,

Figure 7-51. Boat inspection station at the entrance to a state park in Colorado, United States. Strict inspection of boats and fishing equipment is the only effective way to protect a water body from the spread of zebra mussels. Photo by J.S. Aber.

but may live in other kinds of habitats just as well. And finally, some animals are strictly dry, upland dwellers that rarely, if ever, enter wetlands.

Insects comprise the largest and most diverse group of wetland invertebrates. Insects have numerous adaptations for life on, under, and above the water, and the ways in which they cope may change substantially as they undergo metamorphosis in life stages. Among all wetland insects, mosquitos are most hated, because they transmit human diseases, particularly malaria. The primary sources for mosquitos that impact human populations are, in fact, mostly man-made structures that provide mosquito-breeding habitats. Out of thousands of mosquito species, only a few are truly dangerous for humans; most mosquitos are either relatively harmless or even beneficial.

Coral reefs are developed in shallow, clear, tropical and subtropical seas around the world. Reefs are organically constructed wetlands that are among the most biologically diverse environments, even though they occupy only about 0.1% of the world's oceans. Coral reefs are sensitive to environmental conditions, especially water temperature, and rising ocean water temperature in recent decades has led to several episodes of coral bleaching.

Amphibians are completely dependent on wetlands, and many reptiles are also closely associated with wetlands. Around the world, amphibians face a crisis of extinctions. The reasons for this precipitous decline are many and varied; disease caused by a chytrid fungus is the leading explanation in most parts of the world. Reptiles, including alligators, turtles, lizards and snakes, are generally less dependent on aquatic or wetland habitats for their life cycles than are amphibians. Many people have an aversion to amphibians and reptiles, especially snakes, but these animals all play important roles within wetland ecosystems and should be protected.

For many people, birds are the most obvious and attractive wildlife in wetlands; the sheer number of birds and diversity of species demonstrate the biological richness and productivity of wetlands. Waterfowl include those birds that typically swim or dive. Shorebirds, on the other hand, walk on beaches, mudflats, sand banks, and wet meadows or wade in shallow water. Many of these birds are highly specialized for the way in which they procure food and how they move about in water or on land. Fishing birds include pelicans, cormorants, the anhinga, several raptors, and penguins. Their eating habits range from the opportunistic great blue heron to the snail kite that dines almost exclusively on apple snails.

Mammals, like other animals, have adapted to diverse niches within wetlands. Besides humans, beavers are the most effective hydraulic engineers in the animal world. They may quickly transform a stream into a wetland complex that benefits many other plant and animal species. Along with muskrat, mink, otters and seals, they were hunted throughout Eurasia and North America for their high-value fur. Fur farming and declining demand for natural furs combined with conservation efforts have allowed remaining populations to stabilize, and some species are even expanding their ranges.

For all wetland animals, human encroachment and continuing loss of viable habitats are the greatest threats to their existence. Human hunting for feathers, skin, fur, food, or sport has severely impacted many wetland animals and driven some species to extinction. Pollution from human sources – pesticides, herbicides, fertilizer and heavy metals – may weaken animal immune systems, interfere with reproduction, or cause death. Humans have directly or indirectly allowed many wetland species to migrate beyond their native habitats and, thus, to become invasive in other areas. Once established, such invasive animals may alter local habitats, adversely impact native species, impair human land use, and be extremely difficult to control or eradicate.

8 Wetland change

8.1 Introduction

All aspects of the Earth's environmental system change. Some features change quickly, others imperceptibly slowly. Some changes take place in a constant or cyclic fashion; other changes are irregular, episodic, or even chaotic in nature. Wetlands are no exception to this dynamic situation. As parts of the greater environmental scheme, wetlands respond to outside influences and may in turn affect those outside conditions. Wetland changes may be considered under two broad headings (Charman 2002):

- Autogenic – changes that take place within a wetland habitat caused by natural growth and evolution due to internal forcing factors. These changes may be brought about by vegetation succession, soil development, sediment burial, and other processes that take place within the wetland.
- Allogenic – changes induced by external forcing factors, which are of sufficient magnitude to influence or alter internal processes. External factors include changes in climate, rising sea level, tectonic uplift or subsidence, volcanism, fire, modifications of drainage, and so on.

The balance between autogenic and allogenic effects depends on magnitudes and rates of various processes working within and upon wetlands. If the external (allogenic) environmental framework remains relatively stable for extended periods, then internal (autogenic) developments dominate wetland evolution. Somewhat stronger and more variable external factors could either accelerate or inhibit internal developments, still with relatively continuous, but changing wetland conditions. In extreme cases, major changes by external factors may bring such instability that the wetland functions break down and the wetland ceases to exist. In regard to peatlands, Charman (2002) identified five primary factors for their development and change, namely climate, geomorphology, geology and soils, biogeography, and human activities. These five factors apply to all other kinds of wetlands as well.

All wetlands are subject to both autogenic and allogenic processes which are operating constantly at different rates and changing through time. These various factors often interact with each other in feedback relationships. As a simple example, consider the role of cloud cover for the Earth's climatic system. During an episode of global warming, more water would evaporate from the oceans, lakes and wetlands, thus leading to increased atmospheric moisture and greater cloud cover. As cloud cover increases, more incoming solar radiation is reflected into space resulting eventually in a cooler climate and a return to prewarming conditions. This scenario illustrates a negative feedback relationship, which leads to long-term stability.

Wetland Environments: A Global Perspective, First Edition. James Sandusky Aber, Firooza Pavri, and Susan Ward Aber.
© 2012 James Sandusky Aber, Firooza Pavri, and Susan Ward Aber. Published 2012 by Blackwell Publishing Ltd.

The opposite kind of feedback is demonstrated by glaciers and snow cover. During a period of global climatic cooling, glaciers, pack ice, perennial snow, and ice sheets expand in coverage, leading to an increase in the Earth's albedo, greater reflection of incoming solar energy, and a cooler climate. In this case, further climatic cooling favors still more snow and ice coverage and still more global cooling. This positive feedback relationship may lead to a prolonged ice age, which has happened several times in the Earth's past (Hambrey and Harland 1981; Caputo and Crowell 1985; McCay et al. 2006). Of course, the Earth's climatic system is much more complicated than just clouds and ice sheets – there are many other processes in operation that contribute to the net result. Wetlands are important components of this environmental system, as they both store carbon and are sources for greenhouse gases (carbon dioxide and methane).

8.2 Hydroseral succession

Hydroseral succession refers to the gradual transition involving both water and vegetation beginning with open-water habitat and ending with a raised bog or forest cover. The concept was developed by Clements (1916) and promoted by Tansley (1939). The general sequence starts with open water of a shallow lake, pond or estuary and progresses to aquatic vegetation, emergent vegetation and terrestrial fen, and culminates with a raised bog or dry woodland habitat. This theme has many variations and diversity of development. According to Charman (2002, p. 145), "hydroseral succession is by far the most important and commonly encountered example of autogenic change in peatlands." Two main pathways may initiate the development of peatlands:

- Terrestrialization – begins with pre-existing lakes or other shallow water bodies that were created by various geomorphic processes. The fate of all lakes is to infill with sediment and eventually be transformed into wetlands. This scenario is especially common in regions of former glaciation, such as Scandinavia, northern Russia, the northern United States, and much of Canada. Lakes and estuaries may be formed in many other settings by ground-water solution, volcanism, meandering rivers, sea-level change and crustal movements.
- Paludification – happens when ground water rises to the surface and creates anaerobic soil conditions. A change in climate, for example, could lead to increased ground-water recharge and reduced evapotranspiration with the net result that the water table rises toward the surface. Changes in surface drainage, caused by crustal movements, stream captures, beavers, soil development or human activities, may increase ground-water levels. Once the soil becomes saturated, wetland conditions are established and may lead eventually to peat accumulation.

Both pathways initiated bogs in Scandinavia. Many bogs developed from lakes left by the retreating ice sheet at the close of the last glaciation, some 14,000 to 10,000 years ago (Fig. 8-1). Other bogs appeared when crustal rebound took place as the ice load was removed; regions below sea level rose into the terrestrial and fresh-water environment. Infilling by lacustrine sediment led to fen conditions, which set the stage for further hydroseral succession into bogs The general sequence of individual bog development took place in five stages (Fig. 8-2; Masing 1997):

- Stage 1 – Centrally thinned wooded bog or open fen, surrounded by a bog pine forest (Fig. 8-3). Lake sediment and sedge peat have infilled the lower part of the basin to the level where the upper surface is cut off from mineral soil or surface runoff from the surrounding terrain. At this point, *Sphagnum* moss may colonize the surface.
- Stage 2 – Centrally open raised bog with beginnings of bog hollows (Fig. 8-4). *Sphagnum* peat accumulation has raised the bog surface above the surrounding moat and created a nutrient-poor, acidic environment in which few trees could survive. Moss

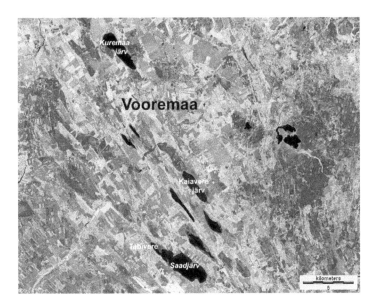

Figure 8-1. Satellite image showing the Vooremaa drumlin field north of Tartu, Estonia (see Color Plate 8-1). Drumlins are elongated hills that were molded by glaciation into smoothed, streamlined landforms. Long, narrow lakes occupy the troughs between drumlins, and bogs have developed in several of these troughs (dark green). Active vegetation is green; fallow fields are purple-maroon; red patch in left-center is a cut-over peat mine at Visusti Soo. Landsat false-color composite; TM bands 2, 4 and 5 color coded as blue, green and red; image date 28 May 2007; processing by J.S. Aber.

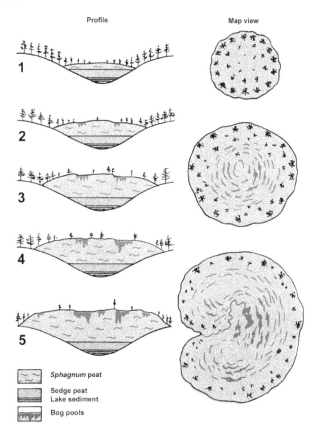

Figure 8-2. Stages in bog development beginning with initial appearance of *Sphagnum* (1) to a mature bog complex (5). Schematic profiles and map views; not to scale; adapted from Masing (1997, p. 45).

Figure 8-3. Valgesoo (White mire), a young bog in southeastern Estonia. Valgesoo is in an early stage of development (1) and is just starting to become convex (raised in the center). The central portion, seen here, shows small pine trees, moss, and grass cover. Ground water is only 20–50 cm deep (Aaviksoo, Kadarik and Masing 1997). Photo by J.S. Aber.

Figure 8-4. Teosaare Bog (Snail island mire) displays sparse cover of dwarf pines and beginning development of hummocks and hollows with small pools (stage 2). It is surrounded by pine forest. Kite aerial photograph by J.S. Aber and S.W. Aber; east-central Estonia.

growth begins to build hummocks separated by hollows.
- Stage 3 – Concentric rings consisting of well-defined hummock ridges and linear hollows following contours on the bog surface. The mature raised bog begins to show its age when the surface develops this characteristic wrinkled pattern of hollows and hummock ridges (Fig. 8-5). Standing water forms pools in hollows, and a pine forest grows on the edge of the bog.
- Stage 4 – Narrow bog pools surrounding irregular hollows and small lakes in the center (see Color Plate 2-2A). The hollows may contain pools of standing water that are perched at levels above the water table and surface drainage in the surrounding terrain. The water in the pools inhibits growth of peat, and the pools gradually expand, deepen and merge into small lakes that grow outward through time.
- Stage 5 – Water in central lakes increases in volume and eventually overflows and erodes outlet channels that drain to the surrounding, lower terrain. The original bog is split by the outlet streams; this process leads to mire complexes with multiple bogs, internal lakes, and intervening streams (Fig. 8-6). The pattern of a mature bog may be quite complicated, depending on its raised surface, presence of mineral islands, and distribution of pools, lakes, and streams (see Fig. 2-25).

This process of hydroseral succession generally takes several millennia to progress from stage 1 to stage 5. The rate of bog formation may be determined from thickness and age of the preserved peat. In Estonia, for example, the main phase of bog development began in the mid-Holocene, 7000 to 4000 years ago, peat accumulated at an average rate of 1.0 to 1.5 mm per year, and most bogs have peat 5-7 m in thickness (Masing 1997).

Bogs grow outward as well as upward (Weckström, Seppä and Korhola 2010). Acidic runoff spills from the raised or blanket bog onto the adjacent terrain, increases water levels, kills any trees, and promotes the spread of *Sphagnum* moss. This is one of the main causes of mire expansion, but people have stopped this process in many places by digging boundary ditches to divert bog runoff and protect adjacent forest and cropland.

Paludification and successional changes may also result from long-term soil development. Consider a recently deglaciated terrain or a region uplifted above sea level by tectonic

Figure 8-5. Linear hummock ridges covered by dwarf pine and intervening pools at the center of Männikjärve Bog (stage 3). Boardwalk crosses the bog with observation tower near the center. The surrounding pine forest can be seen in the background. Kite aerial photograph by J.S. Aber and S.W. Aber; east-central Estonia.

movement. In both cases, the surficial material consists of sediment or rock that is little weathered from its original mineral state. As physical and chemical processes begin to work, the minerals are broken down into smaller particles, and soluble compounds are leached out and transported downward. Microorganisms and plant roots accelerate these processes, and organic matter begins to accumulate. In some situations, a hardpan layer may form within or below the soil that inhibits downward water drainage. Retention of surface water and leaching of base cations lead to anaerobic and acidic conditions that favor the establishment of *Sphagnum* moss and further accumulation of organic matter. This sequence of events creates a positive feedback for expanding bogs, which has happened in many places across North America from Alaska to Newfoundland (Charman 2002).

8.3 Sea-level change and crustal movements

Glacial isostasy is the process of lithospheric depression beneath the weight of an ice sheet and subsequent rebound when the ice mass is reduced or removed. Glacial eustasy, on the other hand, refers to worldwide changes in sea level as a consequence of changing volume of glacier ice on land. Both glacial isostasy and eustasy, thus, are related to the volume – thickness and areal coverage – of ice sheets, but the relationship is neither simple nor fully understood at present.

8.3.1 Glacial eustasy

At the end of the last major glaciation, the water held in continental ice sheets in North America and Eurasia was released into the ocean and sea level rose by approximately 120 m worldwide. If the modern ice sheets in Greenland and Antarctica were to melt, sea level would rise by another 80 m (Hughes et al. 1981). Thus the total possible glacioeustatic range of sea level is on the order of 200 m. During the past one million years, the Earth has experienced multiple cycles of glaciation and interglaciation in which sea level varied by >100 m. The records for these sea-level cycles are preserved in deep-sea sediments, ice cores from Greenland and Antarctica, cave formations, and other geological sources.

The margins of many continents slope quite gently into the sea forming wide, shallow continental shelves. The Atlantic and Gulf coasts of the United States are good examples. In such situations, shoreline positions migrated over tens to hundreds of kilometers as sea level rose and fell during glacial cycles. These sea-level cycles could be preserved as intervals of marine deposits (high sea level) separated by surfaces of erosion (low sea level). A complete sedimentary record of past sea-level changes ideally would be developed in stable coastal areas far removed from glaciation. Much of this sedimentary record is, unfortunately, not directly accessible; it is underwater on the continental shelves. An excellent record is preserved in South Florida (Enos and Perkins 1977). Six episodes of high sea level (interglaciation) are marked by coral reefs and other marine deposits:

1. Recent (Holocene) sea level, last few thousand years.

Figure 8-6. Meenikunno, a large bog complex south of Tartu, Estonia. Meenikunno is a mature, raised bog that encompasses several lakes and mineral islands (stage 5). A. Overview toward the center of the bog, which is a nearly treeless plateau (light zone in middle distance). B. Dwarf pine trees, grass, and moss cover in the southern portion of the bog. Notice the wooden board path to right. Photos by J.S. Aber.

2. Sangamon sea level, peak ~130,000 years ago.
3. High sea level with peak ~180,000 years ago.
4. High sea level with peak ~236,000 years ago.
5. High sea level with peak ~324,000 years ago.
6. Earliest preserved high sea level, age uncertain.

The highest sea level achieved during any of these cycles was by the Sangamon (Eemian) Sea (Fig. 8-7). Sea level stood 5–8 m higher than at present; all of southern Florida was submerged, and the shoreline was located approximately 250 km (150 miles) farther north across the Florida peninsula. In many other parts of the world, the Sangamon Sea is marked by marine deposits or erosional beach terraces a few meters above present sea level (Fig. 8-8). In the Bahamas, sea level stood about 2 m above present during the time interval 132,000 to 118,000 years ago. Near the end of this period, sea level rose rapidly to +6 m; this highest stand of sea level was quite brief, lasting perhaps only a few centuries before declining rapidly (Neumann and Hearty 1996). The mechanism for such a large and rapid change in sea level is thought to be related to possible ice-sheet surging into the ocean.

Until recently, the Sangamon sea level was considered to be the highest of the Pleistocene. However, sea level may have stood even higher

Figure 8-7. Miami Oolite, the limestone formation that underlies much of Miami, Florida. The limestone was deposited as carbonate sand on a shallow marine bank, much like the modern Bahama Bank. The Miami Oolite is a northern continuation of the Key Largo Limestone trend; they were deposited simultaneously during high sea level ~125,000 years ago (Enos and Perkins 1977). Photo by J.S. Aber.

about 420,000 to 400,000 years ago (Hearty et al. 1999; Poore and Dowsett 2001). Sea level may have exceeded 20 m above present in the Caribbean Sea, Bahamas and Bermuda, which implies that both the Greenland and West Antarctic ice sheets were absent or greatly reduced. Considering the modern concerns for sea-level rise, it is wise to keep in mind the magnitudes and rates of entirely natural eustatic sea-level

Figure 8-8. Wembury Bay, near Plymouth, southwestern England. Bedrock platform in foreground is a strandflat a few meters above present sea level. It was eroded by wave action during the Eemian (Sangamon) high stand of sea level ~125,000 years ago. Photo by J.S. Aber.

fluctuations in the not-too-distant past (see below).

8.3.2 Glacial isostasy

Beneath thick ice sheets the amount of lithospheric (crustal) depression ranges from several hundred meters to more than a kilometer. In order to compensate for such a large depression beneath a crustal load, the surrounding area may rise creating a forebulge (Anderson 1988), although the amount of marginal uplift is generally much less than the amount of central depression. When the ice mass disappears, the depressed region rebounds and the forebulge collapses. These movements take place slowly, and millennia are required for complete adjustment to isostatic equilibrium.

Since the last glaciation, extensive coastal regions of North America and Eurasia have uplifted tens to hundreds of meters and some regions are continuing to rebound, particularly the Hudson Bay and Great Lakes areas in Canada and the northern Baltic Sea vicinity in Scandinavia (Andrews 1970; Tushingham 1992). In like manner, the glacial forebulges have subsided. For example, the Atlantic Ocean is encroaching on the Chesapeake Bay region at a rate of about 30 cm per century (Colin 1996), the southern portion of Lake Michigan is sinking (Tushingham 1992), and similar subsidence is taking place in southern portions of the Baltic Sea and North Sea regions in Europe.

In glaciated coastal regions, both postglacial rise in sea level and crustal rebound took place simultaneously. A typical sequence of events is demonstrated in the fjord district of southwestern Norway. As crustal rebound lifted the land faster than rising sea level, fjord connections to the open ocean were broken and marine water was replaced by fresh water (Fig. 8-9). Eventually smaller basins filled in with sediment and turned into bogs, which may be cored for sediment samples and radiocarbon dating (see Fig. 3-30). Such cores reveal the basin's hydrologic history, and the transition from brackish to fresh-water sediment marks the point at which the threshold was uplifted above sea level (Fig. 8-10). The positions of former shorelines are well marked in the landscape by raised beaches and deltas (Fig. 8-11). On this basis, the pattern and timing of crustal uplift may be reconstructed (Fig. 8-12).

8.3.3 Complicated responses

Changes in relative sea level may become quite complicated, where the effects of both crustal depression/rebound and eustatic sea level are involved. The coastal area of western British Columbia shows the possibilities. Most of the shelf of Queen Charlotte Sound was ice-covered and depressed during maximum glaciation, more than 15,000 years ago. When northern Vancouver Island was deglaciated by 13,000 years ago, the shelf and coastal areas were submerged under the Pacific (Clague 1989). Marine shorelines were in some places as much as 200 m higher than today. Crustal rebound then took place; the coastal region was uplifted, and parts of the present shelf were eventually exposed as dry land (Luternauer et al. 1989). Soil development and forest growth took place about 10,500 years ago. Meanwhile fjords to the east were still depressed well below sea level. Between 10,500 and 9000 years ago, most remaining glaciers on the mainland melted and eustatic sea level rose. The results were rebound

Figure 8-9. Bolstadfjord, near Bergen in western Norway. A. Overview looking toward the fjord entrance under the bridge at center distance. Water depth in the foreground is several hundred meters. B. Close-up view of bridge over the fjord entrance. Tidal flow creates ripples in the water surface over the threshold, which is only a few meters deep. This fjord basin is barely connected to the marine environment at present. Photos by J.S. Aber.

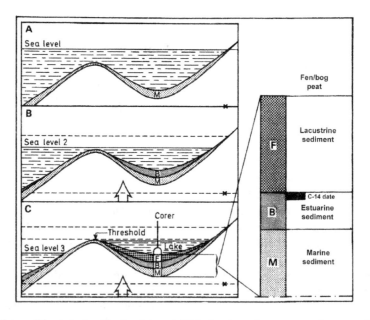

Figure 8-10. Use of a small basin in hard bedrock to establish the time of emergence above sea level during postglacial rebound of the crust. M = marine, B = brackish, F = fresh water. Date of the top brackish sediment marks the time of threshold uplift above sea level. Adapted from Anundsen, K. 1985. Changes in shore-level and ice-front position in Late Weichsel and Holocene, southern Norway. *Norsk Geografisk Tidsskrift* 39, p. 205–225, Fig. 2.

of the fjord-head region (emergence) and drowning of the shelf area (submergence).

The Mississippi River delta exhibits another complicated situation for isostasy and eustasy (Fig. 8-13). During times of major glaciation, when sea level was low, the Mississippi River incised a deep valley across the delta and transported sediment into the Gulf of Mexico well beyond the delta. During interglacial times, in contrast, high sea level allowed sediment accumulation, infilling of the valley, and growth of the delta. These substantial changes in sediment mass of the delta induced crustal subsidence and rebound exceeding 9 m (30 feet) along

Figure 8-11. Kame terrace, Eidslandet, western Norway. The terrace surface (farmstead) is underlain by stratified sand and gravel of a glaciomarine delta. The terrace scarp (foreground) represents the ice margin position and the terrace surface marks sea level in the fjord at the time the delta was deposited by glacial melt water. Photo by J.S. Aber.

the Mississippi valley (Blum et al. 2008). Lesser depression and rebound extended >100 km to the east and west along the coast. The timing of rebound coincided with release of melt water at the end of each glacial cycle from continental ice sheets to the north. At present, high sea level combined with sediment accumulation and compaction are causing crustal depression and delta sinking. This example demonstrates the importance of glacially related processes for understanding delta behavior far beyond the actual limits of ice sheets.

8.3.4 Modern sea-level rise

Modern eustatic sea level is rising at a rate of about 0.3 cm per year due to slow melting of glacier ice[1] and thermal expansion of sea water (Pilkey and Young 2009). While 3 mm per year may not seem like much, this rate adds up to 30 cm (one foot) per century. For low-lying coastal wetlands this becomes significant. From the Maldive Islands in the Indian Ocean (see

---

[1] Glacier ice refers to land-based glaciers and ice sheets. The melting of marine pack ice and ice shelves does not affect sea level, as these ice bodies are already floating.

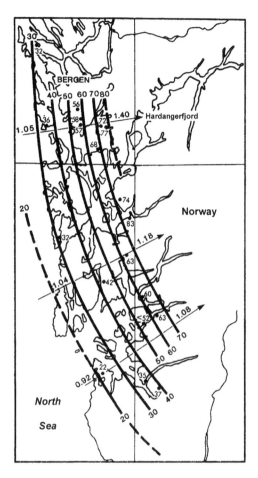

Figure 8-12. Reconstruction of crustal uplift in southwestern Norway since the Younger Dryas phase of glaciation, approximately 11,000 corrected radiocarbon years ago (Denton and Hendy 1994). Contours indicate amount of uplift (in m relative to modern sea level); uplift gradients are given in m/km along profiles. Map adapted from Anundsen, K. 1985. Changes in shore-level and ice-front position in Late Weichsel and Holocene, southern Norway. *Norsk Geografisk Tidsskrift* 39, p. 205–225, Fig. 8.

Fig. 5-21), to Kivalina, a barrier island on the Alaskan Chukchi Sea coast, to the venerable city of Venice, Italy, populated places and fragile wetlands are at risk of disappearing during the twenty-first century. In many cases, these places will be simply abandoned, as engineering solutions to rising sea level are either too costly or environmentally unsound (Pilkey and Young 2009).

Figure 8-13. Natural-color satellite image of the Mississippi Delta region showing distinct sediment plumes in the Gulf of Mexico (see Color Plate 8-13). The Mississippi River delivers about 500 million tons of sediment into the gulf annually via the main stem of the river and its Atchafalaya distributary. A. Atchafalaya Delta, B. Lake Pontchartrain, C. Mississippi Delta, D. Mobile Bay. New Orleans is the light patch on the southern side of Lake Pontchartrain. MODIS image acquired 5 March 2001; adapted from NASA Visible Earth <http://visibleearth.nasa.gov/>.

As noted above, the relative position of sea level along any particular coast depends on both eustatic changes and local land movements. Many coastal environments are sinking or retreating due to a combination of local factors – sediment compaction, shore erosion, reduced sediment supply, crustal subsidence, subsurface solution, etc. Local land sinking plus global eustatic rise add up to substantial rates of sea-level increase in many localities. Galveston, Texas, for example, experienced a relative sea-level rise of 6 m (20 feet) during the twentieth century at a relatively constant rate of 6 cm per year (Pilkey and Young 2009).

For low-lying coasts that slope gently into shallow seas, a small increase in sea level would result in a large inland migration of the shore, and substantial erosion would take place in unconsolidated sediments. The North Sea coast of northwestern Denmark illustrates these consequences dramatically (Fig. 8-14). Coastal cliffs stand tens of meters high and are composed of unconsolidated mud, sand and gravel sediments left by the last glaciation of the region (Fig. 8-15). Historic structures – farms, villages, churches – built during the Middle Ages hundreds of meters back from the cliff are now falling into the sea (Fig. 8-16). The long-term rate of cliff retreat is on the order of one meter per year (H. Lykke-Andersen, pers. com. 2005). A large volume of eroded sediment is transported northward by longshore drift and has, over millennia, built out the Skagen peninsula as a huge sand spit (Fig. 8-17).

Clearly, coastal processes are removing sediment and land in some sectors and building new land in other places. The debate about what to do has been settled in Denmark with the decision that nature should run its course in most places. Nonetheless, some historical structures have been protected from further coastal erosion (Fig. 8-18). Similar situations exist in many other coastal settings. Another famous case is the Cape Hatteras Lighthouse built on the Outer Banks of North Carolina in 1870. At that time, it was half a kilometer inland and well above mean sea level (Pilkey and Young 2009). It took just over a century of erosion and land loss

Figure 8-14. Satellite image of Vensyssel region, northern Jylland, Denmark (see Color Plate 8-14). Note the milky blue color of water along the North Sea coast derived from beach and cliff erosion in contrast to the clean water of the Kattegat. Landsat false-color composite based on TM bands 1, 3 and 7 color coded as blue, green and red; date of acquisition 4 June 2010. Location of Mårup Kirke indicated by asterisk (*) west of Hjørring. Image from NASA; processing by J.S. Aber.

Figure 8-15. Vertical view of North Sea coast, cliff, and beach at Mårup Kirke, near Skallerup Strand, northwestern Denmark. Note shadows of two people walking on beach. Kite aerial photograph by I. Marzolff, S.W. Aber and J.S. Aber.

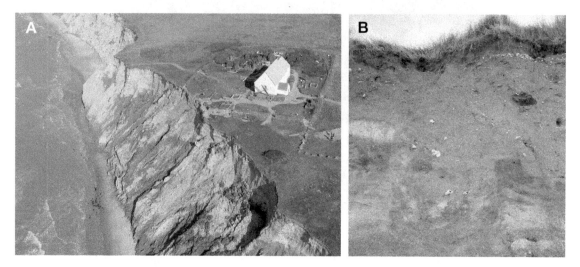

Figure 8-16. A. Mårup Kirke and graveyard. The church was built in the first half of the 1200s well back from the cliff; as erosion approached the church, it was closed in 1926. Note that a corner of the churchyard (cemetery) has already fallen down. Kite aerial photograph by I. Marzolff, S.W. Aber and J.S. Aber. B. Eroding cliff face in cemetery. Note pair of human leg bones exposed in the cliff (lower center of view). Photo by J.S. Aber.

Figure 8-17. Northern tip of Skagen Peninsula, Denmark; Kattegat Sea to right, Skaggerak Sea in left background. Notice rock structures parallel to shore designed to limit beach erosion. Kite airphoto by S.W. Aber and J.S. Aber.

Figure 8-18. Højerup Gamle Kirke (old church) on the edge of Stevns Klint, south of Copenhagen, eastern Denmark. The church was built in 1250–1300 a good distance from the cliff. Erosion of chalk at the base of the cliff has undermined the more resistant upper layer of limestone, and a landslide in 1928 brought down a portion of the church building. A. Overview of cliff and remaining church building. Note riprap placed at base of cliff to prevent further erosion below the church. B. Close-up view of remaining tower and nave of church; the choir and apse fell down in the landslide. Photos by J.S. Aber.

Figure 8-19. Cape Hatteras Lighthouse on the Outer Banks, North Carolina, United States. Seen here following flooding caused by Hurricane Isabel in September 2003. The original lighthouse location is the circular sand pad in the foreground. Adapted from original photo by M. Wolfe; obtained from Wikimedia Commons <http://commons.wikimedia.org/>.

before waves threatened to topple it during a storm in 1980. Following a long and bitter debate, the National Park Service reached a decision to move the lighthouse 500 m back from the shore, which would both preserve the structure and allow natural coastal processes to continue. The move was accomplished in 1999 (Fig. 8-19). The present location should be safe for another century (National Park Service Cape Hatteras 2010).

During the twenty-first century, eustatic sea level is likely to rise by at least 30 cm, perhaps by one meter, and perhaps even two meters according to various projections (Church and White 2006; Pilkey and Young 2009). This eustatic rise will be tempered by local crustal movements, shoreline erosion and deposition, and other factors, such that some regions will experience little change in relative sea level, while others will suffer large increases of several meters. These scenarios are based on current rates and trends, which show no signs of slowing or reversing and may, in fact, accelerate. A sea-level rise of just one meter would cause immense damage to human coastal structures, displace millions of people, and threaten the survival of coastal ecosystems, particularly wetlands, many of which are already under assault from human activities.

8.4 Climate change

8.4.1 Climate basics

The water supply in most wetlands is a delicate balance between gains and losses, which are controlled to a considerable extent by climate. The idea that climate is constant is simply wrong; climate varies continually in response to events on Earth as well as extraterrestrial influences. The time scales of climatic change range from decadal variations to shifts over millions

Figure 8-20. The whole Earth photographed during the Apollo 17 mission showing Africa, Arabia, and Antarctica (see Color Plate 8-20). Clouds, ice, and deserts have high albedo; oceans and land vegetation have low albedo. Image courtesy of K. Lulla, NASA Johnson Space Center. Photo ID: AS17-148-22727, December 1972.

Figure 8-21. Eruption of Mt. Kliuchevskoi volcano (snow covered) on the Kamchatka Peninsula of Russia. Ash plume drifted toward the southeast over the Pacific. The 1991 eruption of Mt. Pinatubo in the Philippines created a significant ash effect that is credited with measurable climatic cooling during 1992–93. Image courtesy of K. Lulla, NASA Johnson Space Center. Photo ID: STS68-150-045, September 1994.

of years. Just as sea level rose and fell during several glacial cycles of the past one million years, so the Earth's climate experienced large oscillations from glacial to interglacial conditions. Several factors play primary roles within the climatic system.

- Solar energy – may vary in the quantity or quality (spectrum) of radiation, and oscillations in earth/sun orbital geometry affect the seasonal and latitudinal distribution of sunlight received on Earth.
- Greenhouse gases – carbon dioxide, methane, water vapor, and ozone absorb thermal-infrared radiation emitted from the Earth's surface, thus warming the lower atmosphere.
- Albedo – refers to the reflectivity of the Earth's surface and atmosphere (Figs. 8-20 and 8-21). Increased albedo – clouds, volcanic dust, snow, ice, deserts – causes more incoming radiation to reflect back into space, thus having a cooling effect. Decreased albedo – water bodies and vegetation – has the opposite effect.
- Oceanic circulation – oceans, atmosphere and ice sheets are coupled in terms of the distribution of heat around the world. Oceanic currents move heat and salt in a thermohaline circulation system.

The Sun is the source of all energy that drives the Earth's climatic system; solar energy is the ultimate control of climate. The 11-year cycle of sunspots is so well known to astronomers and the public alike that it is usually considered to be a permanent fixture of solar activity. However, the sunspot cycle is variable in length and strength. For example, the most recent sunspot cycle (23) reached a peak in 2000, which was comparable to cycles of the late twentieth century. However, the decline to the low ebb of the cycle extended until 2008–09, and a new cycle (24) began late in 2009, about four years

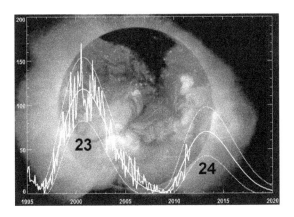

Figure 8-22. Recent solar activity during sunspot cycle 23 and a prediction for cycle 24 (see Color Plate 8-22). Adapted from Hathaway (2011). <http://solarscience.msfc.nasa.gov/SunspotCycle.shtml>

behind schedule (Fig. 8-22). The next sunspot cycle is forecast to reach a level much lower than the 2000 peak, and if this comes to pass it will have a significant impact on world climate (NASA 2011b). In fact, some scientists are now forecasting the possibility of significant global cooling by the mid-twenty-first century.

The so-called Little Ice Age (*c.* AD 1200–1900) was most likely a consequence of prolonged lows in solar activity. Sunspots all but disappeared during the Maunder minimum, 1645–1715 (Eddy 1977). Several independent lines of evidence support the existence of this interval of reduced solar output (Le Roy Ladurie 1971, 2004), and global cooling was reinforced by numerous volcanic eruptions (Nesje and Dahl 2000). The net result was global climatic cooling on the order of 1-2 °C (Grove 1988), which is documented worldwide from the Pacific islands (Cook et al. 1991; Nunn 2003), to the tropical Andes (Thompson et al. 1986), to western Canada (Campbell et al. 1998), to Greenland (Fig. 8-23). Similar, even larger, climatic excursions driven principally by solar activity took place during previous millennia and decidedly influenced wetland development (see chapter 9).

The Earth's climate is unique among planets of the solar system. The temperature range allows for the existence of water simultaneously in solid, liquid, and vapor states and is just right for supporting life. Since the beginning of life

Figure 8-23. Viking settlement in southwestern Greenland started in the late tenth century during the Medieval climatic optimum. The settlements prospered for two centuries, but then fell on hard times when climate began to cool after about AD 1200. A. Stone walls of Hvalsey church, the best preserved of any Viking building in southwestern Greenland. A wedding in 1408 at this church is the last recorded event in the history of Viking Greenland. B. Viking grave remains. Lower ^{18}O values in Viking and Inuit teeth for the interval AD 1400–1700 correspond to the cooler temperatures of the Little Ice Age (Fricke, O'Neil and Lynnerup 1995). Photos courtesy of P. Jensen.

on Earth at least 3½ billion years ago, the range of climates on Earth has remained within tolerable limits for life, a most remarkable fact. Regulation of the Earth's climate is due in large part to certain heat-trapping gases in the atmosphere. Carbon dioxide is most important overall. At present, CO_2 makes up only about 0.3‰ of atmospheric gases, but this amount has increased dramatically since the Industrial Revolution. Methane (CH_4) is another strong greenhouse gas, the abundance of which is connected with

Figure 8-24. Buried carbon derived from biological activity. A. Huge coal strip mine in eastern Germany. Miocene age brown coal (lignite) is mined here for power production nearby. Photo by J.S. Aber. B. Chalk monuments in western Kansas, United States. Chalk is composed of microscopic coccoliths (produced by marine algae), which were deposited in great abundance worldwide during the late Cretaceous. Kite airphoto by S.W. Aber and J.S. Aber.

Figure 8-25. Rounded edges in this bedrock outcrop result from rapid weathering under a spruce-aspen forest. Such weathering releases cations that are delivered to the ocean. Photo by J.S. Aber at Cordova Pass, south-central Colorado, United States.

northern hemisphere wetlands (Korholaa et al. 2010). Carbon is added to the atmosphere by organic weathering of soil minerals and removed by burial of excess organic carbon and carbonate (Fig. 8-24).

The amount of CO_2 present in the atmosphere is linked to the amount of CO_2 dissolved in sea water. The ocean is the major reservoir for CO_2 at the Earth's surface, and CO_2 exchanges freely between ocean and atmosphere. The CO_2 content of sea water is determined, in turn, by buffering cations such as K^+, Ca^{2+} and Mg^{2+} derived from weathering of rocks on land (Fig. 8-25) and by the rate of burial of organic carbon as carbonate and hydrocarbon-rich sediment. Fresh-water and marine wetlands play key roles in the carbon cycle both as carbon sinks and sources of CO_2 and CH_4. "Climate change will likely affect the ability of wetlands to sequester carbon, but the results will vary and are difficult to predict" (Kusler 1999).

8.4.2 Climate and wetlands

Climate change has both regional and global implications for wetlands. Mires that are expanding, vertically or horizontally, accumulate carbon in the peat and muck mass. Likewise, healthy coral reefs that are growing upward or outward sequester carbonate in their hard structure. On the other hand, factors that may reduce water supplies, alter nutrients or temperature, and negatively influence hydrophytic plants and aquatic animals could lead to shrinkage of wetland environments and the release of greenhouse gases. In a general way, warmer and drier climate would reduce fresh-water wetlands, while a cooler, wetter climate would favor wetland expansion.

One could imagine a simple relationship in which cooler, wetter climate leads to expansion of wetlands, burial of organic carbon, and drawdown of atmospheric greenhouse gases. In turn, this would result in still cooler climate and further growth of wetlands, which would amplify

Figure 8-26. Devils Lake vicinity, northeastern North Dakota, United States. Landsat TM band 5 (mid-infrared) satellite image showing expanded extent of lakes, 8 September 2000. EDL = East Devils Lake, which was formerly a separate lake (see Fig. 8-28). Image from NASA; processing by J.S. Aber.

climatic cooling – a positive feedback relationship. A shift toward warmer, drier climate would have the opposite consequence, again a positive feedback situation. However, neither climate nor wetlands behave in such simplistic ways. Consider, for example, the northern Great Plains of the United States. Climate has become gradually warmer during the past half century in the Dakotas and Minnesota. However, precipitation has also increased, with major consequences for wetlands, as demonstrated by frequent flooding of the Red River of the North and expansion of Devils Lake in the northeast of North Dakota (Fig. 8-26).

Devils Lake is a system of connected, large lake basins scooped out during the last glaciation of the region, some 14,000 to 12,000 years ago. Lakes in the enclosed drainage basin are subject to sizable changes in their elevation, area, capacity, chemistry, and biomass (Aber et al. 1997). These changes take place mainly in response to climatic fluctuations. During the past half century, Devils Lake has experienced a dramatic increase in its surface elevation on the order of 40 feet (12 m) with corresponding increases in surface area and volume (Fig. 8-27).

Lake elevations above 1440 feet have led to significant flooding of adjacent cities, parks, roads, farms, sewage treatment plants, a military base, a Native American reservation, and other human structures. This elevation was surpassed in 1997, and Devils Lake reached 1446 feet elevation during the summer of 1999, at which point the lake overflowed and began draining eastward into Stump Lake. Despite this outlet, lake level has continued to rise and now exceeds 1450 feet (U.S. Geological Survey 2010). Countless smaller lakes and marshes in this vicinity have responded in a similar fashion, which demonstrates that global warming may have unexpected consequences on a regional basis, in this case leading to greatly expanded wetland habitats (Fig. 8-28).

One may look also at the micro-scale impact of climate change (see Tables 3-1 and 2), for example, in bogs of Estonia. During warm/dry periods, hummocks and ridges are stable or increasing, whereas bog hollows expand during cool/wet climatic intervals (Karofeld 1998). Frenzel and Karofeld (2000) demonstrated at Männikjärve Bog (see Color Plate 3-10) that hummocks and ridges are sinks for methane; in

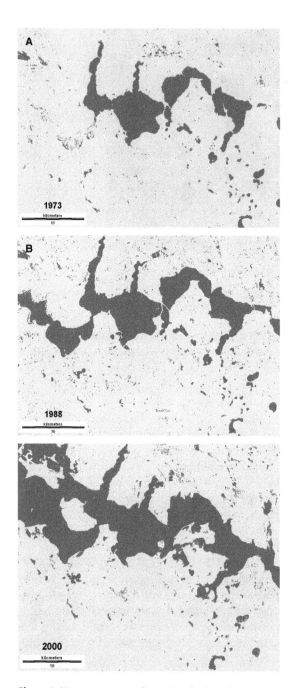

Figure 8-27. Expansion of Devils Lake based on Landsat MSS and TM imagery. Devils Lake itself (excluding smaller, unconnected lakes) more than tripled in area from 122 km² in 1973, to 186 km² in 1988, to 372 km² in 2000. Image from NASA; analysis by J.S. Aber.

contrast, hollows and pools are sites of methane emission. Gradual changes in hummock-hollow distribution, thus, affect methane flux within the bog system. A negative feedback relationship exists, such that warm/dry weather causes an increase in hummocks and ridges and thus a reduction in methane emission. Reduced atmospheric methane would, in turn, lead to a decreased greenhouse effect and cooler climate. On this basis, it appears that bogs of the Baltic region have the potential to dampen changes in climate brought about by other factors.

These examples illustrate that connections between climate and wetlands may become quite complex. When multiple autogenic and allogenic processes interact, the results may be unexpected. Many natural and human factors may influence the ability of wetlands to store or release carbon, and predicting the impact of climatic change on wetlands is a particularly daunting task. The numbers of variables and feedback relationships are simply too many, and too little is known about them to reach any convincing opinions about the future of wetlands and climate. Even the most basic climatic influences, such as solar energy and volcanic eruptions, are impossible to predict. While it appears likely that the Earth's climate will continue to warm during the twenty-first century, any attempt to model future global climate change and its impact on wetlands is fraught with uncertainty.

8.5 Fire

Fire is an integral part of many ecosystems, and humans have set fires since ancient times for hunting, warfare, clearing forests, burning agricultural waste, and other purposes (Fig. 8-29). Prescribed burning is used routinely by the U.S. National Forest Service and other agencies for habitat maintenance. Controlled fire is essential for preserving the tallgrass prairie in the Flint Hills of eastern Kansas (Applegate, Flock and Finck 2003). Prairie fire renews the grass, inhibits trees and woody vegetation, and returns nutrients to the soil (see Color Plate 6-37). Prescribed burning has likewise been used to

Figure 8-28. Panoramic northward overview of the eastern margin of East Devils Lake (see Color Plate 8-28). The former shoreline stands with dead trees on left side; shallow lake and marsh toward right were created by rising water of the last few years. East Devils Lake was a separate lake in the past, but is now joined with Devils Lake. Blimp aerial photograph by J.S. Aber, W. Jacobson and S. Salley, October 2003.

Figure 8-29. Smoke plumes drift toward the southwest from oil-field fires started during the Kuwait War. Image courtesy of K. Lulla, NASA Johnson Space Center. Photo ID: STS37-73-047, April 1991.

control cattails and other emergent vegetation in prairie marshes (Fig. 8-30).

In other situations, fire is quite destructive for wetlands, as demonstrated by recent peat fires in Indonesia. Peat fires spread during the 1997 El Niño event and recurred in 2002 and 2004, particularly in Central Kalimantan (Indonesian portion of the island of Borneo). These fires were exacerbated by ongoing conversions of forested peatland to cropland for the Mega-Rice Project, which began in 1995 and involved deep drainage and the use of fire for quick forest clearance. The amount of carbon released by Indonesian peat fires in 1997 alone is estimated to represent 1000 to 2000 years of carbon accumulation and was the greatest single emission of atmospheric CO_2 since records began in 1957 (Page et al. 2002).

The potential climatic impact of fire is mixed. Release of carbon dioxide from burning forest and peat enhances the greenhouse effect, whereas increased soot and ash in the atmosphere increases the albedo with a possible cooling effect. Certainly warm, drought conditions promote wetland fires, as seen during the summer of 2010 in Russia, when smoldering

Figure 8-30. Smoke plume from prescribed burning of cattails and other emergent vegetation at the state Cheyenne Bottoms Wildlife Area, central Kansas, United States. Such burning requires dry conditions and is usually carried out in late winter or early spring in order to remove standing thatch from the previous year. Blimp airphoto by J.S. Aber.

peat fires covered Moscow with smoke for weeks. In this case, the problem was compounded by past drainage and mining of peat, which left the remaining peat dry and especially susceptible to fire (Pollowitz 2010). In the western United States, extreme fire years are related to Pacific Ocean circulation anomalies (Littell, Peterson and McKenzie 2010).

Regarding the peatlands of Alaska and northern Canada, warmer and drier climate could lead to lower water levels and more frequent fires, which "could accelerate greenhouse gas emissions and cause much of the stored soil organic matter to be released back to the atmosphere" (Turetsky 2010). On the other hand, removal of forest cover by fire leads to a long-term increase in albedo, as the exposed ground is more reflective than forest canopy. In the case of boreal forest fire, Randerson et al. (2006) found that long-term increase in surface albedo was more significant climatically than the release of CO_2, and they concluded that increased boreal forest fires might not affect global warming.

In a comprehensive investigation of boreal peatland in northern Manitoba, Canada, Camill et al. (2009) compared carbon accumulation rates, fire severity, and vegetation composition over the past 8000 years. They found that shifts in fire severity and carbon accumulation lagged behind changes in climate and vegetation and concluded that fire severity and carbon accumulation are mediated by changes in vegetation as a consequence of climatic shifts. On the basis of these examples, fires may have dramatic impacts on wetland vegetation and soil, and burning peat may generate spikes in greenhouse gases. Over the long term, however, the roles of fire, wetlands, and climate change are less well understood and may have distinct regional variations. What is abundantly clear is that human-caused fires have accelerated the release of greenhouse gases and conversions of land cover that increase albedo. The climatic consequences of these contradictory factors remain to be seen.

8.6 Summary

Wetlands are dynamic environments that change through time in response to internal (autogenic) processes and external (allogenic) factors. The balance between autogenic and allogenic effects depends on magnitudes and rates of various processes working within and upon wetlands. Five primary factors for wetland development and change are climate, geomorphology, geology and soils, biogeography, and human activities. These processes and factors often interact with each other leading to positive or negative feedback relationships on local, regional, and global scales.

Hydroseral succession refers to gradual transitions involving both water and vegetation in the life cycle of a wetland under relatively stable conditions. This is the most common type of autogenic change brought about by vegetation succession, soil development, accumulation of sediment, and other processes operating within a wetland. Two means for initiating the development of peatland are terrestrialization and paludification. Once started, bogs grow upward and outward through distinct stages of development, which lead after several millennia to mature mire complexes.

Many allogenic factors may influence wetland development. Among the more important are changes in sea level, crustal movements, climatic variations, fire, and human activities. During the past million years, the Earth has experienced multiple cycles of glaciation and interglaciation in which sea level varied by >100 m. Major changes in coastal wetlands resulted as the shoreline migrated. Areas of direct glaciation were depressed hundreds of meters by ice sheets, then rebounded when the glaciers disappeared. Immediately surrounding ice sheets, the opposite happened. The combined effects of changing sea level and crustal movements may have complicated effects in many regions, with direct impacts on coastal wetlands. Current projections are for rising sea level during the twenty-first century with major consequences for coastal wetlands and human populations.

Water supply and wetland vegetation are controlled to a large extent by climate. Several factors play primary roles within the Earth's climatic system, namely solar energy, greenhouse gases, albedo, and oceanic circulation. Fresh-water and marine wetlands are key parts in the global carbon cycle both as carbon sinks and sources of greenhouse gases. In a general way, warmer and drier climate would reduce fresh-water wetlands and release greenhouse gases, while cooler, wetter climate would favor wetland expansion and carbon storage. However, neither climate nor wetlands behave in such simplistic ways. In fact, the connections between climate and wetlands may become quite complex.

Fire is an integral part of many ecosystems and may be used beneficially for management of some wetland habitats. In other situations, fire is quite destructive. The potential climatic impact of fire is mixed. Release of carbon dioxide from burning forest and peat enhances the greenhouse effect, whereas soot and ash in the atmosphere increase the albedo. Furthermore, removal of forested wetland cover by fire leads to an increase in albedo with a possible cooling effect. Over the long term, the roles of fire, wetlands, and climate change are poorly understood and may have distinct regional variations.

When multiple autogenic and allogenic processes interact, there may be unexpected effects for individual wetlands as well as globally. Given that even the most basic climatic influences, such as solar energy and volcanic eruptions, are impossible to predict, forecasting the impact of climatic change on wetlands is a particularly daunting task that requires a great deal more interdisciplinary scientific investigation. What is abundantly clear, however, is that human impacts on wetlands have accelerated the release of greenhouse gases and conversions of land cover that increase albedo. The climatic consequences of these contradictory anthropogenic factors remain to be seen.

Wetlands through time 9

9.1 Introduction

Wetlands have existed since the earliest records of fossil life on Earth. The most common fossils from the Archean Eon (>2.5 billion years ago) are stromatolites, which were essentially reefs built by cyanobacteria (blue-green algae) that proliferated in shallow marine environments. In fact, the oldest known fossils dating from 3.5 billion years ago are stromatolites (Virtual Fossil Museum 2010). Stromatolites pumped their waste product, namely oxygen, into the early anaerobic ocean of the world, which led to deposition of massive banded-iron formation and later of terrestrial red beds (see Color Plate 2-10), as atmospheric oxygen began to increase. It is safe to say that stromatolites living in shallow marine and coastal wetlands utterly transformed the global environment prior to the appearance of multicellular life (c. 700 million years ago). Then, as now, prokaryotic life forms the bulk of the biosphere, maintains biogeochemical cycles, and renders the Earth habitable for all other life as we know it today, and wetlands are essential ingredients in this environmental scheme.

Stromatolites continue to survive in the modern world in the Bahamas and at Shark Bay, a World Heritage Area in western Australia, where they live in hypersaline tidal pools (Fig. 9-1). High salinity excludes most predators and competitors, which allows the stromatolites to thrive as they did in ancient times (Nature of Shark Bay 2010). The stromatolites of Shark Bay became established there about 2000 to 3000 years ago when sea level and water salinity reached modern conditions. The story of stromatolites illustrates the deep time dimension that is necessary to understand the Earth's environmental system and how it came into being.

Geologists apply a principle known as "uniformitarianism," which states that the present is the key to the past. In other words, understanding modern earth processes, such as volcanoes, glaciers, earthquakes and so on, gives necessary analogies for interpreting the past. This approach may be turned around; the present is the consequence of the past. Indeed, in order to interpret modern environmental conditions, it is often necessary to know about past developments that led to current circumstances. It would be impossible, for instance, to understand why the southern portion of Lake Michigan is sinking today while the northern part is rising without knowing about the impact of the last glaciation more than 10,000 years ago (see chapter 8.3.2).

Given their situations in low-lying landscapes that contain anaerobic water and soil, wetlands tend to accumulate organic-rich sediment through time. Both macro- and microfossils of flora and fauna are well preserved and abundant in such sediments. Many of the most important fossil assemblages have been found in strata of ancient wetland deposits. Well-known American

Wetland Environments: A Global Perspective, First Edition. James Sandusky Aber, Firooza Pavri, and Susan Ward Aber.
© 2012 James Sandusky Aber, Firooza Pavri, and Susan Ward Aber. Published 2012 by Blackwell Publishing Ltd.

Figure 9-1. Stromatolites exposed at low tide in the Hamelin Pool Marine Nature Reserve, Shark Bay World Heritage Area, western Australia. Modified from original photograph by P. Harrison; obtained from Wikimedia Commons <http://commons.wikimedia.org/>.

examples include the Eocene Florissant Formation, a former lake in Colorado with spectacular insect remains (Veatch and Meyer 2008), and the Upper Carboniferous Hamilton Quarry, an estuary channel-filling in eastern Kansas with all types of invertebrate and vertebrate fauna and flora (Mapes and Mapes 1988). These and other examples demonstrate that wetlands served as harbors and refugia for biodiversity in the past (Greb and DiMichele 2006).

Among the organic deposits of wetlands, peat is of major economic importance, because upon burial it may be transformed into coal. Thus, ancient wetlands produced the coal that generates about one-third of the world's electricity supply (Greb and DiMichele 2006). It comes as no surprise, then, that a tremendous amount of research has been conducted on the origins, disposition, and energy potential of coal around the world. Amber is ancient tree resin found in association with low grades of coal, and it provides unique evidence regarding life and environmental conditions in past wetlands. Approaching recent times, peat contains a record of environmental and climatic conditions spanning the past few millennia. This record is crucially important for understanding the recent evolution of the Earth's climatic system as well as human impacts on the environment.

9.2 Coal

The majority of coal preserved in sedimentary strata of the continents was created during two major pulses of accumulation. The first was the Permo-Carboniferous interval in the late Paleozoic (c. 360 to 250 million years ago; Geological Society of America 2009). The second interval began in the Cretaceous (late Mesozoic) and continued well into the Cenozoic (c. 145 to 20 million years ago). These intervals were extremely favorable for flourishing land vegetation and the preservation of peat under humid climatic conditions. Peat and coal also accumulated in lesser amounts at other times between and since these two main coal-forming intervals.

9.2.1 Paleozoic coal

Massive accumulation of coal during the late Paleozoic was preceded by a long developmental phase of terrestrial vegetation that began in the late Ordovician and early Silurian, when the first land plants appeared about 450 to 430 million years ago. These were mosses and lichens that were obligate wetland dwellers, but not peat-forming plants (Greb, DiMichele and Gastaldo 2006). Land plants diversified during the Devonian, and marsh and forest swamp environments emerged in which plants were adapted for low-oxygen and low-nutrient conditions. This set the stage for peat marshes and peat swamps by the end of the Devonian (c. 360 million years ago), and peatlands became widespread during the following Carboniferous Period. Ombrotrophic tropical mires arose in the middle Carboniferous, and *Glossopteris* flora spread into temperate climates during the Permian (Greb, DiMichele and Gastaldo 2006).

Permo-Carboniferous coal is found on all modern continents in a seemingly haphazard geographic distribution. This pattern becomes clear when the continents are reassembled into the late Paleozoic supercontinent, known as Pangaea (Fig. 9-2), which was first proposed by Alfred Wegener in 1912. He relied on paleoclimatic evidence to support his reconstruction of Pangaea (Schwarzbach 1986). Coal deposits were found in two primary climatic zones,

Figure 9-2. Reconstruction of Pangaea during the early Permian Period about 280 million years ago. The supercontinent consisted of two main sectors, Laurasia in the north and Gondwana in the south, which had joined together through a series of continental collisions that created the Appalachians, Urals, and other mountain ranges. Modified from original map by Kieff; obtained from Wikimedia Commons <http://commons.wikimedia.org/>.

Figure 9-3. Surface mining of the Weir-Pittsburg coal bed (Cherokee Group, middle Pennsylvanian, upper Carboniferous). This was the most economically important coal bed mined in southeastern Kansas, United States (Brady and Dutcher 1974). Coal mining has now ceased in this vicinity. Photo by J.S. Aber.

equatorial and temperate, which were separated by subtropical deserts. The equatorial coal zone is mainly upper Carboniferous and stretched across Laurasia in what are now parts of eastern North America (Fig. 9-3), western and eastern Europe, and northern and eastern Asia. These coal deposits were formed in swamps dominated by lycopods and tree ferns that were the most widespread tropical mires in earth history (Greb, DiMichele and Gastaldo 2006). Arthropods, fish, amphibians and reptiles were abundant in these wetlands.

In the Gondwanan sector of Pangaea, coal deposits are found in parts of what are now eastern South America, southern Africa, India, Australia, and Antarctica. Gondwana was located in the southern hemisphere spanning climatic zones from temperate to polar, and ice-sheet glaciation was centered on the South Pole (Caputo and Crowell 1985). Gondwanan coals are mainly Permian in age and are distinctly different floristically from coals of Laurasia. Gondwanan mires were dominated by gymnosperms, such as *Gangamopteris* and *Glossopteris*. The Gondwanan mires were the first widespread non-tropical peatlands in earth history, and some at high latitude may have formed under permafrost conditions similar to modern palsa mires (Greb, DiMichele and Gastaldo 2006).

During the Permian, Pangaea's climate began to dry out, wetlands became increasingly restricted, and desert zones expanded greatly (see Color Plates 2-10 and 3-11). These climatic trends culminated with a massive extinction event at the end of the Permian, perhaps the greatest extinction in earth history, which caused a nearly total collapse of wetland ecosystems worldwide (Greb, DiMichele and Gastaldo 2006). The following Triassic Period was dominated by widespread deserts. Wetlands began to recover in some places by the middle Triassic, particularly in high southern latitudes of Gondwana – thick and extensive coal in modern Antarctica, and the earliest known salamander, frog, turtle, and mammal fossils come from this period.

9.2.2 Cretaceous–Tertiary coal and lignite

The Cretaceous Period represents the next great expansion of wetlands in general and peatlands

Figure 9-4. Stratotype for the boundary between the Cretaceous (K) and Tertiary (T) systems; Stevns Klint, south of Copenhagen, Denmark. A. Cliff profile displays uppermost Cretaceous chalk in the lower portion and lowermost Tertiary limestone above. Note people on beach for scale. B. Close-up view of the K/T boundary, which is marked here by a thin dark seam known as the fish clay (Håkanson 1971). Scale pole marked in 20-cm intervals. Photos by J.S. Aber.

in particular. Angiosperms (flowering plants) appeared during the early Cretaceous (c. 145 to 100 million years ago), including many now-common hydrophytes such as waterlilies (Nymphaeaceae) and lotus (Nelumbonaceae). Likewise, salt-tolerant mangroves related to extant species appeared during the Cretaceous, as did sea grasses. Nonetheless, forested mires continued to be dominated by conifers with an understory of ferns; mosses and ferns were important in some raised mires (Greb, DiMichele and Gastaldo 2006). Cretaceous coal is found in many places around the world: western North America (see Fig. 2-9), Central America and northwestern South America, eastern and central Asia, and New Zealand. This was the acme of dinosaur development, and their fossils are commonly found in coal and other wetland strata along with aquatic and flying reptiles, aquatic birds, and other creatures. Blood-sucking insects (black flies and mosquitos) made their first appearance in late Cretaceous amber (see below).

The Cretaceous culminated with another massive extinction that affected all categories of life – micro and macro, plant and animal, marine and terrestrial. The cause of this extinction is widely attributed to an asteroid or comet impact, which spread a distinctive iridium-rich so-called boundary layer in sedimentary deposits around the world (Fig. 9-4). However, other tectonic and climatic events may have contributed to this extinction. In any case, among those plants and animals that survived, many were obligate wetland inhabitants, which suggest that wetlands may have served as refugia during the extinction event (Greb, DiMichele and Gastaldo 2006).

Following the end-Cretaceous extinction, global changes in flora and fauna took place early in the Tertiary; mammals and angiosperms quickly rose to dominance in most parts of the world, and widespread mires developed. Tertiary coal deposits are found in many regions, particularly western North America, western South America, central Europe (see Fig. 8-24A), and southeastern Asia (Greb, DiMichele and Gastaldo 2006). In general, Tertiary peat experienced less burial and chemical conversion compared with older coals. Tertiary coal, thus, usually has lower carbon content and is more properly called brown coal or lignite. Tertiary brown coal is up to 90m thick in some basins, which suggests multiple or stacked mire accumulations (Fig. 9-5).

The breakup of Pangaea, which had begun during the Mesozoic, reached its apex with several completely separate or semi-isolated continents during the middle Cenozoic. This led to increasing climatic zonation and high biodiversity, especially for angiosperms, such that wetland habitats and resulting coal deposits

Figure 9-5. Thick brown coal (lignite) exposed in a surface mine at Lubstów near Konin, west-central Poland. Excavating machines are working at multiple levels beneath a relatively thin cover of overburden. Photo by J.S. Aber.

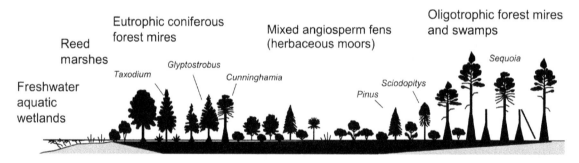

Figure 9-6. Schematic illustration of wetland types during the Miocene of central Europe (see Fig. 8-24A and Fig. 9-5). Coniferous trees are labeled; adapted from Greb, DiMichele and Gastaldo (2006, Fig. 19B).

varied widely through time and space. As an example, the Miocene lignite of central Europe was laid down in a complex succession of habitats that included many types of eutrophic to oligotrophic wetlands (Fig. 9-6).

9.3 Amber

Amber is a treasury of past flora and fauna. Amber is found primarily in Cretaceous and Tertiary coal and lignite deposits and provides further evidence for reconstructing wetland paleohabitats. Resin oozes from trees primarily as a defensive mechanism and upon death; ancient tree resin was the sticky trap that captured plant debris, small vertebrates, invertebrates, and air. Amber preserves extinct species, and these fossils possess lifelike forms in three dimensions with original textures and color patterns (Poinar and Poinar 1999). Millions of years later, these exquisite inclusions reveal geologic history and paleoenvironments. Engel (2001) wrote that amber allows scientists to "peer directly into past eons with far greater clarity than is possible from other kinds of fossils" (p. 161). Amber is also a beautiful gemstone; early gem use of amber amulets and bead necklaces dates from the Stone Age.

Resin is a viscous, amorphous hydrocarbon produced in specialized structures within plant tissue. Although all woody plants produce resin, not all tree resin becomes amber. Certain factors favor the transformation such as chemical composition, resistance to decay, ability to polymerize (harden), and potential for burial.

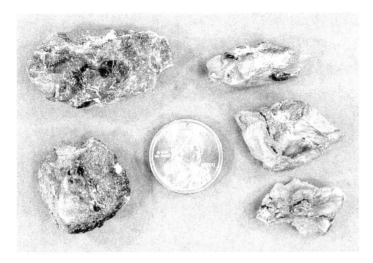

Figure 9-7. Specimens of Cretaceous amber from Kansas, known as jelinite. Dollar coin is 1 inch (25 mm) in diameter. Photo by S.W. Aber.

Polymerization is a chemical fossilization that may be relatively rapid when the resin is exposed to sunlight and air (Langenheim 2003). These criteria explain why it is believed amber was created from copious resin-producing trees in moist tropical or subtropical environments; these conditions favor photosynthesis throughout the year and provide abundant carbon in spite of low-nutrient tropical and wetland soils. When the hardened, polymerized resin is buried with vegetation, soil and sediment, it undergoes additional change over geologic time. The maturation from polymerized resin to amber is a continuum wherein modern resin progresses to ancient resin, and this in turn changes to subfossil resin or copal, and then to amber. Hardened resins such as copal have not yet completed polymerization and preserve extant species of flora and fauna, whereas amber preserves extinct species.

The oldest known amber is from the Carboniferous, 320 million years old, found in U.S. coal deposits in Illinois (Bray and Anderson 2009). Amber is known also from the Triassic (Grimaldi et al. 1998) and Jurassic (Philippe et al. 2005; Azar et al. 2010). Relatively abundant Cretaceous amber is found in Canada (McAlpine and Martin 1969; McKellar et al. 2008; McKellar and Wolfe 2010; Zobel 1999), the United States, the Middle East, and Asia (Zherikhin and Ekov 1999). Some of the oldest reported wetland insects come from early Cretaceous amber in Lebanon (Azar, Prokop and Nel 2010; Poinar and Milki 2001). Cretaceous amber in the United States is located in Wyoming (Grimaldi et al. 2000; Kosmowska-Ceranowicz, Giertych and Miller 2001), Kansas (Fig. 9-7; Aber and Kosmowska-Ceranowicz 2001a, b), and North Carolina (Fig. 9-8).

Cretaceous amber from New Jersey is some of the best known in North America and most significant in the world because of the abundant and diverse flora and fauna (Grimaldi 1996). This amber contains more than 250 species of plants and animals with some of the oldest primitive fossil ants (Engel and Grimaldi 2005), mushrooms, potter wasp, bee, and flower. New Jersey amber was likely derived from the conifer *Pityoxylon* in the Pinaceae family (Grimaldi, Shedrinsky and Wampler 2000). This amber tree grew in coastal swamps and resembled modern *Pinus, Picea* or *Larix* (Greb, DiMichele and Gastaldo 2006). In addition to preserving life, amber contains fossil air bubbles that have been analyzed to show that Earth's atmosphere some 67 million years ago (end of the Cretaceous) contained 35 percent oxygen compared to present levels of approximately 21 percent (Landis 2009). The consequence of this elevated oxygen level is unclear, but is another example of how amber reveals the past.

Figure 9-8. North Carolina amber. A. Lignite and clay beds in the Cape Fear Formation (Cretaceous) from which the amber was recovered. Scale is 12 inches (30 cm) long. B. Amber specimens in matrix; ~4 inches across. Photos courtesy of V. Krynicki.

Tertiary amber comes from many locations including Australia (Bickel 2009; Hand et al. 2010), but the most abundant Tertiary amber is from the Baltic region of central Europe, the Dominican Republic and Mexico. Regardless of locations, identifying the origin of the resin or the amber-producing tree may be a puzzle, because chemical studies are not corroborated by flora inclusions, and amber tree descendants occur in different geographic areas. For example, it is often suggested that the most common amber trees were araucarian (family Araucariaceae) in origin; however, pine cones, needles, and wood fragments are missing from Baltic amber (Engel 2001), and these gymnosperms are no longer abundant in the northern hemisphere where the amber is found (Setoguchi et al. 1998). Modern-day equivalents to this amber tree resemble Kauri pine (*Agathis australis*), monkey puzzle (*Araucaria araucana*), and Norfolk Island pine (*A. heterophylla*), which occur primarily in the southern hemisphere.

Tertiary amber from the Baltic region, known as succinite, is the most abundant amber in the world. The depositional environment of the Baltic amber, according to Kosmowska-Ceranowicz (2009), included an ancient River Eridanus, which ended at the Eocene Sea and created a delta in the vicinity of northern Poland and Kaliningrad, Russia. The identity of the Baltic amber tree is still being debated with several contenders (Szwedo and Szadziewski 2009), but it is accepted that the amber trees grew in wetlands such as swamps and lowland forests (Greb, DiMichele and Gastaldo 2006). Preserved insects, spiders, and other small animals including larvae of caddisflies, mayflies, and water bugs suggest standing water in the ancient swamps (Larsson 1978).

Less controversy surrounds the amber tree in the Dominican Republic and Mexico. These ambers derive from an "algarrobo forest," and the extinct species *Hymenaea protera* (Poinar 1995, 1999; Corday and Dittrich 2009). This species resembles the present *Hymenaea*, which is closely related to the modern algarrobo or carob tree, and *Hymenaea verrucosa* Gaertner, known as the East African Copal (Poinar 1992; Huber and Langenheim 1986). In general, Dominican and Mexican ambers contain more floral and faunal inclusions than does Baltic amber, including small flies such as midges and mosquitos, marsh beetles, mayflies, water striders and caddisflies; all of which lived in wetlands such as ponds, rivers, intertidal mudflats, and shorelines (Poinar and Poinar 1999). The first anopheline mosquito is preserved in Dominican amber, signaling that these potential malaria-carrying vectors were present in Neogene times.

Primary sources of amber are mainly Cretaceous coal and Tertiary lignite deposits; however, quite often amber is eroded from these parent sources and redeposited into younger sedimentary strata. For example, ambers from the Ukraine and the Bitterfeld region in Germany are found in Miocene sediment, yet have similar

Figure 9-9. Selection of Danish amber pebbles retrieved from the North Sea coast (see Color Plate 9-9). From the collection of P. Laursen; photo by J.S. Aber.

Figure 9-10. Buried wood in the lower till of the Independence Formation, northeastern Kansas, United States. Abundant wood in this till represents a spruce forest that was overrun by an ice sheet about 700,000 to 600,000 years ago. From Aber, J.S. 1991. The glaciation of northeastern Kansas. *Boreas* 20, p. 297–314.

infrared absorption spectra to Baltic amber, which is Eocene in age (Kosmowska-Ceranowicz 1999, 2009). Amber continues to erode from Eocene deposits nowadays and is transported onto beaches around the Baltic Sea and North Sea, where it is collected by many people (Fig. 9-9). Baltic amber is also mined commercially from a layer of marine glauconitic sand referred to as "blue earth" (middle Eocene; Engel 2001) in Poland, the Ukraine, and Germany, but the largest operation by far is at Yantarny, Kaliningrad, Russia (Krielaars 2010). Fashioned and polished amber objects and jewelry remain popular today because of the gem's beauty and rarity.

9.4 Pleistocene and Holocene wetlands

The late Cenozoic was a time of global cooling that culminated in widespread, repeated, ice-sheet glaciations during the Pleistocene Epoch (*c.* 2.6 million to 10,000 years ago). The associated frequent changes in sea level and climatic fluctuations placed strong adaptive pressures on wetland communities, and *Sphagnum* moss came to dominate mires in northern latitudes. *Sphagnum* was preadapted to exploit oligotrophic habitats (Greb, DiMichele and Gastaldo 2006):

- Extensive aerenchyma allows it to survive in low-oxygen conditions.
- Compact form, overlapping leaves and rolled branch leaves help it to retain water.
- Ability to acidify water may retard bacterial decomposition and allow organic matter to accumulate.
- Release of acid may weather underlying mineral soil and lead to formation of a clay pan that retains water.

Taken together, these capabilities favored *Sphagnum* expansion and exclusion of most other vegetation that could not tolerate low Eh and pH and lack of nutrients. *Sphagnum* mires, co-inhabited by spruce, tamarack, birch, heaths, sedges, and pitcher plants, spread across vast regions of northern North America and northern Eurasia during interglacial and postglacial periods. The last major glaciation (late Wisconsin) took place between 25,000 and 10,000 years ago, and this advance of ice sheets destroyed most evidence for earlier wetlands in the glaciated regions.

A few sites with older wetland strata were preserved in protected spots, for instance in northeastern Finland (Engels et al. 2010). In the central United States, a spruce forest was overrun by an advancing ice sheet in eastern Nebraska and northeastern Kansas approximately 700,000 to 600,000 years ago (Aber 1991). Remains of spruce wood as large as fence posts are preserved regionally in glacial deposits (Fig. 9-10). The nearest comparable spruce forest today is located 1200 km to the north in central

Manitoba, Canada. For the most part, still, our knowledge of pre-late Wisconsin wetland vegetation and paleoclimate is quite fragmented in time and space. Following is a short review of climatic conditions for the latest Pleistocene and Holocene from selected regions of the world.

9.4.1 Nordic region

Much of what we know about paleoclimate and environmental conditions of the latest Pleistocene and Holocene was first discovered in the Nordic countries of northern Europe based on wetland deposits and fossils, and this region continues to be at the forefront of paleoclimate research (Seppä et al. 2010). Fennoscandia has abundant lakes and mires of many sizes and types, and the region experiences unusually warm and remarkably variable climate for its latitude due to influence from the North Atlantic. Fundamental discoveries, beginning in the late nineteenth and early twentieth centuries, include the Blytt-Sernander scheme for late glacial and postglacial climate change (Boreal, Atlantic, Sub-boreal, Sub-atlantic), recognition of the Holocene thermal maximum (HTM) by Gunnar Andersson, and establishment of modern pollen analysis by Lennart von Post (Fig. 9-11). Already in these early studies, an attempt was made to link vegetation and climate with phases of the Baltic Sea and the archaeological record of early humans (Birks and Seppä 2010).

During the twentieth century, pollen analysis of peat and lacustrine sediment became a standard approach (see Fig. 3-33), and all kinds of paleoclimatic proxies have since been investigated ranging from oxygen isotopes (e.g. St. Amour et al. 2010) to growth patterns in pearl mussels (e.g. Helama et al. 2010). Various dating techniques have become quite sophisticated in order to provide high-resolution temporal control (e.g. Snowball et al. 2010). Any convincing reconstruction of paleoclimate depends upon multiple techniques that are complementary to each other yet independent in their sample materials and methods. For the Nordic region, "the main biological and physical proxy techniques show a remarkably consistent signal for Holocene temperature trend" (Seppä et al. 2010, p. 651). Three main climatic phases are well documented (Fig. 9-12):

- Late-glacial/early Holocene climatic amelioration (c. 11,000 to 8000 years ago) – rising temperature, high lake levels, and ice sheets shrinking into alpine glaciers.
- Holocene thermal maximum (c. 8000 to 4000 years ago) – high temperature, pronounced drought, low to dry lakes, and alpine glaciers disappeared (Nesje and Kvamme 1991).
- Neoglaciation of late Holocene (last 4000 years) – general cooling, high lake levels, and renewed or expanded alpine glaciation.

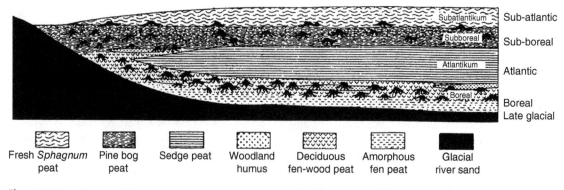

Figure 9-11. Schematic illustration of peat layers from a bog in Närke, central Sweden. Tree stumps occur in two dark peat layers (Boreal and Sub-boreal), and tree remains are absent from two light peat layers (Atlantic and Sub-atlantic). Modified from Birks, H.J.B. and Seppä, H. 2010. Late-Quaternary palaeoclimatic research in Fennoscandia – A historical review. *Boreas* 39/4, p. 655–673, Fig. 1; adapted from von Post (1946).

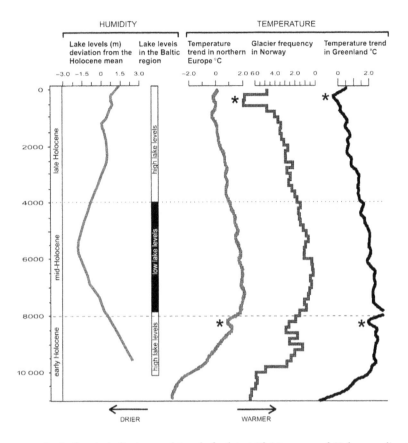

Figure 9-12. Summary chart of main indicators and trends for latest Pleistocene and Holocene climate in northern Europe and Greenland. Asterisks (*) indicate the 8.2 ka cooling event and the Little Ice Age. Dates given in calendar years ago. Compiled from many sources; adapted from Seppä, H., Birks, J.B., Bjune, A.E. and Nesje, A. 2010. Current continental palaeoclimatic research in the Nordic region (100 years since Gunnar Andersson 1909) – Introduction. *Boreas* 39/4, p. 649–654., Fig. 1.

These long-term trends were interrupted by "abrupt events" during which the climate became markedly colder, namely the cooling at 8200 years ago (8.2 ka event) and the Little Ice Age (c. 800 to 100 years ago). Changes in wetlands and vegetation generally were caused by climatic fluctuations. In the subarctic tundra region of northwestern Finland, for example, peatlands initiated and expanded rapidly in the early Holocene, then expansion slowed to a modest pace in the mid-Holocene (Weckström, Seppä and Korhola 2010). Rapid expansion again took place in the late Holocene (4000–3000 years ago), then gradually slowed toward present rates. This scenario follows the overall trends for Holocene climate change seen throughout Fennoscandia.

Wetland and upland vegetation did not depend solely on climatic conditions, however. When attempting to interpret ancient vegetation, one should recognize that "no similar plant communities exist today," as Iversen (1973, p. 26) emphasized. The vegetation succession that took place in Fennoscandia during the Holocene was driven by changes in climate, but also depended on several other factors, such as plant migration rates, species competition for light, changes in soil quality, and human impact. During the late-glacial phase, for instance, changes in aquatic plants generally preceded changes in upland trees, which indicates the ability of wetland vegetation to respond quickly to climatic warming, as seeds and plant fragments are carried by migrating waterfowl. As climate continued to warm during the early Holocene, the arrival of tree species often did not correspond to their temperature requirements. The timing of tree immigration was instead determined by rates of dispersal and locations of glacial refugia from which they spread (Iversen 1973).

In southern Scandinavia, the Atlantic Period represents the Holocene thermal maximum (HTM), and vegetation consisted of a climax deciduous forest. The following Sub-boreal and Sub-atlantic periods correspond to the Neoglaciation, in which climate became gradually cooler and wetter. Certainly vegetation responded to this climatic shift as well as to declining soil quality as nutrients were depleted. The impact of early humans becomes more and more evident during the latter half of the Holocene. Forests were cleared with axe and fire for grazing cattle and growing crops, beginning in the Neolithic (Fig. 9-13) and intensifying during the Bronze and Iron ages. By the Iron Age, southern Scandinavia was largely deforested.

Bogs have yielded many archaeological treasures in northern Europe (Fig. 9-14). Among the most spectacular finds are so-called bog bodies, which are, in fact, mummified human bodies in which soft tissues are preserved including skin, hair, and stomach contents (Fig. 9-15). The best-preserved bodies come from raised bogs in which low oxygen and high acidity limited decomposition and performed a natural tanning process. For preservation to be successful, the bodies had to be submerged below water or buried in peat during the cold season and covered quickly; the bodies were typically placed in old peat-digging holes and held down with sticks or turf (National Museum of Denmark 2010b). Most of the Danish bodies date from the late Bronze Age or early Iron Age. Many other bog bodies have been discovered in Ireland, England, Scotland, the Netherlands, Germany, and Sweden. It appears that "powerful forces were at work in the Iron Age society of northern

Figure 9-13. Kong Asger's høj, a Neolithic burial mound, island of Møn, southeastern Denmark. The Neolithic was the time when agriculture began in southern Scandinavia. Population grew, society became more complex, and monumental building projects were undertaken. A. Overview of mound. B. Interior view of stone-lined passage that leads to a stone-roofed burial chamber in the center; height of passage ~1 m. Danish passage graves date from the middle Neolithic, c. 5000 years ago (Rud 1979). Photos by J.S. Aber.

Europe, and the bog bodies of the period must have held a special place" (Coles and Coles 1989, p. 196).

As we approach the historical era, the Medieval climate optimum (c. AD 700–1200) was a time of extremely favorable climate in northern Europe. Harvests were good, fishing was abundant, sea ice stayed far to the north, vineyards existed 500 km north of their present limits, and famine was rare (Le Roy Ladurie 1971). This was the period of great Viking expansions from Scandinavia. The basis of Viking success was mastery of wetlands, in particular their ships which could negotiate inland rivers, estuaries, and shallow coastal waters as well as the open ocean, which allowed them to colonize new lands (Iceland, Greenland) and overpower other people in the British Isles and continental Europe as far south as the Mediterranean Sea (Savage 1995).

The importance of wetlands is demonstrated by Viking settlements, which were often placed on estuaries, fjords, or coastal embayments. For example, a fleet of Viking ships was discovered and excavated from the floor of Roskilde Fjord, a shallow estuary, on the island of Sjælland, Denmark in 1962 (Fig. 9-16A). Five ships were scuttled about AD 1070 in a channel of the fjord in order to protect Roskilde, then the capital of Denmark, from potential attack from the sea (Viking Ship Museum 2010). The aquatic way of life was so important that Vikings even departed life in ships (Fig. 9-16B).

As the preceding archaeological examples demonstrate, the human presence in Nordic lands became increasingly important during the late Holocene. People utilized wetlands directly and impacted them indirectly through agriculture, building settlements and other activities, all of which modified drainage and vegetation. Most of these changes took place in prehistory, but were nonetheless substantial in altering wetland habitats, particularly in the more populated southern and western portions of the Baltic region.

Figure 9-14. Stone monument marking the discovery site of the Sun Chariot, one of the most famous Bronze Age artifacts from Denmark, dating from about 1400 BC. The Sun Chariot depicts a horse-drawn, six-wheeled wagon in cast bronze with a gold-covered sun disk that displays a spiral motif. The Sun Chariot (Solvognen) was found when the Trundholm Mose (bog) in northwestern Sjælland was plowed for the first time in 1902 (National Museum of Denmark 2010a). Modified from original photograph by Jom; obtained from Wikimedia Commons <http://commons.wikimedia.org/>.

9.4.2 North America

As the Laurentide Ice Sheet shrank toward central Canada at the end of the Pleistocene and in the early Holocene, a multitude of proglacial lakes developed along the retreating ice margin.

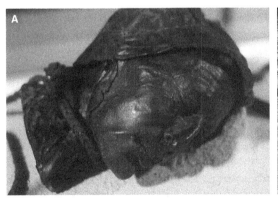 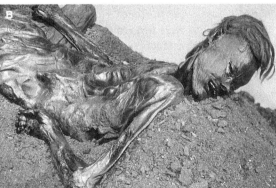

Figure 9-15. Danish bog bodies. A. The head of Tolland Man is restored and displayed at the Silkeborg Museum. This person was likely a holy man, who was sacrificed, judging by the condition of his skin and stomach contents. B. Grauballe Man is displayed at the Mosegård Museum near Århus. It is likely that he was executed or the victim of crime, as suggested by the slash across his throat. Photos by J.S. Aber.

Figure 9-16. Danish Viking ships. A. Viking Ship Museum at Roskilde. Original wood is reconstructed on iron frames. In the foreground is a large, ocean-going cargo ship built initially of pine in western Norway and later repaired with oak in southern Norway and Denmark (Viking Ship Museum 2010). Image modified from original obtained from Wikimedia Commons <http://commons.wikimedia.org/>. B. Restored Viking cemetery with ship graves at Lindholm on a prominent hill overlooking the Langerak estuary, northwestern Denmark. Photo by J.S. Aber.

Many were short lived, lasting only until they overflowed and outlet channels were eroded or until the ice margin retreated and uncovered a lower drainage route. Others survived for millennia. Among the latter was glacial Lake Agassiz, which is considered to be one of the largest fresh-water lakes that ever existed. Lake Agassiz covered during different phases most of Manitoba, and parts of Saskatchewan, Ontario, North and South Dakota, and Minnesota (Fig. 9-17). Overflow from Lake Agassiz and other large proglacial lakes cut spectacular spillway channels across the northern Great Plains, and similar spillways were formed in the Great Lakes region (Fig. 9-18). Much of the modern Mississippi, Ohio and Missouri drainage systems, as well as the Hudson and St. Lawrence systems, were created by such melt-water floods during the last deglaciation.

The exact timing of the three main Holocene climatic phases differed somewhat in North America and other parts of the world, but the overall pattern is broadly similar globally – climatic amelioration in the early Holocene (Preboreal and Boreal), climatic optimum in the mid-Holocene (Atlantic), and climatic cooling in the late Holocene (Neoglaciation; Sub-boreal and Sub-atlantic). A major change in vegetation took place in eastern North America during a brief period around 10,000 years ago (Fig. 9-19). From northeastern Kansas to Nova Scotia, spruce forest was replaced abruptly by mixed conifer/deciduous forest (to the east) or prairie grassland (to the west). This rapid transition represents the end of the Pleistocene and beginning of Holocene climatic conditions.

The Holocene thermal maximum (HTM) is known as the Altithermal or Hypsithermal in North America. During this phase, sand dunes were especially active in the Great Plains from Texas to Saskatchewan (Miao et al. 2007). In the southern Great Plains drought was severe, surface- and ground-water sources dried up, and bison herds were diminished (Meltzer 1999). In the following Neoglaciation phase, climate was wetter and cooler in the southwestern United States (Polyak et al. 2001), and many wetlands reappeared in the Great Plains (Fig. 9-20).

In central Canada, wetland habitats shifted from wetter fen in the early Holocene to drier, forested bogs during the HTM. Fire severity increased to a peak after 4000 years ago, following the HTM rather than during it (Camill et al. 2009). Vegetation and permafrost were also affected by the Medieval climatic optimum and

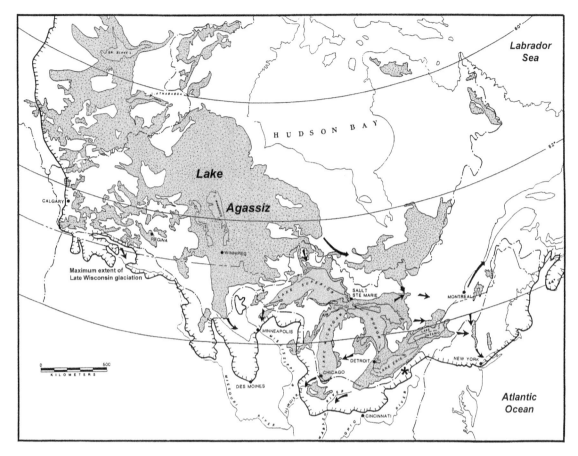

Figure 9-17. Extent of proglacial lakes (gray) in the central and eastern portions of North America during the retreating phase of the late Wisconsin glaciation, maximum limit of late Wisconsin glaciation, and major drainage routes (arrows). Asterisk (*) indicates location of Conneaut spillway (next figure). Marine areas not shown; adapted from Teller (1987, Fig. 2).

Figure 9-18. Conneaut Marsh is among the largest marshes in the state of Pennsylvania, United States. It is situated in a glacial spillway valley that carried melt water south of Lake Erie during the final phase of the last glaciation of the region (Van Diver 1990). A drainage ditch runs along the center of the marsh; location shown in previous figure. Superwide-angle view toward northwest; blimp aerial photograph by J.S. Aber and S.W. Aber.

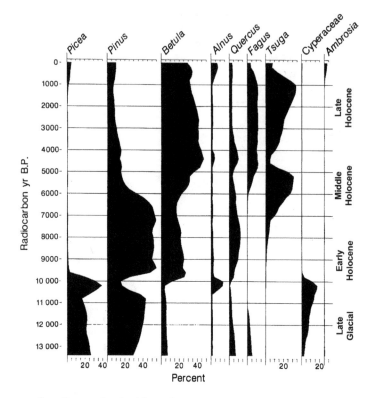

Figure 9-19. Summary pollen diagram for Gould Pond, Maine, United States. Note the peak in spruce (*Picea*) and other large shifts in species abundance just before 10,000 radiocarbon years ago, which reflect rapid changes in climate at the end of the Pleistocene. Adapted from Jacobson, Webb and Grimm (1987, Fig. 2).

Figure 9-20. Sand hills surround Clear Lake in the Valentine National Wildlife Refuge, north-central Nebraska, United States. These dunes are now stabilized and lakes are abundant in the Nebraska Sand Hills, but during the mid-Holocene Altithermal the dunes were active and lakes were much reduced. Kite aerial photo by J.S. Aber and S.W. Aber.

Table 9-1. Late glacial and Holocene climatic cycles of North America. * Date is approximate beginning of each climatic cycle in calendar years before present. Adapted from Viau et al. (2002).

Date*	Period	Climatic conditions
110	Modern	Modern climatic optimum
600	Little Ice Age	Coldest climate of Holocene
1650	Neo-atlantic	Medieval climatic optimum
2850	Sub-atlantic	Continued cooling
4030	Sub-boreal	Beginning Neoglaciation
6700	Atlantic II	Holocene climatic optimum
8100	Atlantic I	Continued warming
10,190	Boreal	Early Holocene warming
12,900	Younger Dryas	Cold late-glacial interval
13,800	Allerød	Warm late-glacial interval

the Little Ice Age. Evidence has accumulated for several climatic cycles during the latest Pleistocene and Holocene of North America (Viau et al. 2002). Climatic oscillations, each lasting a period of roughly 1650 ± 500 years, took place and caused changes in vegetation across the continent (Table 9-1). These cycles may represent changes in atmospheric circulation with global climatic consequences, which are documented in ice cores and marine sediments. The origin of millennium-scale cycles is uncertain, but many scientists consider solar forcing a likely mechanism (see below).

9.4.3 Tropics and Antarctica

Moving to low latitudes, ice-core records from tropical alpine glaciers provide unique climatic data that may be compared with other climatic proxies. Ice cores from Kilimanjaro glaciers, for example, span the entire Holocene with high-resolution oxygen-isotope and dust data that corroborate other climatic indicators for equatorial eastern Africa (Thompson et al. 2002). Warmer and wetter conditions marked the first African Humid Period (~11,000 to 4000 years ago), when lakes rose as much as 100 m above current levels. Lake Chad, for instance, was comparable in area to the modern Caspian Sea. Lake levels dropped abruptly 8300 years ago, which marks a dry period and corresponds with the 8.2 ka event in northern Europe. Another abrupt change took place from about 6500 to 5000 years ago with cooling that coincided with a second humid period. A third abrupt event was a severe drought that occurred about 4000 years ago. This drought lasted three centuries and was the greatest recorded historically in tropical Africa; it coincided with the First Dark Age, when many early civilizations collapsed (Thompson et al. 2002). This climatic record has major implications for Holocene wetland conditions across tropical Africa, the Middle East and western Asia, especially in regard to lakes and their associated features.

The Quelccaya ice cap in the Andes Mountains of southern Peru preserves another unique climatic record. Ice cores provide direct physical evidence for the past 1500 years. Relatively warm climate prevailed before about AD 1550 and after 1900. The intervening cold period represents the Little Ice Age (Thompson et al. 1986). This record compares favorably with late Holocene temperature fluctuations in North America (e.g. Campbell et al. 1998). Farther south, Lake Titicaca rose significantly during the sixteenth through nineteenth centuries as a result of more humid, cooler conditions (J. Argollo, pers. com.).

Still farther south, coastal Antarctica hosts numerous penguin nesting sites. The number and distribution of these sites through time is a proxy for paleoclimate (Baroni and Orombelli 1994). Nesting sites were particularly widespread during the Holocene thermal maximum, 4300 to 3000 years ago (Fig. 9-21). Timing of the Antarctic HTM is noticeably later than for the northern hemisphere, however, which reflects a significant lag time for the southern ocean to respond to global climatic change.

9.4.4 Holocene climate and early man

From this brief review of Holocene climate in representative parts of the world, it should be clear that significant changes in climate did occur during the past several millennia. These climatic shifts were broadly comparable worldwide, but with distinct regional variations, which had major impacts on local wetland habitats and

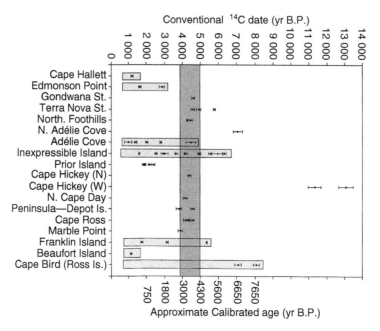

Figure 9-21. Age distribution of modern and abandoned penguin nesting sites in Victoria Land, Antarctica. Modern nesting sites are shown by horizontal rectangles (light gray). Nesting sites were particularly widespread during the mid-Holocene climatic optimum (dark gray). Such favorable environmental conditions have not been repeated since. Adapted from Baroni and Orombelli (1994, Fig. 2).

life. A popular concept exists, at least among the general public, that climate should be stable and that any change in climate is undesirable and must somehow be related to human activity. On the other hand, most paleoclimatologists have come to the conclusion that the major shifts in climate during the Holocene were driven primarily by solar forcing, including variations in sun/earth orbital geometry (Thompson et al. 2002) and fluctuations in solar output (Eddy 1977), as well as by episodic volcanism (Nesje and Dahl 2000).

Humans have played increasingly disruptive roles in terms of habitat alteration during the Holocene. Ruddiman (2005) argued that significant human impact on climate began with the advent of agriculture some eight millennia ago. Agriculture has a profound effect on natural vegetation – forest, prairie, and wetland are replaced in favor of various annual crops. The net result is transfer of carbon previously stored in the biomass into the atmosphere. Carbon dioxide is derived mainly from forest clearing; methane is a byproduct of growing rice and raising cattle. Increasing levels of atmospheric carbon cannot be absorbed into the oceans at rates as fast as they were created by agricultural expansion. Ruddiman (2005) concluded that the natural trend for climatic cooling in the late Holocene may well have been slowed or even reversed by increasing levels of greenhouse gases derived mainly from agriculture.

Certainly early man had an effect on local and regional wetland conditions, primarily through forest clearance, agriculture and irrigation, but it is not yet clear that this human activity had significant influence on global climate prior to the Industrial Revolution. In fact, Nesje and Dahl (2000) concluded that pulses in volcanic eruptions played a major role in cooling events in the northern hemisphere during the Little Ice Age, and that rapid glacier retreat and climatic warming in the late nineteenth and early twentieth centuries were more related to a quiescent phase of volcanism than to human-caused emissions of greenhouse gases. This subject is exceedingly complex, and considerable scientific uncertainty remains as to the role of human activities for past events and trends in Holocene climate.

9.5 Summary

Wetlands have existed since the earliest records of fossil life on Earth, and wetland life has exerted a profound influence on the world's environment and climatic system. Beginning 3.5 billion years ago, stromatolites (blue-green algae) in coastal wetlands pumped oxygen into the early ocean and atmosphere, which eventually led to oxygen-rich environments for most of the world. Given their situations, wetlands tend to accumulate organic-rich sediment including diverse fossils and peat. Upon burial these sediments may be transformed into lignite, coal, and amber. The majority of coal preserved in sedimentary strata of the continents was created during two major pulses of accumulation – the Permo–Carboniferous (late Paleozoic) and Cretaceous (late Mesozoic) through Tertiary (Cenozoic).

Permo-Carboniferous coals accumulated on the supercontinent Pangaea in two primary climatic zones, equatorial and temperate, which were separated by subtropical deserts. A massive extinction event at the end of the Permian caused a nearly total collapse of wetland ecosystems worldwide. The Cretaceous Period represents the next great expansion of wetlands, and it culminated with another massive extinction that affected all categories of life. Among those plants and animals that survived, many were obligate wetland inhabitants, which suggest that wetlands may have served as refugia during the extinction event. Following this extinction, widespread and diverse mires developed during the Tertiary.

Amber is a chemical fossilization of tree resin, and ancient resin was the trap that captured plant debris, small animals, and air bubbles. The hardened resin was buried and polymerized, changing into amber with exquisitely preserved extinct organisms in three-dimensional, lifelike positions. Most amber is found among carbonaceous sedimentary rock associated with coal and lignite deposits that date from Cretaceous and Tertiary times. Amber provides the perfect time capsule with which to examine extinct flora and fauna, and reconstruct past paleoclimates and wetland habitats. Amber is valued as a beautiful gemstone today and as far back as the Stone Age.

The late Cenozoic was a time of global cooling that culminated in widespread, repeated, ice-sheet glaciations. The associated frequent changes in sea level and climatic fluctuations placed strong adaptive pressures on wetland communities, and *Sphagnum* moss came to dominate mires in northern latitudes. Much of what we know about paleoclimate and environmental conditions of the latest Pleistocene and Holocene was first discovered in northern Europe based on wetland deposits and fossils. Three main climatic phases are well documented: 1) late-glacial/early Holocene climatic amelioration, 2) mid-Holocene thermal maximum, and 3) Neoglaciation of late Holocene. These long-term trends were interrupted by "abrupt events" during which sharply colder climate returned ~8200 years ago and during the Little Ice Age.

The exact timing of the three main Holocene climatic phases differed somewhat in North America and other parts of the world, but the overall pattern is broadly similar globally. In North America, vast proglacial lakes formed during deglaciation, and sizable regions were depressed below the marine limit. A major change in vegetation took place rapidly in eastern North America about 10,000 years ago, which marked the shift from late glacial to early Holocene climate. Continued climatic warming led to the Altithermal of the mid-Holocene. Across the continent, cooler and wetter climatic conditions developed during the late Holocene.

Ice-cores from tropical alpine glaciers in Africa and South America provide unique records of climate change that confirm other types of paleoclimatic proxies. In combination these records demonstrate significant changes in tropical climate during the Holocene which had major impacts on lakes and other wetland habitats as well as early civilizations. Even in

Antarctica, the global extent of the mid-Holocene climatic optimum is shown by fossil penguin nesting sites. On this basis, it is clear that significant changes in climate did occur during the past several millennia. These climatic shifts were comparable worldwide, but with distinct regional variations, which had major impacts on local wetland habitats and life.

Humans have played increasingly disruptive roles in terms of habitat alteration during the Holocene. Some have argued that significant human impact on climate began with the advent of agriculture some eight millennia ago. On the other hand, most paleoclimatologists have come to the conclusion that the major shifts in climate during the Holocene were driven primarily by solar forcing and by episodic volcanism. Certainly early man impacted local and regional wetland conditions, but it is not yet clear that this human activity had significant influence on global climate prior to the Industrial Revolution. This subject is exceedingly controversial, and considerable scientific uncertainty remains.

10 Environmental cycles and feedback

10.1 Biogeochemical cycles

10.1.1 Wetland elements

Biogeochemical cycles involve all elements and their chemical compounds at the Earth's surface. These elements are contained in various portions of the biosphere, atmosphere, hydrosphere, cryosphere, and lithosphere (Fig. 10-1). Their residence times vary from minutes to millions of years, and their movements and transformations involve energy. All these aspects interface in wetlands, which support high biodiversity and are important storage points for water and carbon. The chemistry of wetlands depends upon many factors including the influence of bedrock and soil, inflow and outflow of surface and ground water, climate and vegetation, characteristics of surrounding terrain, and human impacts. Complex and often poorly understood feedback relationships are involved in all these cycles. Mitsch and Gosselink (2007) summarized the roles wetlands have in biogeochemical cycles:

- Wetlands may be sources, sinks, and transformers of elements and their compounds.
- Chemical reactions may display strong seasonal patterns related to hydroperiods, temperature, and organic activity.
- Wetlands are often coupled with adjacent environments through chemical exchanges and feedback relationships.
- Wetland productivity varies from quite high to rather low depending on availability of nutrients.
- Storage and cycling of nutrients in wetlands are different from drylands or deep-water environments.
- Human activities have greatly altered chemical cycles and nutrients in many wetlands.

Key wetland elements include nitrogen, potassium, iron and manganese (see chapters 4.5 and 5.3), sulfur, phosphorus, and carbon. The chemical status and transformations of these elements depend primarily on the presence of oxygen within wetland water and soil. Three zones are typically developed from the surface downward: aerobic, facultative, and anaerobic (Fig. 10-2). The vertical position of these zones may fluctuate seasonally, depending on the hydroperiod, and over longer time frames due to changes in drainage, vegetation cover, climate, and other factors.

10.1.2 Nitrogen

Nitrogen is a major nutrient and is often a limiting factor in flooded wetlands or peatlands. The ammonium ion (NH_4^+) is the common reduced form of nitrogen, called mineralized nitrogen, which is biologically produced by decay and decomposition of organic matter. Once ammonium is formed, it may be used directly by plants or anaerobic microbes, which convert it

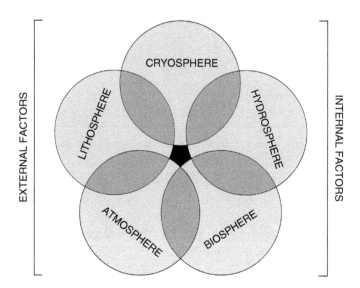

Figure 10-1. Schematic relationships of the lithosphere (mineral), cryosphere (ice), hydrosphere (water), biosphere (life) and atmosphere (air) to each other and to other factors that influence the Earth's surficial environment. External factors include solar energy, cosmic radiation, meteorite impacts, etc. Internal factors are related to plate tectonics, heat flow, geomagnetism, etc.

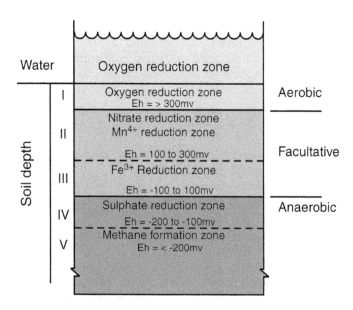

Figure 10-2. Schematic sequence of wetland chemical transformations with decreasing Eh conditions at depth. Adapted from Charman, D. 2002. *Peatlands and Environmental Change*. J. Wiley & Sons, London and New York, 301 p, Fig. 3.12b.

Figure 10-3. Nitrogen oxides (NO_x) are released by burning gasoline and diesel fuel in vehicles and contribute to acid rain. Photo by J.S. Aber; Warsaw, Poland.

again into organic matter. In marshes with algal blooms, pH may exceed 8, in which case ammonium is converted into ammonia (NH_3) and released into the atmosphere (Mitsch and Gosselink 2007). Under aerobic conditions, ammonium may be converted into nitrite (NO_2^-) and then nitrate (NO_3^-) respectively by *Nitrosomonas* and *Nitrobacter*. Denitrification is the further conversion of nitrate into nitrogen (N_2) or nitrous oxide (NO_x) gases by facultative bacteria under anaerobic conditions. Nitrogen gas makes up the largest fraction of the atmosphere (*c*. 80%), so its production has no impact on climate. Nitrous oxide, however, is a greenhouse gas that could affect climate.

Human activities have essentially doubled the amount of nitrogen entering terrestrial and wetland ecosystems, as result of fertilizer runoff, nitrogen-fixing crops, and burning fossil fuels (Fig. 10-3; Mitsch and Gosselink 2007). This extra nitrogen is a nutrient in wetlands, leading to eutropic conditions and algal blooms. When the algae die, oxygen is consumed by the decay process, and hypoxic (<2 mg/L dissolved oxygen) conditions result in fish kills and widespread "dead zones" in coastal marine waters around the world. Off the coast of Louisiana, for instance, the hypoxic zone appears annually and extends over 15,000 to 20,000 km² in many years (Fig. 10-4).

10.1.3 Phosphorus, potassium and sulfur

Phosphorus, like nitrogen, is also often a limiting nutrient in wetland habitats, particularly in northern bogs, fresh-water marshes, and deep-water swamps (Mitsch and Gosselink 2007). For example, Everglades saw grass (*Cladium jamaicense*) communities are limited by phosphorus (Charman 2002). Phosphorus is typically bound in minerals as well as organic litter and peat, but does not occur in wetland gases. It is not directly affected by redox conditions, but does interact with several other elements that affect its availability for plant growth (Reddy et al. 1999):

- P is fixed as aluminium and iron phosphates under acidic conditions.
- P bound by calcium and magnesium under alkaline conditions.
- P is most bioavailable under neutral to slightly acidic conditions.

Phosphorus is considered to be an essential element for life in the DNA molecule and adenosine triphosphate, which stores chemical energy. Arsenic is chemically similar to phosphorus, but is a well-known poison for most life. Recently a bacterium from alkaline soil of dry Lake Mono, California demonstrated the ability to substitute arsenic for nearly all phosphorus in its body. The GFAJ-1 strain, a member of the common

Figure 10-4. Oxygen concentration in Gulf of Mexico, southern Louisiana and southeastern Texas (see Color Plate 10-4). 2006 Seamap hypoxia map overlaid with June 2006 chlorophyll OC3 MODIS-derived image. Red and orange colors indicate hypoxic conditions in coastal waters. A. New Orleans, B. Mississippi River delta. Adapted from NASA (2010a).

Figure 10-5. Pyrite crystals on massive pyrite. Specimen measures ~8 × 4 cm; from Eagle County, Colorado. Adapted from original photo by R. Lavinsky; obtained from Wikimedia Commons <http://commons.wikimedia.org/>.

Gammaproteobacteria, thrived and reproduced when fed an arsenic-rich solution (Wolfe-Simon et al. 2010). This remarkable wetland discovery has major implications for understanding the chemical basis of life as well as the possibility of life on other worlds.

Potassium and sulfur are generally abundant in wetland environments and are not likely to be limiting factors for plant growth. In the anaerobic zone, sulfur and sulfate are reduced by *Desulphovibrio* bacteria into hydrogen sulfide (H_2S), which gives the distinctive smell of swamp or marsh gas, or to solid iron sulfides, depending on the presence of iron (Charman 2002). On the other hand, hydrogen sulfide may be oxidized under aerobic conditions by *Thiobacillus* bacteria into iron sulfate or sulfuric acid. The latter augments the acidity of many wetlands, particularly after sulfide-rich peatlands are drained (Charman 2002). Iron sulfides are quite dark and give the black color typically seen in anaerobic soils. One common iron-sulfide mineral is pyrite (FeS_2), popularly known as fool's gold (Fig. 10-5). Pyrite is often found in

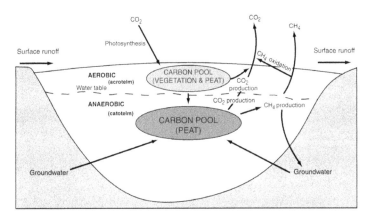

Figure 10-6. Basic scheme for carbon storage and transformations in wetlands. Carbon movements may take place as gases in air, dissolved in water, or as particulate matter. Adapted from Charman, D. 2002. Peatlands and Environmental Change. J. Wiley & Sons, London and New York, 301 p, Fig. 5.2.

coal and is the main source for sulfur dioxide released upon burning coal (see section 10.3.1).

10.2 Carbon cycle

10.2.1 Carbon reservoirs

Carbon is the chemical basis for life and exists in gases, is dissolved in water, and forms solid organic compounds and inorganic minerals essentially everywhere on Earth. Published values for global soil and peat carbon storage vary widely, but most estimates are around 1500 Gt[1] for total soil carbon, of which about 500 Gt are peat carbon (Adams 1999; Roulet 2000). Thus peat carbon represents about one-third of soil carbon. In comparison, atmospheric carbon is about 750 Gt, carbon stored in the world's land vegetation is in the range of 500–1000 Gt, and ocean surface water has some 700–1000 Gt carbon (Adams 1999). Based on these approximate values, it is apparent that wetlands, particularly peatlands, are major storage points for carbon as well as sources and sinks for greenhouse gases (CO_2 and CH_4).

---

[1] Gigatons of carbon (1 Gt = 1 billion tons = 1 Petagram = 1×10^{15} g).

Carbon held in air, biomass, and surficial water has relatively short residence time – days to decades, whereas soils and peat may retain carbon for centuries or millennia. Deeper carbon reservoirs include methane clathrates in seafloor sediment and permafrost, deep ocean water, fossil fuels, and carbonate sediments (see Fig. 8-24). Residence time in these carbon pools ranges from centuries to hundreds of millions of years.

Carbon is taken into wetlands primarily via plant photosynthesis in the aerobic zone and lost mainly by organic decay and decomposition in both the aerobic and anaerobic zones (Fig. 10-6). The boundary between these zones often fluctuates on a seasonal basis as the water table rises and falls. The conversion of organic matter into carbon dioxide is accomplished by respiration in the aerobic zone. In the anaerobic zone, fermentation produces acids, alcohols and carbon dioxide, and methane is derived from *Archaea* bacteria (methanogens).

10.2.2 Carbon balance

The accumulation or release of carbon in peatlands and other types of wetlands depends on the balance between productivity and decay. Wetlands are, for the most part, among the most productive ecosystems in the world,

Figure 10-7. Oligotrophic bog in Finland showing hollow in foreground and hummock with sparse pines in the background (see Color Plate 10-7). Both productivity and decay are limited by the short growing season, anaerobic conditions, low pH, lack of nutrients, and cold temperature. Photo courtesy of E. Volkova.

although the rates of productivity and decay are difficult to estimate accurately or measure directly. The exception is northern peatlands dominated by *Sphagnum* mosses with low productivity, but decay rate is also quite slow in this setting (Fig. 10-7).

In general, the rate of decay is of greater importance than productivity for long-term accumulation of peat (Charman 2002). Decay rate in peat is determined by temperature, water content, presence of oxygen, microbes and animals, and the type of plant material. Decay is usually faster in the aerobic zone (acrotelm) and decreases with depth in the anaerobic zone (catotelm). *Sphagnum* has much slower decay rates than other hydrophytes; thus, bogs generally have lower decay rates than fens. Two fundamental aspects apply to the peat accumulation model of Clymo (1984):

- Acrotelm decay is fast and considerable material is lost by the time it enters the catotelm through burial.
- Peat decay continues, albeit at a much slower rate, in the catotelm.

Given constant rates of productivity and decay over the long term (millennia), the total amount of decay increases as the thickness of peat increases. Eventually the total rate of decay would match the rate of productivity so that no further peat could accumulate, reaching a steady-state equilibrium. This model depends on the ideal assumption of constant conditions and rates, and while the general concept of this model appears valid, climatic changes, human impacts, and other wetland factors have varied significantly during the Holocene (see chapter 9).

10.2.3 Carbon gases and climatic feedback

Of the two carbon greenhouse gases, methane has a far greater thermal effect, some 20 times stronger than carbon dioxide, but methane has a much shorter lifespan in the atmosphere. Thus, the release of methane into the atmosphere has a strong but short-lived greenhouse impact, whereas carbon dioxide emission has a weaker but longer-lasting greenhouse effect (Charman 2002). Gas production depends to a large extent on the position of the water table; a high water table favors methane over carbon dioxide, and methane production increases greatly with higher temperature. Temperate wetlands display clear seasonal patterns in their methane and carbon dioxide emissions, whereas tropical wetlands are much less seasonal (Mitsch and Gosselink 2007). Thus, the balance of gas emissions has climatic implications and partly depends, in turn, on climatic conditions (see chapter 8.4).

In a general way, peatlands exert a negative feedback influence on climate (Charman 2002). In warm interglacial times, as during the Holocene, peatlands grow upward and spread outward, thus storing carbon removed from the atmosphere and reducing the greenhouse effect. Much of the 500 Gt or so of carbon held in peatlands has accumulated during the past few millennia. This drawdown of greenhouse gases, in fact, may have limited potential global warming and led to late Holocene cooling (see chapter 9.4). Franzén (1994) proposed that this mechanism was responsible for initiating glacial cycles during the Pleistocene (Fig. 10-8). Conversely, during glacial episodes peatlands were much reduced both by cold climate and expansion of ice sheets, so carbon was released into the atmosphere, strengthened the greenhouse effect, and limited potential global cooling. Glaciation sculpted vast, poorly drained landscapes in which peatlands could develop again during each subsequent warm interval (Fig. 10-9). "In this respect, comparisons between the global distribution of wetlands with areas subjected to glaciation during ice ages show a remarkable correspondence" (Franzén 1994, p. 300).

This theory remains controversial. Certainly links exist between continental glaciation, distribution of peatlands, carbon storage and greenhouse gases, and these connections result in a negative feedback influence for climate change. However, this cannot be viewed as the only or primary driving force for Pleistocene climatic/glacial cycles. Much larger carbon reservoirs (ocean, permafrost), stronger climatic factors (solar energy, albedo), and ocean-land-glacier interactions were equally, if not more, important for modulating Pleistocene climate. Nevertheless, "the size and functioning of any large carbon pool are of importance to past and future climate change" (Charman 2002, p. 192).

Figure 10-8. Organic-rich peaty soil exposed in the Lancer moraine, southwestern Saskatchewan, Canada. The peat has yielded an uncorrected radiocarbon date of 31,300 ± 1400 years old. This material was presumably derived from mid-Wisconsin interstadial deposits that were deformed and overrun by late Wisconsin glaciation. Photo by J.S. Aber.

10.3 Fossil fuels

10.3.1 Fossil-fuel consumption

The large-scale extraction and burning of coal ushered in the Industrial Revolution and continues to be a primary energy source for modern industry and society in many parts of the world (Figs. 10-10 and 10-11). Coal represents ancient, so-called dead carbon that accumulated in peat-forming wetlands and then was effectively removed from the carbon cycle and stored for tens to hundreds of millions of years. Megamining and massive burning of fossil fuels (coal, oil and natural gas) have multiple direct and indirect impacts on modern wetlands both locally and globally. In addition, natural gas is the feed stock for most inorganic ammonia-based fertilizers (see section 10.1.2).

Burning fossil fuels, forest clearance, and other human activities are causing atmospheric carbon to increase by about 3 Gt per year (Adams 1999). In comparison, current carbon burial rate in all northern peatlands is only 0.07 Gt per year (Charman 2002). As a result, atmospheric carbon dioxide has grown from <320 ppm in the mid-twentieth century to >390 ppm in the early twenty-first century, an increase of >20% in only half a century (Tans 2010). The mean

182 ENVIRONMENTAL CYCLES AND FEEDBACK

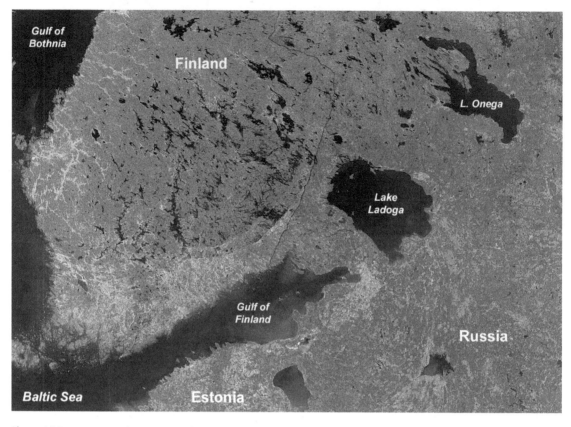

Figure 10-9. Eastern Baltic region including parts of southern Finland, northwestern Russia, and northern Estonia. This terrain was glaciated repeatedly during the Pleistocene. Lakes, seas and wetlands of diverse types and sizes abound throughout the area, and mires are widespread. MODIS panchromatic (visible) image, 29 May 2002; adapted from NASA Visible Earth <http://visibleearth.nasa.gov/>.

Figure 10-10. Railroad loading and transportation of coal to be used for electricity generation in the United States. A. Train loading facility at a mega-mine near Gillette, Wyoming. B. Transport of coal across the Great Plains at Alliance, Nebraska to energy-consumption markets in the central and eastern U.S. Photos by J.S. Aber.

annual growth rate for the most recent decade (2000–09) was around 2 ppm. Beyond the enormous, but as yet uncertain, climatic implications for greenhouse warming, many other wetland consequences result from mining and burning fossil fuels.

10.3.2 Coal mining and acid rain

Coal mining historically was done mostly underground in relatively small workings. However, since the mid-twentieth century large open-pit strip mines have become increasingly common for economic and mine safety reasons. Early strip mines were often simply abandoned after removing the coal and left to overgrow with brush and trees. Those areas are marked by ridges of spoil and intervening troughs. As an experiment, trees were planted on several thousand hectares of mined land in southeastern Kansas, United States in the late 1930s (Muilenburg 1961), and these sites are now wildlife refuges known for excellent deer hunting and fishing. However, acid water leaching from coal-mine debris, known as gob piles, has degraded downstream surface and ground water, and only the most acid-tolerant organisms have survived (Arruda 1992). Since 1969, Kansas state law requires that mined land must be reclaimed for productive agricultural use (Fig. 10-12). Mining

Figure 10-11. Concrete plant associated with a large coal strip mine in eastern Germany (see Fig. 8-24A). Manufacturing concrete releases carbon dioxide from two sources, first burning fossil fuel for thermal energy, and second the conversion of carbonate rock into lime and CO_2. Photo by J.S. Aber.

Figure 10-12. Big Brutus restored at the mining heritage visitors' center near West Mineral, Kansas, United States. One of the world's two largest electric-power shovels, it stands 50 m tall with a working weight of 5000 metric tons. Reclaimed mined land on right side. Blimp airphoto by J.S. Aber.

for other valuable ores – copper, gold, diamonds, lead and zinc – has similar deleterious impacts on wetland habitats (see Fig. 1-15).

Mining is now strictly regulated for safety and environmental reasons in most developed countries. However, mining remains a "wild west" activity in many developing and poorly developed countries of the world. In order to fuel its rapid industrial growth, China has become the world's leader in coal use, overtaking the United States, European Union and Japan combined, and India is following close behind (Bradsher and Barboza 2006). Chinese coal mining has increased dramatically in recent years with little attention to environmental consequences, which has led to increased wasteland and desertification, declines in ground water, land subsidence, and salt-water intrusion in coastal regions (Yang 2007). Untreated drainage from mines and tailings has greatly polluted surface and ground water.

Most upper Carboniferous coal in eastern North America has high sulfur (pyrite) content. Sulfur dioxide gas emitted from burning such coal is converted by a catalyst, such as NO_2, into sulfuric acid (H_2SO_2), which is a major component of acid rain. Acid rain and snow across southeastern Canada and the eastern United States has caused acidification of many thousands of lakes and streams, particularly in regions with crystalline bedrock and soft water that have little buffering capacity, such as the Adirondack Mountains and northern Appalachians (US Environmental Protection Agency 2007). Acid deposition on soils releases aluminium, and both low pH and aluminium are directly toxic for many fish species, few of which can survive a pH less than 5. Because of this problem, low-sulfur coal and lignite from western Canada and the United States are now preferred for electric power plants.

Acid precipitation is also a serious problem in northwestern Europe, where many historic buildings have been damaged (Fig. 10-13). The disappearance of *Sphagnum* mosses as a consequence of atmospheric pollution was described more than two centuries ago in Great Britain (Paal et al. 2009), and high concentration of SO_2 is the main cause for *Sphagnum* decline in other industrial areas. Bog experiments in Scotland demonstrated that deposition of sulfate from acid rain leads to a decline in methane emission, presumably because sulfate-reducing bacteria outcompete methanogens (Gauci, Dise and Fowler 2002).

Strict pollution controls on coal-burning power plants and automobile exhaust have helped to reduce the acid rain problem in North America and western Europe in recent years, but uncontrolled emissions continue in many other parts of the world (Marquardt 2012). In China, for example, massive air pollution from coal burning has caused serious issues for human health, and widespread acid rain has severely damaged rivers and related wetland habitats (Yang 2007).

10.3.3 Estonian oil shale

Burning oil shale in northeastern Estonia heavily damaged nearby bogs through deposition of calcium-rich fly ash (Karofeld 1996; Paal et al. 2009). The Ordovician oil shale is extracted from several large open-pit and underground mines and fuels power plants and chemical factories that represent significant parts of the Estonian economy. During Soviet times, industrial development was emphasized, and large power plants began operating in the 1950s and 60s with uncontrolled emission of fly ash. *Sphagnum* mosses began to disappear from nearby bogs in the 1970s due to a combination of increased pH and calcium in bog water and upper layers of peat. In the most polluted bogs, pH values rose above 5 to >6, compared with pH <4 in unaffected bogs. *Sphagnum* mosses completely disappeared from some bogs, and Scots pine (*Pinus sylvestris*) showed increased radial growth. As *Sphagnum* mosses died out, bogs lost their buffering ability, and a rapid shift took place in which other bryophyte species and vascular plants became established on less acidic to near-neutral soils with higher nutrient levels. In effect, the oligotrophic bogs underwent a conversion into mesotrophic fens.

Since the collapse of the Soviet empire and Estonian independence, atmospheric pollution from burning oil shale has been reduced drastically because of decreased production of

Figure 10-13. Restoration of Marmor Broen (Marble Bridge) in the historic royal district of central Copenhagen, Denmark. This restoration was undertaken because decades of exposure to urban and industrial pollution had caused the stone to decay badly. Stone from the original quarry in eastern Germany was imported in order to carry out an authentic restoration. A. Bridge during restoration with heavily damaged stones, B. Master stonemason finishes new stone sculptures, C. Completed bridge restoration with a mix of new and old stones. Photos by J.S. Aber and S.W. Aber.

electricity and greatly improved pollution controls (Fig. 10-14). Already by the mid-1990s the first signs of bog recovery were evident (Karofeld 1996), and bog renewal has continued with the spread of mosses, increased number of *Sphagnum* species, decreased radial growth of Scots pine, lower contents of heavy metals in mosses, and decrease in pH (Paal et al. 2009). *Sphagnum* mosses reappeared first on hummocks and pool banks, places that depend solely on atmospheric precipitation, and once established the mosses released acids that further lowered water pH. This self-recovery process has been most remarkable and relatively quick. Still, the most damaged sites have not yet reached the status of intact bogs, and air pollution remains a problem close to the largest power plants.

10.4 Human experiment

Homo sapiens are conducting a global experiment involving immense use of fossil fuels, vast land conversions, and intense water consumption in order to support and raise the living standard for a rapidly growing human population, which has now reached seven billion people (US Census Bureau 2010). The experiment is uncontrolled; in other words, no standard procedures or protocols are in place. Measurements of processes and results are insufficient to monitor the progress of the experiment, and our theoretical understanding of the various biogeochemical cycles is incomplete. Most decisions for this experiment are made for immediate individual survival, gain or

Figure 10-14. Modern oil-shale power plant produces electricity and hot water near Narva, northeastern Estonia. Fly ash is captured by dust filters, and pipelines in the foreground carry fly-ash slurry to holding ponds. Processed fly ash is sold for various construction, agricultural, and industrial applications. Photo courtesy of E. Karofeld.

comfort rather than the greater good of society or long-term environmental sustainability.

The greatest impacts of this human experiment on wetlands involve the carbon and nitrogen cycles as well as biodiversity. In all three, results of human activities have already exceeded so-called boundary limits, beyond which consequences are uncertain or perhaps irreversible (see Preface, Table 1). Other major biogeochemical impacts on wetlands include land conversion, water consumption, phosphorus loading, ocean acidification, and chemical pollution. For many people, wetlands are still regarded as wastelands, and for the majority of humanity daily survival has a much higher priority than protecting natural resources for a distant future. Meanwhile, citizens of developed countries are unlikely to give up their relatively rich lifestyles, residents of developing countries will seek to achieve the living standards of richer countries, and global population will continue to grow, perhaps reaching a peak around 10 billion people late in this century (Lutz, Sanderson and Scherbov 2001). Given this point of view, human encroachment, alteration, and destruction of wetlands is likely to continue, particularly in developing and poorly developed countries.

The human experiment has been costly for environmental resources in general and wetland habitats in particular. Nonetheless, the conservative nature of wetland flora and fauna gives them the capacity to survive under adverse conditions, as happened during previous extinction events (Greb, DiMichele and Gastaldo 2006). As the Estonian oil-shale example demonstrates, many wetlands still retain the ability for internal self-renewal once external disruptions are reduced or removed. A central problem facing wetland maintenance and restoration for the twenty-first century, then, is how to reduce human environmental impacts, particularly for the carbon and nitrogen cycles. The human factor for both cycles derives in large part from massive use of fossil fuels, which are burned for energy and processed for fertilizer.

Before assigning any blame for greenhouse gases, however, it would be wise to consider the question: who is responsible for carbon emissions? China has become the world's largest energy consumer with coal supplying some 70% of its energy needs (Webber 2010). In spite of huge increases in domestic coal production, China is the world's largest importer of coal from other countries. In effect, developed nations have exported their dirty manufacturing, coal, and carbon emissions to China, which is now also the largest carbon emitter. Just as offshore banks are used to hide cash, developed countries employ offshore manufacturing to hide their true carbon usage. This may look good for the short-term carbon accounts of developed countries, but this practice has decidedly negative consequences (Webber 2010):

- Efficiency and pollution – Instead of burning the coal in highly efficient power plants, it is burned in less efficient plants that have little pollution control. Thus, more nitrous oxides, sulfur oxides, particulate matter, and carbon dioxide are emitted per megawatt-hour of electricity produced.
- Mining impact – Countries that export coal to China bear the environmental impacts of coal mining without benefiting from the energy production.
- Transportation – Shipping coal overseas adds transportation costs and emits still more greenhouse gases from burning diesel fuel.

Any realistic hope of mitigating human impact on the carbon and nitrogen cycles on a global basis means a combination of actually reducing the consumption of fossil fuels and effectively lowering carbon and nitrogen emissions from their use. Many schemes, regulations, incentives, treaties, and means of valuation have been proposed or enacted to promote energy conservation and wise use of natural resources (see chapter 11). Smil (2011) analyzed global energy usage and technology. Comparing developed and developing countries, he reached two main conclusions.

- Most people in the world need to consume much more energy if they are to enjoy reasonably comfortable, healthy and productive lives.
- The United States and Canada consume twice as much energy per capita as do the richest European Union countries or Japan, but they do not have better standards of living.

Smil argued that per capita energy consumption in the U.S. and Canada could be reduced substantially without a decline in quality of life, and this would promote innovation, strengthen the economy, and help the environment. However, "any calls for restraint or reduction of North American energy use are still met with rejection" (Smil 2011, p. 214). He cited the serial failures of international agreements to reduce carbon emissions – Kyoto, Bali, Copenhagen and Cancún. He rejected numerous technological fixes to energy production and carbon pollution as logistically or financially unfeasible, predicted that atmospheric CO_2 would exceed 450 ppm, and emphasized the need for conservation and moderation of energy consumption.

In terms of renewable energy, wind power emerged during the late twentieth century as the most promising technology for relatively rapid development on a viable financial basis with minimal impacts on wetlands or other environments. A milestone was achieved in 2008, when global wind-energy capacity surpassed 100,000 MW (megawatts), which represented 1% of total electricity use worldwide (Musgrove 2010), and this share had increased to 2% by 2011 (Smil 2011). Denmark now produces the highest portion (20%) of its national electrical energy from the wind (Fig. 10-15), and the United States has a feasible goal to reach this level by 2030 (Fig. 10-16). Corresponding amounts of fossil-fuel consumption and carbon emissions, thus, could be avoided.

10.5 Summary

Biogeochemical cycles involve all elements and their compounds at the Earth's surface, and wetlands serve as storage points, transformers and sources for many elements. Key wetland elements include nitrogen, potassium, iron and manganese, sulfur, phosphorus, and carbon.

Figure 10-15. Wind farm in the rural landscape of western Denmark at Ramme Dige. These wind turbines are situated near archaeological remains in the foreground including Neolithic and Bronze Age burial mounds (a) and a wall from the Iron Age (b). Kite aerial photo by S.W. Aber, J.S. Aber and I. Marzolff.

Figure 10-16. Flat Ridge Wind Farm located on the High Plains of south-central Kansas, United States. This wind farm was constructed in 2008 and became operational in 2009 with a nominal generating capacity of 100 MW. The wind farm is compatible with traditional agricultural land use including crops and cattle grazing (foreground). Kite aerial photo by S.W. Aber and J.S. Aber.

Their chemical status depends to a large extent on the presence of oxygen in wetland water and soil. Nitrogen is a major nutrient that is often a limiting factor for plant growth in wetland habitats. Human activities have greatly increased nitrogen input, and excess nitrogen leads to hypoxic conditions in coastal marine environments. Phosphorus is another limiting nutrient in many wetlands. Although phosphorus is considered to be essential for life, some bacteria have the ability to substitute arsenic for nearly all phosphorus in their bodies. Potassium, sulfur, iron and manganese are generally more abundant and, thus, are not limiting factors in most wetland environments.

Carbon is the chemical basis for life, forms greenhouse gases, and is a primary component of fossil fuels. Wetlands, particularly peatlands, are major carbon reservoirs as well as sources for carbon dioxide and methane. Carbon is taken into wetlands primarily via plant photosynthesis in the aerobic zone and lost mainly by organic decay and decomposition in both the aerobic and anaerobic zones. The accumulation or release of carbon depends on the balance between productivity and decay, and the rate of decay is generally of greater importance than productivity for long-term accumulation of peat. In a general way, peatlands exert a damping influence on climate change. During warm intervals peatlands expand, accumulate carbon, draw down greenhouse gases, and minimize global warming; the opposite takes place during cool periods. It is possible that this mechanism may have initiated glacial cycles during the Pleistocene and led to late Holocene cooling.

Burning fossil fuels, forest clearance, and other human activities are causing atmospheric carbon to increase by about 3 Gt (~2 ppm) per year. Beyond the enormous, but as yet uncertain, climatic implications for greenhouse warming, many other consequences for wetlands result from mining and burning fossil fuels. Mining has direct negative impacts on wetland habitats; burning fossil fuels generates acid rain that alters downwind water and soil chemistry and releases greenhouse gases that may influence climate. Burning oil shale in Estonia during the Soviet era released fly ash that caused severe impacts for nearby bogs; however, these bogs have demonstrated remarkable self-recovery following large reductions in fly-ash pollution during the past two decades.

Homo sapiens are conducting an uncontrolled global experiment involving large-scale use of fossil fuels, land conversions, and water consumption in order to support and raise the living standard for a rapidly growing human population. The greatest impacts of this human experiment on wetlands involve the carbon and nitrogen cycles as well as biodiversity. In all three, results have already exceeded so-called boundary limits. The conservative nature of wetland flora and fauna gives them the capacity to survive under adverse conditions. However, any realistic hope of mitigating human impact on the carbon and nitrogen cycles means a combination of reducing the consumption of fossil fuels and effectively lowering carbon and nitrogen emissions from their use, particularly in the United States and Canada.

Wetland services, resources and valuation 11

11.1 Human use of wetland ecosystems

Throughout history wetlands have sustained human populations by acting as sources of food, water, fiber, shelter and solace. Examples range from ancient Sumerian and Egyptian cultivations of reed beds for fiber and building materials, to Aztec experimentation with chinampas wetland agricultural methods and water harvesting techniques (Boule 1994). Archaeological evidence also indicates that Mesopotamians valued the aesthetics of wetland landscapes and sought to replicate them by creating gardens that mimicked and incorporated wetland pools and pond features (Boule 1994). Similar gardens indicative of later Islamic influence were common across Mughal India during the sixteenth through eighteenth centuries. Many examples of these still survive today and include the famous gardens of Srinagar and those found within the complex of the Taj Mahal in Agra, India.

Even so, early agriculturalists, like their more modern counterparts, recognized the economic value of modified wetlands as cropland. Scarborough (2009) documented the significant alterations to wetland landscapes in the ancient Mayan lowlands and highlands of Mexico and other parts of Mesoamerica. Here, the construction of raised fields and drained platform planting areas, and other engineering modifications such as ditches, earthen berms, and terraces, were not uncommon. The horticultural gardens of the Dal Lake of Srinagar, India provide a comparable modern version. Carefully maintained raised earthen beds along the shallow shores of the Dal Lake provide space for vegetable harvests. Similar floating cultivation techniques called *dhap* using floating mats of dead aquatic plants are seen in the wetlands and waterlogged areas of rural Bangladesh.

The push for agricultural expansion led to wetland conversions across various parts of the world well into the twentieth century. The industrial age sped up this process, as increased population pressures led to the development of untouched wetlands for settlement and profit. Policy initiatives from colonial India to North America pursued sustained wetland conversion strategies. Often these were not based on scientific knowledge about the role of wetlands within the larger biosphere, so that they were erroneously represented as wastelands (Mitsch and Gosselink 2000).

Beginning in the second half of the twentieth century, however, numerous studies began documenting the valuable economical, hydrological and ecological functions provided by wetland habitats at local, regional and global scales (Table 11-1). Scientific journals, such as *Wetlands, Estuaries, Journal of Wetlands Ecology*, and *Wetlands Ecology and Management* among others, were founded specifically to study important aspects of this ecosystem, its characteristics, and its functioning. Scientists examined

---

Wetland Environments: A Global Perspective, First Edition. James Sandusky Aber, Firooza Pavri, and Susan Ward Aber.
© 2012 James Sandusky Aber, Firooza Pavri, and Susan Ward Aber. Published 2012 by Blackwell Publishing Ltd.

the complexity of wetland ecosystem services and functions and the interactions between their intertwined terrestrial, hydrological, biological, chemical, and atmospheric components. This chapter examines some of the significant biophysical and socio-economic contributions provided by wetland ecosystems.

Table 11-1. Ecosystem, hydrological and economic goods and services provided by wetlands.

Ecosystem Services	Hydrological Services	Economic Services
• Provide habitat • Biogeochemical cycling • Nutrient storage & cycling • Carbon storage • Climatic influences • Erosion protection • Pollution amelioration • Storm surge and coastal protection	• Flood abatement • Water quality improvement through biological, physical & chemical processes • Waste water treatment • Water storage & diversion	• Extractive activities • Agriculture & ranching • Aquaculture • Forestry • Peat extraction • Salt production • Pearl production • Transportation services • Irrigation • Hydroelectric power generation • Ecotourism, recreation & birding

11.2 Ecosystem services

Ecosystem services provided by wetlands are broad ranging in scope. These services are not just localized to specific wetland sites, but are beneficial at regional and global scales as well. Examples include contributions to global carbon storage, organic and inorganic nutrient cycles, and protection from storm surges and severe flood events. Wetland protection and conservation efforts are also often framed by the need to preserve these sites as important repositories of biodiversity. As habitats for flora and fauna, wetlands provide yet another major ecosystem contribution.

11.2.1 Habitats

Wetlands provide habitats for a diversity of plant and animal life in a wide range of geographic zones from the coldest tundra to the flood zones of the equatorial Amazon. Climatic, terrestrial, and hydrologic conditions collectively influence the diversity and density of species within any given wetland type. Thus, while tropical wetlands like the Okavango or the Pantanal support great species richness, just one or two dominant species might be more typical of other areas. For example, *Spartina alterniflora* and *S. patens* often dominate the coastal marsh ecosystems of northern New England (Figs. 11-1 and 11-2). Overall, however, the astonishing biodiversity and productivity supported by these ecosystems

Figure 11-1. *Spartina* sp. in salt marshes of the Plum Island estuary, northeastern Massachusetts, United States. A. Extensive stands of *Spartina alterniflora* (short form). B. *S. alterniflora* (short form) along creek edge with *S. patens* (salt hay) in the foreground with wetland pools in the background. Photos by Firooza Pavri.

Figure 11-2. Vertical aerial view of salt marsh of the Plum Island estuary, Massachusetts, United States. This area was settled in the 1640s and ditched for improved hay production, which continued until the 1960s. The distinctive wavy pattern, known as cow lick, is typical of *Spartina patens* (compare with previous figure). Blimp aerial photograph by S.W. Aber, J.S. Aber, and V. Valentine (visible in upper left corner).

is maintained through the dynamic and complex cycling of nutrients and energy transfers between biotic and abiotic components within the system.

Reliably quantifying global estimates of wetland habitats is difficult due to the lack of systematically collected data from many parts of the world (Scott and Jones 1995; Finlayson et al. 1999; Zedler and Kercher 2005). Recent ongoing efforts have attempted to fill this data gap through inventorying satellite imagery (Rebelo, Finlayson and Nagabhatla 2009). However, much work remains to be done. Past studies estimated the total expanse of wetlands globally to be somewhere in the region of 12.8 million km^2 based on a variety of international inventories (Finlayson et al. 1999). This makes up close to 12 percent of the Earth's total land area. However, other estimates suggested totals as low as 5.3 million km^2 (Matthews and Fung 1987). While such wide ranges leave much uncertainty as to the exact extent, at present the international Ramsar Convention on Wetlands provides the most comprehensive repository of information on wetland habitats globally (Ramsar 2010c).

The Convention on Wetlands was signed in Ramsar, Iran in 1971 and currently has 160 Contracting Parties worldwide. Contracting countries to the Ramsar Convention are obliged to provide detailed information on wetlands listed as part of the wetlands conservation treaty and these data are made available through the Ramsar Sites Information Service (RSIS). Systematically gathered site-specific geographic and biophysical data along with management and conservation issues pertaining to each site are provided and regularly updated. While the RSIS may not list all wetland habitats found within individual countries, it goes a long way toward providing a comprehensive database on at least the more important and well-recognized wetland sites.

For the United States, a recent comprehensive study conducted using satellite imagery and digital photography indicates 43.6 million total hectares of wetlands across the conterminous states (Dahl 2006). Ninety-five percent of this area comprises fresh-water wetlands while the rest is represented by marine or coastal habitats. Between 1998 and 2004, the study found an increase of approximately 12,950 hectares of wetlands annually for the United States, and these net gains were attributed to wetland restoration and enhancement activities as a result of regulatory requirements.

Detailed and comprehensive inventories such as the one mentioned above are, however, both expensive and time consuming. Most scientific studies have opted to focus on specific wetland sites, particular wetland types, or target groups of organisms such as fish, birds, and amphibians (Brinson and Malvárez 2002; Tockner and Stanford 2002; Deil 2005; Keddy et al. 2009). Such detailed analyses provide a wealth of information on unique habitat conditions, migratory and threatened species, non-native invasive species, and changing land-use conditions within and around wetland sites.

Wetlands are crucial for supporting fish populations. They provide food, shelter, spawning, and nursery areas for both marine and inland fresh-water species. Riverine wetlands are home to a wide diversity of fish species and these are often uniquely adapted to survive fluctuations in water levels and nutrient inputs based on flood events. For example, the seasonal flood cycle of southern Africa's Okavango River starting in November drives a proportion of its 80 fish species from the main river channels into ephemerally flooded areas (Merron and Bruton 1995; Scott 2010). Similarly, the Sudd region of the upper Nile River floodplain spans across Sudan, Ethiopia and Uganda and records some 118 species of fish. Many of these species migrate into nutrient-rich seasonally flooded plains and coincide their breeding and spawning with the floods to ensure adequate food for juvenile fish (Peck and Thieme 2010). Such flood pulses initiate ecological processes between different components of a riverine ecosystem. Other adaptations to variable salinity levels allow migrating fish such as salmon and trout to move through fresh-water, estuarine and marine habitats. The Chinook salmon (*Oncorhynchus tshawytscha*), for instance, migrates hundreds of kilometers from the Bering Sea to the upper tributaries of the Yukon River in North America to spawn (Burridge and Mandrak 2010).

Wetlands are also home to both resident and migratory species of birds. Marsh habitats that serve as migration stopover points for birds traveling between wintering and breeding areas provide vital food reserves. The conservation of interconnected wetlands along major flyways is critical to maintaining successful populations of migrating bird species. This may be particularly difficult as the three major global flyways: the Americas, the African–Eurasian and the East Asia–Australasian, each span dozens of individual countries with unique sets of conservation challenges.

The case of the Siberian crane (*Grus leucogeranus*), which once wintered in northern India, provides one example. The migrating population of the endangered Siberian crane had been reduced to critical numbers in recent decades. In most years just a handful of mating pairs reached their wintering grounds in Keoladeo National Park in Rajasthan, India. However, the crane has not been spotted there since 2002. The loss of wetland stopover sites along its migratory route from central Siberia through central Asian countries and Afghanistan into northern India has been identified as one of the main reasons for its disappearance (BirdLife International 2010a). Other factors that have been suggested as possible contributors to the problem include the crane hunting tradition in Afghanistan (a flyover country) and the decline in forage areas within Keoladeo due to buffalo grazing (Boojh, Patry and Smart 2008). With vanishing wetlands, a few remaining areas provide relief to large numbers of migrating birds and are soon unable to sustain large visiting populations. Moreover, such wetland oases, in the absence of protection and active management, may be particularly vulnerable to and influenced by surrounding land-use change, seasonal rainfall patterns, habitat fragmentation, and changes to hydrological conditions. Organizations such as BirdLife International target these issues by working with local partners in different countries to raise awareness, promote conservation efforts, identify threatened bird species, and develop comprehensive conservation action plans (BirdLife International 2010b).

Numerous physical, economic, political, and socio-cultural factors contribute to the loss of wetland habitat and consequentially a reduction of their biodiversity. For any given region, it is usually the combination of several different factors that leads to habitat loss. These include the following:

- the draining and filling of wetlands for agriculture or development
- habitat fragmentation
- aquaculture
- alteration of river regimes through diversions, dams, impoundments, and flood-control mechanisms
- overharvesting of wetland products
- clearing of wetland vegetation
- contamination through agricultural and industrial runoff
- eutrophication and algal blooms
- introduction of invasive species that crowd out native plants and animals
- changes in land tenure and ownership regimes
- changes in government policy
- encroachment by small-scale farmers and ranchers
- political uncertainty
- absence of conservation management
- poverty and political conflict
- variability in regional and global precipitation and climate patterns
- sea-level rise or fall

Monitoring, education and conservation efforts involving multiple stakeholders could be critical to the sustainability of these habitats.

11.2.2 Wetlands and biogeochemical cycles

Biogeochemical cycles play critical roles in maintaining ecosystem functions and processes. The cycling of nitrogen (N), phosphorus (P) and carbon (C) through wetland ecosystems occurs through a series of complex processes (see chapter 10). Each of these is necessary for the viability of wetlands and contributes to their global cycles. Nitrogen is made available to wetlands through biological fixation, precipitation, and through human point and non-point source discharges (White and Reddy 2009). Wetlands provide vast reservoirs of N in inorganic and organic forms within soils. Plants assimilate nitrogen into plant tissue and once incorporated, N moves through the food chain from producers to consumers and then decomposers (Cutter and Renwick 2004). N is also leached from the soil into ground and surface waters. Human interference in the nitrogen cycle has resulted in the overloading of N from human waste, sewage and agricultural runoff into wetland areas. Such overloads may cause eutrophication, and the excessive growth of algae in wetland pools and ponds may choke out native wetland plants.

Like nitrogen, phosphorus is considered a limiting nutrient in wetland ecosystems (Richardson and Vaithiyanathan 2009). Phosphorus is found in terrestrial sinks and is made available in the soil from rock weathering and erosion processes. P makes its way through the food chain via uptake by plants and onto consumers until it is returned to the soil by decomposers. Phosphorus is also made available through the excessive use of fertilizers and runoff from agricultural fields, and through human waste, leading to problems similar to those observed with the overloading of N in wetlands.

Carbon enters the wetland ecosystem through the photosynthesis process whereby plants take in CO_2 in the presence of sunlight and water to produce sugars. While carbon dioxide is returned to the atmosphere through respiration and decomposition, a significant amount of organic matter made up of carbon is sequestered in dead material within wetland soils (Dise 2009). Due to slow decomposition processes in the presence of water, wetlands play an important role in the atmospheric carbon cycle serving as an important repository or sink of carbon. As Kusler (1999) pointed out, carbon sequestration and storage in wetlands is dependent on their size and type, the vegetation present, soil depth, pH, temperature, precipitation, nutrients, and ground-water levels. Hence, different wetland types have different carbon accumulation and storage capacities (Dise 2009). For instance, peatlands may sequester carbon at a slow rate for many thousands of years (Gorham et al. 2007).

The focus on carbon due to its critical role in climate change has added urgency to understanding the role of wetlands in carbon storage. Studies have suggested that significant amounts of the world's soil carbon could be sequestered in wetlands. Carbon released from wetlands through conversion to other land uses, burning, melting permafrost, and drying of peatlands, may therefore contribute to significant increases in global atmospheric carbon concentrations. Climate change and higher temperatures could also cause changes within wetland ecosystems. High-latitude wetlands under permafrost and peatlands across North America and northern Eurasia are the focus of numerous studies interested in observing how rising temperatures may lead to melting permafrost and accelerate the release of greenhouse gases. Particular focus is placed on CO_2 and methane (CH_4) releases from these sites. Biogenic CH_4 releases from wetlands in thawing permafrost regions and other tropical regions are of particular concern, because CH_4 is identified as a more potent greenhouse gas than CO_2 (Denman et al. 2007). Changing climates would undoubtedly impact wetlands globally. How they respond to these changes and what roles they might play in the future remain under investigation.

11.2.3 Storm surge and coastal flood protection

Nowhere are the protective services rendered by wetlands more obvious than along coastlines. Coastal wetlands act as effective barriers protecting inland areas from storm surges. They absorb and dissipate the most devastating first impacts of wave action. Coastal wetlands are generally built on alluvial materials brought down and deposited by rivers, loose sand and gravel, and dead plant material called wrack, which accumulate over long periods of time. The accretion of such material maintains and expands the marsh, creates sand dunes and barrier islands, and counters the inevitable erosion due to wave action.

The root systems of plant species in coastal marshes are also uniquely adapted to withstand constant wave action. For instance, studies found that the densely matted and deep root systems of high-salt-tolerant marsh species found in the more saline zones of the Louisiana marshes were better able to withstand the onslaught of waves from the devastating Hurricane Katrina (2005),

Figure 11-3. Dense prop root system of *Rhizophora apiculata* (red mangrove) supplying air and stability found along coastal western India. Photo by Firooza Pavri.

than their low-salinity counterparts (Howes et al. 2010). Similarly, mangroves in tropical regions are uniquely adapted with dense and buttressed prop-root systems to withstand constant and severe wave action and may recover relatively quickly even from larger, more powerful surges (Fig. 11-3). Moreover, the intricate network of tidal creeks and fresh-water channels that typify delta marshes such as the Sundarbans of South Asia or the estuarine backwaters of India's southwestern Kerala coast further work to dissipate the destructive power of waves on the shoreline. Coral reefs perform a similar protective role (see chapter 7.2.3).

The channeling and impoundment of rivers for flood control, navigation and irrigation, and dredge-and-fill activities have hindered the transport and deposition of sediment and the natural accretion processes necessary for maintaining coastal marsh complexes. Such activities may dramatically reduce the size of coastal wetlands. Without the annual deposition of materials, the marsh loses its nourishment and ability to keep up with coastal erosion or natural subsidence. From coastal Louisiana and the Mississippi Delta to the Brahmaputra, Ganges, Mekong, and Red River deltas, the consequences of coastal marsh losses are similar. The vulnerability to coastal flooding increases dramatically, inland fresh-water systems are threatened by salt-water intrusions, and losses in biodiversity have long-term negative ecosystem and economic consequences.

11.3 Hydrological services

Water comprises an essential element of all wetlands. Besides providing an obvious source for fresh water, numerous additional hydrological services make wetlands some of the most functionally productive ecosystems in the world. Wetlands improve water quality through toxin removal, enhance ground-water recharge, and protect against erosion from coastal storms and flood events.

11.3.1 Flood abatement

Beyond coastal regions, wetlands also provide flood abatement services along flood-prone river plains. Wetlands with large storage capacities act as sponges holding excess water, gradually releasing the water over extended time periods, and minimizing the damage caused by sudden snow melts and extreme precipitation events. Flood-control services also reduce runoff velocity and prevent stream- and river-bank erosion during peak water discharges. Runoff is, however, influenced by climatic factors, seasonality and soil characteristics. In some cases, such as upper latitude wetlands in Alaska and Siberia, runoff velocity may be high due to permafrost conditions which hinder storage and percolation (Carter 1997).

With urban, industrial and agricultural development and wetland conversions, the economic costs associated with flood events may rise astronomically. The catastrophic 1993 floods of the upper Mississippi River basin due to excessive precipitation resulted in over 40 deaths and close to US$20 billion in economic costs (Perry 2000). More than 150 rivers and tributaries were in full flood, affecting vast sections of the central United States (see Color Plate 4-16; Johnson, Holmes and Waite 2010). At the time, it was the most expensive flood ever, devastating towns, inundating millions of farmland hectares, and destroying transportation links and urban services across nine states. While not all of these costs were related directly to the conversion of floodplains and wetlands to other land uses, flood impacts are exacerbated by the lack of naturally occurring flood abatement systems such as wetlands, which could store large quantities of water and reduce the height of flood peaks (Zedler and Kercher 2005). A similar major flood episode took place on the lower Mississippi River in 2011, which required opening the Morganza spillway and consequent flooding of the Atchafalaya region in southern Louisiana. This flood has renewed calls for significant changes in river and wetland management.

11.3.2 Water quality

Most experts agree that wetlands serve as buffers or filters, which trap excess nutrients, heavy metals, particulate matter, suspended sediments, and toxic contaminants (DeBusk and DeBusk 2001; Verhoeven et al. 2006; Rodriguez and Lougheed 2010). In recent decades, constructed and natural wetlands have been used globally to treat waste water effectively and aid in the incorporation of toxic metals like lead (Pb), copper (Cu), zinc (Zn), mercury (Hg) and cadmium (Cd), among others (Vymazal 2008). As DeBusk and DeBusk (2001, p. 255) identified, a variety of processes active in wetland environments improve water quality:

- Biological – wetland plants, some types of algae, and microbes remove excessive nutrients and toxic metals through uptake and incorporate them into their tissues. Nutrients such as nitrates and phosphates are readily taken up and stored by hydrophytic vegetation including *Typha*, *Phragmites*, and other reeds before eventually being released through the process of decomposition or leaching. In some cases, nutrients may remain stored for the long term through the deposition of partially decomposed material and the formation of peat.
- Physical – the slow movement of water through wetlands, the physical structure of plant roots and stems, and the presence of emergent and submerged plants all promote the trapping of suspended solids and particulate matter or even, in some cases, larger non-plant litter items. This matter is often deposited on the surface of the soil or

remains trapped among the exposed roots and stems (Fig. 11-4).

- Chemical – adsorption, ion exchange, and precipitation may aid in the removal and storage of heavy metals from water traveling through wetlands. Metal accumulations and chemical processes of removal are dependent on soil conditions including pH, cation exchange capacity, redox potential and the presence of organic matter (Laing et al. 2008).

Point and non-point introductions of pollutants including nutrients and heavy metals from agricultural runoff, urban sewage, storm-water overflows, and industrial effluents remain a threat to fresh- and salt-water quality the world over (Fig. 11-5). Such nutrient overloads are often responsible for hypoxic dead zones at the mouths of major rivers, estuaries, and enclosed bays, leading to the widespread disruption of aquatic life. A recent study identified more than 200 such dead zones worldwide, with the largest covering 120,000 km^2 in the Baltic Sea (Schrope 2006). These areas cause significant economic disruptions to vital fisheries, as seen both in the case of the blue crab industry in the Chesapeake Bay of the eastern United States and shellfish beds and coastal fisheries at the mouth of the Mississippi River in the Gulf of Mexico. While wetlands alone cannot ameliorate this problem, they provide an important service in improving water quality through the processes outlined above.

11.3.3 Water storage and diversion

Wetlands have for millennia served as reservoirs of fresh surface and ground water to meet human and ecosystem demands. The storage capacities of wetlands depend on a host of factors including their size and their watersheds, which comprise the wetland basin and catchment areas. Storage capacity is also influenced by the height of the water table. Generally, the

Figure 11-4. Dense stand of *Avicennia marina* (gray mangrove) along the west coast of India. Notice debris and litter which gets entangled and stranded among the branches and roots during low tide. Photo by Firooza Pavri.

Figure 11-5. The highly polluted Mithi River estuary flowing through suburban Mumbai along coastal western India at low tide with a stand of *Avicennia marina* (gray mangrove), a highly salt-tolerant species which also withstands relatively high levels of pollution. Photo by Firooza Pavri.

higher the water table (during wet periods) the lower the storage capacity and vice versa (Carter 1997). Storage capacity also influences the abundance of flora and fauna available within a given wetland, serves to replenish streams and lakes during dry periods, and helps with groundwater recharge.

In rural parts of the developing world without piped water systems, communities are often dependent on wetlands for household and agricultural needs. For instance, beels are expansive low-lying seasonally flooded regions adjacent to major floodplains in eastern India and Bangladesh that provide water supplies and fishing opportunities to resident populations for many months of the year. When seasonal floodwaters recede, remnant deeper beels may still dot the landscape and provide water resources while their drier counterparts are used for agriculture. Similarly, hoars (another local wetland term), are found in northeastern parts of Bangladesh and are much deeper bowl-shaped depressional wetland features which may contain fresh water almost year-round. They can be quite large and during severe monsoonal flooding may inundate vast areas of adjacent land. In other instances tanks or reservoirs built into depressions serve as artificially created wetlands all across South Asia. They are used as sources of water by resident communities, serve bird and resident wildlife and cattle populations, and often support aquatic plant communities. Such natural and artificial wetland features provide vital freshwater resources to rural communities and their significance cannot be emphasized enough.

From the earliest hydraulic civilizations of Mesopotamia and Egypt, the development of water as a resource for agriculture, urbanization, and industrial use has led to significant alterations to the flow regimes of rivers. Through the construction of storage and hydroelectric dams, water-diversion schemes, river channelization for navigation and flood control, floodplain and deltaic wetlands have reduced dramatically in extent. This chapter has already considered some of the consequences of such wetland reductions. Perhaps one of the most striking examples of large-scale anthropogenic changes to river regimes and wetland deltas comes from the 1950s water diversion schemes built on the Amu Darya and Syr Darya rivers that feed into the Aral Sea (Fig. 11-6). The Aral Sea basin covers 2.2 million km^2 across central Asia. A landlocked sea in an arid region, it is fed principally by the Amu Darya and Syr Darya, which originate in the Pamir and Tien Shan mountains and meander across a distance of 2500 km before emptying into deltas along its southern (Amu Darya) and northeastern (Syr Darya) shores (World Bank Report 1998).

Today the Aral Sea is shared by the five Central Asian countries of Uzbekistan, Kazakhstan, Tajikistan, Turkmenistan, and Kyrgyzstan, which were all once part of the former Soviet Union. Agricultural policies of nineteenth-century Russia targeted the development of this region for cotton and rice cultivation. The Soviet period saw a further expansion of irrigation through large-scale water diversion schemes and a network of canals, which dramatically reduced water flows into the Aral, shrank its volume by more than 75 percent, and increased salinity levels from 10 g/L in 1960 to 100 g/L by 2004 (NATO 2005). Figure 11-6 illustrates this startling reduction in area with an outline of the 1960 shoreline of the sea.

The consequences of such radical reductions in water volume have altered the landscape of the region and the configuration of the sea itself. The desiccation of the Aral Sea has resulted in its breakup into several smaller water bodies. The salinization of rivers, ground water and agricultural land due to the mobilization of salts from the expansion of irrigation and inefficient water use has had serious economic consequences. The Amu Darya and Syr Darya deltas were once important migratory stopping points for birds and nourished vibrant fisheries. Today, only a small fraction of these ecosystems remain, yet their re-establishment through concerted international cooperation could yield benefits to the local and regional ecology and economy. Recent cooperative agreements between the central Asian countries concerning integrated water basin management and conservation efforts provide some hope.

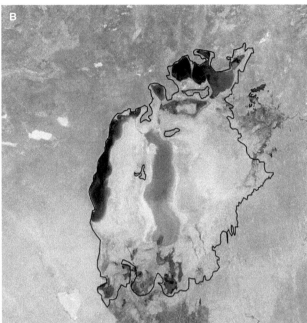

Figure 11-6. A. Map of the Aral Sea region and drainage in central Asia. Produced by A. Dailey. Map made with data from Natural Earth, Global Administrative Areas, and World Wildlife Fund Terrestrial Ecoregions Database (Olson et al. 2001). Accessed online <www.naturalearthdata.com> <http://www.gadm.org/> and <http://www.worldwildlife.org/science/data/item6373.html> February 2011. B. Dramatic Terra MODIS image from 26 August 2010 showing the desiccation of the Aral Sea due to water diversion schemes on the Amu Darya and Syr Darya (see Color Plate 11-6). The deltas of the Amu Darya (southern shore) and Syr Darya (northeastern shore) are visible on the image in green. The 1960s shoreline of the sea is marked. Image adapted from NASA's Earth Observatory. Accessed online <http://earthobservatory.nasa.gov/IOTD/view.php?id=46685> October 2010.

11.4 Economic services

Wetlands provide obvious economic benefits to society, and their contributions to sustained human settlement and expansion are well documented. Their economic contributions fall largely within the primary or extractive sector of the economy, but they also contribute to recreation, tourism, transportation, and the services industry in general. Increasing population pressures and the rising demand for food products, materials, and economic development have led to the over-exploitation of a number of wetland sites around the world. Such cases are only compounded by the politics and unequal power relations of competing interests encroaching upon these dwindling resources.

11.4.1 Extractive industries

Extractive activities and industries that rely on wetland resources include hunting and trapping, fiber and building materials, wetland agriculture, cattle ranching, fisheries, commercial aquaculture, pearl production, forestry, hydroelectric energy, peat mining, and salt production (Fig. 11-7), among others. Industry dominance within any given wetland site depends on a number of factors. These may include the following issues:

- a region's development status
- its history of resource use
- external market factors
- local demand for products
- policies governing resource use

Figure 11-7. Salt production pools at the southern end of the Dead Sea, Israel and Jordan. The evaporation ponds are separated by a central dike running more or less north–south along the international border. Potash, bromine, magnesium and other compounds are produced from the brine and salt. Image courtesy of K. Lulla, NASA Johnson Space Center. Photo ID: STS28-96-065, August 1989.

- surrounding land use and settlement patterns
- land ownership and tenure patterns.

Hunting, trapping, foraging, and small-scale fishing and agriculture are some of the oldest activities carried out by humans in wetland environments. Indigenous communities in many parts of the world still depend on these activities and maintain their traditional knowledge and technologies when extracting resources. For instance, the Hudson and James Bay lowlands of Canada include bogs, fens, and permafrost peatlands (Abraham and Keddy 2005). Here, indigenous communities of Cree, Dené and Métis dominate this sparsely populated region and still engage in fishing and hunter-gathering activities (Birkes et al. 1994).

Small-scale fishing is common all along India's coastal regions (Fig. 11-8). Often local fishing communities clear sections of coastal mangroves and construct low stone walls along shorelines which serve to trap fish during low-tide conditions. These simple fishing techniques require minimal capital inputs except for maintaining the stone walls. Individuals pick stranded fish off the sand beach once tides have receded. In most cases, such fishing techniques yield small subsistence catches, and often are just supplements to output from agricultural plots. In other cases, fishing using small boats may yield fairly large catches and serve local markets.

Small-scale rural farmers often use seasonally flooded wetlands as livestock grazing sites that provide large quantities of fodder during the dry season. Examples of these are observed in the elevated shallow dambos of Zimbabwe or the fadama of Nigeria (Turner 1984). Cattle ranching in the várzea adjoining major rivers of central Amazonia concentrates in the low-lying pastures during drier periods and shifts to higher-elevation cleared floodplain forests during wet periods (Junk and Piedade 2005). Hay production is another traditional use of wet meadows, salt marsh, and seasonally flooded sites (Fig. 11-9).

Figure 11-8. Small-scale fishing on the western coast of India. A. Cleared mangroves in the background make way for small-scale fishing areas. Stone walls, carefully constructed and maintained, serve to strand fish during low tide, which then can be gathered easily. B. Small fishing boats. C. Fisherfolk sort daily catches of tiny shrimp and *Sardinella melanura* (tarli). Photos by Firooza Pavri.

Figure 11-9. Traditional hay production on the Vistula River floodplain at Kaminoski National Park, near Warsaw, Poland. The floodplain is drained with canals to promote hay growth. Photo by J.S. Aber.

Combined rice and fish cultivation in flooded paddy fields of China and other parts of southeastern Asia is a tradition that dates back several centuries. The flood pulse system of northern Thailand's Songkhram River basin provides an example of the complex and beneficial interactions between the seasonally flooded forested wetlands providing important spawning and nursery habitat for fish adjacent to converted rice–fish cultivation fields (Wood and van Halsema 2008).

Mining peat for fuel and horticultural products has taken place since prehistoric times. Drainage is a first fundamental step to dry out the surface peat sufficiently so that it can be worked, and then peat may be extracted in three ways (Charman 2002):

- Cutting of blocks – Traditional hand cutting or mechanical cutting is typically done for local domestic production. The peat bricks or turves are laid out to dry for several weeks then transported home for fuel (Fig. 11-10A). This approach does limited damage to bogs because of its small scale. The top material is not good quality for fuel and is often thrown into previously cut bog, providing a source for vegetation regrowth to take place.
- Extrusion of sausages – Small tractors pull equipment that cuts into the peat and draws peat up to the surface where it is extruded in long round strips or sausages. Although the surface disturbance appears minimal, the bog biomass and height are reduced, and invertebrate populations are impacted. A relatively large bog area is necessary to produce peat in this manner.
- Peat milling – Commercial peat extraction involves heavy equipment and is practiced

Figure 11-10. Peat production. A. Pile of peat turves for domestic use in Ness, Outer Hebrides, Scotland. Modified from original photograph by Maclomhair; obtained from Wikimedia Commons <http://commons.wikimedia.org/>. B. Pile of milled peat ready for shipment off site at Rannu Soo, west of Tartu, Estonia. Photo by J.S. Aber.

on a large scale worldwide. The bog surface is milled with rotating drums fixed with pins that break up the peat. After drying, the milled peat is picked up with a large vacuum and placed in piles for further handling (Fig. 11-10B). The process may be repeated several times each summer, and each year drainage must be lowered to continue harvesting peat. In the end the peatland is almost completely destroyed, although in recent years attempts have been made to restore cutover peatlands (e.g. Quinty and Rochefort 2003).

Examples of large-scale commercial extractive activities using wetland ecosystems are often scrutinized closely due to their more visible impacts. Global market forces and local government policies promoting export-led growth and responding to local development pressures often drive such undertakings. The recently established oil-palm industry of Kalimantan, Indonesia provides one such example (Wood and van Halsema 2008). Beginning in the mid-1990s, in response to the global demand for biofuels, the peat-swamp forests of central Kalimantan were cleared and burned to make way for oil-palm plantations. With international investment, oil-palm production rose from 5 million tons in 1995 to 18 million tons by 2008, generating important export revenue for Indonesia (U.S. Department of Agriculture 2007, 2010c). This conversion of wide swaths of Indonesia's peatlands to oil-palm resulted in significant releases of CO_2 (Wood and van Halsema 2008). Ironically, the call for greener biofuels has inadvertently led to the destruction of peatlands and subsequent release of greenhouse gases.

Another example of commercial extraction comes from aquaculture production or the so-called "blue revolution" (Simpson 2011). Aquaculture refers to the farming of aquatic organisms such as fish, mollusks, crustaceans and aquatic plants from fresh, brackish, and marine waters. Aquaculture is undertaken on a large scale in many countries of the world and often involves the conversion of coastal and inland wetlands to fish farms (Alongi 2002; Seto and Fragkias 2007). A recent Food and Agriculture Organization report (FAO 2009) suggested that close to 44 percent of the seafood (including fish, crustaceans and mollusks) consumed globally now comes from aquaculture production. Aquaculture production has increased at a brisk 6.5 percent annually from approximately 28 million tons in 1999 to 50 million tons in 2007 (Table 11-2).

Asia leads the world in aquaculture production – supplying a staggering 88 percent of global demand. As Table 11-3 indicates, in 2007 China alone contributed over 60 percent of the world's aquaculture at 31 million tons, with India (3.4 million tons) and Vietnam (2.2 million tons) trailing a distant second and third. Several factors explain China's large market share in aquaculture. The country has a long history of aquaculture production dating back several millennia. Since the 1950s the Chinese government has provided many incentives for aquaculture production and invested large research funding into developing more efficient techniques (Guo 2000). These factors, coupled with the widespread traditional practice of

Table 11-2. Aquaculture production for fish, crustaceans and mollusks by region. Source: Food and Agriculture Organization. 2009. *Yearbook of Fishery Statistics 2007*. Accessed online <ftp://ftp.fao.org/docrep/fao/012/i1013t/i1013t.pdf> October 2010.

Region	1998 (tons)	1998 (%)	2002 (tons)	2002 (%)	2007 (tons)	2007 (%)
World	28,412,656	100	36,781,779	100	50,329,007	100
Africa	186,362	0.7	453,638	1.2	824,762	1.6
Americas	1,261,893	4.4	1,799,643	4.9	2,430,546	4.8
Asia	24,922,366	87.7	32,358,284	88.0	44,565,579	88.6
Europe	1,921,404	6.8	2,042,398	5.6	2,339,515	4.7
Oceania	120,631	0.4	127,816	0.4	168,605	0.3

Table 11-3. Aquaculture production for fish, crustaceans and mollusks by top 5 producing countries. Source: Food and Agriculture Organization. 2009. *Yearbook of Fishery Statistics 2007*. Accessed online <ftp://ftp.fao.org/docrep/fao/012/i1013t/i1013t.pdf> October 2010.

Country	1998 (tons)	1998 (%)	2002 (tons)	2002 (%)	2007 (tons)	2007 (%)
China	18,721,938	66.0	24,141,658	65.6	31,420,275	62.4
India	1,908,485	6.7	2,187,189	6.0	3,354,754	6.7
Vietnam	338,920	1.2	703,041	1.9	2,156,500	4.3
Indonesia	629,797	2.2	914,071	2.5	1,392,904	2.8
Thailand	594,579	2.1	954,696	2.6	1,390,031	2.8
Total	22,193,719	78.1	28,900,655	78.6	39,714,464	78.9

pond fish culture along river floodplains and in flooded rice paddies, have contributed to China's leading position in aquaculture production.

With increasing demand, aquaculture is poised to grow even more rapidly in the coming years. Shrimp cultivation is the largest by value, accounting for roughly 15 percent of the traded volume of all seafood. It is often undertaken in coastal brackish wetlands or constructed ponds pumped with sea water. For many countries, the export value of seafood now far outstrips traditional commodities like rice, tea, etc. In the United States alone, aquaculture production was valued at US$87 billion in 2007 (Food and Agriculture Organization 2009). In comparison, the total retail equivalent value of the United States beef industry was US$76 billion for 2007 (U.S. Department of Agriculture 2010d).

Given these trends, scientists have voiced concern over the ramifications of this growth industry on wetland ecosystems worldwide. Studies have documented the potential negative effects of large commercial-scale aquaculture undertakings on ecosystems including wetlands (Alongi 2002; Seto and Fragkias 2007). These include the:

- Clearing of coastal mangroves to make way for shrimp farms.
- Pollution from waste and other chemicals and pharmaceuticals used in intensive aquaculture, which may lead to eutrophication and algal blooms.
- Upsurge of fish diseases and parasites such as sea lice that thrive under contained conditions often found at large aquaculture sites or in fish pens and cages.
- Dilution of wild fish stock with the accidental introduction of genetically modified farmed species.

As this section suggests, primary sector extractive industries may have potentially quite damaging consequences for wetland ecosystems. While it would be impractical to prevent the use of these resources, the focus should be on more sustainable extraction and wise-use practices, which has been the call of organizations like Ramsar.

11.4.2 Pearl production

Pearls are beautiful gemstones created by mollusks located in fresh- and salt-water wetlands. Chapman (2009) recognized 85,000 extant species of mollusks. However, only 20 mollusk species are pearl-bearing, and of those six are responsible for the overwhelming majority of today's cultured-pearl production (Table 11-4). The majority of marine pearls come from the akoya pearl (*Pinctada fucata*), golden- and silver-lipped pearl (*P. maxima*), and black-lipped pearl (*P. margaritifera*) from Japan and tropical South Sea regions; whereas the majority of fresh-water pearls come from the triangle shell (*Hyriopsis cumingii*) and wrinkle shell (*Cristaria plicata*) in Asia; the washboard shell (*Megalonaias nervosa*) is from North America.

Natural pearls are rare, and gem quality ones especially rare. The Abernethy Pearl, for instance, is a large gem-quality fresh-water pearl from *Margaritifera margaritifera*, discovered in 1967 in the River Tay in Scotland. One of the most threatened fresh-water species worldwide, *M.*

Table 11-4. Main pearl-producing mollusks, water chemistry, geographic locations, typical pearl sizes and colors. Adapted from Gemological Institute of America (2000), Matlins (2006) and Wise (2006).

Water	Type	Location	Size	Color
Salt	*Pinctada fucata* "akoya" pearl oyster	Japan, China, Australia, South Korea, New Guinea	6–7 mm	Gray, white, yellow
	Pinctada margaritifera "black-lipped" pearl oyster	Cook Islands, French Polynesia south of equator, Hawaii	8–14 mm	Steel gray to black
	Pinctada maxima "silver- and golden-lipped" pearl oyster	Indonesia, Malaysia, Vietnam, Philippines, Thailand, Australia	8–15 mm	Cream, golden, silver
Fresh	*Cristaria plicata* "wrinkle" or "river" shell	China, Vietnam, Japan, Korea	Most 3–5 mm Up to 14 mm	Bronze, peach, plum, champagne
	Hyriopsis cumingi "Unio" or "triangle" shell	China		
	Megalonaias nervosa "washboard" shell	North America		White, cream

margaritifera is commonly known as the pearl mussel; it has a circumpolar distribution with a lifespan that commonly exceeds a century (Ziuganov et al. 2000; Helama and Valovirta 2008). In addition to modern pearls, ancient pearls exist as a testament to wetland environments in the past; the oldest known fossil pearls date from the Triassic, some 230–210 million years old (American Museum of Natural History 2002).

Pearls are secretions that form within mollusks in response to an irritant that enters the shell and lodges in the mantle tissue. The mineral composition of pearl and mother of pearl is aragonite ($CaCO_3$). When pearls form within the mollusk, the surface skin is coated in nacre, which consists of layers of aragonite crystals joined by conchiolin, a binding protein (Wise 2006). Nacre is added in concentric layers that are microns-thin and resemble an onion viewed in cross section. In the pearl-producing mollusks, undulations in nacre layers produce tiny grooves that act as a diffraction gradient to break up light into component colors, which create a rainbow or iridescence and a warm glow that seems to radiate from inside (Fig. 11-11).

Early in the twentieth century, demand for pearls exceeded the supply of natural pearls, which had become scarce as a result of overfishing mollusks. At this time, a British biologist working in Australia, William Saville-Kent, is credited with creating the process used to culture pearls (Matlins 2006). Two periculturists, Tatsuhei Mise and Tokichi Nishikawa, applied this technique and received a patent in 1907 for the process of producing cultured akoya pearls in Japan (Ward and Ward 1998). K. Mikimoto, an entrepreneur, adapted the scientific process, using the Mise–Nishikawa method for commercial production and made his name synonymous with cultured round pearls sold to global consumers. Japan dominated the pearl industry with a half-century head start, but more recently South Sea pearls are becoming a popular jewelry choice, and 2011 was the fiftieth anniversary of the French Polynesia-Tahitian cultured black-pearl industry (*Jewellery News Asia* 2011a).

Pearl production is a simple form of aquaculture with positive impacts on wetlands and the economy. The production of South Sea pearls serves to demonstrate the main procedures as described by J. Branellec (2011, pers. com.), a pearl farmer whose family helped initiate cultured golden-pearl production in the Philippines. Culturing pearl starts with the capture or harvest of wild *Pinctada maxima* mollusks, which are needed to create a breeding stock. This initial harvest was carried out by the Badjao, an indigenous tribe from Malaysia but mostly living a nomadic life on boats. After mollusks reach a certain age of maturity, they are separated into pearl hosts and donors. Donors are chosen based on depth of gold, silver, or black color in the shell; they provide the mantle tissue that is grafted behind the bead nucleus

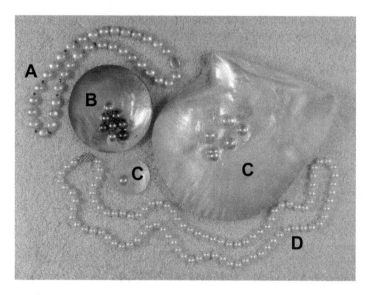

Figure 11-11. Selection of fresh- and salt-water pearls and mother-of-pearl shells (see Color Plate 11-11). A. Chinese, multi-colored fresh-water, probably *Hyriopsis cumingi*. B. Black-lipped pearl oyster, *Pinctada margaritifera*. Diameter of shell = 85 mm. C. Golden-lipped pearl oyster, *Pinctada maxima*. D. Japanese akoya pearls, *Pinctada fucata*. White with rose overtone. Photo by S.W. Aber and J.S. Aber.

implant in the host. The bead is mother-of-pearl, usually from a fresh-water mollusk, such as the Mississippi River pigtoe clam (*Pleurobema beadleianum*), which is commonly cultured on North American pearl farms (Wise 2006).

The mantle tissue and bead stimulate the host oyster to secrete layers of nacre around the implanted irritant; the final thickness is determined by length of time in the water before harvest. Culturing may also be accomplished without a bead, using a tissue implant only; the tissue alone creates the *activation*, not a nucleation process, to stimulate the host oyster to secrete nacre, and the resulting pearl is all nacre (E. Strake, pers. com. 2011). For salt-water pearls, only one nucleus is implanted; however, cultured fresh-water pearls most often use implantations that are nucleated with up to 50 pieces of tissue (Wise 2006). The size of implant bead in pearl oysters depends on the size of mollusk; South Sea pearls are among the largest in size as compared to akoya and fresh-water pearls.

Pearl farming also takes place in fresh-water environments such as ponds, lakes, and rivers. China harvested fresh-water pearls in the thirteenth century and today is a major supplier of all sizes, shapes, qualities, and quantities of fresh-water pearls in various natural colors (Pearl-Guide 2011). The commercial fresh-water cultured-pearl industry in the United States was started by John and Chessy Latendresse, who pioneered American pearl production in the 1950s, created the "all-American pearl," and made the American Pearl Company a tourist destination in Tennessee (Burch 1995). Each pearl takes some 3–5 years to grow; in contrast, many Asian fresh-water pearls are completed in only one year. The Latendresses showed that the Japanese akoya would accept North American mussel shell nuclei, and exporting farm-grown mussel shells overseas lessened the pressure on natural mussel populations.

Cultured-pearl farming is a business that may be in balance with wetland ecosystems. It is an operation that neither pollutes nor drains wetlands because the essence of pearl farming is in generating and sustaining life. Although wild mollusks are captured, bred, and grown in captivity, they are not used as a food source, but rather as pearl hosts and donors. Cultured-pearl farming provides a sustainable alternative to overharvesting wild mollusks for natural pearls or shells. Each year more mollusks are released back into the ocean by pearl farmers than were

Figure 11-12. Lock and dam 15 on the Mississippi River at Rock Island, Illinois. A. String of grain barges, 3 wide and 4 long, entering lock 15 from upstream end. B. Towboat *Crimson Glory* pushes barges out of lock 15 (downstream). The barge immediately in front of the towboat carries diesel fuel. Bridge has rotated sideways to allow passage of the towboat. Photos by J.S. Aber.

originally harvested, thereby increasing the wild population. In fact, Branellec (2011, pers. com.) described pearl production as a non-extractive industry and eco-friendly business (*Jewellery News Asia* 2011b).

Salt-water pearl farmers, thus, have a keen interest in protecting waters and life in coastal bay, lagoon, atoll, and reef areas especially against the greatest threats to pearl farming and marine biodiversity, namely illegal fishing and poaching through the use of explosives and poisons (DeVantier, Alcala and Wilkinson 2004). Philippine government regulation, for example, protects the interests of pearl-farm cultivation areas by creating "no-take" zones, where fish and other marine life extraction is legally banned. South Sea pearl farms are located in the Coral Triangle, which includes the Philippines, Indonesia, Malaysia, and other countries in the Sulu-Sulawesi, South China, Celebes, and Philippine seas. The Coral Triangle contains half of the worlds' coral reefs and is heralded for its rich marine biodiversity. The World Wildlife Federation (WWF 2011) reported that all nature-based tourism in the Coral Triangle earned over US$12 billion. Black-pearl production in Mexico from the Sea of Cortez is another popular venue for ecotourism (Sea Cortez Pearl Blog 2011). Global pearl production statistics are closely guarded, and numbers differ among pearl associations; nonetheless, pearl farming and ecotourism are co-existing multi-billion dollar industries that depend upon clean and viable wetland habitats.

11.4.3 Services industries

Water has been used as a means of transportation since the earliest civilizations. Humans settled close to water bodies for food, water and fertile soils, as well as the ability to transport themselves and their goods easily (see Fig. 1-6). In some wetlands, water levels are actively managed and rivers are dredged and channelized to facilitate water transportation throughout the year. The Mississippi River is a good example of a river serving as a lifeline for commerce and a vital economic link. Despite the popularity of alternative means of transport, including rail and road, decades of channelization, dredging, and the construction of locks and dams have allowed the Mississippi to continue functioning as an important means of transportation connecting the coast to the deep interior of the continent (Fig. 11-12). Typically, petroleum products such as gasoline and fuel oil are shipped upstream from Louisiana, while agricultural products from the Midwest and Great Plains, including wheat and corn destined for export markets, are shipped downstream.

Similar to the Mississippi corridor, the ambitious Hidrovia transportation project on the South American Paraná–Paraguay river system connecting Buenos Aires in Argentina to Cáceres in Brazil was proposed to expand the upstream navigation of large vessels (Alho 2005). This project would have had a significant environmental impact on the Pantanal wetlands. After much discussion and opposition from environmental and other groups, the project has been set aside for now due to the many ramifications on the flow regime of the river and its adjoining wetlands (Alho 2005).

Canals have been constructed since ancient times for irrigation and navigation. Canals and navigable rivers are subject to the same laws of nature that govern uncontrolled streams. Water flow, sediment erosion, flooding, transport of pollutants, and related issues are of great importance for long-term maintenance and operation of canals. Canals also have special circumstances, as they often require dedicated water sources, and they may connect drainage basins of dissimilar type. Migration of exotic aquatic species, such as the zebra mussel (see chapter 7.4) has occurred in some canals.

The old Erie Canal was the first major waterworks project in the United States, built in 1817–1825. The canal connected Lake Erie to the New York harbor tidewater in a multi-level route that followed the local terrain and was fed by local water sources. Water flow in the canal was required for several purposes (Langbein 1976):

- Filling the canal at the beginning of each spring season.
- Water for lockage to replace water lost from higher to lower levels.
- Water lost by seepage through the berm and towpath banks.
- Water diverted for industrial power usage.

The canal was initially a great economic success, which led to a boom in canal building. The Erie Canal was enlarged twice in 1836–1862 and 1905–1917. It is now known as the New York State Barge Canal, and is used today mainly for recreation (Fig. 11-13). In general, canals in the

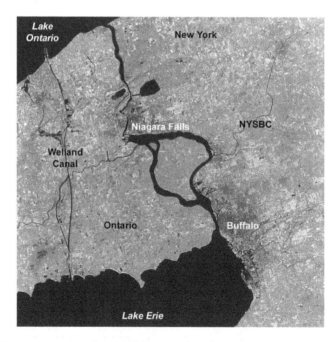

Figure 11-13. Canals in western New York, United States and southern Ontario, Canada (see Color Plate 11-13). The Welland Canal is nowadays the primary shipping connection between lakes Ontario and Erie. The New York State Barge Canal (NYSBC) is utilized mainly for recreational purposes. Landsat false-color composite; TM bands 2, 3 and 4 color coded as blue, green and red. Active vegetation appears red and pink. Acquired June 1983; adapted from NASA Goddard Space Flight Center.

Figure 11-14. Imperial Valley, southern California and northern Mexico (see Color Plate 11-14). The All American Canal (AAC) enters right side of scene and flows westward just north of the U.S.–Mexican border. This is the last tap from the Colorado River on the U.S. side of the border. Bright red irrigated cropland is evident in the Imperial Valley of the United States, but crops have less irrigation water in Mexico. Landsat false-color composite; TM bands 2, 3 and 4 color coded as blue, green and red. Active vegetation appears red and pink. Acquired April 1983; adapted from NASA Goddard Space Flight Center.

eastern United States were built for transportation purposes, whereas canals in the western part of the country are primarily for irrigation (Fig. 11-14).

A similar canal-building boom took place in Europe during the eighteenth and nineteenth centuries. The result is an interconnected system of waterways that allows shipping and pleasure boating throughout most of western and central Europe. Locks are the conventional means to change water levels, but many other ingenious techniques have been used to lift or lower boats from one level to another (Figs. 11-15 and 11-16).

Recreational uses of wetlands provide significant economic development opportunities and incomes to local regions. Involving local and indigenous populations in tourism may provide dual benefits in terms of generating local employment opportunities and promoting conservation goals (Scheyvens 1999). This also serves to alleviate often fraught relationships between marginalized indigenous populations and top-down government conservation and management efforts. Studies have documented successes of participatory management in ecotourism efforts (Cater 1994; Drumm and Moore 2005; Boojh, Patry and Smart 2008). However, as Scheyvens (1999) cautioned, not all ecotourism efforts may end up empowering local people or recycling revenue through local economies. Even so, economic multiplier effects suggest that wetland sites support additional service sector activities including hotels, inns, guides, restaurants, and recreational venues beyond income generated through fees and permits to these sites.

Figure 11-15. Elbląg canal, northern Poland. A. Tour boat is riding on a trolley carriage that travels up a ramp from the lower canal to a higher level. The trolley is pulled by a cable that runs over the large wheel in the background. B. Power to move the cable and trolley is provided by a water wheel. Photos by J.S. Aber.

Figure 11-16. Falkirk Wheel in Scotland, United Kingdom. The rotating wheel connects the Forth and Clyde Canal (below) with the Union Canal (above), a 24-m (~80-foot) vertical difference at the wheel. These canals had been linked by a series of locks that had fallen into disrepair and been abandoned. The wheel was opened in 2002. Photo courtesy of S. Jones.

Figure 11-17. Tour group travels in small boats to view plants and animals on the Amazon River near Manaus, Brazil. Photo courtesy of K. Buchele.

The popularity of wildlife and ecotourism in recent decades has meant that wetlands have become particularly desirable sites for visitors interested in birding, wildlife, and other recreational activities. The recreation industry and tourism have always been significant economic earners for the grand wetland sites of the world including the Everglades, Pantanal, Sundarbans, Amazonia (Fig. 11-17), and Okavango (Fig. 11-18), among others. These areas boast a great variety and abundance of flora and fauna and also market key megafauna such as the Bengal tiger (*Panthera tigris tigris*) of the Sundarbans, the white rhino (*Ceratotherium simum simum*) in the Okavango, and the more elusive jaguar (*Panthera onca*) of the Pantanal.

Smaller and perhaps lesser known wetland sites also market themselves as niche areas for particular activities such as birding and wildlife watching. For instance, the comparatively small 2800-hectare Keoladeo National Park in western India is a famous World Heritage site that draws large numbers of international tourists during the winter bird migration season by marketing itself as the home of the majestic Sarus crane

Plate 1. Aerial overview of the Rachel Carson National Wildlife Refuge along the Atlantic coast of southeastern Maine, United States. The salt marsh, pools and tidal channels intervene between the beach front (right) and mainland (left), both of which have dense residential development. The human presence here has strong influence on the wetland water supply, vegetation and wildlife. View toward north; blimp airphoto by J.S. Aber, S.W. Aber and V. Valentine.

Plate 1-9. Presque Isle is a sandy spit that extends from the mainland into Lake Erie in northwestern Pennsylvania, United States. The transition from sandy shore, to shallow water, to deep lake is depicted in this panoramic view looking toward the northeast. Kite aerial photo by J.S. Aber and S.W. Aber.

Wetland Environments: A Global Perspective, First Edition. James Sandusky Aber, Firooza Pavri, and Susan Ward Aber.
© 2012 James Sandusky Aber, Firooza Pavri, and Susan Ward Aber. Published 2012 by Blackwell Publishing Ltd.

Plate 2-2A. Water is the primary ingredient for wetlands. Nigula Bog, southwestern Estonia. Water fills numerous shallow pools of irregular size and shape. *Sphagnum* moss (reddish brown) surrounds each pool, and in between the pools, low hummocks are covered with heather and dwarf pines. A wooden walkway about half a meter wide is laid directly on the bog surface and runs across the bottom and right sides of the scene. Kite aerial photo (Aber et al. 2002).

Plate 2-10. Sangre de Cristo Formation exposed near Cuchara in south-central Colorado, United States. Several thousand meters of red sandstone, shale and conglomerate accumulated as alluvial fans during the Permian when the Ancestral Rocky Mountains were uplifted. The red color indicates oxidizing conditions in the depositional environment. These strata were tilted upward later, when the modern Rocky Mountains were deformed in the Eocene. Photo by J.S. Aber.

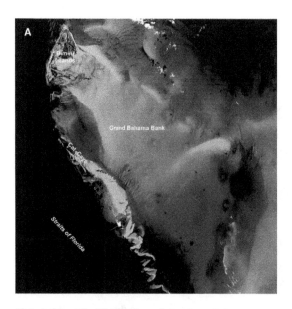

Plate 2-5A. Bimini Islands and Cat Cays, Bahamas. Shoals of carbonate sediment show distinctly as pale blue through the shallow, clear water of Grand Bahama Bank. False-color Landsat image in which vegetation on islands appears in red. Field of view ~75 km across; image courtesy of NASA Goddard Space Flight Center.

Plate 2-12. Watercress (*Rorippa nasturium-aquaticum*), an emergent wetland plant. A member of the mustard family (Cruciferae), it is a succulent, long-stemmed plant growing in tangled masses or low mounds up to ~30 cm (1 foot). The leaves have a strong peppery taste; watercress is highly valued for food flavoring and medicinal uses (Tilford 1997). Watercress absolutely requires clear, flowing water with temperatures <18 °C (65 °F), which means it favors spring-fed streams. Flint Hills of east-central Kansas, United States; photo by J.S. Aber.

Plate 2-16. Salt marsh and swamp at the Wells National Estuarine Research Reserve. Panorama looking toward the southwest along the Atlantic coast of southeastern Maine, United States. A bridge on Drakes Island Road (*) is part of a water-control structure that limits tidal flow between the marine lagoon in the background and marsh in the foreground. Notice the distinct vegetation zones, which reflect variations in water depth and salinity. Blimp airphoto by J.S. Aber, S.W. Aber and V. Valentine.

Plate 2-25B. Close-up view of tree-covered mineral island developed on a drumlin. Note distinct vegetation zones: (a) *Sphagnum* moss, (b) pine, (c) birch (partly bare), and (d) ash, elm, maple and other deciduous hardwoods, some of which display fall colors. Kite aerial photo; Aber et al. (2002).

Plate 3-4. The man-made pond in the foreground trapped recent runoff and has a high content of yellowish-brown suspended sediment. The next pond downstream did not receive this sediment and has a clean, dark blue appearance. Central Kansas, United States. Kite aerial photo by J.S. Aber and S.W. Aber.

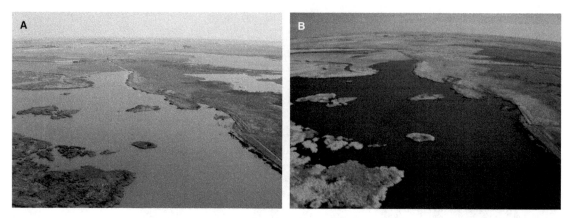

Plate 3-8. Color-visible (A) and color-infrared (B) digital images of marsh at the Nature Conservancy, Cheyenne Bottoms, central Kansas, United States. Active vegetation appears in bright red-pink colors in the latter. Kite aerial photographs from Aber et al. (2009, Fig. 5).

Plate 3-9. Close-up view of marsh and shore at Luck Lake, Saskatchewan, Canada. Distinctive vegetation zones are revealed by differences in plant texture, pattern, and color. The maroon plant is red samphire (*Salacornia rubra*), which grows on saline mudflats. Kite aerial photo by J.S. Aber and S.W. Aber.

Plate 3-10. Vertical kite aerial photograph in visible light showing water pools and vegetated hummocks in the central portion of Männikjärve Bog, Estonia. *Sphagnum* moss species display distinctive green, gold, and red autumn colors along with pale green dwarf pine trees on hummocks. These dramatic colors are not displayed so clearly at other times of the year. Field of view ~60 m across. Based on Aber et al. (2002, Fig. 2).

Plate 3-11. Red Hills in Barber County, southern Kansas, United States. Alluvial sediment accumulated in an ancient desert floodplain environment and is now eroding in a badlands topography. Bright red color indicates oxidizing conditions of deposition. Kite airphoto by S.W. Aber.

Plate 3-12. Exposed mudflat on the shore of Luck Lake, Saskatchewan, Canada. Color analysis of this photograph gives light gray to pale blue-green colors for the mud, indicating reducing conditions in the sediment. Kite aerial photo by J.S. Aber and S.W. Aber.

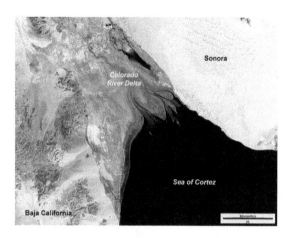

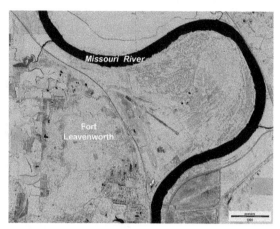

Plate 3-14. False-color composite Landsat TM image of the Colorado River delta in the Sea of Cortez, Mexico. Bands 3 (red), 4 (near-infrared) and 5 (mid-infrared) color coded as blue, green and red. Irrigated crops are bright yellow-green; dry mud/salt flats are bright cyan; sand dunes are near white. Landsat 5, March 2004. Image from NASA; processing by J.S. Aber.

Plate 3-17. Ikonos false-color composite of Fort Leavenworth, northeastern Kansas, United States. Green, near-infrared and red bands color coded as blue, green and red; active vegetation appears dark green to yellow-green colors. Dataset acquired August 2000; compare with Figure 3-16. Image from NASA; processing by J.S. Aber.

Plate 3-23. Playa basins on the nearly flat, featureless High Plains in west-central Kansas, United States. Ephemeral lake in a playa depression during a wet period. Water fills a shallow basin in a fallow field with green winter wheat fields in the background. Kite aerial photograph; after Aber and Aber (2009, Fig. 17).

Plate 3-28. Schematic illustration of the Munsell Color system. Hue is the spectral color (circumference); chroma is the intensity of color (radius); value is the brightness (vertical axis). Modified from original illustration by SharkD; obtained from Wikimedia Commons <http://commons.wikimedia.org/>.

Plate 3-29. Distinctive colors are displayed in this wetland soil. Dark gray/brown indicates a high content of organic matter, and orange mottles show oxidized iron. Taken from Vasilas, Hurt and Noble (2010, Fig. 7).

Plate 4-3. Overview of Dry Lake, an ephemeral lake at the terminal point of an enclosed basin on the High Plains in west-central Kansas, United States. A. May 2007, a wet year with the lake full of water; note small overturned row boat in lower left corner for scale. B. May 2008 displays a wet mudflat surrounded by salty soil. Similar views toward the southwest; kite airphotos by S.W. Aber and J.S. Aber.

Plate 4-6. Marsh-pond complex in the Sand Hills near Lakeside in western Nebraska, United States. Bright maroon and golden-orange colors in this pond are presumably caused by carotenoid pigments of invertebrates, such as brine shrimp, brine flies and rotifers, in the hyperalkaline water typical of the western Sand Hills region (Bleed and Ginsberg 1990). Kite airphoto by S.W. Aber and J.S. Aber.

Plate 4-12A. Finney Wildlife Area in southwestern Kansas, United States. Originally designed for a much larger lake, upstream ground-water pumping has reduced spring flows, and diversion of surface water has rendered this artificial "lake" into a wetland wildlife area. During rare wet years, a tiny puddle of water is held behind this dam. Asterisk indicates position of the outlet tower. Kite airphoto by J.S. Aber.

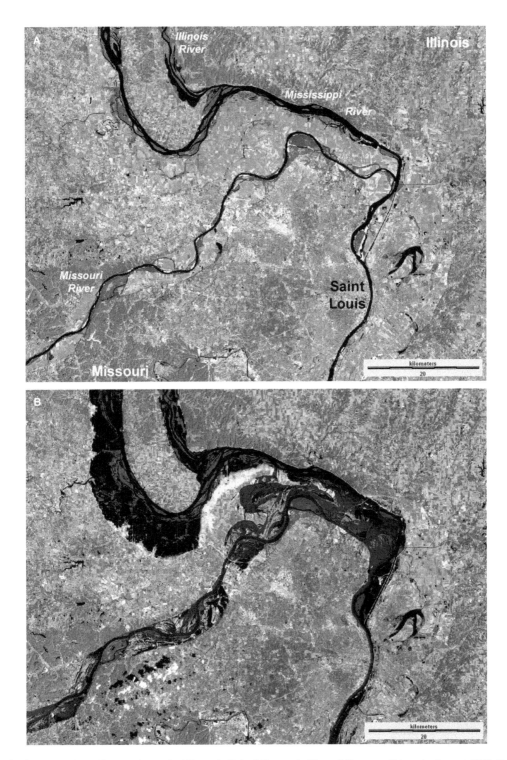

Plate 4-16. Mississippi River at St. Louis, Missouri, United States. A. Normal flow conditions in August 1991. B. Waning flood conditions in August 1993. Landsat TM bands 2, 5 and 7 color coded as blue, green and red; active vegetation appears in green colors. Image from NASA; processing by J.S. Aber.

Plate 4-23. Iron-cemented rødsten (redstone) formed where ground water emerged from a gravel bed exposed in the cliff. Photo by J.S. Aber; Ristinge Klint, Denmark.

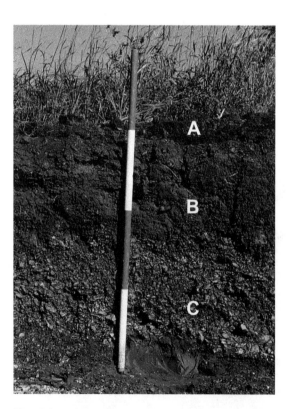

Plate 5-11. Hydric soil profile from a playa mud flat at Dry Lake, west-central Kansas, United States. A. Freshly dug pit, ~30 cm deep, showing distinct horizons from the surface downward. Note the redoximorphic iron accumulation (*) and dark gray color at depth. B. Clump of greenish-gray (5 GY 6/1), clayey silt with ferric iron accumulation (*). Photos courtesy of B. Zabriskie.

Plate 5-3. Olpe Soil is a compound upland soil. The lower portion (C) is a paleosol developed in highly weathered (leached and oxidized), older (Neogene), alluvial gravel. The upper part (A & B) is composed of younger (Holocene) loess. Scale pole marked in feet (30-cm intervals). Flint Hills, Kansas, United States. Photo by J.S. Aber.

Plate 5-12. Close-up view of *Sphagnum warnstorfia* moss. This moss forms red tussocks at the edge of oligotrophic bogs. Karelia, Russia; photo courtesy of E. Volkova.

Plate 6-11. Close-up, vertical view of *Azolla cristata* on marsh margin. Green-colored algae is below water level; maroon-colored *Azolla* is above water. *Azolla* spreads rapidly by vegetative reproduction and may form extensive mats (CAIP 2008). Field of view ~25 cm across; taken from Aber et al. (2010, Fig. 3).

Plate 6-8. Red samphire, also known as saltwort (*Salicornia rubra*). Succulent stems turn red in summer and may form bright maroon carpets on saline mud flats (see Color Plate 3-9). Cattle relish the high salt content; as human food, it may be eaten in salads and stews or used for pickling (Lahring 2003). Comb is ~5 inches (12.5 cm) long; photo by J.S. Aber at Frederick Lake, Saskatchewan, Canada.

Plate 6-14. Red admiral butterfly (*Vanessa atalanta*) feeds on the dazzling white flower cluster of the buttonbush (*Cephalanthus occidentalis*). Photo by J.S. Aber and S.W. Aber, Lake Kahola, Kansas, United States.

Plate 6-21. Pickerel weed (*Pontederia cordata*). The spade-shaped leaves and delicate violet flowers are quite attractive, and pickerel weed is a popular ornamental plant. It may form dense cover in shallow marsh, pond and lake environments. Photo by J.S. Aber; Conneaut Marsh, northwestern Pennsylvania, United States.

Plate 6-29. Old Wives Lake, Saskatchewan, Canada. Notice distinct and continuous vegetation zones marking high shorelines along the margin of the lake. Kite aerial photograph by J.S. Aber and S.W. Aber.

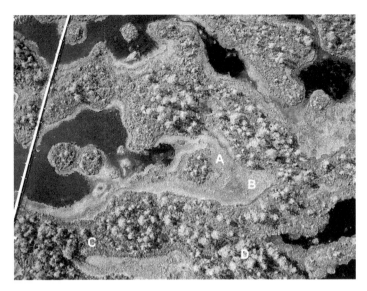

Plate 6-30. Männikjärve Bog, east-central Estonia. Autumn vegetation zones follow irregular pools, hollows and hummocks in the raised bog. A. *Sphagnum cuspidatum* floating in water. B. *S. cuspidatum* around pool shore. C. *S. rubellum* above water. D. Pine trees on hummocks along with dwarf shrubs. Elevated boardwalk is ~2 feet (60 cm) wide. Photo by J.S. Aber.

Plate 6-34. Purple loosestrife (*Lythrum salicaria*). Close-up view of flower stalk. Photo by J.S. Aber; Conneaut Marsh, northwestern Pennsylvania, United States.

Plate 6-37. Controlled prairie burning at a small spring in the Flint Hills, Kansas, United States. Such burning is typically done in latest winter or earliest spring to remove dead thatch of previous year. Nutrients in the thatch are returned to the soil to enrich spring growth. Photo by J.S. Aber.

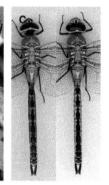

Plate 7-5. Damselfly and dragonfly comparison. A. Eastern forktail female damselfly (*Ischnura verticalis*) resting with wings folded together over her back. B. Common green darner dragonfly (*Anax junius*) sitting on foliage with its wings extended to the sides. Wing span ~10 cm. C. Temperature-dependent color change of the common green darner: left at 7 °C, right at 24 °C. Photos A and C courtesy of R. Beckemeyer; photo B by J.S. Aber.

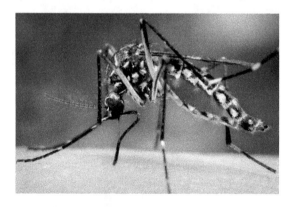

Plate 7-6. Close-up view of a female *Aedes aegypti* mosquito taking a blood meal from a human host. This is the most common species of *Aedes* and a major vector for West Nile Virus as well as Dengue and Dengue hemorrhagic fever. Image by J. Ganthany (2006), courtesy of U.S. Centers for Disease Control and Prevention (CDC).

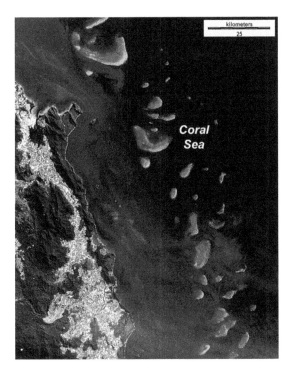

Plate 7-7. Portion of Great Barrier Reef and Coral Sea near Cairns off the coast of northeastern Australia. Altogether, this reef covers approximately 30,000 km² (UNEP 2004). Landsat TM satellite image composite derived from TM bands 1, 3 and 5; TM dataset acquired 16 July 2009. Image from NASA; processing by J.S. Aber.

Plate 7-9. Longear sunfish (*Lepomis megalotis*) displays brilliant orange and turquoise colors. This male is quite possibly guarding a nest. Native to eastern North America, the longear sunfish prefers shallow, clear water with slow to moderate current in streams with clean gravel beds (Ohio Department of Natural Resources 2010). Photo by J.S. Aber and S.W. Aber; Kahola Creek, eastern Kansas, United States.

Plate 7-8. Florida Keys region south of the Everglades on mainland Florida, United States. The modern coral reef forms a shallow, well-defined ridge (light blue color) along the edge of the continental shelf east and south of the keys. Landsat TM satellite image composite made from TM bands 1, 3 and 7 color coded as blue, green and red; TM dataset acquired 23 April 2008; processing by J.S. Aber.

Plate 7-12. Adult American bullfrog (*Rana catesbeiana*) sits in a patch of mosquito fern (*Azolla* sp.) in a prairie marsh. This frog is a game species; the largest native frog in North America, its body may reach 20 cm in length and weigh nearly 1 kg (Elliot, Gerhardt and Davidson 2009). Photo by J.S. Aber; central Kansas, United States.

Plate 7-13. Plains gartersnake (*Thamnophis radix*) has caught a small American bullfrog (*Rana catesbeiana*) at the edge of a prairie marsh. The snake has already swallowed the entire right hind leg of the frog, which has ceased to struggle. Photo by J.S. Aber and S.W. Aber; central Kansas, United States.

Plate 7-15. Northern painted turtle (*Chrysemys picta*), ventral (bottom) view. The brilliant colors form an intricate pattern on the plastron and bottom edges of the carapace. This species thrives in shallow marsh pools and eats both plant and animal foods; adults prefer plants and juveniles are more carnivorous (Collins and Collins 2006). Photo by S.W. Aber; central Kansas, United States.

Plate 7-18A. Laguna Atascosa National Wildlife Refuge, southern Texas, United States. This refuge is a key site on the central flyway; migrating birds avoid crossing over the Gulf of Mexico to the east and inland deserts to the west. Overview looking toward the northeast with Laguna Madre, a hypersaline lagoon, on the right and the mainland on the left. Kite aerial photograph by S.W. Aber and J.S. Aber.

Plate 7-24B. Scarlet ibis (*Eudocimus ruber*) seeks crabs and mollusks in mangrove thickets and coastal mudflats. Photo courtesy of M. Buchele, Brazil.

Plate 7-25. Dabbler ducks. A. Mallard (*Anas platyrhynchos*) drake displays its distinctive color pattern with a green head and white collar. B. Cinnamon teal (*A. cyanoptera*) has a broad bill. Both are native to North America. Photos by J.S. Aber.

Plate 7-26. Diving ducks. A. Redhead (*Aythya americana*) drake. Redheads usually feed at night and rest during the day. B. Canvasback (*A. valisineria*) drake; note the long bill and large webbed feet. Both are native to North America. Photos by J.S. Aber.

Plate 7-27. The Mandarin duck (*Aix galericulata*) drake is among the most colorful and beautiful of all birds. Native to eastern Asia, it breeds in eastern Siberia, China and Japan, and winters in southern China and Japan (Harris 2009). Photo by J.S. Aber.

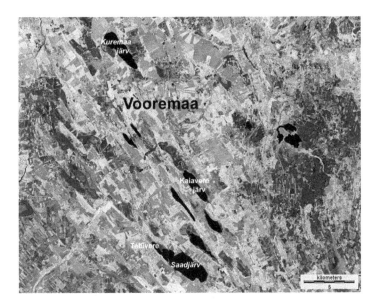

Plate 8-1. Satellite image showing the Vooremaa drumlin field north of Tartu, Estonia. Drumlins are elongated hills that were molded by glaciation into smoothed, streamlined landforms. Long, narrow lakes occupy the troughs between drumlins, and bogs have developed in several of these troughs (dark green). Active vegetation is green; fallow fields are purple-maroon; red patch in left-center is a cut-over peat mine at Visusti Soo. Landsat false-color composite; TM bands 2, 4 and 5 color coded as blue, green and red; image date 28 May 2007; processing by J.S. Aber.

Plate 8-13. Natural-color satellite image of the Mississippi Delta region showing distinct sediment plumes in the Gulf of Mexico. The Mississippi River delivers about 500 million tons of sediment into the gulf annually via the main stem of the river and its Atchafalaya distributary. A. Atchafalaya Delta, B. Lake Pontchartrain, C. Mississippi Delta, D. Mobile Bay. New Orleans is the light patch on the southern side of Lake Pontchartrain. MODIS image acquired 5 March 2001; adapted from NASA Visible Earth <http://visibleearth.nasa.gov/>.

Plate 8-14. Satellite image of Vensyssel region, northern Jylland, Denmark. Note the milky blue color of water along the North Sea coast derived from beach and cliff erosion in contrast to the clean water of the Kattegat. Landsat false-color composite based on TM bands 1, 3 and 7 color coded as blue, green and red; date of acquisition 4 June 2010. Location of Mårup Kirke indicated by asterisk (*) west of Hjørring. Image from NASA; processing by J.S. Aber.

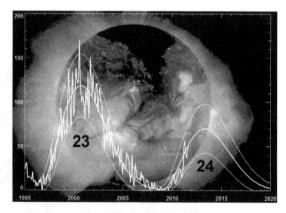

Plate 8-22. Recent solar activity during sunspot cycle 23 and a prediction for cycle 24. Adapted from Hathaway (2011). <http://solarscience.msfc.nasa.gov/SunspotCycle.shtml>

Plate 8-20. The whole Earth photographed during the Apollo 17 mission showing Africa, Arabia and Antarctica. Clouds, ice, and deserts have high albedo; oceans and land vegetation have low albedo. Image courtesy of K. Lulla, NASA Johnson Space Center. Photo ID: AS17-148-22727, December 1972.

Plate 8-28. Panoramic northward overview of the eastern margin of East Devils Lake. The former shoreline stands with dead trees on left side; shallow lake and marsh toward right were created by rising water of the last few years. East Devils Lake was a separate lake in the past, but is now joined with Devils Lake. Blimp aerial photograph by J.S. Aber, W. Jacobson and S. Salley, October 2003.

Plate 9-9. Selection of Danish amber pebbles retrieved from the North Sea coast. From the collection of P. Laursen; photo by J.S. Aber.

Plate 10-4. Oxygen concentration in Gulf of Mexico, southern Louisiana and southeastern Texas. 2006 Seamap hypoxia map overlaid with June 2006 chlorophyll OC3 MODIS-derived image. Red and orange colors indicate hypoxic conditions in coastal waters. A. New Orleans, B. Mississippi River delta. Adapted from NASA (2010a).

Plate 10-7. Oligotrophic bog in Finland showing hollow in foreground and hummock with sparse pines in the background. Both productivity and decay are limited by the short growing season, anaerobic conditions, low pH, lack of nutrients, and cold temperature. Photo courtesy of E. Volkova.

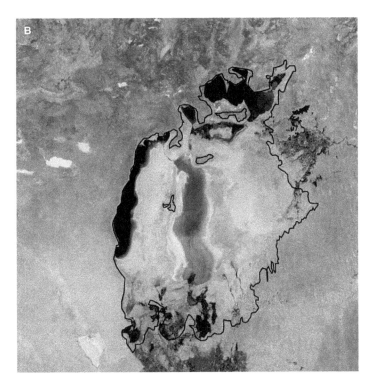

Plate 11-6B. Dramatic Terra MODIS image from 26 August 2010 showing the desiccation of the Aral Sea due to water diversion schemes on the Amu Darya and Syr Darya. The deltas of the Amu Darya (southern shore) and Syr Darya (northeastern shore) are visible on the image in green. The 1960s shoreline of the sea is marked. Image adapted from NASA's Earth Observatory. Accessed online <http://earthobservatory.nasa.gov/IOTD/view.php?id=46685> October 2010.

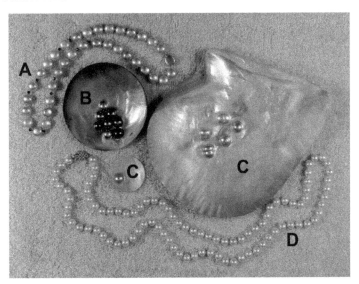

Plate 11-11. Selection of fresh- and salt-water pearls and mother-of-pearl shells. A. Chinese, multi-colored freshwater, probably *Hyriopsis cumingi*. B. Black-lipped pearl oyster, *Pinctada margaritifera*. Diameter of shell = 85 mm. C. Golden-lipped pearl oyster, *Pinctada maxima*. D. Japanese akoya pearls, *Pinctada fucata*. White with rose overtone. Photo by S.W. Aber and J.S. Aber.

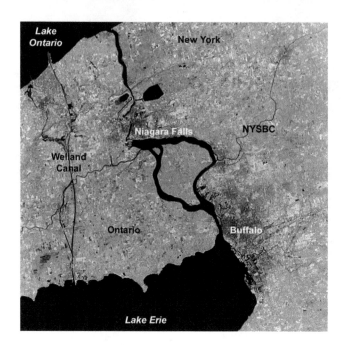

Plate 11-13. Canals in western New York, United States and southern Ontario, Canada. The Welland Canal is nowadays the primary shipping connection between lakes Ontario and Erie. The New York State Barge Canal (NYSBC) is utilized mainly for recreational purposes. Landsat false-color composite; TM bands 2, 3 and 4 color coded as blue, green and red. Active vegetation appears red and pink. Acquired June 1983; adapted from NASA Goddard Space Flight Center.

Plate 11-14. Imperial Valley, southern California and northern Mexico. The All American Canal (AAC) enters right side of scene and flows westward just north of the U.S.–Mexican border. This is the last tap from the Colorado River on the U.S. side of the border. Bright red irrigated cropland is evident in the Imperial Valley of the United States, but crops have less irrigation water in Mexico. Landsat false-color composite; TM bands 2, 3 and 4 color coded as blue, green and red. Active vegetation appears red and pink. Acquired April 1983; adapted from NASA Goddard Space Flight Center.

Plate 13-9. Tar Creek carries highly polluted mine runoff from the Treece-Picher-Cardin vicinity, as shown by its typical rust-orange color at low flow. According to the U.S. Army Corps of Engineers, "Tar Creek is highly toxic and, for all intents and purposes, dead" (USACE 2005); however, continued stream monitoring has recently revealed the presence of some fish and macro-invertebrates. View downstream just east of Commerce, Oklahoma. Photo by J.S. Aber.

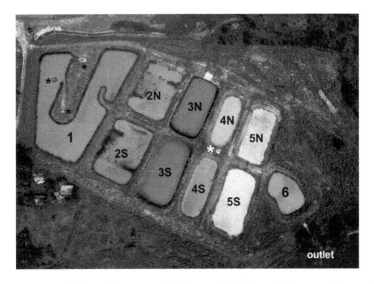

Plate 13-12. Passive treatment facility at Commerce, Oklahoma. Vertical aerial photograph annotated with pool numbers. Inlet pool (1) with three artesian wells (*) upper left; outlet channel lower right. Two treatment trains identified as north (N) and south (S). Notice change in water color from the inlet cell (1) to the outlet pond (6). White asterisk (*) indicates position of wind and solar power for cells 4. Vehicles in upper left corner for scale; helium-blimp airphoto by J.S. Aber and S.W. Aber.

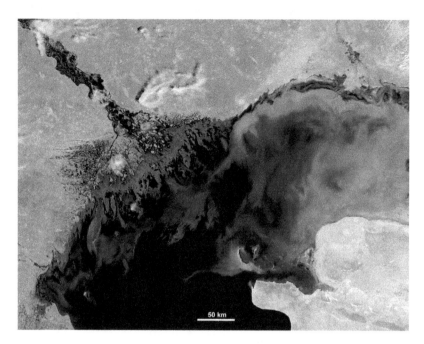

Plate 14-1. Volga River delta in the northern Caspian Sea, southern Russia. MODIS true-color image, acquired 11 June 2005. The river divides into >500 distributary channels forming a wetland web that supports hundreds of plant and animal species. It is among the most productive fisheries in the world, including the caviar-producing sturgeon. Image adapted from NASA's Earth Observatory <http://earthobservatory.nasa.gov>.

Plate 14-20. Saddle-billed stork (*Ephippiorhynchus senegalensis*), a large and wide-ranging African stork. This stork inhabits fresh, brackish, and saline marshes, wet meadows, and other open wetlands. Seen here in a marsh of the Okavango River delta region. Photo courtesy of M. Storm.

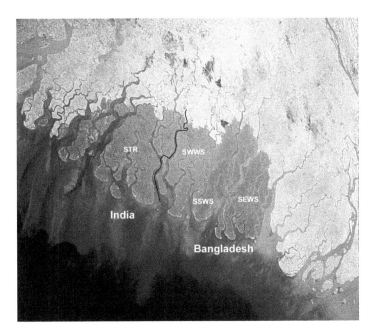

Plate 15-2. Ganges Delta, Sundarbans, India and Bangladesh. The protected mangroves are highlighted in red in this color-infrared photograph. Black line marks the national boundary. STR – Sundarbans Tiger Reserve, SWWS – Sundarbans West Wildlife Sanctuary, SSWS – Sundarbans South Wildlife Sanctuary, and SEWS – Sundarbans East Wildlife Sanctuary. Image courtesy of K. Lulla; STS061B-50-007, December 1985.

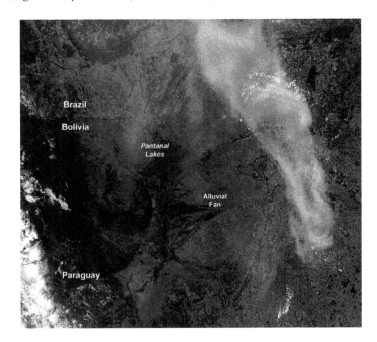

Plate 15-9. Pantanal region, South America. The image clearly shows one of the largest alluvial fans at the center, while the Pantanal Lakes are visible to the north and west of the alluvial fan as black dots. The Brazilian landscape to the east and north of the fan indicates agricultural activities. Healthy vegetative cover is observed in Bolivia to the west of the Pantanal. Red dots indicate active fires. Moderate Resolution Imaging Spectroradiometer (MODIS) natural-color image from 16 June 2003; adapted from NASA's Earth Observatory <http://earthobservatory.nasa.gov/>.

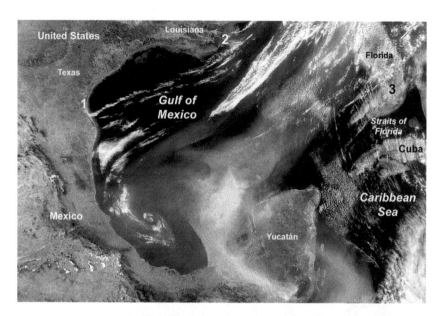

Plate 15-10. Natural-color, SeaWIFS satellite image of the Gulf of Mexico region with minimal cloud cover. Selected case study sites: 1. South Texas, 2. Mississippi River delta, 3. Florida Everglades. Smoke from forest fires covers part of the Yucatán Peninsula and southern Gulf. Image date 24 April 2000; adapted from NASA Visible Earth <http://visibleearth.nasa.gov/>.

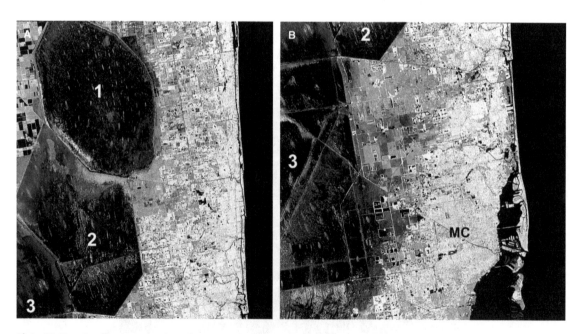

Plate 15-13. Satellite images of southeastern Florida showing surface-water conservation areas 1, 2 and 3. Note how vegetation patterns reflect the direction of surface-water flow in the conservation areas. The brighter red patches are hammocks with hardwood trees. A. Region of Pompano Beach, north of Miami. B. Miami vicinity. The Miami Canal (MC) begins at Lake Okeechobee and runs across central Miami into Biscayne Bay. Landsat TM bands 2, 3 and 4 color coded as blue, green and red; active vegetation appears bright red and pink. Date of acquisition February 1983; adapted from NASA Goddard Space Flight Center.

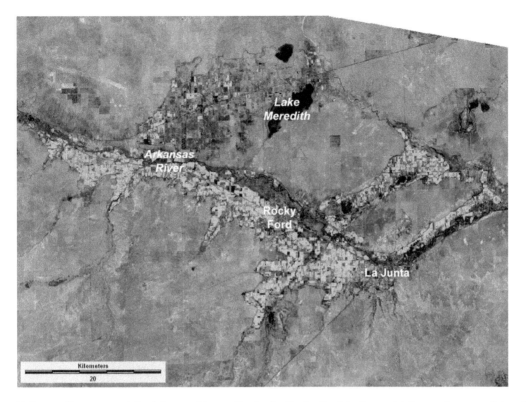

Plate 16-4. Satellite image of the Arkansas River valley in the Rocky Ford–La Junta vicinity, southeastern Colorado, United States. Bright green indicates irrigated crops; maroon-pink shows dry upland areas. Lake Meredith is a major storage point for water diverted from the Arkansas River. Landsat TM bands 2, 4 and 5 color coded as blue, green and red; 7 August 2009. Image from NASA; processing by J.S. Aber.

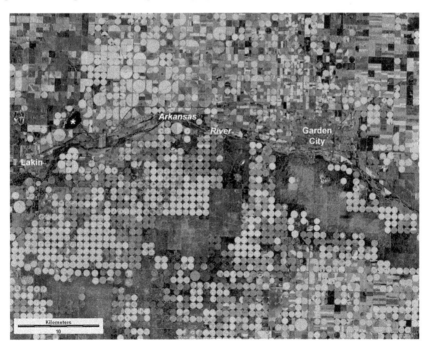

Plate 16-6. Satellite image of the Arkansas River valley in the Lakin-Garden City vicinity, southwestern Kansas, United States. Bright green indicates irrigated crops; maroon-pink shows dry-fallow areas. Small irrigation circles are one-half mile (0.8 km) in diameter; large circles are one mile (1.6 km) in diameter. Asterisk (*) indicates Lake McKinney, a holding point for water diverted from the Arkansas River via the Amazon Ditch. Landsat TM bands 2, 4 and 5 color coded as blue, green and red; 4 August 2007. Image from NASA; processing by J.S. Aber.

Plate 16-10. Overviews from the middle of Dry Lake, Kansas looking toward its eastern end. A. Lake completely full of water following heavy snowmelt and spring rains, May 2007. B. Pool of rust-colored water surrounded by mud/salt flat, May 2010. Water color is presumably a result of suspended sediment as well as saline invertebrates. C. Salt flat completely covers the dry lake basin, October 2010. Kite and blimp aerial photos by J.S. Aber, S.W. Aber, G. Corley, D. Leiker, C. Unruh and B. Zabriskie.

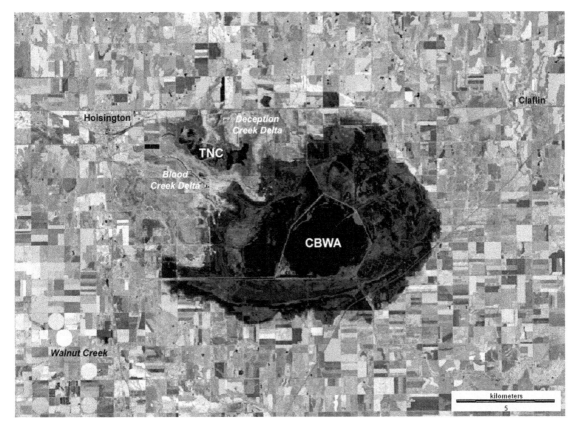

Plate 16-20. Multitemporal Landsat image based on TM band 4 (near-infrared) for 2006, 2007 and 2009, color coded respectively as blue, green and red. Bright colors represent significant changes in land cover from year to year; dull-gray colors indicate little change in land cover. The broad maroon-purple zone shows the extent of high water in 2007; black and dark blue show perennial water bodies; compare with Fig. 16-16. CBWA - Cheyenne Bottoms Wildlife Area; TCN - The Nature Conservancy. Image from NASA; processing following the method of Pavri and Aber (2004), central Kansas, United States.

Plate 16-21. *Azolla* bloom in TNC marshes, Cheyenne Bottoms, Kansas, United States. Comparable views both taken in mid-October. A. 2009. *Azolla* indicated by maroon color; cattle in lower right corner. B. 2010. Cattail (green) is prominent; *Azolla* is not evident in this aerial view, but small patches were observed on the ground at the edges of the marsh. Blimp and kite airphotos by J.S. Aber, S.W. Aber, G. Corley and B. Zabriskie.

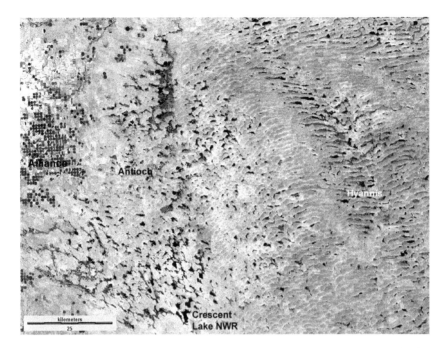

Plate 16-24. Satellite image of the western Nebraska Sand Hills, United States. The eastern side of this scene displays massive transverse dunes, and smaller barchan dunes are found to the west. The most active vegetation appears dark green; less active vegetation is pale green. Numerous lakes (blue to black) occupy the troughs and swales between dunes. Landsat TM bands 2, 5 and 7 color coded as blue, green and red; image date 10 August 2010. Image from NASA; processing by J.S. Aber.

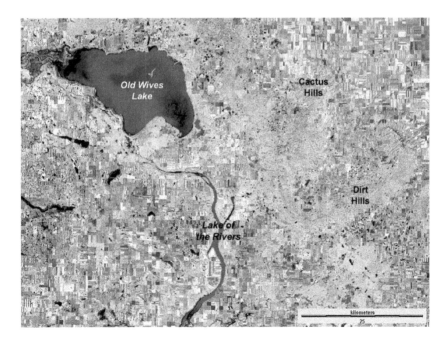

Plate 16-29. Satellite image of the Missouri Coteau, southern Saskatchewan, Canada. Dirt Hills and Cactus Hills are large ice-shoved ridge complexes. Old Wives Lake and other water bodies are full during a wet episode. Active vegetation appears in green and yellow-green colors. Landsat TM bands 1, 5 and 7 color coded as blue, green, and red; image date 9 September, 2002. Image from NASA; processing by J.S. Aber.

Plate 16-33. Satellite images of Old Wives Lake in wet and dry phases. A. Wet phase with lakes full of water, September 2002. Old Wives Lake displays swirling patterns of suspended sediment driven by wind and waves as well as an algal mat (green) in the southeastern portion.

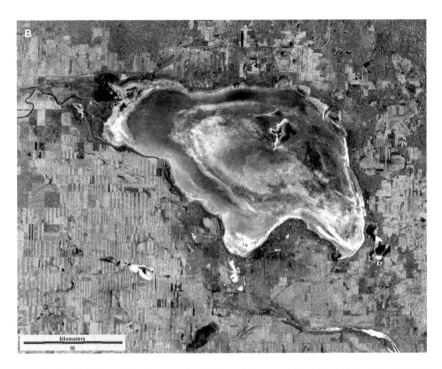

Plate 16-33. cont'd   B. Old Wives Lake is completely dry with exposed mud/salt flats, July 1988. Some water remains in Wood River and Frederick Lake. Landsat ETM/TM bands 3, 4 and 5 color coded as blue, green and red. Active vegetation appears in green and yellow-green colors; fallow/bare ground is pale purple to red; water bodies are light blue to black. Image from NASA; processing by J.S. Aber.

Plate 16-40. Near-infrared image (April 2005) of West Creek within the Parker River basin showing extensive drainage ditches across the salt marsh. Data Source: 1:5000 digital orthoimagery from Massachusetts Office of Geographic Information (MassGIS 2011). Adapted from Pavri and Valentine (2008).

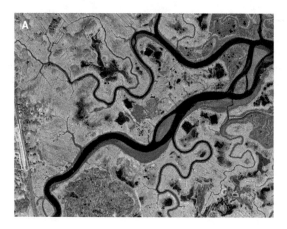

Plate 16-41A. Near-infrared image (April 2005) of the Rowley River salt-marsh system with a number of pools and pannes in black, active vegetation in red across upland areas and marsh areas yet to green-up in pale shades of pink and gray. Data source 1: 5000 digital orthoimagery from Massachusetts Office of Geographic Information (MassGIS 2011). Adapted from Pavri and Valentine (2008).

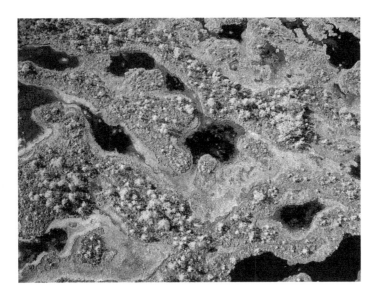

Plate 16-50. Vertical view over central portion of Männikjärve Bog, east-central Estonia. Striking color zonation and intricate spatial patterns of pools, moss, and other vegetation are displayed in this autumn scene, which is ~100 m across. Kite aerial photo (Aber and Aber 2005).

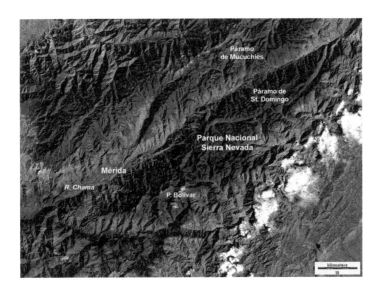

Plate 17-4. Rare, nearly cloud-free satellite image of Andes region in western Venezuela. Bright green indicates evergreen and deciduous forest; dark olive-green is alpine shrub of the páramo. Landsat MSS bands 1, 4 and 2 color coded as blue, green and red. Acquired 13 January 1979. Image from NASA; processing by J.S. Aber.

Plate 17-20. Satellite image of the central portion of San Luis Valley. The region north and east of the Rio Grande is an internal basin draining into several lakes. GSD - Great Sand Dunes, RL - Russell Lakes, SLL - San Luis Lake. Landsat TM bands 2, 5 and 7 color coded as blue, green and red. Active vegetation appears green. Acquired 21 August 2009. Image from NASA; processing by J.S. Aber.

Plate 17-26. Close-up view of a pool at Russell Lakes State Wildlife Area. Dark green emergent vegetation is mainly bulrush, and algae forms a mat on water surface. Blood-red patches presumably reflect carotenoid pigments of invertebrates in saline water. Blimp aerial photo by S.W. Aber and J.S. Aber.

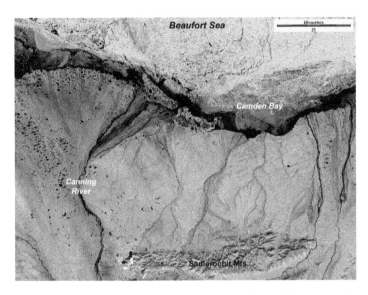

Plate 17-28. Satellite image of a portion of the North Slope region, including the Canning River and its delta, north of the Sadlerochit Mountains, northeastern corner of Alaska, United States. Land vegetation is just beginning to green up (yellow-green), and sea ice (cyan) has pulled away from the shore in this early summer scene. Small lakes and streams are ice-free (black), but larger lakes and rivers still have partial ice cover. Most of this area is part of the Arctic National Wildlife Refuge, including the coastal plain 1002 Area, as well as Kaktovik Inupiat Corporate Lands. Landsat ETM+ bands 1, 4 and 5 color coded as blue, green and red. Image date 6 July 2000. Image from NASA; processing by J.S. Aber.

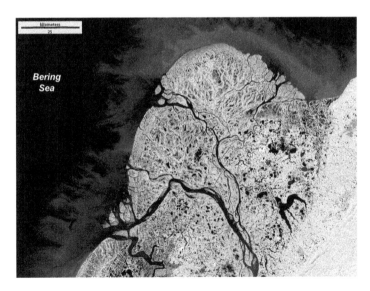

Plate 17-30. Satellite image of the Yukon River and Yukon Delta National Wildlife Refuge, Alaska, United States. Land vegetation is just beginning to green up. Lakes, rivers and the sea are ice free in this late spring image; suspended sediment is highlighted in light blue colors. Intricate patterns are displayed in the delta and coastal environments. Landsat ETM+ bands 1, 4, and 5 color coded as blue, green and red. Image date 18 June 2002. Image from NASA; processing by J.S. Aber.

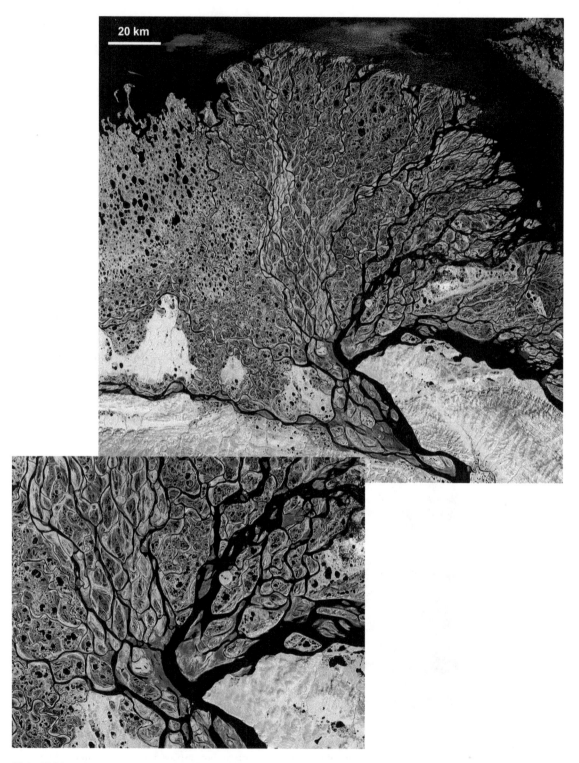

Plate 17-31. The Lena River delta on the Arctic coast of eastern Siberia, Russia. This is one the most beautiful false-color Landsat images ever produced; it shows the intricate network of meandering waterways in the classic fan-shaped delta. The inset image depicts details of the distributary system. The delta is a highly protected wetland wilderness in a permafrost environment and is an important refuge and breeding ground for many wildlife species. Adapted from NASA; obtained from Wikimedia Commons <http://commons.wikimedia.org/> October 2011.

Figure 11-18. Tourists load luggage in preparation for a charter flight to a wildlife observation camp in the Okavango Delta region of Botswana. Photo courtesy of M. Storm.

Figure 11-19. Information kiosk for tourists on the Wetlands and Wildlife National Scenic Byway. The byway route follows state highways as well as county roads and includes Cheyenne Bottoms and Quivira National Wildlife Refuge in central Kansas, United States. Photo by J.S. Aber.

(*Grus antigone*), which often reaches up to 1.8 m in height, among 360 other recorded and rare bird species (Boojh, Patry and Smart 2008). In other cases, wetland sites market themselves as stopover points on intercontinental bird flyways, promising the viewing of large numbers of migratory birds and waterfowl. The North American mid-continent wetlands complex of Cheyenne Bottoms and Quivira National Wildlife Refuge in central Kansas provides one such example (Fig. 11-19).

The U.S. Fish and Wildlife survey for 2006 suggests that 47.7 million individuals across the United States engaged in bird watching activities near their homes or on trips, and nearly 20 million individuals took special trips away from home to observe wild birds (USFWS 2006). While clearly not all these individuals visited a wetland site to do so, the numbers suggest a potentially large birding community that wetland sites could target for recreational activities. Beyond wildlife watching and birding, fishing and hunting also generate significant revenues. Studies estimate that in the U.S. alone, 2.3 million migratory bird hunters spent approximately US$1.3 billion on hunting trips and equipment in 2006 (USFWS 2006).

As this section demonstrates, the economic services provided by wetland habitats are substantial and wide ranging. Balancing economic needs with habitat conservation and wise use is critical to ensure the sustainability of ecosystems and the cultures that depend on them.

11.5 Wetland valuations

Previous sections of this chapter have elaborated on the numerous ecosystem services and products that wetlands provide to society. Examples also have illustrated that the use and value placed on a wetland resource are always context-dependent and based on numerous locally contingent and in some cases regionally and internationally relevant factors. This is complicated by the fact that in addition to resources such as water, fiber, and food, which may be valued monetarily or via other means, wetlands are multi-functional and provide habitats, cycle nutrients, store carbon, protect against flood and storm destruction, and improve water quality. Moreover, they are appreciated for their aesthetic and intrinsic qualities. Quantifying and assigning values to such a plethora of services is a complex undertaking.

11.5.1 Why value wetlands?

Environmental economists have argued that placing a value on environmental resources ensures that decisions about their use, allocation, and management are made more judiciously. Valuation requires the measurement and monitoring of a resource to account for and compare all of its costs and benefits to society and the ecosystem. In the case of wetlands, studying these ecosystems enables us to understand them more thoroughly, and monitoring them allows for assessments of their current status and how that may change over time. Moreover, valuation also creates awareness of the biophysical, economic, social, and cultural benefits of such ecosystems. It identifies stakeholders, user groups, and those who may benefit from wetland services and products, thereby eliciting interest in wise and sustainable-use practices (de Groot et al. 2006).

Valuations give decision makers the tools with which to assess policy and management alternatives. While traditional economic approaches have used market-based techniques to assign a value to environmental resources, ecological and other socio-cultural approaches provide techniques for non-market valuations as well. Wetlands, like other natural resources, serve multiple functions, provide multiple uses, and engage multiple stakeholders. The failure to value these resources appropriately may often result in their degradation or overuse. Valuing wetlands using integrated market and non-market approaches could allow for a fuller multidimensional appraisal of their contributions to the ecosystem and society.

11.5.2 Property regimes and externalities in wetland use and valuations

How wetlands are used, who uses them, and what rules regulate their use depend on a number of factors including the geographic and socio-political context in which they are found. Institutional economists and theorists examining property rights provide us with a starting point to answer some of these questions (Ostrom 1990; Bromley 1991; Hanna and Munasinghe 1995). They suggest that formal property rights institutions guide the use of resources, such as wetlands, and generally reflect the larger social and environmental goals of a society. While policy and management decisions play important roles in wetland health, theorists argue that wetlands with clearly established property regimes stand the best chance of sustainable use. Property regimes detail legal protections accorded, specify resource rights and rules of use, and clearly define the monitoring and enforcement mechanisms for misuse. Hanna, Folke, and Maler (1995, p. 17) identified the four broad property rights institutions or regimes observed globally (Table 11-5). These are:

Table 11-5. Four categories of property rights and their conditions.

Private property	• Individual or corporate ownership, control & access • May include some restrictions on resource use based on prevailing environmental laws
State/public property	• Managed by federal, state or local government • Balances environmental, economic & social responsibilities • Multifaceted rules of use • Public access
Common property	• Collectively owned & managed • Shared responsibilities • Rights of access & use exclude non-members
Open access	• No established or enforced property rights or rules of use • Constraints on misuse absent • First-come, first-served conditions

Private property is commonly observed across most parts of the world. Here, an individual or business is assigned ownership of a resource or property and controls access to it. The individual has considerable leeway to use resources on this property and may exclude non-owners from use. However, it is not uncommon to see certain constraints on use even for private property. In the case of wetlands in the United States, for instance, Section 404 of the Clean Water Act requires a private property owner to be granted a permit before any dredge-and-fill activity affecting a wetland site may be undertaken on his/her private property.

State/public property includes resources that fall under the jurisdiction of the state, which acts as caretaker and manages the resource based on a set of socially accepted principles. A little over 30 percent of the total land area of the United States is public property, with management authority vested in a variety of federal agencies, including the U.S. Forest Service, the Bureau of Land Management and the National Park Service, among others (Platt 2004). Rules that govern the use of resources on these lands may be multifaceted. For instance, recreational and other uses of the Everglades National Park are governed by the rules set by the National Park Service.

Rules are generally socially and environmentally responsive in that they are formulated to provide the community with access to this resource, yet ensure that environmental concerns for the wetland's health are not compromised. Usufructuary rights are often observed on state property and include only use (not ownership) rights granted to a resource. As an example, fishermen are allowed to extract fish from coastal waters, yet they do not own those waters. Or, in the Great Plains region of the United States, ranchers are permitted to graze cattle and use prairie wetlands and watering holes on state-owned public land, but they are not assigned any ownership rights to those lands.

Going back to our example from Keoladeo National Park in India, the state which manages the park primarily as bird habitat also provides usufruct rights to local communities to collect and sell fuel wood and timber from the rapidly expanding *Prosopis juliflora* (common name: *Vilayati babul*) invasive species that has spread across the grasslands and drier pool areas of the park. Here, by providing usufruct access we see the state balancing environmental management concerns with local needs for fuel wood by adjusting rules of use within the boundaries of the park (Boojh, Patry and Smart 2008).

Common property comprises resources that are held collectively by a group of people. Members of this group reserve rights of access and may exclude use by non-members. However, the inadequate articulation of rules of use and exclusion, or the insufficient monitoring of non-member use could lead to overuse and result in Hardin's (1968) "tragedy of the commons" prediction (see below). Common property regimes are often observed in the case of coastal and inland fisheries, which due to their dynamic nature moving from one area to another can pose significant management challenges. Under the right conditions, however, common property regimes may be quite effective institutions to manage such resources sustainably (Crean 2000). Traditional societies may often control and manage resources in common. Indigenous groups in the Amazon or the Aborigines of Australia provide examples of this.

The acequia (ah-say-kya) irrigation system at San Luis, Colorado is another example of common ownership and operation. Derived from practices in Spain and northern Africa, the acequias are long ditches that divert water from the Culebra Creek into irrigated fields (Fig. 11-20). Each user shares in access to water and responsibility for maintenance, and an elected mayordomo (superintendent) enforces the rules. The initial acequia was constructed by hand in 1851, and the system continues to function in a traditional manner today (Fig. 11-21).

Open access refers to property regimes without established rights or rules governing their use. Often such instances lead to the overuse of resources because of the lack of constraints on use and the absence of monitoring or enforcement measures (Fig. 11-22). Moreover, users have few incentives to engage in sustainable resource-use practices, given that

Figure 11-20. City of San Luis, Colorado in the foreground and irrigated fields of the Culebra Creek valley behind. Irrigation water is delivered in a common system of acequias (ditches) that are jointly operated and maintained by all water users in the district. Blimp airphoto by S.W. Aber and J.S. Aber.

Figure 11-21. Cattle graze in acequia-irrigated wet meadows of the Culebra Creek valley with the Sangre de Cristo Mountains in the background. Photo by J.S. Aber; near San Luis, south-central Colorado, United States.

there are no guarantees that others would not act in their own self-interest. In the event of significant pressures on a particular resource, open-access regimes may be misused to the point of degradation. Hardin's tragedy of the commons thesis more accurately describes these situations. The open oceans are often used as a classic example of open-access regimes. In contrast with continental waters, few monitoring mechanisms exist to supervise fish catches in the open ocean, and international fishing policies are weakly enforced at best.

Accounting for externalities is another important aspect of any resource valuation exercise. Hackett (2006, p. 66) pointed out that "externalities are positive or negative impacts on society that occur as a by-product of production and exchange..., they are not included in the factors that underlie market supply and demand." Externalities may be thought of as spillover effects, which reflect the full cost or benefit of any alterations to an ecosystem or resource-use activity (Cutter and Renwick 2004). In the past, the cost of the construction of a hydroelectric dam rarely included the cost of habitat destroyed or agricultural land lost, or changes to local climate regimes. Yet, these are real ecosystem and societal costs that are associated with the project, but were never accounted for in the decision to construct the dam.

The example of the oil-palm industry of Kalimantan used earlier illustrates this point well. The price of oil-palm on global markets does not reflect the cost of carbon emissions to the

Figure 11-22. *Avicennia marina* (gray mangrove) appear in the background along with shanties that have built illegally along a rocky shore in suburban Mumbai, India. If such settlements are allowed to persist unchecked, residents may engage in illegally clearing mangroves for fuel wood and other materials. Photo by Firooza Pavri.

atmosphere from peatland clearing and its future influence on climate change. Conversely, beneficial externalities occur when, for instance, an artificial wetland is created primarily to treat waste water, but also provides habitat to bird and animal species, flood protection, and carbon sequestration services. Ecologists have argued for the inclusion of externalities in any ecosystem valuation exercise to reflect the true total costs and benefits of any market exchanges of resources or habitat alterations (Cutter and Renwick 2004).

11.5.3 How to value wetlands?

Various approaches and methods have been offered to assess the economic value of resources such as wetlands. The most familiar of these is cost–benefit analysis (CBA). Jules Dupuit presented this technique back in the nineteenth century, and during the early twentieth century it was used to evaluate large-scale hydro projects in the United States. Since then, CBA has been put to use in various fields including transportation, public health, and the environment (Portney 2008). CBA provides a comparative assessment of various alternatives, so that informed decisions can be made before adopting resource development strategies. As the name suggests, this and other methods such as contingent-evaluation (CV) techniques allow one to measure the gains and losses of any proposed activity for those affected by it. As Portney (2008) explained, benefits or gains are often measured by using the "willingness to pay" principle, while the costs or losses are calculated based on the amount of compensation needed to offset negative consequences. Field and Field (2006, p. 116) outlined the basic structure of a cost–benefit analysis, which includes:

- Clearly identifying the issue or project.
- Identifying the different input and output components.
- Calculating the benefits and costs of these components.
- Comparing and assessing the benefits and costs.

Cost–benefit analysis is most often measured in quantifiable economic terms. However, not all benefits or costs are so easily summarized. In those cases, economists have devised ways to assign shadow prices using proxy values or replacement costs to enable quantification (Cutter and Renwick 2004).

Researchers have pointed out problems with using the cost–benefit approach to ecosystem valuations (Hanley and Spash 1995). For instance, it is difficult to assign monetary values to ecosystem services such as nitrogen and carbon storage. As this book has indicated, ecosystem functions are complex and interconnected. Changes to one system are bound to have ramifications on other parts of the same system or other interconnected systems. How do we anticipate all of these and account for them? Similarly, wetlands do not have clearly demarcated spatial boundaries and their benefits might be accrued over wide areas. This may pose problems when accounting for benefits and costs. For example, the flood abatement benefits of a wetland might be felt far downstream from the actual wetland site. Furthermore, making changes to ecosystems today could have ramifications in a century's time that we may not even anticipate today.

How do we assess and account for impacts in the future? Economists have provided discounting techniques to account for costs and benefits accruing at different time periods. This involves summing up future benefits or losses in present-day terms using a specified discount rate. The choice of discount rate used for this process invariably influences the outcome and may in turn affect the economic viability of a project. Even so, CBA and CV are widely used across the world. Depending on the specific goals of a valuation exercise and the resources and time available to undertake it, other economic methods used in the past have included cost-effectiveness, hedonic pricing, travel-cost, restoration- and replacement-cost methods.

In responding to criticisms of economically focused methods, some have offered alternative approaches to valuing environmental resources while still retaining the methodological contributions of CBA and mainstream economic techniques. Turner, Georgiou and Fisher (2008, p. 9) proposed the ecosystem-services approach, which in addition to linking human welfare to ecosystem health, seeks to identify non-monetized services provided by wetlands and incorporate these services into the valuation process. Such comprehensive approaches attempt to evaluate ecosystem functions and services, alongside their social contexts, governance and management arrangements. Within the valuation exercise, the ecosystem services approach also simulates different futures based on ecosystem alterations to provide policy makers with a wide range of options from which to choose (Turner, Georgiou and Fisher 2008).

Finally, stakeholder involvement at various stages of the valuation process, including establishing management objectives, configuring governance structures, and determining values, are essential components of such endeavors.

A similar approach is advocated by the international body on wetlands, the Ramsar Convention, and is summarized here (de Groot et al. 2006). This approach is a five-stage process. In the first stage, management goals and objectives are assessed to gauge the parameters of the endeavor and make informed decisions about how to proceed with a valuation. Second, a comprehensive stakeholder analysis is initiated, where various stakeholders and beneficiaries of a wetland site are identified and their roles made explicit. Considerable attention is paid to this stage of the process, with the acknowledgement that stakeholders contribute significant knowledge and are beneficiaries of wetland services and products and hence may play important roles in the success or failure of proposed changes to a wetland site.

The third stage includes an exhaustive functional assessment of hydrological, biological, chemical, physical, and ecological contributions and an inventory of products and services offered by a wetland. In the fourth stage, the actual valuation and analysis of benefits in economic, environmental, and social terms using market and non-market based methods are undertaken. The fifth and final stage includes communicating the results of the valuation exercise to all stakeholders and decision makers. The Ramsar-supported approach, thus, places equal emphasis on the market and non-market elements of wetland ecosystems, prioritizes the role of stakeholders, and engages in the widespread dissemination of results.

11.6 Summary

This chapter provides a general overview of wetland services, functions and uses with examples from across the world. Wetlands provide habitat refuges for flora and fauna, and they aid in the capture, storage and cycling of nutrients through the Earth's terrestrial, hydrological and atmospheric components. They offer water purification services by filtering organic and inorganic impurities and provide protective services, which guard against destruction from flood and storm events. By sequestering large amounts of carbon in soil and plant matter, wetlands also play vital roles in regulating the Earth's climate. These diverse services are coupled with their roles as fresh-water storage areas and transportation links to ferry goods and people.

Human ingenuity and technological innovations have allowed for a diverse set of economic goods to be extracted from wetland sites. Communities in parts of the developing world in particular depend on resources from wetland sites to meet their everyday livelihood needs of food, fuel, shelter, fiber, and water. Meanwhile, the extractive sector of the economy has benefited from products such as animal pelts, peat, fish and shellfish, fuel, timber, agricultural and ranching land, and hydropower. Pearl production is a simple form of aquaculture with positive impacts on wetlands and the economy, and pearl farmers have a keen interest in protecting waters and life in coastal and reef areas. Finally, the aesthetic appeal of these landscapes has also provided economic benefits. Recreation and tourism are major economic drivers providing important sources of local employment and often leading conservation and protection efforts.

Given these numerous functions, services and products, the sustainable management of wetland environments is critical. A first step in this endeavor is the establishment and protection of well-defined and socially relevant property-rights regimes. Identifying the most appropriate policies or management strategies requires a careful valuation of this resource. Valuation techniques have transitioned from focusing primarily on the economic benefits accrued, to more holistic considerations that include ecological, social, cultural, and economic benefits and attempt to account for externalities. Valuing wetlands in this way helps account for their full ecosystem and societal services and benefits.

Conservation and management: Wetland planning and practices

12.1 The conservation movement

The Merriam-Webster English dictionary defines conservation as "a careful preservation and protection of something; especially: planned management of a natural resource to prevent exploitation, destruction, or neglect." Scientists might dispute the inclusion of the word "preservation" in the definition above. Preservation implies setting land aside for its intrinsic and natural beauty, which is somewhat different from the usual intent of conservation. In general, however, the definition for conservation captures the term's essential elements and its general usage over the past century and a half.

Tracing the development of conservation thought within the United States leads one to several early influential works. These have played enduring roles in molding the public's collective understanding of the environment and our place within it. The Lewis and Clark expedition (1803–6) was the first significant venture to chart and record the western lands of the United States. It was also the last great pre-industrial journey in the United States. Lewis and Clark's meticulously detailed natural history contributed substantially to our knowledge of the dynamic American landscape prior to large-scale European influence. Their travel was accomplished entirely with human, animal, and natural power to move the expedition, dubbed the Corps of Discovery, across the continent and back (Fig. 12-1).

Later, the ideas of key nineteenth- and early twentieth-century American literary figures such as Ralph Waldo Emerson, Henry David Thoreau, John Muir, and Aldo Leopold among others contributed to intellectually articulating an environmental ethic. Emerson's essay *Nature*, Thoreau's essay *Walking* and more famous work *Walden*, John Muir's passionate defense of a preservationist approach, and Leopold's much later *Sand County Almanac* roused the idea of nature requiring protection. These universally acclaimed works provided ethical arguments for the more careful use of the environment, with some advocating the need for setting aside nature to counterbalance developmental and urban ills witnessed elsewhere.

The most explicit systematic scientific account of human impact on the environment from this early period was geographer George Perkins Marsh's work *Man and nature; or, Physical geography as modified by human action* (Marsh 1864). Marsh's careful and detailed analysis using European and North American examples of human influence in environmental modification and his elaboration of conservation guiding principles provided the first comprehensive exploration. Adopting a systems approach, he was able to identify how elements within the environment interacted with each other and how changes to one

Wetland Environments: A Global Perspective, First Edition. James Sandusky Aber, Firooza Pavri, and Susan Ward Aber.
© 2012 James Sandusky Aber, Firooza Pavri, and Susan Ward Aber. Published 2012 by Blackwell Publishing Ltd.

Figure 12-1. Lewis-and-Clark bicentennial reenactment at Kaw Point, where the Kansas River enters the Missouri River. A. Overhead view with the 55-foot (17 m) keelboat at center tied up with smaller pirogues above. Blimp airphoto by S.W. Aber and J.S. Aber. B. Replica keelboat and pirogues on display for the public with the Kansas City, Missouri downtown skyline in the background. Photo by J.S. Aber.

system might have repercussions elsewhere. For instance, Marsh built connections between deforestation and soil degradation, the draining of marshes and lakes and the decline of dependent flora and fauna, and the channeling of rivers and the loss of fertile floodplain soils (Marsh 1864). He was not the first to make these connections. Yet, through his work, Marsh provided the scientific rationale for tempering the exploitation of natural resources for human development with an approach that recognized the finite nature of resources and the need for their conservation.

These philosophical arguments along with burgeoning scientific evidence amplified public awareness, and by the end of the nineteenth century subtle changes became apparent in United States' policy toward natural resource use. Theodore Roosevelt's presidency (1901–8) served to entrench more firmly a conservationist approach to resource management, advocating a "planned and orderly development" of natural resources to benefit the citizens of the United States (Library of Congress 2002). Among other initiatives, Roosevelt worked to expand the National Parks Service, established what would later become the National Wildlife Refuge system with the first wildlife refuge at Pelican Island in Florida as a breeding ground for native birds, and expanded National Monument sites through Presidential Proclamation (Library of Congress 2002).

During Roosevelt's term, the United States Forest Service was established in 1905 with Gifford Pinchot serving as its first chief. Pinchot was also a key advisor to Roosevelt on forest and natural resource policy and played an influential role in establishing a more scientific approach to resource management. His focus on wise use within the forestry sector promoted a utilitarian philosophy of resource management. He recognized the vital role natural resources played in the development of the nation, but also acknowledged the need for their protection and enhancement for future generations. Despite the significant industrial development pressures of the twentieth century, early ethical ruminations and guiding principles built a conservation philosophy that attempted to balance the preservation of certain lands and resources from use, while promoting multiple-use and sustained-yield policies for others (Cutter and Renwick 2004).

Other countries experienced similar conservation trajectories and early European ideas of conservation clearly influenced the United States' approach. Germany's conservation tradition also identifies with key works including Wilhelm Heinrich Riehl's *Naturgeschichte des Volkes* and Ernst Rudorff's essay in the *Preußische*

Jahrbücher, both influential treatises in the establishment of later institutions that promoted public education and awareness of the natural environment (Uekötter 2004). Establishment of museums of natural history, zoos, and botanical gardens across Europe and America during the nineteenth century were examples of such undertakings.

The era of colonialism was then at its height, so that European ideas of conservation spread to colonies across Africa and Asia. For example, in India the British employed Dietrich Brandis, a German forester and close mentor of Gifford Pinchot, who implemented a strategy for scientific forest management. Brandis returned to Europe with lessons learned in India and shared these with his European and American colleagues. At the time, a scientific approach to resource management meant closely monitoring resources, and in the case of forests, determining their total volume, their rates of yield and harvest, and outlining strategies for their extraction. In India, Brandis put into practice a hierarchical system of protected lands that prohibited human use in the more restricted zones.

The hunting tradition in Britain, likewise, accorded protective status to tracts of land. Similar traditions in British India allowed for exclusive hunting grounds that limited human access. The establishment of parks for wildlife conservation and big-game hunting, as seen in British East Africa, provided yet another example of how conservation strategies were implemented across the globe (Hingston 1931). Imperial Russia also established hunting reserves during the time of Peter the Great in the 1700s, with later developments in Russia's conservation philosophy leading to the formation of a nature reserve system of which the zapovednik was to be the most restrictive and excluded all uses except those for scientific study (Strebeigh 2002).

This discussion traces the evolution of the modern conservation movement from a largely Eurocentric perspective. Yet, it is important to recognize that conservation beliefs, attitudes, and practices persisted and continue to persist in the non-European world. Indigenous ideas of conservation persevered in pockets despite the imposition of conservation and management ideas from outside. For instance, sacred forests including those on forested wetland sites across Asia and Africa were protected from overuse because of their association with local religious beliefs and deities. Similarly, permit systems to so-called royal hunting forests were to control access and use by the general population and served a dual conservation purpose. In other instances, through the granting of usufruct rights, sustainable indigenous resource-use practices continued side by side with European conservation strategies.

Wetland protection and management during this period was given scant attention. In fact, one would argue that wetland conversions to other land uses accelerated through the nineteenth century and well into the twentieth century. Even so, it is important to recognize this early history of conservation for the lasting influence it had on the management of natural resources across the globe. Moreover, and as chapter 14 shows, with the global awareness of wetlands as critically important ecosystems, a new era of management and protection starting during the 1970s sought to correct the policy missteps of the past.

12.2 Wetland resource management

Natural resource management provides a scientific and systematic approach to protecting resources so as to meet ecosystem and societal needs. Over the past few decades, scientists have generally agreed that natural resource management requires an integrated systems approach to adequately address relevant issues (Kent 2000a; Turner, van den Bergh and Brouwer 2003; Ramsar Convention Secretariat 2007b). A systems approach recognizes the complex and interconnected character of natural and human systems and implicitly acknowledges the advantages of studying the whole by simultaneously considering its constituent parts and the interactions therein.

An integrated systems approach requires a broad understanding of the biophysical and socioeconomic aspects of natural resources, their availability, use, and allocation. In the case of wetlands, it may involve coordinating and addressing the biological, hydrological, chemical, terrestrial, engineering, socio-cultural and economic aspects of sustainably managing this resource. Moreover, wetlands are generally part of a larger watershed or drainage basin and hence both upstream and downstream impacts of any alterations to a wetland site must be considered.

The Ramsar Convention has similarly advocated an integrated wise-use strategy when managing wetlands, arguing for the "maintenance of their ecological character, achieved through the implementation of ecosystem approaches, within the context of sustainable development" (Ramsar Convention Secretariat 2007b, p. 12). Ramsar uses the Brundtland Commission Report's definition of sustainable development as that which meets the needs of the present without compromising the ability of future generations to meet their own needs (World Commission on Environment and Development 1987). Recognizing the vital role wetlands play in providing services and products to dependent communities, Ramsar advises a management approach that tempers development with the maintenance and enhancement of this ecosystem to ensure its long-term sustainability.

12.3 Wetland management plans

A wetland management plan, like any comprehensive land-use plan, outlines a vision for the future stewardship and development of a particular wetland site. It defines goals and outcomes and provides the framework for sustaining an outlined conservation and development management strategy. In general, wetland management plans are stewardship plans that consider the sustainability of a wetland site not in isolation, but in relation to its surrounding geography, land uses, and communities.

As planners suggest, several factors influence the success of a management plan in realizing its objectives (Creighton 1992; Steiner 2000). Plans need careful preparation and clearly defined implementation strategies and administrative responsibilities with realistic budgets and timelines to meet desired goals and objectives. They need to be adaptive or responsive to feedback received during the entire planning process. Wetlands are multifaceted and dynamic ecosystems and plans should be flexible enough to incorporate both expert and stakeholder feedback by fine-tuning strategies and practices. Doing so would more appropriately reflect changing conditions and result in productive outcomes. Finally, management plans developed in consultation with a wide range of stakeholders and receiving their broad support have the greatest chances of success. The general steps in the wetland management planning process are outlined below (Fig. 12-2) and follow suggestions from planners and wetland managers (Steiner 2000; Ramsar Convention Secretariat 2007b).

- **Vision** – All resource management plans require an initial vision statement. This

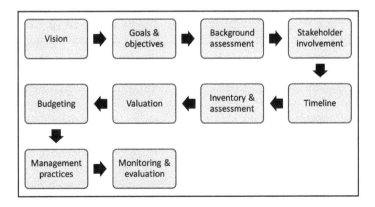

Figure 12-2. Schematic illustration of steps involved for wetland management planning.

statement contextualizes the wetland management issue, identifies the general threats and challenges, and outlines an expected future, which maintains or improves the ecological health and character of a wetland.

- **Goals and objectives** – In addressing a problem or issue that presents itself to a particular community, a management plan must include goals and outcomes and list expected long-term ecosystem and societal benefits through its implementation. Goals must be detailed, unambiguous, and measureable, so that a plan's success or failure may be assessed and lessons learned can be transferred to other situations.
- **Background assessment** – Having a clear idea of the current situation is critical in defining future pathways for improvement. Scoping the geographic context of the wetland site and surrounding land uses, gauging wetland-use practices and management strategies, identifying key stakeholders, determining legal parameters, zoning regulations, and other protection ordinances impacting the site are all essential before initiating a project.
- **Stakeholder involvement** – Studies have demonstrated that citizen and stakeholder involvement at the planning stage of each step improves its chances of success. Involving community members and user groups in identifying issues and proposing solutions which involve their future builds consensus and provides credibility and legitimacy to a plan (Creighton 1992). Moreover, involving stakeholders could provide innovative solutions to particularly vexing wetland management problems that include multiple community or user groups.

  To work, stakeholder involvement needs to be transparent and the benefits of participation should be made obvious. Surveys, public workshops, field trips, educational forums, and focus groups may all help include a diverse group of stakeholders, represent their views and opinions and tap into their knowledge. Stakeholder feedback could be incorporated into an adaptive management strategy whereby new knowledge or findings may readjust or clarify goals and strategies. Similarly, a plan must also involve other players including local government, non-profit agencies and public land-trust institutions in the process. Groups such as Ducks Unlimited, The Nature Conservancy, and similar other organizations within the United States possess a wealth of information that could be beneficial during the inventorying and implementation phases of a project.
- **Timeline** – Developing a realistic timeline that specifies deadlines for the completion of particular components of a plan is essential to its overall success. The time required to complete different aspects of a project depends on its scale and scope. Providing enough time for different tasks is necessary to ensure that important components are not overlooked.
- **Inventory and assessment** – This step involves a systematic data collection effort to document and characterize the biophysical and socioeconomic characteristics of a site. Generally, this includes field surveys and other mapping, remote sensing, and socioeconomic inventorying techniques. Due to financial and time constraints, however, a rapid assessment program which may be completed within a day or less provides a suitable alternative. Organizations such as Ramsar and the U.S. Environmental Protection Agency recommend several different approaches for the rapid assessment of ecological conditions and anthropogenic impacts within a wetland site (Fennessy, Jacobs and Kentula 2004). These have been tested at various sites and under various conditions with a wide variety of metrics that produce robust and verifiable results.
- **Valuation** – A cost–benefit or other evaluation technique as explained in chapter 11 could be implemented to present alternative strategies and help in their selection.
- **Budgeting** – Developing a financial plan to pay for the costs incurred in implementing the different components of the management plan is an important step in the process. Identifying specific sources of money,

whether they are from federal, state, local government, non-profit or private sources is undertaken at this stage.

- **Management practices** – Based on the overall objectives as well as the inventory, valuation and budgeting steps completed earlier, the management plan selects the most appropriate combination of practices to implement at the project site. A review of some of these practices is provided starting in the next section of this chapter.

- **Monitoring and evaluation** – Implemented management strategies need to be monitored and periodically evaluated for their effectiveness in meeting stated goals. Adjustments to plan components may become necessary as new issues emerge or unexpected problems are encountered.

12.4 Wetland management practices

The resource management practice used at a particular site would depend on what type of wetland is being managed and for what specific purpose. Management practices vary depending on whether one is working in a riparian wetland, peatland, coral reef, or salt marsh, and would be conditional upon whether the wetland is being managed for bird habitat, to provide a source of fresh water and sustenance to a local community, or to enable transportation and navigation. In each of these cases, the management strategies involved would be somewhat different. Yet, in each case, a wetland manager must be mindful of the impacts that different strategies could have on other components of the wetland ecosystem and its dependent human community.

Wetland management strategies are generally based on current scientific assessments and the collective wisdom of successfully implemented past strategies. However, improvements in technology and advances in the knowledge of wetland functioning require a regular re-assessment and fine-tuning of strategies used. A survey of past and current terrestrial-hydrologic, biochemical, and socio-economic strategies to manage wetlands is provided in Table 12-1.

Table 12-1. Wetland management strategies and practices.

Terrestrial & Hydrologic	Biological & Chemical	Socio-economic
• Ditching	• Grazing	• Moral suasion
• Dredging & channeling	• Stocking	• Mutual enforcement
• Agricultural tiling	• Propagation	• Incentive and disincentive mechanisms
• Vegetative buffering	• Species introductions	
• Water level manipulation	• Chemical applications	• Taxes, fees, subsidies, tradable credits, bonds, covenants, easements, transfer of development rights
• Vegetation manipulation		
• Terrestrial modifications		
• Prescribed burning		• Stakeholder incentives
• Mowing, disking, plowing		• Green labeling & certification

12.4.1 Terrestrial and hydrologic-based strategies

Terrestrial modifications to maintain wetland habitats or alter them into other land uses have included ditching, dredging, filling, levee construction, agricultural field tiling, and impoundment works. Depending on the extent of modifications undertaken, these may preserve basic ecosystem processes – allowing the wetland to sustain itself – or alter the hydrological regime of a wetland, cause unnatural water fluctuations and flows, and destroy its ecological balance. Modified wetland ecosystems have sustained human populations for centuries and today one would be hard-pressed to find wetland sites entirely free of human influence. Some of the commonly observed methods to manage wetlands are elaborated below.

Ditching involves the construction of channels through marshes and peatlands with the aim of draining pool and pond features and controlling the accumulation of surface water (Resh 2001). Historically, this process was carried out to aid land-use conversions and control mosquito populations. Along the northeastern coast of the United States, for instance, deep parallel ditches were constructed in the early twentieth century to lower water tables, avoid the accumulation of standing surface water, and thereby reduce mosquito breeding sites (Anon 1997; Resh 2001). Sediment removed in the process was spread across the marsh surface to create elevated areas, which were then used for agriculture or cattle grazing.

Dredging and channeling are undertaken to excavate bottom sediment of river banks and floodplains and deepen open standing water. Dredging operations may be quite large in scale and involve the use of underwater mechanical excavators connected to pipes that transport bottom sediment to barges or to shore banks. Such operations are regularly undertaken to ease navigation through major inland and coastal waterways.

Biotic communities adapt to the natural dry and wet cycles experienced by wetlands. Anthropogenic interferences with these flow cycles, including the timing of peak flows, continuity of flows and rapidity of change in flow patterns, have consequences on dependent communities (Welcomme et al. 2006a). For wetlands whose hydrologic flows have been altered, mimicking natural low- and high-water cycles through controlling water levels could serve an important function in supporting dependent communities and biotic integrity (Snodgrass and Burger 2001). Active water-level management through a system of locks and sluices may influence the composition and structure of wetland vegetation, and provide suitable breeding, foraging and spawning habitat for bird, fish, and other animal populations (Fig. 12-3). Furthermore, water-level management also serves to control nutrient concentrations, dissolved oxygen, water clarity and quality, and may help manage the spread of invasive species (Laubhan and Roelle 2001; Snodgrass and Burger 2001).

The semi-natural pond wetland system observed across parts of China and elsewhere in Asia provides an example of managed water levels (Yin, Shan and Mao 2006). The shallow pond, small dam, weir-and-sluice landscape found in agriculturally intensive parts of China demonstrates an actively managed water irrigation and distribution system. By providing water-storage capacity, these maintained pond wetlands act as buffers during heavy seasonal precipitation events and provide an environment for fish farming. A sluice system that connects individual ponds to the regional drainage system also channels water to fields for

Figure 12-3. A dam and outlet spillways at Little Salt Marsh function to store water and regulate downstream flow in the Quivira National Wildlife Refuge, south-central Kansas, United States. Twenty-one miles (34 km) of canals along with many levees and other drainage-control structures are used to divert, hold, and distribute water throughout the refuge. Kite aerial photo by J.S. Aber and S.W. Aber.

Figure 12-4. Borders of riparian forest provide buffers between the river channel and agricultural fields on the floodplain. A. Neosho River valley, east-central Kansas, United States. B. Hornád River valley on the border between Slovakia (foreground) and Hungary (left background). The river here forms a large meander loop; two overflow flood channels (c) cut across the narrow neck of the loop. Kite airphotos by S.W. Aber, J.S. Aber, and J. Janočko.

irrigation. Thus, this ingenious long-standing practice not only provides protection from storm water, but also services dependent agri-fishing communities (Yin, Shan and Mao 2006). Similarly, ponds excavated into floodplains across parts of West Africa and South America provide fish culture environments stocked with fish trapped by retreating floods or through aquaculture (Welcomme et al. 2006b).

Agricultural tiling involves the laying of perforated drainage tubes or pipes approximately 1 m below the soil surface to drain excess water from wet fields, ephemeral basins and other low-lying areas into ditches. Agricultural tiling was undertaken to help improve crop yields by removing excess soil moisture and provide more uniform growing and harvesting conditions throughout an agricultural field. During the nineteenth and twentieth centuries, extensive sections of the United States' agricultural belt were tiled to improve yields (Busman and Sands 2002). Conservation practices today recommend the removal of these systems to recreate earlier wetland conditions.

Vegetative buffer strips constructed and maintained along riparian zones or protectively encircling wetland sites provide some defense from adjacent land-use stressors (Gregory et al. 1991). Buffers provide habitat transition zones and may serve as corridor connectors for wildlife moving between wetland sites (Kent 2000b). They also help reduce the velocity of runoff into streams and rivers, guarding against bank erosion, and may aid in pollution abatement through the uptake of nitrogen and phosphorus (Fig. 12-4).

The construction of nesting platforms and sites for wetland fauna and bird life is an essential management strategy employed in numerous wetlands. Bird nest islands (Fig. 12-5), artificially controlled water levels, and manipulated vegetation assemblages provide conditions necessary to sustain migrating and nesting populations.

The role of natural fire regimes in maintaining ecosystem health has been well established. Wetlands also experience natural fires which help stimulate new growth, clear dead organic matter and undesirable vegetation, release nutrients, and open up bird nesting habitat (Kent 2000b). Prescribed and controlled burns within wetland sites may be initiated for similar objectives (see chapter 8.5). Additionally, controlled burns are also undertaken to alter species compositions and often used to assist in eliminating invasive species such as cattail (*Typha* sp.). The timing of prescribed burns is an important consideration and depends on the purpose of the burn. Generally, burns to remove invasive species are timed to coincide with seed germination cycles, while those undertaken to provide bird habitat need to be completed well before the breeding and nesting season (Laubhan and Roelle 2001).

Figure 12-5. S-shaped artificial island for bird habitat in flooded pools at the state Cheyenne Bottoms Wildlife Area. Levees in the background are part of an extensive system to control water levels in various pools for habitat management. Kite airphoto by J.S. Aber and P. and J. Johnston; central Kansas, United States.

Supplementary mechanical methods of controlling plant species distributions and cover types in wetlands include mowing, disking, and plowing (see Fig. 6-36). Such practices are generally undertaken to mimic more natural processes such as grazing and burning, and to create suitable foraging and nesting bird and animal habitat. Disking breaks up sod and low vegetation and aerates the soil surface (Kent 2000b). Mechanical methods may also be effective in controlling problematic vegetation and altering vegetation assemblages (Laubhan and Roelle 2001).

12.4.2 Biological and chemical strategies

Domesticated and wild animals graze in wetlands across the world and provide an important contributing factor in maintaining vegetation structure in terms of height, density, and plant species distributions (Laubhan and Roelle 2001). Grazing may also be effective in managing invasive species, helping control the spread of aquatic plants, and maintaining the biological diversity of wetlands. *Spartina patens* (salt hay) dominates coastal marshes along the northeastern coast of the United States that traditionally served as cattle-grazing areas (see Figs. 11-1 and 11-2).

Floodplain grasslands have also been invaluable to dependent ranching communities. For example, in the Pantanal, seasonally wet grasslands occupy close to 30 percent of the total area and low-intensity cattle ranching has been a part of this landscape for more than 200 years (Mourao et al. 2005). Increasing population pressures, however, could rapidly change this scenario as deliberate manipulation of water levels and more sustained and intensive grazing practices promote wetland conversions into pasture land. Overgrazing may also be detrimental to wetland habitats, due to trampling and soil compaction, and when forage areas are significantly depleted dependent bird communities experience shortages (Boojh, Patry and Smart 2008).

Flora and fauna are often actively managed at wetland sites across the world. For instance, stocking is undertaken to support local fisheries, maintain adequate supplies for recreational purposes (Fig. 12-6) or ecological balance, and sustain threatened or endangered species (Welcomme et al. 2006b). Propagation is another method whereby sections of wetlands are artificially seeded through transplants, cuttings, tubers, or rhizomes (Kent 2006b). Local species are generally selected for this purpose and their suitability to the biotic and hydrologic characteristics of the site are closely assessed before propagation is undertaken (Kent 2006b). As chapter 6 illustrates, species are also managed to control troublesome and persistent invasive populations.

Figure 12-6. Fly fishing in a shallow subalpine lake that is stocked for sport fishing in the Culebra Range, Sangre de Cristo Mountains of southern Colorado, United States. Photo by J.S. Aber.

Herbicides are used to temporarily manage invasive vegetation when all other methods fail to control its spread. The problems with non-selective herbicide use are widely reported. Targeted herbicide applications need to be undertaken with care, under calm wind conditions and where treatment is limited to smaller regions. Herbicide applications are often undertaken with other control measures such as burning, disking, or flooding to improve success rates.

12.4.3 Socio-economic strategies

The strategies outlined above describe specific techniques to manage wetland ecosystems across the world. Social scientists have also reflected on what sorts of mechanisms could influence human behavior to realize desirable environmental ends, whether those include reducing pollution or engaging in sustainable resource extraction practices. Socio-economic strategies encompass a broad range of mechanisms or instruments that are used to guide and regulate human access to and use of environmental goods and services. These instruments reflect the broader geographic and policy context of the region within which they are implemented and generally direct behavior through a series of incentives or disincentives.

Incentive- or disincentive-based mechanisms may be informally or formally established and implemented. Informal mechanisms are those that are generally agreed upon and observed by different parties even in the absence of any formal rules that require them to participate. The most common form of an informal mechanism is moral suasion. Urging individuals to adopt conservation practices by "doing the right thing" and appealing to their civic and ethical responsibilities toward the environment is a classic example of moral suasion. Rather than coercive measures that enforce rules of environmental use, moral suasion relies on the voluntary adoption of best practices (Weersink 2002).

The 1960s environmental movement provides a classic example of moral suasion changing the attitudes and practices of people worldwide with respect to the environment, natural resource use, recycling, the ethical treatment of non-humans, and so on (Handy 2001). Positive spillover effects tend to be high when moral suasion is applied, as it is not limited to targeting just one environmental practice; rather it works to voluntarily change a particular mindset (van Kooten and Schmitz 1992). The strategy followed by the Canadian Ministry of Agriculture vis-à-vis the adoption of sustainable agricultural practices provides a good example (Weersink 2002). The Ministry hopes to persuade farmers to adopt more sustainable practices by conducting education and awareness campaigns, rather than implementing more restrictive mandates. Chapter 14 discusses how Canada has used a similar approach when implementing wetlands policy.

Some social scientists disagree, however, that such voluntary measures are effective. They argue that not all individuals are swayed into adopting conservation practices based on civic, ethical, or moral arguments and hence the potential for the "free rider" problem is high with voluntary measures. Moreover, moral suasion suggests that individuals have few other constraints placed on them when adopting conservation practices. This is not the case, particularly in the developing world where communities are entirely dependent on wetland resources to meet basic needs. Poverty or other circumstances may, in those cases, force individuals to engage in unsustainable resource extraction strategies. In those cases, arguments appealing to an individual's civic responsibility are unlikely to be taken into consideration when engaging in resource use and extraction.

Another informal strategy to change behavior toward natural resource use through voluntary means is by applying pressure on individuals to adopt and abide by conservation practices that are in the long-term interest of all stakeholders. The mutual enforcement of informal rules amongst members of a community dependent on a particular resource provides a case in point. For instance, as Acheson and Brewer (2003) detailed in the case of the lobster fishery off the coast of Maine, informal rules and an aggressively protected territorial system, guard against incursions by newcomers and ensure a sustainable livelihood for well-established lobstering communities. In another example, fisherfolk along the western coast and estuaries of India abstain from fishing during the religiously auspicious month of *shraavan*, which also coincides with the monsoon season and fish-spawning cycles. In this case, religious tradition melds with a conservation practice that allows for the sustainability of local fish stocks and livelihoods.

Formal conservation management strategies are more ubiquitous, and some might argue more effective at promoting conservation behavior through either a regulatory approach or a system of incentives and disincentives. Command-and-control approaches attempt to change the environmental behavior of individuals or companies directly through legislation. These approaches include enforcement mechanisms through a system of fines and penalties that ensure compliance. Chapter 14 more fully explores environmental laws, policy and standards explicitly targeted toward wetland conservation.

Other formal incentive-based wetland conservation strategies provide more flexibility in meeting environmental objectives than their command-and-control counterparts. These strategies seek to employ market approaches through taxes, fees, charges, subsidies, tradable credits, bonds, covenants, easements, transfer of development rights and the like to urge individuals to adopt conservation practices that work for the larger common good (Hackett 2006). Economists have argued that such approaches provide adequate incentives for conservation-seeking behavior and encourage companies to innovate by coming up with greener technologies that make them more competitive.

Examples of direct and indirect incentives-based strategies for wetland conservation are found at both the federal and state levels in the United States. For instance, several states charge a special tax on the sale of fertilizers and pesticides. This money is often set aside for water-pollution abatement programs. Non-point source pollution such as runoff from agricultural fields is particularly difficult to track and monitor. Hence, such indirect taxes seek to instill more judicious pesticide and fertilizer use by farmers. In other cases, point sources of pollution from factories are easier to monitor and are assessed discharge fees.

At the federal level several programs exist to incentivize wetland conservation programs directly. The U.S. Natural Resources Conservation Service (NRCS) oversees the Wetlands Reserve Program (WRP) created through the 1990 Farm Bill, which currently provides payments to farmers who restore and protect once farmed wetlands on their property (NRCS 1996). Over 900,000 hectares are currently enrolled in the WRP across the United States (Fig. 12-7; NRCS 2010b). In another program intended to reverse the decline in waterfowl populations the U.S. Fish and Wildlife Service, through the North American Waterfowl Management Plan, gives farmers economic incentives for incorporating farming practices that could benefit waterfowl habitat (NRCS 1996).

Similar tax and incentive programs in countries across the world attempt to balance development needs with wetland conservation practices. For instance, the European Union (EU) imposes environmental taxes and fees on pesticides, landfills, and water pollution (Hackett 2006). The EU provides authorization to individual countries to promote wise-use farming practices that incorporate the

Figure 12-7. Playa-marsh landscape in the foreground is enrolled in the U.S. Wetland Reserve Program. Wind farm in the background has minuscule environmental impact on the wetland habitat. Photo by J.S. Aber, southwestern Kansas.

conservation of environmentally sensitive areas through incentive programs that are funded jointly by the EU and member nations (Shine and de Klemm 1999). These programs have been implemented widely across EU nations. As Shine and de Klemm (1999) outlined, the Countryside Stewardship Scheme implemented in the United Kingdom provides payments to landowners for wetland-based rehabilitation and restoration efforts. In Australia, programs offer property tax rate rebates and concessions to landowners who voluntarily conserve wetlands and rehabilitate native vegetation (Whitten et al. 2002).

Beyond these formal and informal mechanisms to promote conservation best practices applicable to wetland settings, institutional ecologists have suggested that strengthening stakeholdership in a resource and making property regimes relevant and responsive to dynamic local conditions are important considerations. More recently, green labeling and certification aim to connect the supply and demand side of products by recognizing those producers that intentionally engage in biodiversity preservation or practice sustainable extraction methods. Such efforts may have indirect impacts on wetlands, although the reliability of certification and labeling programs has been questioned by some. The peat producers Fafard et Frères and Premier Tech Horticulture in Québec, Canada, became the first companies worldwide to obtain VeriFlora certification in 2010 for responsible management of peatlands (*Peat News* 2010). This certification assures that good management practices are followed in all phases of peat extraction and bog restoration in order to achieve sustainable development.

12.5 Summary

This chapter provides a brief history of conservation thought in the United States, Europe, and elsewhere. The review suggests that there was a great deal of cross-fertilization of conservation ideas between different countries during the nineteenth century. Moreover, European colonialism spread these ideas and practices across parts of Africa and Asia. While wetlands did not figure prominently in discussions on conservation during this early period, the wetland conservation policies and practices implemented today rest on this shared history and knowledge.

Scientists have recognized the importance of approaching wetland management from an integrated systems perspective. Such an approach ensures that the multifaceted nature of wetland ecosystems and their complex interactions are recognized and taken into account. To this end, wetland management or stewardship plans need to consider a wetland site in relation to its surrounding geographic and socio-economic context. A plan's design elements must reflect locally specific conditions, incorporate the views and knowledge of a broad range of stakeholders, and be mindful of the essential role these resources play in providing livelihoods for dependent communities. In other words, wetlands must be managed not just for their ecosystem services but also their significant contributions to humans.

Wetlands are dynamic and ever-changing systems that have been altered through human use and ecological processes. As Botkin (1995) argued, acknowledging and understanding the complexity of change in nature, whether induced by humans or not, allows us to arrive at sustainable solutions to confront modern-day environmental problems. Over the centuries, our knowledge and management strategies have shifted from converting wetland ecosystems into other land uses, to integrating conservation practices that may ensure their sustainability. A survey of commonly used terrestrial, hydrological, biological, chemical and socio-economic management practices reveals that wetlands require multiple management techniques – physical, chemical, biological, and social. The unique combination of practices implemented at any given wetland site depends upon the goals and objectives of management and the characteristics of the site itself.

13 Wetland restoration, enhancement and creation

13.1 Introduction

The contemporary field of restoration ecology has relatively recent origins, dating back to the early 1980s. It draws theoretically from disciplines such as ecology and conservation biology. While ecology focuses on the interactions between organisms and their environment, conservation biology studies and monitors individual species and ecosystems with the aim of providing appropriate management strategies for their conservation. Both these fields provide the scientific and theoretical basis for restoration ecology.

Restoration ecology focuses broadly on restoring habitats that may have been degraded or converted to other land uses through human or other modifications. The Society for Ecological Restoration International (SERI), an organization that promotes the restoration of ecosystems, defined it as "the recovery of an ecosystem that has been degraded, damaged, or destroyed" (SERI 2004). A more specific definition is provided by Cairns (1988, p. 3), who suggested that restoration ecology comprises the "full or partial placement of structural or functional characteristics that have been extinguished or diminished and the substitution of alternative qualities . . . with the proviso that they have more social, economic, or ecological value than existed in the disturbed or displaced state."

To provide sustainable solutions for degraded ecosystems, restoration ecology is firmly grounded in scientific understandings of how ecosystems function and their resilience and vulnerabilities. It also incorporates an understanding of ecological engineering or bio-engineering techniques to provide suitable alternatives (Mitsch and Jørgensen 2003). Furthermore, restoration ecologists are attentive to the critical role played by communities and stakeholders in the long-term success of such undertakings and are focused on enhancing the joint socio-ecological value of restored ecosystems.

In general usage, the term "restoration" implies a return to some previous state. Yet, what should this state be? For some, it invokes the notion of a wilderness state prior to human intervention. However, not only is this impractical across most of the world today, but it also negates the vital role humans play in shaping our environments, including wetlands. The Society for Ecological Restoration International suggests that restoration activities must be targeted to accelerate ecosystem recovery with respect to its health (functional processes), integrity (species composition and community structure), and sustainability (resistance to disturbance and resilience) (Clewell, Rieger and Munro 2005, p. 2). Thus, restoration projects should direct attention to process and function rather than recapturing the conditions present

---

Wetland Environments: A Global Perspective, First Edition. James Sandusky Aber, Firooza Pavri, and Susan Ward Aber.
© 2012 James Sandusky Aber, Firooza Pavri, and Susan Ward Aber. Published 2012 by Blackwell Publishing Ltd.

at a historical moment in time for any given site. Given the complexity of such tasks, restoration projects likewise require long-term commitments in management, monitoring and evaluation.

Restoration projects necessitate the use of a reference ecosystem that could serve as a model or point of comparison and evaluation. For instance, to restore a wholly altered coastal marsh, information on hydrology, soil conditions, species compositions and vegetation assemblages prior to degradation and alteration are important and necessary starting points. However, these data may not be easily available from the altered wetland site itself. In those instances, restoration ecologists refer to a site or, ideally, multiple sites that exhibit similar geographic and ecological conditions. Collectively, these sites may provide information on the expected range of ecosystem conditions and characteristics. In so doing, they fill the information gap and provide a direction to work toward (IWWR 2003). Finally, reference sites also are used to measure the long-term success or failure of restoration undertakings (Kusler 2006). Without such comparative assessments, the successes of capital and time investments would be difficult to gauge.

13.2 Terminology

The terminology associated with wetland repair often varies by geographic region. Restoration, enhancement, creation, rehabilitation, reclamation, repair, and mitigation are just some of the more commonly used terms (Fig. 13-1; van der Valk 2009). It is important to distinguish between these terms to ensure consistency of usage

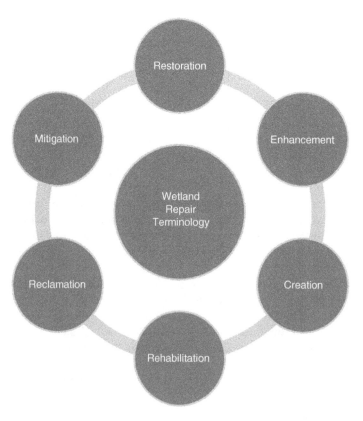

Figure 13-1. Schematic illustration of terminology commonly used in wetland repair.

across agencies working to protect and enhance wetland resources. The Interagency Workgroup on Wetland Restoration (IWWR) within the United States, which includes the National Oceanic and Atmospheric Administration, Environmental Protection Agency, Army Corps of Engineers, Fish and Wildlife Service and Natural Resources Conservation Service, provided the following definitions for standard wetland repair projects undertaken (IWWR 2003). Variations of these definitions are used in other geographic contexts and by other non-governmental and conservation organizations.

Wetland restoration is the return of a "degraded wetland or former wetland to a pre-existing condition or as close to that condition as is possible" (IWWR 2003). In restoration, efforts are made to repair damage to the structure and functions of natural ecosystems. For instance, if hydrologic changes brought about by the construction of dams or ditches have caused the desiccation of a wetland, then a restoration project might attempt to re-establish wetlands through engineering efforts that remove or adjust impoundment structures, fill in ditches, or take out agricultural drainage tiles. **Reclamation**, on the other hand, usually involves conversion of wetland to another type of land use such as agriculture or forestry (Quinty and Rochefort 2003).

Restoration projects typically emphasize restoring function and process within a wetland ecosystem. For example, the Nature Conservancy has worked at the Cheyenne Bottoms wetland complex in central Kansas by restoring former agricultural fields into wetlands through contouring shallow seasonal water-accumulating depressions, prescribed burns, grazing, removing fences, and allowing wetland plants from adjacent wetland sites to re-colonize through natural regeneration processes (Fig. 13-2). Restoration projects are undertaken across the world. However, due to substantial capital and resource investments, such projects are concentrated mostly in developed countries and for the most part are limited in geographic scale. Even so, some restoration projects including those undertaken in the Florida Everglades and within the deltaic marshes of the Tigris and Euphrates rivers and the Mississippi Delta are notable for their size and scope.

Wetland enhancement includes increasing one or more of the functions performed by an existing wetland beyond what currently or previously existed in the wetland. There is often an accompanying decrease in other functions (IWWR 2003). Enhancements may be undertaken on functioning wetland sites or those that have witnessed some degradation. Enhancements serve to deliberately augment certain wetland functions for specific purposes. The construction of bird habitat through the building of nesting sites and bird islands or the deliberate control of water levels to deter invasive species and encourage native wetland vegetation are examples of such measures. These efforts may yield unexpected benefits. At Quivira National Wildlife Refuge, for instance, water level in the Big Salt Marsh was lowered in 2010 in order to control invasive plants (Fig. 13-3). When much of the marsh dried up, many carp (*Cyprinus carpio*) died, which provided a fish feast for birds, including whooping cranes (*Grus americana*) and more than 200 bald eagles (*Haliaeetus leucocephalus*), the most ever seen at the refuge (Pearce 2010).

Wetland creation involves converting a non-wetland (either dryland or unvegetated water) to a wetland (IWWR 2003). In this case, wetland conditions including hydrology, vegetation and wildlife are established with appropriate soil that may, over time, support wetland ecosystems. Wetland creation is undertaken for a variety of purposes. Within the United States, Section 404 of the Clean Water Act requires wetland creation or enhancement projects to be undertaken whenever wetlands are destroyed or converted for development activity. Wetland creation may serve socio-ecological functions and balance the impacts of development on the environment.

Wetland creation is a significantly larger engineering undertaking than wetland enhancement or even, in some cases, restoration. Studies have indicated that such efforts may not always live up to expectations and result in wetlands that have limited functions or are expensive to

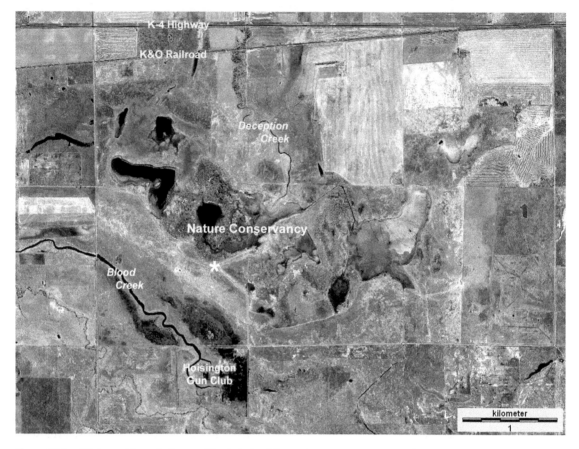

Figure 13-2. High-spatial-resolution satellite image of the Nature Conservancy marshes in Cheyenne Bottoms, near Hoisington, central Kansas, United States. Water bodies are black, and active vegetation is light gray to white. Asterisk (*) indicates site for long-term observations with small-format aerial photography. Ikonos panchromatic (green + red + near-infrared) band, 11 July 2003. Image processing by J.S. Aber.

maintain (IWWR 2003). Many require constant management and long-term oversight. An alternative focus of wetland creation across the world has also been to serve as water-quality treatment systems. These artificially created treatment wetlands serve numerous purposes, including waste-water treatment and managing non-point-source pollution (DeBusk and DeBusk 2001; Mitsch and Gosselink 2000; van der Valk 2009).

Wetland mitigation is a term frequently used in the United States and generally associated with Section 404 of the Clean Water Act of 1972 and as amended in 1977. This act mandated that any development or conversion activity directly impacting wetlands must be balanced with the restoration, enhancement or creation of a wetland or wetland features elsewhere (Lewis 1989). Mitigation is expressly undertaken within the United States to ensure that wetland destruction at one site is suitably compensated with its creation or enhancement elsewhere. Such mitigation wetlands are often seen in connection with highway construction (Fig. 13-4). The United States' No Net Loss Policy similarly endorses wetland mitigation activities to ensure that there is no further decline in the total wetland base of the country.

13.3 Wetland restoration, enhancement and creation design principles

Principles guiding the design and restoration of wetlands incorporate ecological, economic and social considerations to ensure self-sustaining and resilient systems able to withstand natural and human-induced stressors. Humans play an integral role in the health of wetland ecosystems and their participation is essential to ensure productive outcomes. Design principles acknowledge the vital role played by dependent stakeholder communities and accommodate their needs. Promoting sustainable resource-use strategies that meet local needs yet preserve vital wetland functions and processes necessitates carefully crafted adaptive management plans that ensure stakeholder input (Society for Ecological Restoration International and IUCN Commission on Ecosystem Management (SERI IUCN) 2004).

Restoration projects that reflect local geographies and meld into the broader landscape stand the best chance of long-term success. Based on the accumulated knowledge of wetland restoration, enhancement and creation projects from across the globe, an inventory of critical principles and design elements emerge as key to their long-term viability and success (Hammer 1997; Hay and Philippi 1999; Kent 2000a; U.S. Environmental Protection Agency 2000; Zedler 2001; Mitsch and Jørgensen 2003; Quinty and Rochefort 2003; Ramsar 2003; Society for Ecological Restoration International 2004; SERI IUCN 2004; Zedler and Kercher 2005; Mitsch 2006; Money et al. 2009; Ramseier et al. 2009; van der Valk 2009; Natural Resources Conservation Service 2010c). These guiding principles, synthesized from several studies, include the following (Fig. 13-5):

Self-organizing and sustaining systems – The design of wetland restoration and creation projects to mimic natural self-sustaining ecosystems is key to their long-term success independent of human management. Ecosystems may be thought of as flexible, self-organizing units that adapt continually to their changing environments (Mitsch and Jørgensen 2003). Designing

Figure 13-3. Big Salt Marsh wetland complex at Quivira National Wildlife Refuge in south-central Kansas, United States. Saline ground water discharges into springs and streams that feed into the marsh. Water flow is regulated through a series of canals, levees, and control structures. View northeastward; kite aerial photograph by S.W. Aber and J.S. Aber.

Figure 13-4. Mitigation wetland at the intersection of two highways in eastern Kansas, United States. Cattails and spike rush emerge among shallow pools (A) in a wetland that attracts a large variety of waterfowl and shorebirds, including these Canada geese (B). Photos by J.S. Aber.

Figure 13-5. Schematic illustration of design principles for wetland restoration, enhancement and creation.

wetlands to mimic such conditions makes them more resilient to threats. Resiliency allows an ecosystem to maintain its critical functional characteristics despite external natural or human-induced stressors or disturbances. The focus on restoring natural functional processes through the bioengineering of hydrological patterns, nutrient cycles, native floral and faunal communities, and dynamic erosion and deposition processes could ensure such ecological integrity (U.S. Environmental Protection Agency 2000).

Systems approach – As explained in chapter 12, a systems approach is one that adopts a holistic view, considering an ecosystem in its entirety rather than as separate individual component parts. In designing restored or newly created wetlands, such an approach may more effectively address interrelated processes and functions and anticipate the impact of changes made to one element of the system on other areas.

Using a systems approach, design elements may borrow and synthesize the most appropriate elements and technologies from a wide diversity of fields including biology, chemistry, ecology, engineering, hydrology and others. The implementation of a restoration project hence necessitates expertise from a wide variety of fields and from a range of stakeholders that may each contribute to a unique understanding of a system's functional and structural aspects. Moreover, adopting a systems approach makes it possible to anticipate the response of a system to changes implemented and, as such, could be critical to long-term success.

Restoring structure and function – Alterations to the structure and morphological characteristics of wetlands due to their conversion to other land uses may affect ecological functions. For example the draining of wetlands in the American Midwest through agricultural drainage tiles resulted in degraded wetland habitats and changes to water-flow regimes, rendering these areas unsuitable for native wetland biota. Restoration techniques focus on re-engineering morphological site characteristics, which may, in turn, help re-establish wetland functional processes including hydrological regimes, appropriate soil conditions and other beneficial biogeochemical and nutrient cycles (U.S. Environmental Protection Agency 2000). Moreover, recommendations favor the implementation of simple, ecologically sound structural engineering approaches that are less likely to give rise to unanticipated complications down the road.

Employing renewable resources – To mimic natural systems and minimize long-term human maintenance costs, a designed wetland should incorporate as many renewable resources as possible (Mitsch and Jørgensen 2003). For instance, designs could make use of natural sources and flows of water rather than attempt to pump water in from a distance. Instead of building erosion buffers from fabricated materials that might require constant maintenance, planning riparian buffers and vegetated stream banks to minimize erosion and stabilize banks during high-water events would be more sustainable. Similarly, artificial vegetation management techniques that require constant upkeep should be avoided (U.S. Environmental Protection Agency 2000).

Using locally available resources would support ecological integrity and make the restored system more resilient to disturbances and adaptable to changes (U.S. Environmental Protection Agency 2000). Harnessing the power of flowing water to deposit silt and alluvium along riparian or deltaic wetlands and tidal action to flush out nutrient and sediment loads along coastal marshes minimizes the need for artificial equipment to complete such tasks. Such actions ensure a self-propagating system that requires minimal human intervention once an initial set of conditions has been established (Mitsch and Jørgensen 2003). Even so, restored systems do require some long-term management to ensure their sustainability.

Reflecting local geography – A wetland rehabilitation project must be integrated in the context of the broader landscape and watershed within which it is located, either through biotic or abiotic exchanges (Society for Ecological Restoration International 2004). Restoration projects are necessarily localized. Yet, their design must incorporate elements that also are germane to the broader socio-ecological context. Bioengineering of stream banks to prevent erosion would have positive downstream externalities beyond a particular wetland restoration site. Similarly, a restoration project must consider how it interfaces with or complements and augments other related projects in the area that might also be working to expand habitat corridors, minimize pollution, control runoff, improve water quality, and control erosion (U.S. Environmental Protection Agency 2000).

Adaptive management – Adaptive management approaches ensure that adjustments could be made to the overall restoration or rehabilitation plan during any stage of its implementation based on new information received or field observations (Ramsar Convention Secretariat 2007a). Restoration efforts do not always follow a set path and many unanticipated events could necessitate changes to overall plans. Adaptive management allows for adjustments to be made throughout the implementation of a project. Project monitoring is, thus, essential, and progress must be evaluated continually against expected goals.

Balancing ecosystem and human needs – Wetlands provide both human and ecosystem services. Restoring wetlands requires a fine balance between meeting the needs of both. In the developing world, wetland ecosystems have been molded through centuries of human use. In certain cases, resource-use practices have struck a remarkably fine balance between meeting human needs and allowing for the regeneration of the ecosystem and the preservation of its ecological processes. However, increased demand for resources may place pressures on existing ecosystems. Restoration projects in such contexts might incorporate indigenous resource-use strategies and traditional knowledge that benefit the recovery of these ecosystems (Society for Ecological Restoration International and IUCN Commission on Ecosystem Management (SERI IUCN) 2004; Ramsar Convention Secretariat 2007a).

If humans form an integral part of how a wetland is used, their views and needs must be integrated and balanced with the ecological and conservation objectives of a restoration project.

For successful outcomes, the restoration process should be open to community participation, and stakeholders must have their opinions heard and included from the early stages of project inception through post-completion stewardship (Ramsar 2003). Research from the developing world indicates that the long-term participation of locally dependent communities is often a necessary condition for the success of ecological restoration projects (Society for Ecological Restoration International and IUCN Commission on Ecosystem Management (SERI IUCN) 2004).

Planning for disturbance – Restored systems must be resilient enough to withstand and recover from natural and human-induced stresses (Society for Ecological Restoration International (SERI) 2004). As with reference ecosystems, restored ones should be able to endure periodic storms, flood and drought events, fires, and changes in human resource-use patterns, without long-term degradation to structure and function.

Monitoring, evaluation and stewardship – All restoration projects start with a series of goals and objectives. Monitoring progress throughout the project plays an essential role in evaluating whether project objectives have been met and the restoration process has been successful. Monitoring through systematic data collection at periodic intervals helps gauge the performance of structural and functional processes implemented within a wetland site. To allow for temporal comparison, monitoring sites should remain constant and could be pre-determined in the initial planning stages of the project. They may comprise set transect lines, identified bench marks, well sites, stream-flow gauges, and water-quality monitoring stations, all of which could be surveyed using global positioning systems (GPS) and hence easily tracked over time (Clewell, Rieger and Munro 2005).

Evaluation criteria and performance expectations for restored ecosystems are generally spelled out during the initiation of a project. Reference ecosystems that display similar characteristics are often used to evaluate the performance of restored sites. Even though restored wetland ecosystems may be designed to be self sustaining, long-term stewardship mechanisms ensure that investments into restoration are sustained and pay continuing dividends.

It is difficult to predict the outcome of restored or created wetlands with any degree of certainty (Zedler 2000). This is largely due to the myriad conditions that influence the distribution and occurrence of species and the organization of different elements within the system. Trying to anticipate how these might play out beforehand is nearly impossible. Designing self-organizing systems which take advantage of natural processes should incorporate adaptations that could help an ecosystem withstand stresses and ensure long-term productivity.

13.4 Restoration and enhancement considerations

The scope and scale of wetland restoration or enhancement projects necessitate a significant amount of pre-planning. A sequence of steps similar to those outlined in chapter 12.3 of this book is often followed for restoration projects as well. Additional considerations in planning restoration projects are described in brief below and follow recommendations from national and international organizations such as Ramsar, the Society for Ecological Restoration International, the U.S. Environmental Protection Agency and the U.S. Natural Resources Conservation Service.

Selecting a site – The selection of a site for restoration itself is often an involved process and takes into consideration a host of factors (Ramsar 2003). These may include the:

- ecological importance of the site itself.
- expected environmental and social benefits from restoration, both local and those beyond the immediate region.
- economic costs to be expended.
- availability of financial resources.
- long-term economic benefits through increased tourism or habitat protection, etc.
- specific local benefits to communities and stakeholders.
- anticipated technical difficulties in executing a project and possible solutions.

Geographic analysis – Studying a site's geographic present and past, including its topography, surface- and ground-water hydrology, soils, species composition, and other landscape features through field surveys, maps, inventories and past histories is an important part of the restoration planning process and provides a wetland manager an understanding of constraining elements (U.S. Environmental Protection Agency 2000). Information compiled on how land is used at present, including both ownership and usufruct rights, and what resources are extracted from the site provides useful future resource management directions. Similarly information on water rights and water allocation should be inventoried to ensure their adequacy in meeting future needs.

A selected wetland site is understood within its larger geographic context. Regional landscape trends may be monitored not only to understand past changes but also to anticipate what future stressors might affect a wetland site (Society for Ecological Restoration International (SERI) 2004). Moreover, how restoration may augment the ecological services and nutrient and energy flows across the watershed should be considered (SERI 2004).

Understanding degradation – Inventorying conditions that have led to wetland degradation is a necessary part of the initial planning process. Older aerial photographs, topographic maps, planning documents, interviews with informants, or site histories provide information on development activities such as drainage ditches, roads, impoundment features, and impervious surfaces that may have been constructed within a wetland site or in the vicinity (U.S. Environmental Protection Agency 2000).

Selecting a reference ecosystem – As per the recommendation of numerous studies, a reference ecosystem or multiple ecosystems should be identified and used as models through the restoration process and later for evaluation and comparison purposes (IWWR 2003; Ramsar 2003; Society for Ecological Restoration International (SERI) 2004).

Planned enhancements – Restoration plans address how wetland functions and processes are enhanced or restored. Focus areas may address wetland hydrodynamics, soil conditions, vegetation and faunal habitat, and outline actions to be implemented to remove undesirable structures, vegetation or invasive species (Natural Resources Conservation Service 2010c).

Stakeholder input – Planning includes the input and support from stakeholders directly or indirectly affected by a restoration or enhancement project. The long-term success of restoration projects depends on the stewardship commitments of stakeholders and the immediate community. Providing incentives to local communities would help the long-term stewardship of resources (Ramsar 2003; Society for Ecological Restoration International and IUCN Commission on Ecosystem Management (SERI IUCN) 2004; Ramsar Convention Secretariat 2007a). Likewise, the inclusion of traditional knowledge in resource management and sustainable extraction practices may be an important component of a successful restoration project (Ramsar 2003). Publicizing a planned restoration project and soliciting input from the community serve a dual purpose of environmental education and ensuring that local concerns are addressed before implementation.

Implementation steps – A step-by-step plan must be devised outlining what specific techniques would be used in the restoration processes. It should include budgets and timelines for completion. Future operation and maintenance plans must be outlined for restoration projects that incorporate structural features requiring such operation and maintenance. For instance, water-level controls or water-distribution canals would need constant maintenance. Additionally, future maintenance plans must outline mechanical treatment or regular prescribed burns for vegetation health, management of invasive species, or other techniques that may be required to maintain the ecological health of a restored or enhanced system (Natural Resources Conservation Service 2010).

Explanations must be given of how adaptive management strategies would be incorporated into the execution of a project. An outlined plan should also indicate how modifications based on unforeseen conditions could be accommodated.

Monitoring and evaluation – Monitoring and evaluation plans both during and beyond the completion of the project need to be drawn up. Periodic assessments to evaluate the health and condition of a restored ecosystem must be implemented (Society for Ecological Restoration International (SERI) 2004).

13.5 Approaches to wetland restoration and enhancement

Systematic wetland enhancement and restoration projects have a relatively recent history. Even so, a broad range of techniques have been employed to restore and enhance the structure and functions of wetland sites globally. Thus far, resource and cost limitations have confined wetland restoration efforts to North America, Europe and other developed nations. However, emerging research on artificial wetlands from Asian nations suggests such efforts are gaining momentum across the developing world and may be indicative of future trends (Zhang et al. 2010).

The scale and scope of many restoration and creation projects often necessitate involvement by governmental agencies, industry, larger land trusts, and conservation organizations in such undertakings. This is a trend observed across the world and one in which multiple stakeholders play increasingly important roles. Long-term partnerships and scientific exchanges between such groups hold promise in advancing the science of wetland restoration more rapidly. Moreover, the scientific dissemination of successes from individual cases contributes to a more robust understanding of the challenges and solutions in such actions.

Given the complexity of biotic and abiotic interactions dominating wetland environments, scientists generally agree that complete restoration reflecting the diversity of life and processes that typify an original wetland site before disturbance is a most difficult undertaking. Restoration approaches and techniques implemented

Table 13-1. Approaches and methods for wetland restoration and enhancement.

Active Approaches	Passive or Hybrid Approaches
• Landscape contouring • Topographic excavation • Breaching levees • Dam & weir removal • Rerouting roads & embankments • Increasing culvert size • Reconnecting natural creeks and inlets • Restoring the hydro-period • Plugging drainage ditches • Removing spillways • Bioengineering • Species introductions • Connecting wetland patches through corridor connectors	• Restricting human or livestock access to wetland • Natural floral and faunal recolonization • Eliminating water controls • Allowing natural flood events

are selected based on the nature of the problem being addressed, the type of wetland and the unique characteristics of the site itself. The Interagency Workgroup on Wetland Restoration (IWWR 2003) identified two principal approaches, active and passive, to wetland restoration and enhancement (Table 13-1). In many instances, however, it is a combination of the two that works best to restore ecosystem processes.

13.5.1 Active approaches

Active approaches to restoration and enhancement involve physical engineering efforts which may include significant alterations to the topography, soil, and hydrology of a site. Such engineering approaches may involve the use of large earth-moving equipment to breach levees and re-flood marshes, or backfill channels and remove spillways to rebuild floodplains and allow overland flows.

One of the more ambitious examples of such an approach is currently being undertaken to restore some of the original meandering flows of the Kissimmee River, which provides much of the water to the Florida Everglades (see Fig. 3-13). During the 1960s the Kissimmee River basin was channelized through a system of levees, pumps and spillways, which reduced its volume into Lake Okeechobee and overland flows into the Everglades. This dramatically altered the hydrology of sections of the Everglades and reduced its size to make way for agriculture and urban development. By restoring the Kissimmee's natural pre-channelization flows and removing spillways and other water-control measures, the Kissimmee River Restoration project undertaken by the U.S. Army Corps of Engineers (USACE) hopes to restore some 8900 hectares of wetlands by 2015 (USACE 2009).

Plugging ditches to restore tidal flushing or high water tables and saturated or hydric-soil conditions is also often observed with active approaches. The coastal regions of the northeastern United States were extensively ditched to promote agricultural and grazing activities and to control mosquitos (Fig. 13-6). These alterations significantly influenced the configuration of salt-marsh ecosystems and associated bird and wildlife habitat. In the last few decades, concerted efforts have been undertaken to restore natural tidal flushing, reinstate tidal channels and reconnect creeks and inlets. Brackish and estuarine marshes are restored by rerouting restrictions such as roads and embankments, removing unused railway tracks, increasing the size of culverts that allow tidal waters to pass beneath roads, and refilling ditches. Over time this re-establishes habitat and spawning areas, and rebuilds a marsh surface through the accretion of organic matter (Middleton 1999).

Engineering efforts are also used to excavate and contour the land surface to mimic water-accumulating hollows, depressional features and undulating natural topography. These techniques are used in a wide variety of wetland types, and among other things, attempt to restore the hydroperiod, or the duration and frequency of inundation naturally observed at a site prior to degradation. In other cases, a veneer of soil is imported and lain on the surface of a wetland, or organic matter and nutrients are augmented through deliberate additions (Zedler and Kercher 2005).

Dam and weir removal may reinstate the natural flow of rivers and recreate habitat for migrating anadromous fish species. For instance, the 2002 removal of the Smelt Hill dam on the Presumpscot River, which flows through southern Maine in the northeastern United States, has provided some 11 km of potential habitat to the *Salmo salar* (Atlantic salmon), *Alosa sapidissima* (American shad), and *Alosa pseudoharengus* (alewife) (Casco Bay Estuary Project 2003; Maine Department of Environmental Protection 2008). The Presumpscot is one of the few remaining spawning grounds of the

Figure 13-6. Linear ditches across the salt-hay meadow at Rowley, Massachusetts. Such drainage ditches were dug beginning in the seventeenth century and were maintained until the mid-twentieth century. Now they are abandoned and gradually overgrowing and infilling. Blimp airphoto by S.W. Aber, J.S. Aber and V. Valentine.

economically important North Atlantic salmon along the New England coastline of the United States.

In some cases, natural events may inadvertently and unexpectedly speed up the process of restoration. A recent example is the accidental 1990 breaching of the Sieperda Polder in the Scheldt River estuary of the Netherlands (Eertman et al. 2002). Polders date back several centuries and are an example of deliberate human drainage of coastal landscapes. These reclaimed areas are often lower than sea level and are protected from flooding by a series of sea walls and levees, canal-and-lock systems, and draining mechanisms that make them suitable for agriculture or human settlement.

Some five years after the 1990 accidental breaching of the Sieperda Polder during an intense storm, tidal activity quickly re-established a marsh ecosystem with mudflats and brackish vegetation communities where agricultural crops were once grown (Eertman et al. 2002). The expansion of tidal channels through deliberate restoration efforts further enhanced a vibrant coastal marsh system, which now supports a diversity of biota. While this was not a planned restoration project, given the right conditions the Scheldt River estuary demonstrates just how quickly natural tidal processes re-established wetland flora and fauna (Eertman et al. 2002).

Considerable effort has been made in Canada to restore and reclaim cutover peatlands (Quinty and Rochefort 2003). In order to restore functional peat accumulation two steps are essential: a) re-establishing viable plant cover including *Sphagnum* mosses, and b) raising and stabilizing the water table near the surface. These objectives are reached through the following operations: preparing the surface, collecting and spreading plants, spreading straw cover, fertilizing, and blocking drainage.

The Bois-des-Bel site in southeastern Quebec exemplifies this approach. The site had been abandoned for two decades prior to the beginning of restoration in 1999. An area of 8 ha was treated by leveling fields, filling ditches, building berms, and creating pools. Vegetation became re-established rapidly, and fauna returned to the pools. This experimental site is monitored for long-term consequences of restoration particularly on its biodiversity (Peatland Ecology Research Group 2009). Many other sites have been restored or reclaimed in Quebec and New Brunswick as well as the Canadian prairie provinces, with variable results (Quinty and Rochefort 2003).

Other bioengineering or soft engineering approaches include stream-bank stabilization using biodegradable matting or dead tree trunks, constructing habitat and nesting sites, and restoring connections and habitat links between patches of wetlands through riparian buffers or vegetated corridor connectors. These connectors allow animal species to move freely between wetland patches. The actual configuration and close proximity of pools and ponds within a wetland site are also important considerations when attempting to restore habitat for amphibians (Knutsen et al. 1999). Other approaches try to restore patches of wetlands in relatively close proximity to each other even if they are not connected through vegetated corridors. Such patches may serve as important bird feeding areas and stopover points along migratory routes.

Active approaches to restoration and enhancement are time-consuming and financially demanding. They also require considerable expertise and careful long-term planning. The results of active restoration approaches often take a longer time to manifest. Hence, a monitoring and evaluation system must be implemented to ensure that results are in keeping with goals and expectations (IWWR 2003).

13.5.2 Passive and hybrid approaches

Passive approaches involve restoration or enhancement through the removal of factors that reduce a wetland's resilience and viability. This approach allows a wetland to regenerate naturally by removing any problematic factors or reducing the sources of degradation (U.S. Environmental Protection Agency 2000; IWWR 2003). Restricting access to a wetland site either by humans or domesticated animals to allow native flora to re-establish is one example of a passive approach. Cleared coastal mangroves may re-establish across tidal mudflats once human access to these areas is restricted. Floating propagules from established mangroves in the vicinity quickly colonize new areas by anchoring themselves to open mudflats. Allowing rivers and streams to meander through floodplains during natural flood stages by eliminating controls on water flows would help to nourish riparian vegetation with the deposition of silt and alluvial material.

More often, however, combinations of active and passive methods or hybrid approaches are used to achieve long-term restoration goals. An example of this might include breaching a dike or removing a dam that restricts natural water flows to allow for original hydrological conditions to be re-established slowly. Once desired hydrological flows are established, flora and fauna are allowed to re-colonize with minimal further inputs. The restoration of the southern deltaic marshes of the Tigris and Euphrates rivers of Iraq provides an example of such a combined approach (United Nations Environmental Programme 2010).

The Tigris and Euphrates flow through the Mesopotamian region of present-day Turkey, Syria, Iran and Iraq. Both river systems have been actively managed through the development of irrigation canals or impoundments since ancient times. Nearing the mouth of the two rivers in the Persian Gulf, their waters distribute across a wide area of shallow lakes, seasonally inundated floodplains and marshes, giving rise to one of the largest and most important wetland regions of the Middle East, also known as the Mesopotamian Marshes (Fig. 13-7). Since the 1970s, these marshes have reduced dramatically in size due to large-scale upstream impoundment and dam activities and the diversion of their waters for irrigation in Turkey, Syria and Iraq.

Additionally, during the 1980s and 1990s Iraqi leader Saddam Hussein and his regime persecuted political dissidents taking refuge in the dense and tall Mesopotamian reed marshes, and especially targeted the resident Ma'dan or Marsh Arab communities (Scott 1995). As punishment, the regime built dikes and canals to drain the wetlands and divert its waters, which led to the almost complete elimination of the southern marshes (Campbell 1999). Consequently, the Ma'dan communities that lived within these marsh systems for millennia lost access to their ecosystem-dependent economic livelihoods. Unable to fish, or engage in reed cultivation (*Phragmites australis*), harvesting or agriculture, many moved out of the marshes to seek refuge in other regions.

Recent conservation efforts since the end of the Hussein era have breached dikes to re-divert the Tigris and Euphrates waters into the marshes. This re-flooding strategy has resulted in the quick and dramatic comeback of substantial areas of wetlands, with the associated re-establishment of fish, bird and macro-invertebrate communities (NASA 2010b). Images from NASA's Moderate Resolution Imaging Spectroradiometer (MODIS) sensor on the Terra satellite show striking alterations to the southern Iraqi marshes (Fig. 13-8). The changes are clearly visible between the first image from February 2002 at the height of the wetland's drying out and the second from February 2007, when re-flooding efforts had made significant strides in regenerating sections of the marsh complex.

The February 2002 image indicates an almost completely barren landscape in the midst of the desert with a few patches of agricultural activity seen along the sides of canals or fed through irrigation systems. By the same month in 2007, there are noticeable differences. Areas of vegetation have expanded considerably. Wet mud patches and pools in dark shades, indicative of re-flooding efforts, can be seen across the center of the image.

Restoring complex wetland functions and processes is difficult through breaching dikes and re-flooding efforts alone. Concerted floral and faunal restoration efforts in cooperation with local resident communities such as the Ma'dan would be required for their successful re-establishment. Moreover, continued political uncertainty and economic hardship within the region complicate the tenuous nature of restoration efforts, and the sustained long-term health of these marshes remains in question.

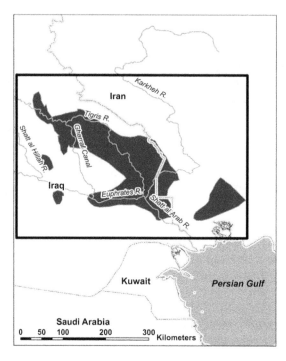

Figure 13-7. Map of the lower Tigris and Euphrates drainage system in Iran and Iraq showing marsh areas (dark gray). Produced by A. Dailey. Map made with data from Natural Earth, Global Administrative Areas, and World Wildlife Fund Terrestrial Ecoregions Database (Olson et al. 2001). Accessed online <www.naturalearthdata.com> <http://www.gadm.org/> and <http://www.worldwildlife.org/science/data/item6373.html> February 2011.

Wetland restoration and enhancement projects may be completed relatively quickly with functions, vegetation and animal species re-colonizing within the span of a few months to years given the right conditions and necessary engineering or other structural changes implemented. The type of wetland generally provides a good indication of how long a restoration project may span. For instance, it may be relatively easy and quick to establish a riparian wetland given the reinstatement of the right topographic and hydrological conditions (Kusler 2006). On the other hand, Zedler and Kercher (2005) suggested that tussock-forming sedge meadows could take decades to establish fully. Meanwhile, forested wetland requiring trees to mature or a peatland necessitating organic matter accumulation might take substantially longer time and management investments and require careful planning and fine hydrological adjustments (Kusler 2006). Similarly, wetland creation projects require substantially greater investments of time than restorations and may take several decades to approximate natural conditions.

13.6 Artificial treatment wetlands

The numerous ecosystem functions offered by wetlands provide the impetus to reproduce such conditions artificially to benefit both humans and the environment. Constructed, artificial or treatment wetlands, as they are interchangeably called, are found across the globe. By 2004, there were some 5000 operational waste-water

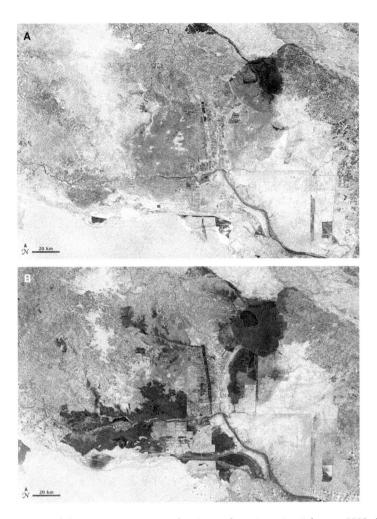

Figure 13-8. MODIS images of the Mesopotamian Marshes in southern Iraq. A – February 2002 shows the marshes almost entirely dried out due to water diversion and impoundment activities. B – February 2007 shows some restored marshes, water bodies and wet surface areas (dark gray) after a few years of re-flooding efforts. Images adapted from NASA's Earth Observatory <http://earthobservatory.nasa.gov/>.

treatment wetlands across Europe and about 1000 in the United States (U.S. Environmental Protection Agency 2004). They are built primarily to take advantage of the water-quality enhancement and purification functions of natural wetlands.

As chapter 11.3.2 elaborates, wetlands act as natural filters and may effectively trap suspended solids, particulate matter, litter and debris. Wetland vegetation and microbes aid in the uptake of nitrogen and phosphorus, and chemical processes facilitate the incorporation of such toxic metals as lead, copper and mercury, among others (DeBusk and DeBusk 2001; Vymazal et al. 2006; Vymazal 2008). Such properties make wetlands particularly suitable to treat waste water or improve water quality and may be more economical and sustainable than maintaining conventional waste-water treatment facilities (U.S. Environmental Protection Agency

2004). As an added benefit, treatment wetlands may also provide habitat to local fauna in some cases.

Treatment wetlands are employed to filter and decontaminate water from agricultural and livestock facilities or treat storm-water runoff (Natural Resources Conservation Service 2010). These facilities are generally located in upland regions away from floodplains and take special precautions in highly permeable soils to minimize the risk of accidental contamination of surrounding areas (U.S. Environmental Protection Agency 2004; Scholz and Lee 2005).

The construction of treatment wetlands requires careful engineering, taking into consideration the soils, hydrology, topography and surrounding land uses. Typically, construction requires excavating the land surface to create a depression or embankments, appropriately grading the surface, transporting in a suitable layer of top soil, if needed, and installing dikes, spillways or other water-control structures to manage water levels and flows (U.S. Environmental Protection Agency 2004). Macrophytes such as *Phragmites* sp. and *Typha* sp. are introduced and allowed to colonize the constructed site (Scholz and Lee 2005). By and large, wetland species that spread easily and are tolerant of high concentrations of nutrients and contaminants are selected for treatment wetlands. Collectively, these features attempt to mimic a natural wetland's structure and provide the conditions necessary for an efficiently functioning system.

13.7 Contaminated mine-water treatment

Mining brings buried minerals to the surface where they undergo weathering and may release acids and toxic compounds into surface and ground water. The deleterious impacts of mining on wetlands are well known and well documented around the world (see chapter 10.3.2). Wetlands may also be utilized to clean up the consequences of past mining. The Tri-state mining district in southwestern Missouri, southeastern Kansas, and northeastern Oklahoma was exploited for lead and zinc beginning in the mid-1800s until 1970, during which time no environmental regulations or controls existed. Mining lead and zinc ore generated a huge volume of waste rock – chat, which was simply dumped on the surface in big piles (see Fig. 1-15). The chat piles, water-filled mines, and former smelting sites are today sources for highly polluted water that contains heavy metals. In addition to lead and zinc, mining byproducts included cadmium, germanium, and gallium (Park 2005). Streams, lakes, and shallow aquifers of the region are severely contaminated (Fig. 13-9).

Figure 13-9. Tar Creek carries highly polluted mine runoff from the Treece-Picher-Cardin vicinity, as shown by its typical rust-orange color at low flow (see Color Plate 13-9). According to the U.S. Army Corps of Engineers, "Tar Creek is highly toxic and, for all intents and purposes, dead" (USACE 2005); however, continued stream monitoring has recently revealed the presence of some fish and macro-invertebrates. View downstream just east of Commerce, Oklahoma. Photo by J.S. Aber.

Such contamination led to the establishment of Environmental Protection Agency (EPA) superfund sites in Missouri, Kansas, and Oklahoma beginning in the early 1980s. After decades of effort, the former mining areas around Joplin, Missouri and Galena, Kansas have mostly been

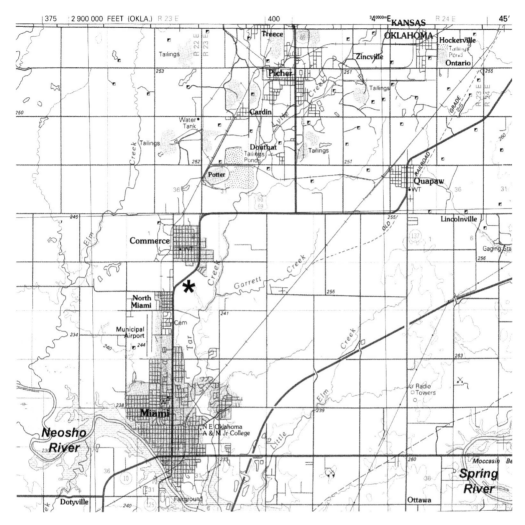

Figure 13-10. Topographic map of the Treece, Kansas and Picher-Miami, Oklahoma vicinity. Passive water treatment system at Commerce indicated by asterisk (*). Adapted from Neosho, Missouri-Oklahoma-Kansas, U.S. Geological Survey, 1:100 000-scale metric topographic map, 30 × 60 minute quadrangle, 1986. Contour interval = 20 m.

reclaimed. However, the overwhelming magnitude of pollution led to EPA buyouts and relocation of residents in Picher, Oklahoma and Treece, Kansas (Fig. 13-10). From a population of some 20,000 people during its mining peak, Picher has become virtually a ghost town. It ceased to exist officially in 2009, but the legacy of ruined land and human health will endure for many decades to come. Efforts now focus on cleanup and remediation (Fig. 13-11).

Soon after active mining ceased, metal-laced mine water began discharging naturally in 1979 through open mine shafts, springs, and artesian wells of the vicinity. Initial discharges were highly acidic; however, water chemistry has become net alkaline with total alkalinity greater than total acidity (Aber et al. 2010). Perennial mine-water flow through two artesian water wells was first identified between Commerce and North Miami, Oklahoma in 1983 (see Fig. 13-10). This location was selected for passive treatment of the toxic water (Nairn et al. 2009). Periodic data collection began in 1998 with monthly sampling since 2004. The targeted

Figure 13-11. Reclamation of mined land in progress at Treece, Kansas. Photo by J.S. Aber.

discharge has pH just <6, total alkalinity ~400 mg/L as $CaCO_3$, flow rates of 400–700 L/minute, and elevated levels of Fe, Zn, Pb, and Cd.

Given the nature of the target discharge, a multi-process unit conceptual design was developed, which included six specific treatment steps (Table 13-2). In addition, an identical parallel treatment train approach was deemed appropriate for at least two reasons. First, the parallel trains allow for simultaneous performance of necessary maintenance and continued treatment. Second, given the research focus of this site, the parallel trains allow experimental manipulations to be conducted. During construction, a third mine-water discharge was discovered and incorporated into the design (Fig. 13-12).

The completed system includes ten distinct process units with a single initial oxidation pond (cell 1) followed by parallel surface-flow aerobic wetland-ponds (cells 2N and 2S), vertical-flow bioreactors (cells 3N and 3S), re-aeration ponds (cells 4N and 4S), horizontal-flow limestone beds (cells 5N and 5S), and a single polishing pond-wetland (cell 6). Cattails are well established in cells 1, 2N and 2S; wind and solar power provide the only necessary energy for the re-aeration ponds (Fig. 13-13). Mine water was diverted into the passive treatment system late in 2008. Monitoring began in January 2009, and preliminary results indicate highly effective reduction or removal of metals from the treated water (Table 13-3). This system represents a state-of-the-art ecological engineering research site for passive-wetland treatment of ferruginous lead-zinc mine waters.

Table 13-2. Summary of final conceptual design process units, primary targeted water-quality parameters and design function. Note the role of artificial wetlands. Based on Nairn et al. (2009).

Process unit	Targeted parameter	Function
1. Oxidation pond	Fe	Oxidation, hydrolysis and settling of iron oxyhydroxide solids and trace-metal sorption
2. Surface-flow wetlands and ponds	Fe	Solids settling
3. Vertical-flow bioreactors	Zn, Pb & Cd	Retention of trace metal sulfides via reducing mechanisms
4. Re-aeration ponds	Oxygen demand and odor	Wind- and solar-powered re-aeration. Stripping oxygen demand and H_2S. Adding O_2
5. Horizontal-flow limestone beds	Zn, Mn & hardness	Final polishing of Zn as $ZnCO_3$ Final polishing of Mn as MnO_2. Adding hardness to offset bioavailability of any remaining trace metals
6. Polishing pond and wetland	Residual solids	Solids settling. Photosynthetic oxygenation. Ecological buffering

Table 13-3. Comparison of system influent and effluent data for selected metals (mg/L) and calculated percentage differences. SE is standard error; BDL refers to data below detectable limits. Adapted from Aber et al. (2010, Table 3).

Metal	Influent Mean	SE	Effluent Mean	SE	Change %
As	0.062	0.0005	BDL	BDL	100
Cd	0.017	0.003	BDL	BDL	100
Fe	172.51	5.54	1.055	0.439	99.4
Ni	0.893	0.012	0.049	0.024	94.5
Pb	0.063	0.009	BDL	BDL	100
Zn	8.093	0.092	0.232	0.062	97.1

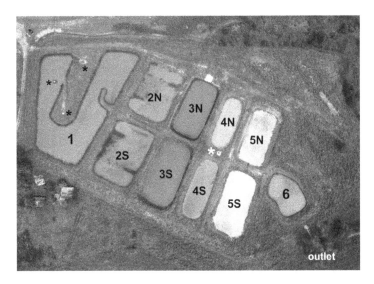

Figure 13-12. Passive treatment facility at Commerce, Oklahoma (see Color Plate 13-12). Vertical aerial photograph annotated with pool numbers. Inlet pool (1) with three artesian wells (*) upper left; outlet channel lower right. Two treatment trains identified as north (N) and south (S). Notice change in water color from the inlet cell (1) to the outlet pond (6). White asterisk (*) indicates position of wind and solar power for cells 4 (see next figure). Vehicles in upper left corner for scale; helium-blimp airphoto by J.S. Aber and S.W. Aber.

Figure 13-13. Ground views of the passive treatment facility at Commerce, Oklahoma. A. Cell 1 surrounded by cattails with an artesian well in the center. B. Wind and solar power run pumps for the re-aeration ponds (cells 4N and 4S). Photos by J.S. Aber.

13.8 Summary

This chapter provides a summary of the application of restoration ecology principles and approaches to wetland repair. The field of wetland restoration is still in its nascent stages. Significant improvements in design and technology are anticipated as our knowledge of wetland functions expands and as we learn lessons from the implementation of repair projects. Restoration projects focus on re-establishing damaged functional processes and enhancing the integrity and resilience of wetland ecosystems. The long-term sustainability of such projects requires significant commitments in management and monitoring efforts. Such commitments are best met by involving multiple stakeholders and members of the community in stewardship activities.

A theoretical understanding of how wetland ecosystems function and results from numerous restoration projects implemented globally provide key design principles to promote their success. Scientists have argued that using a systems approach, restorations should be self sustaining, employ renewable and locally available resources, and focus on rebuilding the structure and functional characteristics of wetlands. Moreover, they should reflect the geographic context and balance objectives to benefit local communities and ecosystem needs. Restoration efforts should adopt principles of adaptive management, ensuring flexibility in implementation based on new or unanticipated information. Clearly defined monitoring and evaluation criteria and performance expectations for restored ecosystems are used to assess whether goals and objectives have been met.

Specific approaches and techniques employed in restoring systems are varied. Active or passive approaches are chosen based on the scope and scale of the problem being addressed, the resources available for restoration, enhancement or creation, and the distinctive characteristics of the geographic site itself. Artificially constructed treatment wetlands provide one example of an active approach to creating a wetland that takes advantage of the water-quality improvement and purification functions provided by these ecosystems.

Even with the most carefully planned restoration and enhancement projects, scientific data suggest that the outcomes of such undertakings are difficult to predict with any degree of certainty. This is, in part, due to the complexity of factors that govern wetland functioning. As Zedler (2001) suggested, dynamic water regimes, topographic and soil conditions, external disturbances, exotic species invasions, failed recolonization efforts, and human involvement all may influence the ultimate success or failure of wetland rehabilitation projects.

14 Wetlands governance and public policy

14.1 Wetlands governance and policy

Wetlands governance refers to the overall legislative framework by national or international institutions, which guides the conservation and management of this resource. The governance and management of wetland ecosystems is, however, complicated by their multifaceted character. Great geographic variability exists in the types of wetlands, they provide numerous services, they are valued differently based on geography and socio-economic context, and they extend across national boundaries so that their management in one region may have direct implications for their health in another. All these factors must be kept in mind as governments and international institutions devise context-relevant ways to maintain and enhance these resources.

14.2 International wetland policy

The emergence of the science of wetland ecology has undoubtedly bolstered conservation legislation across the world. In 1971 the "Convention on Wetlands of International Importance especially as Waterfowl Habitat", also known as the Ramsar Convention or the Convention on Wetlands, was signed in Ramsar, Iran. This intergovernmental treaty outlined global cooperation between individual states in wetland habitat protection and conservation (U.S. Fish and Wildlife Service 1993). While the original intent of the Convention was primarily to provide a habitat for water birds (as reflected by its name), the Convention later broadened its scope to cover all aspects of wetland conservation and wise-use practices. The Convention went into effect in 1975 after seven countries ratified it, and by 2010 there were 160 signatories to the treaty with 1896 protected wetland sites comprising 185.4 million hectares of habitat (Ramsar 2010a).

The Ramsar Convention does not have regulatory mechanisms in place to enforce the treaty. Rather, it relies on member nations to maintain their wetland conservation and wise-use obligations by developing national wetland policies in accordance with their overall natural resource planning strategies. Furthermore, it underscores the importance of establishing wetland nature reserves; member nations are required to designate at least one wetland site within their borders as a wetland of "international importance" in terms of its "ecology, botany, zoology, limnology or hydrology" (Ramsar 2010d).

More recently, the Convention has initiated National Ramsar Committees to provide support for the Treaty's implementation at the national level. These committees are to involve a wide diversity of interested players including local governmental agencies, scientists, and members of non-governmental and community organizations. The Committees are responsible for managing Ramsar sites, identifying new sites, providing expert knowledge, overseeing the implementation of new wetland resolutions, and procuring funds for management through grants disbursed by the Ramsar Convention (Ramsar 2010e).

Cooperation between member countries is fostered through regular progress meetings. These include reviews of wetland sites on the list, conferences on wetland habitat management, and data collection and cooperation with other international conservation bodies. Financial contributions from member states support the Convention and its administrative arm, the Ramsar Bureau, which is located at the headquarters of the World Conservation Union in Gland, Switzerland (Ramsar 2010f).

Criteria for identifying Ramsar wetlands of international importance fall in two categories. Group A Sites include areas containing representative, rare or unique wetlands, and Group B Sites include areas of international importance for conserving biological diversity. Representative examples of wetlands on the List include the more famous sites like the Pantanal in South America, the Everglades in North America, the Volga River delta in Russia (Fig. 14-1), the Okavango in southern Africa, the inland Niger delta in the Sahel, and the Sundarbans in South Asia. However, most sites on the List include smaller and lesser known, yet equally valuable, wetlands important for their local and regional contributions (Ramsar 2010g).

14.3 Wetland policy in the developed world

At the national level, wetlands policy making made significant strides in the latter half of the twentieth century. However, progress in the conservation and sustainable management of wetlands is far from uniform. Countries across the developed world have legislated and enforced stricter regulations for the conservation of these ecosystems, while developing nations, constrained by other societal needs, have been less inclined to promote legislation curbing the development or alteration of wetland environments.

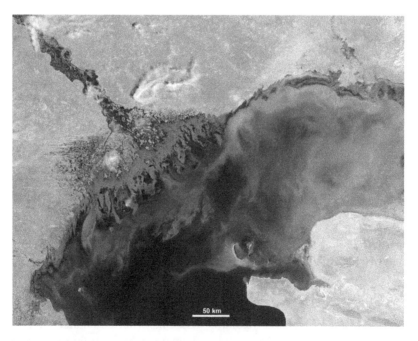

Figure 14-1. Volga River delta in the northern Caspian Sea, southern Russia (see Color Plate 14-1). MODIS true-color image, acquired 11 June 2005. The river divides into >500 distributary channels forming a wetland web that supports hundreds of plant and animal species. It is among the most productive fisheries in the world, including the caviar-producing sturgeon. Image adapted from NASA's Earth Observatory <http://earthobservatory.nasa.gov>.

14.3.1 United States

The United States was a comparatively early entrant into the realm of wetland protection. Today, it ranks among those that closely regulate the development of wetland habitats. Yet, this was not always the case. Nineteenth-century United States policy toward wetlands was to impound, drain and alter them into what were considered more economically productive land uses. This policy was promoted by the Swamp Land Acts of 1849, 1850, and 1860, which allowed for the drainage of wetlands and the expansion of agriculture and urban development (Wheeler et al. 1995; Williams 1990). Many rivers were modified substantially for navigation and hydropower purposes (Fig. 14-2). As Tzoumis (1998) reported, these and other policies were influenced by pro-development philosophies and resulted in dramatic change and acreage declines in wetland landscapes across the country.

Even so, early sustained efforts by duck hunters and other conservation enthusiasts led to the eventual creation of the National Wildlife Refuge system, which promoted the early preservation of habitats including wetlands for waterfowl species (Lewis 2001) and was supported by Migratory Bird Hunting and Conservation Stamps, so-called "duck stamps" (Fig. 14-3). Other early legislative victories included the Migratory Bird Treaty Act of 1918 and the Migratory Bird Conservation Act of 1929, which recognized the importance of sustaining waterfowl and bird populations and in turn their wetland habitats (National Wildlife Refuge System 2010).

The 1960s heralded a new era of environmental legislation and policy initiatives in the United States – not necessarily limited to wetlands. The Federal Water Pollution Control Act Amendments of 1972, now also known as the Clean Water Act, sought to restore and maintain the overall health of the nation's waterways. Section 404 of the Clean Water Act provided federal jurisdiction through a permit system for any dredge-and-fill activity within the "waters of the U.S." Due to the lack of specificity on the part of the U.S. Congress, the U.S. Army Corps

Figure 14-2. U.S. postage stamp commemorating modification of the lower Arkansas River for navigation and hydropower purposes. Issued in 1968; original stamp printed in blue and black. From the collection of J. Vancura.

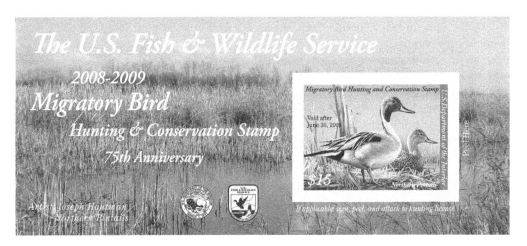

Figure 14-3. U.S. Fish and Wildlife Service 2008–09 Migratory Bird Hunting and Conservation Stamp commemorating the 75th anniversary of the duck-stamp program. Original stamp printed in full color.

of Engineers, in charge of issuing permits, adopted an expanded definition of "waters of the U.S." to contain non-navigable waters including wetlands (Chertok and Sinding 2005). Meanwhile, early cases in front of the courts used the Army Corps of Engineers' expanded definition, thus laying down legal precedent. The definitional confusion was based in part on the failure of Congress to address this issue adequately in 1972 or later in 1977 when it made amendments to the Clean Water Act (U.S. Environmental Protection Agency 2010).

The permit system used by the U.S. Army Corps of Engineers directs that no dredging activity be permitted if alternatives exist. Furthermore, preliminary studies must be carried out on site to minimize the potential impact of the dredging and filling activities on wetlands, avoid the activity altogether where practical, or engage in mitigation activities for unavoidable impacts (U.S. Environmental Protection Agency 2010). Mitigation includes providing compensation for unavoidable impacts through the restoration or creation of wetlands elsewhere. The Army Corps of Engineers cannot issue a permit if the wetland performs important biological and ecological functions, or when the ecological and economic costs outweigh the benefits proposed. The U.S. Environmental Protection Agency (US EPA) has the statutory authority to designate wetlands that are subject to permits and has the ability to veto the Army Corps of Engineers' decisions (US EPA 2010).

The enforcement of Section 404 includes issuing administrative compliance orders requiring violators to cease illegal activity, remove structures to restore sites, and assess civil penalties. For more egregious violations, the U.S. Environmental Protection Agency (US EPA) and the Army Corps of Engineers exercise their criminal enforcement authority, which could lead to jail time and compensation costs (US EPA 2010).

In 1985, the "swampbuster" provision in the Food Security Act took a markedly different approach to wetland protection. It mandated that farmers undertaking alterations that adversely impacted wetlands on their property would be ineligible for many U.S. Department of Agriculture benefits programs, including

Figure 14-4. Boundary sign marking a wetland reserve for a playa on the High Plains of southwestern Kansas, United States. Photo by J.S. Aber.

loans and price support payments (Williams 2005). Thus, rather than a permitting system with civil or criminal consequences, the "swampbuster" provision provided disincentives to discourage wetland alterations. The Wetlands Reserve Program was authorized as part of the Food Security Act of 1985 and amended in Farm Bills of 1990 and 1996 (Natural Resources Conservation Service 2010d). It is a voluntary program to encourage private property owners to restore and protect wetlands (Fig. 14-4). In return for signing conservation easements, landowners receive payments for restoration from the U.S. Department of Agriculture.

Finally, the No Net Loss Policy was an outcome of the 1987 National Wetlands Policy Forum convened by the Environmental Protection Agency to look at wetland management in the United States. Among other issues, the Forum identified that there should be no overall net loss of the remaining wetlands base, and that the restoration of wetlands should be undertaken where feasible (Mitsch and Gosselink 2007). This did not necessarily argue for the cessation of all development, but it made the strong case for remedial and restoration activities at another site to mitigate the alteration of wetlands.

Beyond these rather crucial legislative and policy guidelines, there have also been Presidential executive orders and other state-level regulations that have sought to guide wetland development activities. Counties and townships have also introduced zoning ordinances to regulate activities in wetland sites (U.S. Environmental Protection Agency 2010). These and other measures have undoubtedly influenced wetland conservation and, as reported in recent studies, helped reduce the rate of wetland habitat loss across the United States (Dahl 2000).

In summary, U.S. federal policy on wetlands is carried out under regulations related to land use and water quality rather than an overall national wetland law. Wetland protection is promoted through a multi-pronged approach including regulation, permitting, providing incentives and disincentives, and through acquisition by establishing parks and refuges.

14.3.2 Canada

Canada's foray into wetland policy making is another noteworthy example from the developed world. Canada is distinguished as being first in the world to outline a federal wetland conservation policy. This policy follows the recommendations of the Ramsar Convention on wise use and sustainable management. The Canadian government's 1992 Federal Policy on Wetland Conservation (FPWC) aims to achieve cooperation with the governments of its provinces, territories and private entities to maintain, enhance, rehabilitate, and secure wetlands of significance. It accomplishes these goals by promoting public awareness, developing exemplary wetland conservation practices, stimulating decision making based on scientific and technical factors, and enhancing cooperation between citizens, non-governmental organizations and governmental entities (Government of Canada 1991).

Canada's approach differs significantly from that of the U.S. in that it has adopted a largely non-regulatory policy which relies primarily on the collective participation and cooperation of federal, provincial, and territorial governments to undertake wetland conservation practices (Government of Canada 1991; Rubec 1994). It is difficult to gauge the long-term success of this policy given its recent adoption, but the Canadian government has reported successes through the implementation of mitigation strategies.

Luck Lake in the prairie pothole region of south-central Saskatchewan is a good example of this cooperative approach. Luck Lake is a Saskatchewan Heritage Marsh that includes 6170 acres (~2470 ha) of wetland and 650 acres (260 ha) of adjacent upland (Fig. 14-5). The lake is the terminal point of a small enclosed drainage basin. Water-control structures were built in 1988 by Ducks Unlimited Canada, which continues to manage the lake. Partners in this venture are the Province of Saskatchewan, Saskatchewan Wildlife Federation, Saskatchewan Natural History Society, and Wildlife Habitat Canada. In addition to these partners, the Birsay Water Users Association, Saskatchewan Water Corporation, Rural Municipality of Coteau, local landowners, and federal government were instrumental in organizing the project. Water for Luck Lake comes from nearby Lake Diefenbaker, a huge reservoir on the South Saskatchewan River, via high-pressure pipelines that supply irrigation to local farmers (Fig. 14-6).

Canada's wetland policy has been groundbreaking, as 25 other nations have now implemented similar non-regulatory policy measures. Given that Canada contains nearly one-quarter of total global wetland habitats and is a leading player in the Ramsar Convention with over 30 designated wetland sites, the results of this approach will be keenly studied (Rubec 1994). Furthermore, in the coming decades data will be provided to enable comparative analysis of the pros and cons of regulatory and non-regulatory mechanisms in wetland policy making.

14.3.3 Western Europe

Rather than outline wetland-specific legislation, most European nations have opted to incorporate wetland protection within existing laws on

Figure 14-5. Overview of Luck Lake from the eastern end looking toward the southwest. Extensive mudflats and shallow water are attractive for shorebirds and waterfowl. Kite aerial photograph by S.W. Aber and J.S. Aber.

Figure 14-6. Luck Lake looking from the southern side toward the northwest. A levee divides the lake in half. Pipeline water supply comes into the fenced rectangle (lower left corner) and flows into the western side of Luck Lake. Kite aerial photograph by S.W. Aber and J.S. Aber.

fishing, agriculture, water protection, or industrial development. European Union (EU) countries also align national policies with EU directives. In the case of wetlands, the EU Water, Habitats, and Birds directives have been most relevant. As many western European countries have a federal system, states or counties often play significant roles – not only in enforcing existing federal laws, but also implementing regional wetland conservation regulations.

Germany presents one such example. The German National Federal Conservation Act provides the umbrella policy specifically outlining that wetlands, including fens, marshes, reed beds, meadows and other wetland ecosystems, are protected, and development activity leading to their destruction or impairment is prohibited (Germany Report 2005). More recently, the 2007 National Strategy on Biological Diversity links quantitative targets for improvements in ecological wellbeing to other European Union legislation such as the Water Framework Directive and Birds Directive (Germany Report 2008).

Further, wetland protection is included in German legislation concerning pollution controls and federal forestry, mining, and hunting laws. This provides an additional measure of protection for wetland habitats. German states (or Länder) have the ability to further widen the scope of this protection by enacting supplementary policies that may be implemented regionally and where applicable. Alternatively, states can also provide permits for wetland conversions when this is in the public interest or where mitigation activities are employed; thus balancing development goals with conservation practices (Germany Report 2005).

The United Kingdom, meanwhile, incorporates both site-specific and policy-based mechanisms for wetland management and protection (United Kingdom (UK) Report 2005). To that end, Ramsar-designated wetland sites across the United Kingdom and its territories are monitored and protected under various existing federal laws including, most notably, the Wildlife and Countryside Act of 1981, the 2003 Water Act, and align with European Union Water Framework and Habitat directives (UK Report 2008). A targeted approach adopted by the government identifies key habitats and species and necessitates the development and implementation of action plans for their protection.

Overall, policy adopted across the United Kingdom is largely designed to incorporate Ramsar prescribed best-use wetland practices, monitor significant sites, encourage land-use policies that accommodate wetland habitats, and implement conservation strategies that are holistic (UK Report 2005). All of these goals are supported by regulatory frameworks and implementation plans like the Estuary Plan, Shoreline Plan, or Water Management Plan (UK Report 2005).

In the United Kingdom, governmental agencies have also made forging partnerships with private stakeholders an important aspect of their conservation goals. Beyond these efforts, non-profit conservation organizations are active in the United Kingdom. One example is the Wildflower & Wetlands Trust, which has engaged in wetland habitat and species protection and restoration over the past 50 years. The Trust is now also supporting wetland conservation and education programs in countries across the globe (Wildflower and Wetlands Trust 2008).

14.3.4 Central Europe

Running through central Europe from Estonia to Poland and from Slovakia to Bulgaria are a series of countries that were once part of the Soviet Union or came under Soviet domination and are now sovereign countries. Many are now members, or candidate members, of the European Union and are working to implement EU standards in all aspects including wetlands. During the Soviet period, wetlands, along with other natural environments were neglected or treated as economic resources to be exploited. In Estonia, for example, bogs were drained for forestry and agriculture (Fig. 14-7), and peat was mined for fuel (Fig. 14-8). Bogs in northeastern Estonia were heavily impacted by deposition of fly ash from nearby power plants fueled by oil shale; pH and calcium increased and *Sphagnum* mosses disappeared (Karofeld 1996).

In spite of this bureaucratic situation, some progress was made on bog research and conservation. The Ramsar Treaty led to protected status for several essential wetlands. In 1981, mire protection areas were designated in wetlands that were considered most important for hydrologic aspects and for richness in berries. Additional mire reserves were established in the 1980s and 90s. Estonia therefore has a substantial number of wetlands, with a large total area, in protected status today (Fig. 14-9). *Sphagnum* moss is beginning to grow again in bogs of northeastern Estonia following reduction in fly ash emissions from power plants (Karofeld 1996). The mires of Estonia now attract considerable interest from scientists and ecotourists as exemplary wetlands in Europe, where only

Figure 14-7. Clear cut (left side) for experimental agriculture in conifer forest at Teosaare (snail island) Bog; artificial drainage for improved forestry in left background; Endla Lake in right background. The agricultural experiment was a failure as crop growth could not be sustained. Part of the Endla mire complex in east-central Estonia. Kite aerial photograph by S.W. Aber and J.S. Aber.

Figure 14-8. Peat mine at Ulila near Tartu, Estonia. The peat is burned in an electric power plant (smokestack in right background). Many such operations were shut down after Estonia gained independence from the Soviet Union. Photo by J.S. Aber.

fragments of undisturbed bogs remain (Fig. 14-10; Masing, 1997).

The Estonian experience was repeated in many other parts of central Europe. In northern Poland, for example, Soviet construction of a small fishing harbor at Władysławowo cut off the supply of sand delivered via long-shore drift to the Hel peninsula on the Baltic coast north of Gdańsk (Fig. 14-11). The beach must now be maintained by pumping sand from offshore. Poland's continued reliance on coal for industry and power results in heavy air pollution, particularly during the winter heating season (Fig. 14-12). In some parts of the country, pollution is so severe that water supplies must be protected to maintain quality (Fig. 14-13).

Substantial progress has been made, nonetheless, to preserve key natural monuments, such as the Tatra Mountains in Poland and Slovakia (Fig. 14-14). The formerly glaciated valleys of this range have many fragile alpine wetlands. National parks on both sides of the border provide high levels of environmental protection, and all activities within the parks are strictly controlled. Special permission is required to step off designated paths and for any type of scientific study (Fig. 14-15). These parks have become twenty-first-century meccas for ecotourists seeking nature in a setting that is relatively undisturbed compared with most of Europe.

Figure 14-9. Estonian postage stamp recognizing Matsalu Looduskaitseala (nature protection area; later designated as a national park), which comprises an estuary system at Matsalu Bay, many tiny islets, and the adjacent mainland. *Anser anser* is the wild goose also known as the greylag (graylag in U.S.) goose that was the ancestor of domesticated geese in Europe and North America. Original stamp printed in multiple colors.

Figure 14-10. Tourist observation tower under construction on the margin of Valgesoo (White mire), a small bog southeast of Tartu, Estonia. Photo by J.S. Aber in 2000.

Figure 14-11. Fishing harbor at Władysławowo (left side) cuts off the long-shore flow of sand from the west (left). The beach in foreground must be maintained by pumping sand from the Baltic Sea floor. Northern Poland; kite aerial photo by J.S. Aber and D. Gałązka.

Figure 14-12. Coal stockpiled for winter heating in Warsaw, the capital of Poland. The coal is burned in building furnaces without any pollution control. Photo by J.S. Aber.

Figure 14-13. Dam, reservoir and castle at Dobczyce, southern Poland. Note the iron fence around the reservoir; no public access is allowed in order to maintain high quality of this water supply for the city of Kraków. Photo by J.S. Aber.

14.3.5 Commonwealth of Independent States

Compared with EU countries, the Commonwealth of Independent States (CIS), including Ukraine, Belarus, Russia, Kazakhstan, and several other countries of the former Soviet Union, still requires greater articulation and enforcement of wetland management efforts and regulatory mechanisms. Most of this region inherited a legacy of environmental problems from policies adopted during the Cold War. Pursuing industrial growth and economic development at the expense of environmental concerns was characteristic of the region for most of the twentieth century.

At present, the Ramsar organization is attempting to expand protection of important wetland sites while simultaneously urging the development of greater country-based institutional and legal protections. Even a cursory look at environmental and wetland legislation in the CIS suggests that it is piecemeal at best. Greater openness, however, and the documentation of environmental problems across the region undoubtedly will encourage calls for regulation

Figure 14-14. Tatra Mountains of central Europe. A. Overview of Polish Tatra Mountains looking toward the southwest from near Toporowa Cyrhla; the city of Zakopane is visible in the right background. B. Slovak Tatra Mountains seen from Stará Lesná looking northward. Tourist hotels in the foreground; a ski jump is visible at scene center; the highest peak to right is Lomnicky at 2634 m. Kite aerial photos by J.S. Aber and S.W. Aber.

and protection. Furthermore, ongoing technical and financial assistance from international conservation organizations have helped these countries initiate such discussions (Ramsar 2010a).

As the largest country in the world, Russia has vast wetlands of many different types (Fig. 14-16; see Fig. 14-1). The West Siberian lowlands of Asian Russia comprise some of the most extensive wetland ecosystems of the world. These lowlands range from east of the Ural Mountains to the Yenisei River, covering more than two million square kilometers. Some estimates suggest that approximately half of this area is covered in wetlands (Solomeshch 2005). Peatlands, ranging between 1 and 5 m thick in some cases, occur across the landscape and were probably formed during the end of the last glacial period (Kremenetski et al. 2003). Despite the lack of consistent data, scientists believe these peatlands play an important role in

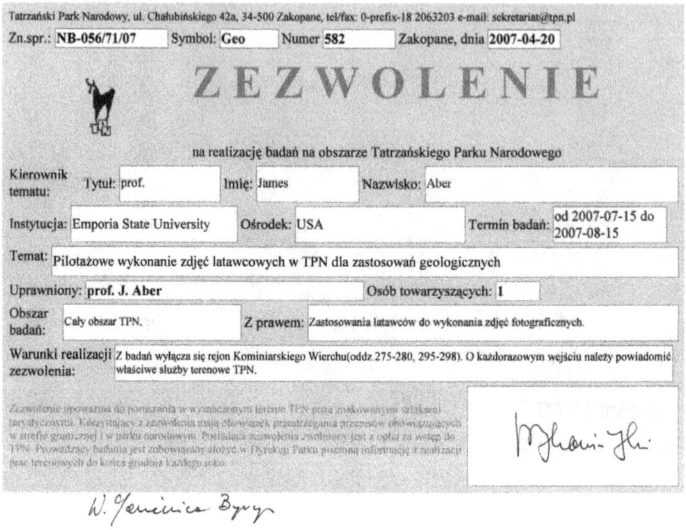

Tatrzański Park Narodowy, ul. Chałubińskiego 42a, 34-500 Zakopane, tel/fax: 0-prefix-18 2063203 e-mail: sekretariat@tpn.pl

Zn.spr.: NB-056/71/07 Symbol: Geo Numer 582 Zakopane, dnia 2007-04-20

ZEZWOLENIE

na realizację badań na obszarze Tatrzańskiego Parku Narodowego

Kierownik tematu: Tytuł: prof. Imię: James Nazwisko: Aber

Instytucja: Emporia State University Ośrodek: USA Termin badań: od 2007-07-15 do 2007-08-15

Temat: Pilotażowe wykonanie zdjęć latawcowych w TPN dla zastosowań geologicznych

Uprawniony: prof. J. Aber Osób towarzyszących: 1

Obszar badań: Cały obszar TPN. Z prawem: Zastosowania latawców do wykonania zdjęć fotograficznych.

Warunki realizacji zezwolenia: Z badań wyłącza się rejon Kominiarskiego Wierchu(oddz.275-280, 295-298). O każdorazowym wejściu należy powiadomić właściwe służby terenowe TPN.

Figure 14-15. Zezwolenie (permission) to conduct kite aerial photography for geological applications in Tatrzański Park Narodowy (TPN), Poland. This official permit was required before any scientific field work could be undertaken during the limited approved time period, 15 July to 15 August 2007. Photo by J.S. Aber.

global carbon sequestration. At present, 19 out of 35 currently designated Ramsar sites in Russia comprise peatlands and encompass approximately nine percent of the total Russian Ramsar area (Russian Federation Report 2005).

While population pressures are not significant in this region, the exploration and extraction of minerals, timber, peat, oil, and natural gas have all contributed to significant habitat alterations and declines in biodiversity. The former Soviet Union engaged in some site-specific wetland habitat protection in western Siberia as far back as the late 1950s (Solomeshch 2005). Based on this system, natural areas across Russia, such as zapovedniks and zakazniks, are categorized based on their size, the degree of protection they are accorded, the flexibility of at-site management observed, and the extent to which sites are off-limits to development (Solomeshch 2005, p. 47).

The Russian government has conferred such status to additional areas across the West Siberian lowland in an attempt to conserve these important resources. Meanwhile, monitoring key wetland ecosystem-health indicators is ongoing at many sites (Russian Federation Report 2005). This will provide a wealth of information on the efficacy of conservation practices. Moreover, the policing and enforcement of management plans ought to yield positive results in these protected areas. Even so, their effectiveness will only become apparent as long-term monitoring efforts yield results.

14.3.6 Australia, New Zealand and Antarctica

Wetlands policy and management are challenged by the generally drier climatic conditions observed across vast expanses of the Australian continent. Even so, Australia has designated over 900 nationally important wetland sites and 64 Ramsar sites that ring the continent (Australian Government 2010a). The density of nationally recognized wetland sites is highest along the wetter east-coast states of Queensland, Victoria, and New South Wales. However, inland wetlands are also found in the drier regions of Western Australia and the Northern Territories. Individual states and territories are responsible for crafting and implementing wetland policy and identifying strategies for protection, which closely involve local stakeholders. As in other developed nations, at the federal level, wetland

Figure 14-16. Wetlands in European Russia. A. Paludified spruce forest in the boreal zone of Karelia (east of Finland). B. Eutrophic birch-*Sphagnum* vegetation of a karst mire, Tula region south of Moscow. Photos courtesy of E. Volkova.

protection is included within broader environment and water-quality protection laws (Australia Report 2008).

Water demands for human consumption, industry and agriculture in Australia have increased stresses on this limited resource and resulted in the over-allocation of rivers and other water sources. The Murray–Darling Basin provides an important case in point. It is one of the largest drainage basins in Australia, covering over more than one million square kilometers across the southeast. A diverse ecosystem of numerous wetlands and marsh habitats including over 16 Ramsar sites, it also meets the needs of approximately 85 percent of Australia's irrigated agriculture and the water demands of close to three million people (Australian Government 2010b). Recent climate patterns have exacerbated water shortages within the basin. Recognizing the necessity for long-term integrated water management, the Australian government implemented the 2007 Water Act to promote more sustainable water-use practices for the basin. In addition, the National Water Initiative emphasizes the need for efficient and balanced water use to meet ecosystem and human needs (Australia Report 2008).

In contrast, New Zealand's wetland management challenges result from the loss of this ecosystem through river channelization, draining, and conversion to other land uses. Studies in the 1990s suggested that only eight percent of New Zealand's original wetlands remained (New Zealand Department of Conservation 1990). The government has initiated measures to mitigate some of these losses through conservation and restoration efforts under the Resource Management Act of 1991 (New Zealand Report 2008).

Finally, some of the most remote regions of the world have wetland ecosystems. Tundra wetlands in coastal eastern Antarctica are frozen for most of the year except for a brief summer period during which soils melt and microbial activity intensifies. Nitrous oxide (N_2O) fluxes have been recorded for these wetland sites built on guano and decaying animal and plant matter (Zhu et al. 2008). With warming climate trends, scientists are calling for the close monitoring of such tundra wetlands as possible sources of methane, nitrous oxide, and other greenhouse gases. For now, limited human interaction with the Antarctic continent and its designation as a space for science to benefit all humankind has accorded it protection. Its protection is also formalized through a complex of agreements falling under the Antarctic Treaty System (Scientific Committee on Antarctic Research 2010).

Successful international cooperation in science, conservation and preservation within this system provides a possible model that could be implemented elsewhere for similar resources that transcend sovereign jurisdictions (Grant 2005).

14.4 National wetland policy in the developing world

Wetland management issues across the developing world are marked by their own distinct set of challenges. Population and development pressures and the lack of institutional protection often exacerbate problems associated with the overuse of wetland resources. Conversion to agricultural land, increased demand for fresh water, coastal aquaculture, inland pisciculture, timber harvesting, and increased flows of agricultural and industrial effluents include just some factors responsible for the destruction of coastal and inland wetlands across the globe (Whigham, Dykyjova and Hejny 1993). The lack of awareness of wetland ecosystem benefits and the tendency in some geographic contexts to classify them as wastelands confound conservation efforts.

Moreover, wetlands management often falls under the jurisdiction of various government ministries and in the absence of national wetland laws and protections, ministerial priorities and development pressures override wetland interests. For instance, in India, while primary wetland management falls under the jurisdiction of the Ministry of Environment and Forests, the departments of Agriculture, Fisheries, Revenue, and Water Resources, among others also exercise some jurisdictional control (Prasad et al. 2002). As an example, coastal mangroves may fall under the control of the Forest Department, while inland wetlands might be the responsibility of the Revenue or Agriculture departments. Furthermore, management is complicated by competing priorities faced by the same department. For instance, should the Revenue Department focus on irrigation infrastructure development or wetlands protection? The Indian Ministry of Environment and Forests has sought to build a National Wetland Inventory to systematize research and aid the management of these resources (India Report 2008).

Mangroves or subtropical and tropical coastal wetlands have witnessed particularly significant declines over the past century. These unique wetland habitats have been shown to provide integral ecological functions, including forming the basis of complex marine food chains. Today, these habitats are disappearing at an accelerating rate across South and Southeast Asia, making way for shrimp aquaculture and other industrial farming practices. For instance, Thailand has lost almost half of its mangrove forests since 1960 to such activities (Earth Island Institute 2010).

Governments are now recognizing the costs associated with this rapid destruction and in certain cases have taken protective measures. The Indian and Bangladeshi governments have made attempts to carefully manage shared mangrove forests covering an area of approximately 10,000 km^2 along the Ganges delta. The Sunderban Biosphere Reserve and the mangrove eco-park in Jharkhali are both recent efforts to protect this ecosystem (Chattopadhyay 2010). Yet, the rise in population within the region has placed significant development pressures on these resources and it remains to be seen if these efforts will sustain themselves or be effective in the long run.

14.5 Shared wetlands

The task of wetland habitat protection is complicated further when two or more countries share these ecosystems. Naturally, political relations between nations pose significant challenges. Major examples of ecosystems ranging over wide geographic extents include the Pantanal shared by Brazil, Bolivia and Paraguay, the Okavango shared by Botswana, Namibia, Angola and Zimbabwe, and the Sundarbans, which span across Bangladesh and India. In each of these cases, national policies alone cannot offer adequate protection. Joint agreements to share the benefits and protect these ecosystems often require the role of mediators to provide equitable solutions to complex management arrangements.

The Okavango River delta is one such case where cross-border troubles between Botswana and Namibia are exacerbated by conflicts over scarce water resources in this dry landscape (International Rivers Network 2010). The designation of the Makgadikgadi Salt Pans (into which the Okavango River eventually drains) as a protected Ramsar site also complicates joint protection efforts. The Okavango wetland is a seasonally filled inland delta for the Okavango River, which has its headwaters in western Angola and makes its way through Namibia before disappearing in the midst of Botswana's Kalahari Desert. The river weaves a maze of lagoons, islands, and channels (Fig. 14-17) before vanishing in the desert (Fig. 14-18). This landscape and the rich soda deposits left behind as the water evaporates or dries into the desert sustain a unique ecosystem famous for wildlife, including elephants (Fig. 14-19) and hundreds of bird species (Fig. 14-20).

Both bilateral and multilateral agreements have played a role in the management of the Okavango River basin. One such tripartite agreement, the Permanent Water Commission on the Okavango reached by Namibia, Botswana, and Angola has so far performed adequately as it seeks to share water resources equitably across borders (Okavango Commission (OKACOM) 2010). In such cases, governmental and non-governmental mediators like the non-profit Green Cross International play important roles in creating the atmosphere for fair solutions or providing technical and scientific assistance. In the case of the Okavango, the Global Environmental Facility has also funded projects and provided management assistance.

The Pantanal provides yet another example of the need for multilateral arrangements for wetland protection. The Pantanal is part of the upper Paraguay River basin (in Bolivia, Paraguay, and Brazil) and Mato Grosso (Brazil). It extends over >360,000 km^2 and comprises riverine, palustrine, and lacustrine wetlands ecosystems (Banks 1991). The larger Paraná–Paraguay River system is to be the main focus for development in the coming decades (Wais and Roth-Nelson 1994). Several hydroelectric projects are either being planned or have already been built across the river system to provide flood controls on the rivers, develop an internal transportation corridor, and boost growth in the interior of these countries. This, along with increased population and agricultural and industrial development pressures will likely change the Pantanal's habitat significantly (Swarts 2000; Junk et al. 2009).

Figure 14-17. Aerial view of a meandering channel in the Okavango River delta during the dry, cool, winter season. Photograph taken from a small plane; courtesy of M. Storm.

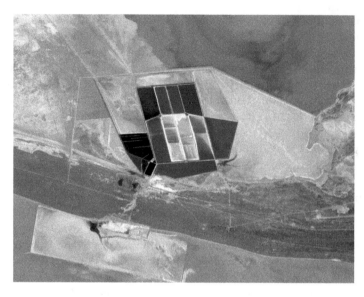

Figure 14-18. View toward southwest of salt extraction pools, Sowa Pan, eastern portion of Makgadikgadi, Botswana. Soda ash (sodium carbonate) and salt are produced from brine. Adapted from NASA Gateway to Astronaut Photography of the Earth; ISS014-E-15732, March 2007 <http://eol.jsc.nasa.gov/>.

Figure 14-19. Elephant taking a bath in shallow water of the Okavango River delta, Botswana. Photo courtesy of M. Storm.

Wetland cooperation between countries in this region has thus far largely focused on the development of the Paraná–Paraguay basin. However, conservation and the sustainable development of this ecosystem were given an impetus through the establishment of the Pantanal Regional Environmental Program in 2002 and the Brazilian government's decision in 2008 to establish a National Wetlands Institute for research and development (Junk et al. 2009). Such systematic scientific monitoring should aid conservation efforts, along with the basin's recognition as a UNESCO Biosphere Reserve in 2000 and the delineation of smaller protected sites by individual countries within the larger basin.

14.6 Summary

This chapter suggests that wetland policy across the globe is far from uniform. While the second half of the twentieth century saw significant steps toward wetland conservation and protection, far more needs to be done to jumpstart these efforts in areas of the developing world including Africa, South America, and parts of Asia. Two emerging philosophies appear to guide wetland management. On the one hand, the regulatory approach adopted by the United States and other nations focuses on laws, enforcement mechanisms, and penalties to protect these habitats. Canada leads a second, largely non-regulatory, approach designed to build partnerships, and through incentives and

Figure 14-20. Saddle-billed stork (*Ephippiorhynchus senegalensis*), a large and wide-ranging African stork (see Color Plate 14-20). This stork inhabits fresh, brackish, and saline marshes, wet meadows, and other open wetlands. Seen here in a marsh of the Okavango River delta region. Photo courtesy of M. Storm.

disincentives to persuade the public and industry to support broad wetland conservation policies.

Human pressures on wetlands in the developing world are becoming more acute. Ever-increasing demands for water to meet agricultural, industrial and domestic needs add severe strains to an already stretched resource. The implications of climate change, increased frequency of severe storm events, and rising sea levels will have ramifications on natural ecosystems, wetlands included. In the absence of effective conservation regulations, increasing land-use pressures to meet food production will result in wetland habitat alterations, as already evident across the world. As a recent Ramsar report suggests, the daunting challenge for nations will be to craft legislation and build conservation strategies ensuring sustainable wetland use in the context of these global threats (Ramsar Strategic Plan 2006). Furthermore, legislation and policy need to be context-specific to account for geographic and developmental factors rather than employing a one-size-fits-all approach.

Low-latitude wetland case studies 15

15.1 Introduction

The low latitudes span between the Equator and 30° latitude across the northern and southern hemispheres. This region receives intense solar radiation throughout the year with variable precipitation patterns ranging from year-round rainfall closer to the equator to distinct wet and dry seasons farther away in the tropics. The tropical air masses that control climates within this geographic belt give rise to tropical moist, tropical wet and dry, and tropical dry climates. For the most part, these temperature and precipitation regimes provide suitable growing conditions year round for a wide diversity of flora and fauna.

Scientists have long since established a correlation between latitudinal gradient and species diversity (Gaston 2007). Empirical evidence suggests that species richness decreases with increasing latitude from the tropics to the poles, with recent analyses supporting the idea of the tropics as both a cradle and museum of biodiversity (Jablonski, Roy and Valentine 2006). This gradient is observed in terrestrial, marine and to a lesser extent in fresh-water realms, with some variation by region and habitat type (Hillebrand 2004).

Several reasons have been provided for this observed species richness gradient. However, an agreed upon explanation remains elusive. Some have suggested that the higher speciation rates in the lower latitudes may be due to a longer and more stable period of diversification in the tropics coupled with lower extinction rates, greater amount of received solar radiation, higher average temperatures, and other biotic and abiotic factors. Reducing the rate of species loss consequently requires equal if not greater attention to be paid to the tropics and lower latitudes (Gaston 2007). However, in the tropics, biodiversity protection is further complicated by human demand and the socio-economic and developmental realities that many countries face.

In this first set of case studies focusing on the low latitudes, we find the selected wetlands supporting a scale of species richness and density seldom observed elsewhere. However, human resource demands from these wetlands also play critically influential roles in their sustainability. This chapter considers some of the large and globally renowned tropical wetland sites including the Sundarbans of South Asia, the Okavango of southern Africa, the Pantanal of South America and the Gulf of Mexico coast of the United States.

These examples cover a wide variety of tropical and subtropical wetland habitats. Coastal mangrove ecosystems are found in the Sundarbans and the Everglades, while a mosaic of fresh-water, intermediate, brackish, and saline marsh environments are found across vast sections of the Pantanal, Okavango, Mississippi Delta, Everglades, and Laguna Madre. Moreover, three of the regions considered in this chapter: the Sundarbans, Okavango and Pantanal, include wetlands that span across country borders,

Wetland Environments: A Global Perspective, First Edition. James Sandusky Aber, Firooza Pavri, and Susan Ward Aber.
© 2012 James Sandusky Aber, Firooza Pavri, and Susan Ward Aber. Published 2012 by Blackwell Publishing Ltd.

complicating issues of conservation and protection. Coordinated management efforts are important to lasting protection. Yet, such coordination is far from simple and requires long-term commitments and recognition of the importance of wetland conservation.

Keddy and Fraser (2005) argued that the size of wetlands does matter. With increasing size, wetland functions such as carbon sequestration, nutrient cycling, flood mitigation and habitat availability operate at much larger scales – with global and regional consequences. We find this to be the case in each of these low-latitude wetland sites. They are impressive not just for the scale and scope of influence they have, but also their ability to support a richness of life seldom seen elsewhere and their capacity to recover from natural and human-induced stressors.

15.2 Sundarbans of South Asia

The Sundarbans coastal mangrove ecosystem stretches across roughly 10,000 km² of deltaic plains formed by the Ganges, Brahmaputra, and Meghna rivers as they enter the Bay of Bengal. It is one of the largest contiguous stretches of mangroves anywhere in the world (Fig. 15-1).

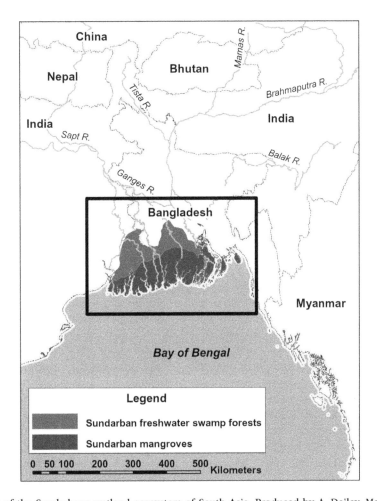

Figure 15-1. Map of the Sundarbans wetland ecosystem of South Asia. Produced by A. Dailey. Map made with data from Natural Earth, Global Administrative Areas, and World Wildlife Fund Terrestrial Ecoregions Database (Olson et al. 2001). Accessed online <www.naturalearthdata.com> <http://www.gadm.org/> and <http://www.worldwildlife.org/science/data/item6373.html> February 2011.

Table 15-1. Percentage of total worldwide mangrove coverage in the top ten countries, which represent approximately two-thirds of all mangroves. Source FAO (2007).

Country	% of total mangroves
Indonesia	19
Australia	10
Brazil	7
Nigeria	7
Mexico	5
Malaysia	4
Cuba	4
Myanmar	3
Bangladesh	3
India	3

The name "Sundarbans" likely arises from the local name for a mangrove species, sundari (*Heritiera fomes*), which was once widespread across the region but is now threatened and in decline.

The Sundarbans is just one stretch of mangroves found at latitudes below 25°N and S. Coastal mangrove systems are observed across tropical regions in Asia and Oceania, Africa, North and Central America, and total some 15.2 million hectares (152,000 km²) (Food and Agriculture Organization 2007). As Table 15-1 shows, the top 10 countries in terms of mangrove area coverage are not limited by geography but spread across the tropical belt encircling the globe. A recent report, however, suggests that some 20 percent of this, or 3.6 million hectares (36,000 km²), have been lost to development and other activities since 1980 (Food and Agriculture Organization 2007). Mangrove ecosystems generally comprise trees and shrubs that have developed special adaptations to saline conditions and variable periods of inundation. They provide numerous economic benefits to resident populations and perform important roles in stabilizing coastlines, mitigating the impact of storms and floods, and providing habitats for numerous invertebrates, crustaceans, fish, and other mammals and birds.

The Sundarbans extends between 21°30′ to 22°40′N latitude and 88°05′ to 89°55′E longitude, straddling the countries of India and Bangladesh with roughly 60 percent of the wetlands contained within Bangladesh and the rest in India. The region experiences typically humid subtropical climate with a mean annual high temperature of 34°C. Both the Southwest summer and Northeast winter Monsoons affect the region. A majority of rainfall is from the Southwest Monsoons and occurs between June and October. It brings an annual average of between 1600 and 1800 mm to the wetlands depending on location.

The Sundarbans may be divided into two broad ecological zones (Fig. 15-1). The Sundarbans mangroves cover the edge zone of the delta immediately adjacent to the Bay of Bengal and comprise mudflats, islands, creeks, channels, estuaries, shallow water bodies, and grasslands that experience periodic inundation. The Sundarbans fresh-water and brackish swamps lie north of the mangroves farther inland within the delta and comprise mixed tropical species. Over the centuries, this latter zone has witnessed widespread clearing to make way for dense human settlement and agricultural activities. A color-infrared space-shuttle photograph clearly reveals the extent of preserved mangroves (Fig. 15-2). The mangroves are indicated in red, and surrounding agriculture is clearly noticed in pale gray tones. Shrimp farming and intensive agriculture are common around the protected mangrove ecosystem. The entire region is dotted with islands and crisscrossed by numerous channels, streams and rivers, which not only provide important influxes of fresh water but also deposit sediment into the Bay of Bengal.

Pollen and sediment analysis from radiocarbon dating indicates a brackish estuarine swamp environment across the delta over the past 9000 years with variable periods of expansion and retreat of mangrove forests possibly due to sea transgression and high sedimentation rates (Hait and Behling 2009). More recently, neotectonic activity causing the delta to tilt eastward and the shifting courses of the Ganges and Brahmaputra rivers over the past several centuries have played a crucial role in the hydrology and morphology of the delta and influenced the sedimentation, accretion, and erosion patterns witnessed across the region (Gopal and Chauhan 2006). Moreover, tidal action, the influence of which may be observed up to 50 km inland,

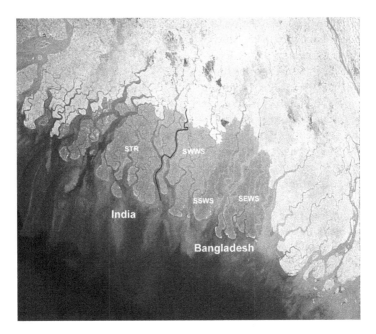

Figure 15-2. Ganges Delta, Sundarbans, India and Bangladesh (see Color Plate 15-2). The protected mangroves are highlighted in red in this color-infrared photograph. Black line marks the national boundary. STR – Sundarbans Tiger Reserve, SWWS – Sundarbans West Wildlife Sanctuary, SSWS – Sundarbans South Wildlife Sanctuary, and SEWS – Sundarbans East Wildlife Sanctuary. Image courtesy of K. Lulla; STS061B-50-007, December 1985.

shapes the land adjacent to the Bay of Bengal and helps build sand bars, dunes, mudflats and islands over time through the accretion of sand and alluvial material deposited by the rivers (Gopal and Chauhan 2006). By the same token, severe storms can wash away islands in a matter of hours or days, and this constantly shifting landscape typifies the processes that have shaped the delta over time.

Cyclonic activity is quite common during the summer and post-monsoon period. The region has experienced enormous loss of life and property in the past due to severe cyclonic events that are coupled with devastating tidal surges followed by widespread flooding. Table 15-2 provides a look at some of the most devastating storms experienced by Bangladesh and the accompanying staggering loss of human life (Hossain 2001). The vulnerability of this coastal zone would only become more acute with predicted sea-level rise, the destruction of protective mangroves due to development demands, and persistent conditions of poverty experienced across the region.

Table 15-2. Cyclonic activity, storm surges, and approximate loss of human life in Bangladesh for selected storms. Adapted from Hossain (2001).

Date	Cause	Casualties
May 1822	Cyclone	40,000
Oct. 1876	Cyclone, storm surge height = 14 m	400,000
Oct. 1897	Storm surge	175,000
May 1965	Cyclone, storm surge height = 4 m	19,270
Nov. 1970	Cyclone, storm surge height = 14 m	500,000
April 1991	Cyclone, storm surge height = 6 m	150,000

Coastal mangrove ecosystems typically observe spatially defined zones based on the topography and configuration of the landscape, substrate, tidal action, and salinity ranges. Scientists have provided classification schemes that divide mangrove forests based on their occurrence along coastlines. Across the eastern coast of India, coastal mangroves are classified into proximal, middle and distal zones depending on their exposure to the open sea, tidal action and influx of fresh water (Mahajan 2010). Studies further indicate that the Bangladeshi section of the Sundarbans exhibits greater species complexity and structure than their Indian counterparts, which may in part be due to variations in salinity levels (Food and Agriculture Organization 2007). Cintron and Novelli's (1984) generalized categorization of fringe, basin, and riverine mangroves characterizes mangrove zonation based on tidal action and salinity levels.

- The fringe zone comprises the front range of mangroves facing the daily onslaught of tidal action and higher salinity levels. Vegetation adaptations include species that exhibit high salt tolerance and buttressed root systems. Such prop root systems serve to anchor trees and dissipate the energy of strong tides and winds as well as for aeration. In the case of the Sundarbans, species such as *Rhizophora* (red mangrove) and *Bruguiera* are often found in this zone, which may experience polyhaline or mesohaline conditions.

  Similarly, *Sonneratia* and *Avicennia* are also common and quite well adapted to diurnal tidal action (Gopal and Chauhan 2006). *Avicennia* develop pneumatophores, which are specialized root adaptations that aid respiration under waterlogged conditions (Fig. 15-3). The roots extend upward between 15 and 20 cm above the soil surface from below-surface lateral root systems. To deal with high salinity levels, species like *Avicennia* also secrete salt through their leaves, with salt deposits often visible on the leaf surface. In other cases, species concentrate salts in the leaves which are then shed, or species may absorb additional water to dilute the concentration of salt in their tissues, giving rise to thick and fleshy succulent leaves (Mahajan 2010).
- Basin mangroves are generally found in sheltered inland spaces behind the front range of fringe-zone mangroves. They are also found in shallow depressions that may be flooded during high tide. Due to inadequate drainage, water may accumulate and stagnate in pools within this zone.

Figure 15-3. Pneumatophores rising 15–20 cm above the soil surface from lateral roots aiding respiration in *Avicennia marina* (gray mangrove). Photo by Firooza Pavri.

- Riverine mangroves occur along the banks of rivers, estuaries, and creeks in coastal regions. They may range far inland from the coast and are generally exposed to regular fresh-water influxes, along with nutrient and sediment loads brought down by rivers. With tidal extents reaching far inland into the Sundarbans, these riverine mangrove zones extend into the interior regions of the Ganges Delta traversed by a dense network of channels and creeks. Generally, this zone experiences far lower salinity levels and a lower incidence of tidal inundation (Mahajan 2010). The famous *Heritiera fomes* may be found in this zone and can grow up to a height of 25 m under ideal conditions.

Human occupation of the Gangetic Delta and Sundarbans region dates back to early human settlement of this part of South Asia. The region also appears frequently in historical accounts which detail the settlement and agricultural fertility of the delta (Chakrabarti 2009). More comprehensive accounts emerge from early nineteenth-century British-led surveys of the region and colonial reclamation efforts that focused on using the forests for timber production, harvesting animals, or converting the delta swamp forests into agricultural land for rice cultivation. The conventional wisdom at the time regarded wetlands as "wastelands" that needed draining and systematic development to become economically productive regions. With the Indian Forest Act of 1878, the British designated sections of the wetlands as Reserved Forests (Gopal and Chauhan 2006). This categorization ensured greater government control over extraction activities and profits.

Today the Gangetic Delta is one of the most densely populated and intensively cultivated regions of the world, and this has placed enormous demands on the Sundarbans ecosystem. Several million people live across parts of the Sundarbans. In addition, the Sundarbans is located at the mouths of two of South Asia's largest rivers, the Ganges and Brahmaputra, draining a combined total of 1.7 million km^2 (Allison 1998). They are influenced by upstream developments occuring across their drainage basins, which span five countries. Dam construction and water-diversion activities, erosion due to upstream land-use conversions, pollution from urban, industrial and agricultural activities, and resource extraction in the form of timber harvesting, fishing, aquaculture, hunting and trapping all play influential roles in shaping the current health and future viability of this system.

Since the end of the British colonial period in 1947 and the subsequent partitioning of the subcontinent, conservation and management of the Sundarbans lie in the hands of two separate governments with two sets of priorities responding to the distinct needs of the region's people. Recognizing the ecological importance of this ecosystem, both India and Bangladesh established nature reserves during the 1970s to protect sections of the Sundarbans from development pressures.

The wetlands are well known for their biodiversity and the Royal Bengal tiger (*Panthera tigris tigris*) is likely their most famous resident. Studies have recorded some 350 species of vascular plants, in excess of 300 species of birds including a large number of migratory species, numerous invertebrates, and just under 50 species of mammals within the Indian and Bangladeshi sections of the wetlands (International Union for Conservation of Nature 1997). Within the Sundarbans, only about 250 Bengal tigers remain (Project Tiger 2011). Hunting, poaching and habitat loss have contributed to a rapid decline from an estimated 40,000 tigers across the subcontinent at the turn of the twentieth century.

India initiated Project Tiger in 1972, which sought to provide protection for the last remaining tiger populations found in small isolated pockets across the country. The Sundarbans Tiger Reserve was established in 1973 with over 2585 km^2 of land and water to protect its key poster species, the Royal Bengal tiger. The reserve later included a national park covering some 1330 km^2 within its boundaries. This designation meant a complete cessation of development activity or local resource use from within the park (Project Tiger 2011). The Sundarbans National Park was listed as a UNESCO World Heritage Site in 1987, while the entire Sundarbans ecosystem was declared a Biosphere Reserve in 1989.

Across the border, Bangladesh similarly established the Sundarbans West (715 km^2), Sundarbans East (312 km^2) and Sundarbans South (369 km^2) Wildlife Sanctuaries. Extending across several deltaic islands, the sanctuaries offered conservation management and protection under the Bangladesh Wildlife Act of the 1970s. There are no villages established within the sanctuaries, while activities such as honey gathering, wood extraction, fishing, and the extraction of golpatta leaves (*Nypa fruticans*) used for thatch material are regulated by the Forest Department (International Union for Conservation of Nature 1997). Even so, illegal extractive activities including hunting and poaching continue to trouble the region.

Despite these conservation efforts, the sheer demand for resources and space in this densely settled delta complicates efforts to balance meeting human and ecosystem needs. Illegal poaching, trapping, fishing and felling of wood continue to pose challenges for officials. During the twentieth century several major animal species saw local extinctions from the Sundarbans. These included the Indian Javan rhinoceros (*Rhinoceros sondaicus inermis*), the gaur (*Bos gaurus*) and the mugger crocodile (*Crocodylus palustris*) among others (International Union for Conservation of Nature 1997).

Few coordinated conservation efforts exist between India and Bangladesh. The two countries have several long-standing issues, which include disputes over the extraction, diversion and allocation of waters from the Ganges and its tributaries upstream from Bangladesh. Meanwhile the diversion of fresh water and migration of the river channels farther eastward have caused higher salinity gradients in the western portion of the mangroves. This has affected local flora and fauna, as reported in the case of localized die-backs of *Heritiera fomes* (International Union for Conservation of Nature 1997). The Sundarbans region clearly faces significant challenges in the twenty-first century. Meeting these will require striking a balance between growing human needs and allowing ecosystem functions to perform. These challenges undoubtedly will be compounded by Indo–Bangladeshi politics and the regional ramifications of global climate change on sea levels, storm activity, temperatures, and precipitation regimes.

15.3 Okavango Delta of southern Africa

North of the Kalahari Desert in southern Africa lies one of the largest seasonally flooded inland deltas in the world. The Okavango Delta is located at 18°45′S latitude and 22°45′E longitude and is formed by the Okavango River, which traverses more than 1000 km from the Angolan highlands – slowly making its way southeastward across the dry interior of southern Africa. The river's flow is dependent on rainfall across its watershed and this water is carried through Angola and Namibia and into Botswana, where it eventually fans out into a bird-foot-shaped alluvial delta (Fig. 15-4). The narrow panhandle of the Okavango River as it enters Botswana is formed by geological fault lines that constrain the water's flow to a contained channel (Fig. 15-5). These faults found across the region are an extension of the East African Rift system. Farther downstream additional faults allow the river greater freedom to branch out, which eventually leads it to distribute its water over a wide and relatively flat basin underlain by deep sand deposits.

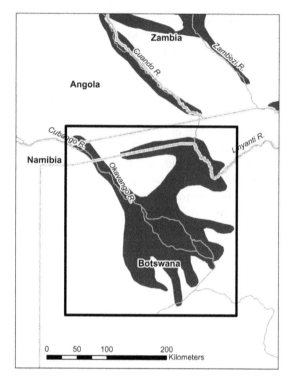

Figure 15-4. Map of the Okavango Delta wetlands and surrounding region of southern Africa. Produced by A. Dailey. Map made with data from Natural Earth, Global Administrative Areas, and World Wildlife Fund Terrestrial Ecoregions Database (Olson et al. 2001). Accessed online <www.naturalearthdata.com> <http://www.gadm.org/> and <http://www.worldwildlife.org/science/data/item6373.html> February 2011.

The river's flood pulse brings life-sustaining water to the delta during the height of the dry season between May and September, when water levels have fallen to critical levels across much of the region. This provides a refuge for wildlife, birds, domesticated cattle, and other foraging animals. Total flows into the delta average 9.4 km^3 per year with annual variations based on precipitation trends (Mendelsohn and el Obeid 2004). The delta is also fed by local rainfall averaging 490 mm during the summer months starting as early as November and especially between January and March (McCarthy et al. 2000).

During particularly wet periods, overflow from the delta may make its way farther southeast via the ephemeral Boteti River to enter the

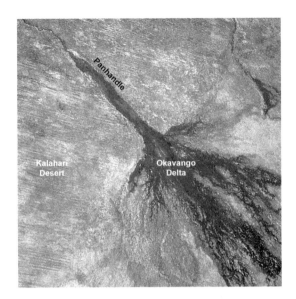

Figure 15-5. Space-shuttle photograph of the Okavango Delta, Botswana, Africa. Dark gray indicates active vegetation. Linear stripes to the north and west are dunes in the Kalahari Desert. Image courtesy of K. Lulla; STS61A-43-29, November 1985.

Makgadikgadi salt-pan depression within the greater Kalahari Basin. The Makgadikgadi salt pans are remnants of a vast ancient lake system that covered this region. Evidence from paleo-floodplains suggests significant fluctuations in alluvial deposition patterns across the region with a delta that could easily be two or three times its current size. Ramberg et al. (2006) suggested a combination of tectonic activity, climate and precipitation changes as possible causes for the delta's variable size through geologic history.

At present, precipitation combined with the Okavango River's waters give rise to a montage of seasonal and permanent wetlands that in full flood may extend over 15,000 km^2, displaying remarkable species richness (Heinl et al. 2007). Ramberg et al.'s (2006) review of recorded species indicated a delta with some 1300 species of plants, 71 species of fish, 444 birds, and 122 mammals. Permanent wetlands are found along the panhandle and the proximal reaches of the delta. Here, reeds (*Phragmites*), papyrus (*Cyperus papyrus*), sedges, bulrushes and other aquatic plants dominate the landscape. The hippopotamus (*Hippopotamus amphibious*) is a permanent resident of these swamps (see Fig. 1-2). Swamp antelope called the lechwe (*Kobus leche*) are also found in large numbers within the permanent swamps. Estimates suggest approximately 40,000 individuals, making them one of the greatest concentrations of large mammals within the delta (Bass 2009). During the flood season, they spread across the delta to take advantage of new foraging sites.

Seasonally or occasionally flooded swamps and grasslands found in shallow depressions are generally observed beyond the permanent swamps (Bass 2009). Elsewhere, mopane or mopani (*Colophospermum mopane*) dominant woodlands are found across the outer edges of the delta and across northeastern Bostwana. These dry open woodlands also include sparse grass and brush cover. Lastly the salt pans of the Makgadikgadi observe high concentrations of salts and minerals and virtually no vegetative cover when dry. During particularly wet seasons when excess water from the delta makes its way to the pans, algal blooms are not uncommon (McIntyre 2007).

The duration of flooding within the delta is dependent on precipitation rates, water-table levels and evapotranspiration rates. The delta acts as an oasis across a particularly arid region of Africa providing a critical resource through the dry season of the year and attracting fauna from great distances. A prime example of this is the elephant (*Loxodonta africana*). Estimates suggest up to 30,000 individuals consider the Okavango Delta home for part of the year (see Fig. 14-19). Once the delta's floods have receded, the elephants migrate farther northeast to seek refuge in the mopane woodlands (Bass 2009). Other commonly found seasonal fauna include large numbers of buffalo (*Syncerus caffer*), zebra (see Fig. 7-40) and impala (*Aepyceros melampus*) (Ramberg et al. 2006, p. 329). Giraffe (Fig. 15-6), kudu antelope (*Tragelaphus strepsiceros*), and African lion (Fig. 15-7) are also found across the delta. Thousands of slightly elevated tree-covered islands dot the landscape and serve as a terrestrial refuge for animals during the full flood season.

The human fossil record dates back several thousands of years across this region of

Figure 15-6. Giraffes (*Giraffa camelopardalis*) in the Okavango Delta. Photo courtesy of M. Storm.

Figure 15-7. Wild adult female African lion (*Panthera leo*) in the Okavango Delta. Photo courtesy of M. Storm.

southern Africa. Humans have undoubtedly had an intermittent, yet long-standing and complex relationship with the delta. As the region experiences greater demands for water from locally expanding human settlements, agro-pastoral development and upstream diversions of the Okavango River, water availability issues for the delta have become quite significant since the end of the twentieth century. Population within the greater Okavango basin was a little over 1.1 million people in 2000, and is expected to rise to 1.6 million by 2020 (OKACOM 2011).

The subsequent demand for fresh water from riparian users is bound to increase over the next decade. Complicating this equation is the fact that the Okavango flows through three extremely arid countries with different development priorities.

Of the waters that enter the Okavango Delta, more than 90 percent originates in the Angolan highlands through rainfall (OKACOM 2011). Approximately 96 percent of this is lost to evapotranspiration or through consumptive use, two percent percolates underground and the remaining flows at the surface (in Ellery and McCarthy 1998). The diversion of water upstream in any of the countries would influence flows into the delta and other downstream users. Recognizing the vital importance of this shared resource, the governments of Angola, Namibia and Botswana accept the need for Integrated Water Resource Management (IWRM), which considers the issue from a whole-ecosystem perspective and includes multiple stakeholders when seeking sustainable management solutions (Heyns 2007).

The Permanent Okavango River Basin Commission (OKACOM), a multilateral institution with representatives from the three riparian states formed in 1994 and seeks to coordinate management and water development efforts. OKACOM expects that shared governance of this resource could help arrive at agreed-upon sustainable use strategies and equitable fresh-water allocations as well as improve water infrastructure and offer water-conservation and pollution-prevention alternatives (OKACOM 2011). Similar multilateral institutions would be critical in managing shared fresh-water resources and adjudicating conflicts that may arise between riparian users in the twenty-first century.

15.4 Pantanal of South America

The Pantanal of South America (derived from the Portuguese word *pântano* and translating as swamp or bog) comprises one of the largest fresh-water wetland sites in the world (Fig. 15-8). It extends over roughly 160,000 km² of seasonally flooded savanna grasslands, forests

Figure 15-8. Map of the Pantanal wetlands and surrounding region of South America. Produced by A. Dailey. Map made with data from Natural Earth, Global Administrative Areas, and World Wildlife Fund Terrestrial Ecoregions Database (Olson et al. 2001). Accessed online <www.naturalearthdata.com> <http://www.gadm.org/> and <http://www.worldwildlife.org/science/data/item6373.html> February 2011.

and permanently flooded lakes and lowland regions across the countries of Bolivia, Brazil and Paraguay (Keddy et al. 2009). The Pantanal is located between 15°S to 22°S latitude and 54°W to 58°W longitude within the upper Paraguay River basin. Roughly two-thirds of the wetlands fall within the country of Brazil with the remaining distributed between Bolivia and Paraguay. As a point of comparison, the Pantanal at the height of the wet season may extend over an area the size of the state of Georgia in the United States.

The wetlands and seasonally flooded grasslands are located within a shallow geographic depression filled in with sediments eroded from surrounding plateaus and highlands since the Quaternary period (Alho 2005). Moreover, as the rivers and tributaries of the Paraguay, which flow across the region, enter the low-relief plains they deposit large quantities of eroded fluvial material. Low topographic relief with a slope of 0.3 to 0.5 m/km from east to west and between 0.03 to 0.15 m/km from north to south ensures slow surface water flows across the Pantanal (Franco and Pinheiro 1982 in Alho 2005). Since the Late Pleistocene, these processes have built geographically expansive fluvial fans, which form a distinctive geomorphological feature of the Pantanal (Assine and Silva 2009). A Moderate Resolution Imaging Spectroradiometer (MODIS) image from 16 June 2003 shows one of the largest fluvial fans quite distinctly in the center of the image, while smoke from agricultural fires in the vicinity provide the image a hazy appearance (Fig. 15-9). Pantanal's lakes are clearly visible to the north and west of the alluvial fan as black dots of variable size. The Brazilian landscape to the east and north of the fan indicates agricultural activities. Healthy vegetative cover is observed in Bolivia to the west of the Pantanal.

The region experiences typically tropical climate, with average annual temperatures at 25°C and distinct seasonal rainfall during the summer months. Rain-fed floodwaters in the northern upper reaches of the Paraguay River system as well as region-wide seasonal precipitation provide the key drivers for life across the region. Rainfall occurs between November and March and ranges between 1200 and 1300 mm per year. Low-slope conditions allow water to overflow river banks starting in the northern reaches and can flood extensive portions of the Pantanal.

The timing, amount of land inundated, and the length of inundation are generally dictated by the level of the river and precipitation patterns. A flood pulse starts across the upper Pantanal in February with southern downstream regions observing delayed floods as late as May and June. As floodwaters recede between August and September during the dry season, isolated pools, watering holes and lakes may retain water for variable amounts of time depending on their depth and evapotranspiration rates. These ephemerally wet areas are also interspersed by narrow rivers and streams and serve as important forage sites for birds and refuges for fish and other mammals during the dry

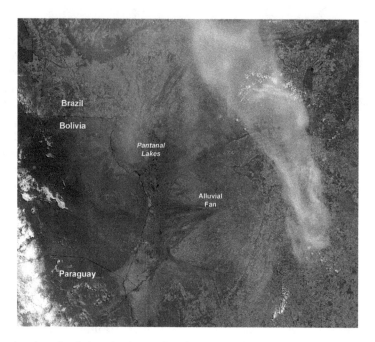

Figure 15-9. Pantanal region, South America (see Color Plate 15-9). The image clearly shows one of the largest alluvial fans at the center, while the Pantanal Lakes are visible to the north and west of the alluvial fan as black dots. The Brazilian landscape to the east and north of the fan indicates agricultural activities. Healthy vegetative cover is observed in Bolivia to the west of the Pantanal. Red dots indicate active fires. Moderate Resolution Imaging Spectroradiometer (MODIS) natural-color image from 16 June 2003; adapted from NASA's Earth Observatory <http://earthobservatory.nasa.gov/>.

season. The annual fish *piracema* (reproductive) migrations are a vivid illustration of the synchronization of fish life cycles to the flood pulse system. Fish first migrate upstream at the start of the rainy season to spawn and then later follow the flood into seasonally inundated grasslands for forage and habitat. During the dry season as the flood retreats they return to perennial rivers, streams and lakes.

Even though the Pantanal is often identified as a single ecosystem, it comprises a diverse mosaic of habitat types based on soil characteristics and the dynamics of inundation (Harris et al. 2005). These conditions give rise to vegetation assemblages that have adapted to periodic flooding or near-drought conditions based on seasonal climate variability. The Pantanal's habitats include savanna grasslands (*campos*), woodlands, gallery or riparian forests lining river banks, elevated forested islands (*capões*), marshes (*baías* and *corixos*) and scrubland (Pearson and Beletsky 2005; Mittermeier et al. 2005).

This montage of habitats contributes to a high level of species richness and density including some 1863 species of vascular plants, 263 species of fish, 463 species of birds and 132 mammal species (Alho 2005, p. 224). Among the more renowned species, the Pantanal jaguar (*Panthera onca palustris*), the world's largest rodent, the capybara (*Hydrochaeris hydrochaeris*), the giant river otter (*Pteronura brasiliensis*), the marsh deer (*Blastocerus dichotomus*), and the abundant Pantanal caiman (*Caiman yacare*) comprise just a few. The wetlands also serve as stopover points for regional bird flyways and provide seasonal foraging and nesting habitat for numerous migratory waterfowl and shorebirds. Moreover, 19 recorded species of parrots, 13 species of heron and egrets, spoonbills, ibis, kingfisher birds, jabiru (*Jabiru mycteria*), the world's largest stork, and rheas (*Rhea americana*) are commonly observed (Alho 2005).

Human influence within the Pantanal dates back to early indigenous groups that settled the region several millennia ago. In comparison to other coastal parts of South America, population numbers have generally been low across the wetlands. A few larger cities encircle the wetlands including Cuiabá in the north, Corumbá in the south, and towns like Bonito, Miranda and Campo Grande which funnel tourists into the wetlands. Current settlement patterns date back to the colonial period and the introduction of cattle ranching across the region during the mid-nineteenth century. Approximately 95 percent of the Brazilian Pantanal is privately owned by ranchers, with large cattle ranches (*fazendas*) commonly observed across the landscape (Mittermeier et al. 2005).

Cattle ranching has coexisted with the native flora and fauna for more than a century. However, changes in agrarian practices brought on by Brazil's global position in agricultural production and the biofuels revolution have meant increased investment in cash crops like soybean, corn and rice within and outside the Pantanal (Junk and de Cunha 2004). The ramifications of deforestation and grassland conversions will undoubtedly affect the ecological integrity of the region. Already ranching activities have been shown to influence fire regimes with devastating short-term consequences for wetland flora and fauna (Alho, Lacher and Gonçalves 1988). The expansion of commercial agriculture and large-scale ranching has introduced exotic species that outcompete native grasses (Harris et al. 2005), while the cultivation of monocultures like soybean may impact long-term soil productivity. The increased use of fertilizers, pesticides and herbicides causes pollution runoff entering streams and rivers, with potentially important implications for fish populations and ecosystem balance. Road development and urban expansion especially across the Brazilian Pantanal require more careful planning and management.

In the 1990s the governments of the La Plata basin comprising Argentina, Bolivia, Brazil, Paraguay, and Uruguay proposed the idea of undertaking the Hidrovia or Parana–Paraguay Waterway project which included the dredging, channelization and straightening of the river to promote upstream transportation and development. The consequences of Hidrovia on sections of the Pantanal ecosystem, its *piracemas* and its flood-pulse system would be nothing short of catastrophic. After consistent opposition from a wide diversity of groups, Hidrovia's future remains uncertain for the time being.

15.5 Gulf of Mexico, United States

The Gulf of Mexico is a semi-enclosed sea that is bounded by the United States, Mexico, and Cuba (Fig. 15-10). It connects eastward via the Straits of Florida to the Atlantic Ocean and southeastward between Cuba and the Yucatán Peninsula to the Caribbean Sea. The United States section stretches from the Florida Keys to the southern tip of Texas in the subtropical range ~25° to 30°N latitude. This region is the tectonically quiet trailing edge of the continent characterized by low-lying, poorly drained landscapes in which geomorphic processes are dominated by stream and coastal erosion and sedimentation. For most of the U.S. Gulf region, clastic sediment – sand, silt and clay derived from inland – makes up coastal deposits and landforms. In South Florida, however, carbonate sediment produced by marine invertebrates and chemical precipitation is predominant in limestone plateaus, sandy shoals, and coral reefs.

Plant hardiness zones are 9 for most of the U.S. Gulf coast and 10 for southern Florida, the Mississippi River delta, and the southern tip of Texas (Arbor Day 2006). Freezing temperatures occur rarely in zone 9 and never in zone 10 (see Table 6-4). Average annual precipitation ranges from approximately 2 m along the north-central Gulf coast to about 0.7 m at South Padre Island, Texas (Anderson and Rodriguez 2008). Gulf surface water average temperature varies from lows of about 10 °C in the winter along the north-central Gulf shore to around 30 °C in coastal southern Florida during the summer (National Oceanographic Data Center 2008).

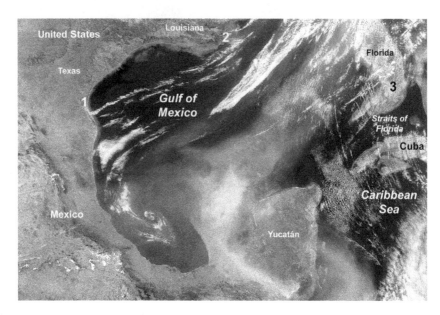

Figure 15-10. Natural-color, SeaWIFS satellite image of the Gulf of Mexico region with minimal cloud cover (see Color Plate 15-10). Selected case study sites: 1. South Texas, 2. Mississippi River delta, 3. Florida Everglades. Smoke from forest fires covers part of the Yucatán Peninsula and southern Gulf. Image date 24 April 2000; adapted from NASA Visible Earth <http://visibleearth.nasa.gov/>.

Table 15-3. Classification of hurricanes and tropical storms based on wind speed. Adapted from Southern Regional Climate Center (2011).

Storm rank		Wind (mph)	Wind (km/h)
Hurricane	Category 5	>155	>250
	Category 4	131–155	211–250
	Category 3	111–130	178–210
	Category 2	96–110	155–177
	Category 1	74–95	119–154
Tropical storm		39–73	63–118
Tropical depression		<39	<63

Table 15-4. Hurricanes striking the United States Gulf of Mexico coast region (2005 to 2010) with category at landfall. No storms of hurricane strength made landfall on the U.S. Gulf coast in 2006, 2009 or 2010. The number and strength of hurricanes were exceptional in 2005. Data from Southern Regional Climate Center (2011).

Year	Hurricane	Cat.	Region
2008	Dolly	2	South Texas
	Gustav	3	Louisiana
	Ike	2	East Texas
2007	Humberto	1	East Texas, West Louisiana
2005	Dennis	4	West Florida, Alabama
	Katrina	4	Louisiana, Mississippi, South Florida
	Rita	3	West Louisiana, East Texas
	Wilma	3	South Florida

The entire Gulf region is subject to hurricanes, which bring widespread strong wind and waves, intense rain, and flooding along the coast and inland (Table 15-3). During the period 2005-10, eight hurricanes made landfall along the U.S. Gulf coast, mostly in 2005 and 2008 (Table 15-4). Hurricanes also may spawn local thunderstorms and tornadoes that wreak additional damage. Even lesser tropical storms and depressions may cause substantial flooding and erosion of coastal areas. Hurricanes are historically most common in southern Florida, moderately frequent in the north-central Gulf region, particularly Louisiana, and least common in the Florida panhandle and southern Texas (Doyle 2009).

Selected case studies include the Florida Everglades, which shares much in common with Caribbean environments; the Mississippi River

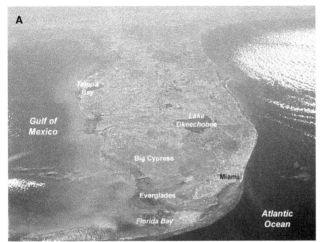
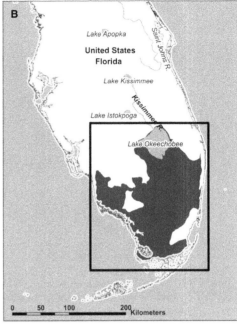

Figure 15-11. Florida peninsula. A. Oblique manned-space photograph looking northward over the Florida Peninsula. Adapted from NASA Johnson Space Center; STS51C-44-26; January 1985. B. Map of southern Florida; the Everglades drainage system is depicted in dark gray. Produced by A. Dailey. Map made with data from Natural Earth, Global Administrative Areas, and World Wildlife Fund Terrestrial Ecoregions Database (Olson et al. 2001). Accessed online <www.naturalearthdata.com> <http://www.gadm.org/> and <http://www.worldwildlife.org/science/data/item6373.html> February 2011. See also Fig. 3-13.

delta, which has suffered much during hurricanes of recent years; and Padre Island and Laguna Madre of South Texas. Water chemistry and sediment input vary greatly among these sites, but all are heavily impacted by past and present human activities. Many portions of these sites are protected in various national parks, seashores, wildlife refuges, and nature preserves.

15.5.1 Florida Everglades

The Florida Everglades is a vast expanse of wet sawgrass prairie that extends from Lake Okeechobee in south-central Florida to the mangroves along the coast of southern Florida (Fig. 15-11). Lake Okeechobee is the upstream source of surface water, which migrates slowly as shallow sheet flow over a flat limestone plateau that slopes imperceptibly southward to the sea. Everglades National Park, established in 1947 (Fig. 15-12), occupies the downstream portion of the water-flow system. Immediately to its north is Big Cypress National Preserve. Continuing to Lake Okeechobee is a series of surface-water storage conservation areas that were established along the Shark River Slough to control and distribute water throughout southern Florida (Fig. 15-13). Prior to human drainage and development, the Everglades was one of the largest fresh-water marsh systems in the world, covering some $10,000\,km^2$ (4000 square miles), but less than half this area remains nowadays (Dugan 2005).

The climate of the Everglades region is semitropical, and palms and tropical plants co-exist alongside warm-loving temperate vegetation. The region experiences two main seasons – wet (summer) and dry (winter). The hurricane season is mainly in summer and early autumn, when sea-surface temperature is warmest, and drought conditions may develop some years in spring. Water drainage from Lake Okeechobee

Figure 15-12. United States postage stamp issued in 1947 to recognize the establishment of Everglades National Park. Original stamp printed in green monotone; from the collection of J. Vancura.

is blocked to the east by the Atlantic Coastal Ridge supported by bedrock of the mixed clastic-carbonate Anastasia Formation to the north and the wholly carbonate Miami Limestone to the south (Bryan, Scott and Means 2008). The Miami Limestone was deposited some 130,000 years ago, when sea level stood 5–8 m higher than present and all of southern Florida was submerged as a shallow bank (see chapter 8.3.1). This limestone surface is exposed now in many places, where it forms subtle ridges and pinnacles. In other places, it is covered by younger wetland sediments including diatomaceous earth, marl, and peat up to 5 m thick in sloughs (Whelan et al. 2009). Peat is mainly of Holocene age and may accumulate rapidly at rates up to 10 cm per century (Bryan, Scott and Means 2008).

The Everglades comprises several types of wetlands. These are arranged in a general progression from north to south, including

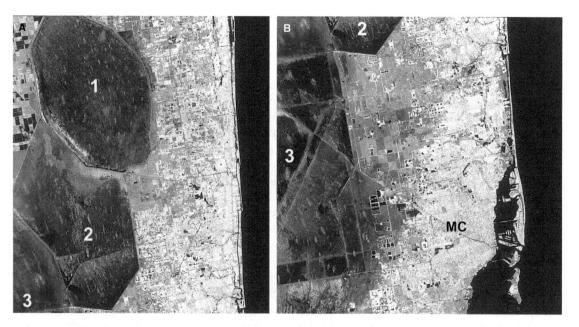

Figure 15-13. Satellite images of southeastern Florida showing surface-water conservation areas 1, 2 and 3 (see Color Plate 15-13). Note how vegetation patterns reflect the direction of surface-water flow in the conservation areas. The brighter red patches are hammocks with hardwood trees. A. Region of Pompano Beach, north of Miami. B. Miami vicinity. The Miami Canal (MC) begins at Lake Okeechobee and runs across central Miami into Biscayne Bay. Landsat TM bands 2, 3 and 4 color coded as blue, green and red; active vegetation appears bright red and pink. Date of acquisition February 1983; adapted from NASA Goddard Space Flight Center.

fresh-water marsh, deep-water swamp, tidal salt marsh, and coastal mangrove. The fresh-water marsh is dominated by Jamaica swamp sawgrass (*Cladium jamaicense*, also known as Everglades saw grass). Also common in this zone are aquatic communities of lilies (*Nuphar* and *Nymphaes* sp.) and bladderworts (*Utricularia* sp.). On limestone pinnacles, so-called hammocks support various hardwood trees. Acidic leaf litter has dissolved moats around the hammocks, and many hammocks have a central solution hole as well (Bryan, Scott and Means 2008). Among the more interesting inhabitants are brightly colored tree snails (*Liguus fasciatus*), which are endemic to southern Florida, Cuba, and Hispaniola (Dugan 2005). Snail populations in each hammock have unique color patterns (Bryan, Scott and Means 2008).

Deep-water swamps of Big Cypress Swamp contain bald cypress (*Taxodium distichum*) and other woody vegetation growing in standing water. Much of this swamp was subjected to heavy logging, but some portions are still well preserved. This region is home to most of the remaining Florida panthers (see Fig. 7-48). Tidal salt marsh and mangrove occupy the coastal zones, in which water chemistry varies from slightly brackish to fully marine (Fig. 15-14). The Everglades contains the world's largest contiguous tract of mangrove swamp in the subtropics (Doyle, Krauss and Wells 2009).

Several major hurricanes have crossed the Everglades in recent decades, including Wilma and Katrina in 2005 and Andrew in 1992 (Fig. 15-15). Hurricane impacts on the mangrove forest include the following (Doyle, Krauss and Wells 2009; Krauss et al. 2009; Smith et al. 2009; Whelan et al. 2009):

- Greatest damage is confined to a narrow coastal zone, 50–500 m wide that may extend for tens of kilometers in length.
- Some heavily damaged zones may recover, but others are permanently converted into intertidal mudflats.
- Mangroves are felled above a critical wind speed of 110 miles per hour (177 km/h), the threshold between categories 2 and 3 hurricanes.
- Substantial sediment is deposited that may raise soil surface elevations by >4 cm. Even after a year of compaction and erosion, the soil surface remains >3 cm above pre-storm level.

Figure 15-14. Red mangrove (*Rhizophora mangle*) exposed at low spring tide in brackish water of Whitewater Bay. Oysters cling to prop roots that are normally submerged, and Spanish moss hangs from upper branches of the mangrove. The mangrove is rooted in peat, which can be seen just above water level. Photo by J.S. Aber; Everglades National Park, Florida.

Figure 15-15. Satellite image of Hurricane Andrew in the Gulf of Mexico just west of Florida after it passed over the Everglades. The eye of the hurricane is well formed and is ~20 km in diameter. Also note thunderstorm clouds along the right edge. Landsat TM band 5 (mid-infrared); 24 August 1992; image processing by J.S. Aber.

- Intact mangrove forest reduces the maximum water height reached inland during storm surges.

The Everglades is a cornucopia of plants and animals; distinctive plants are the royal palm (*Roystonea regia*), West Indian mahogany (*Swietenia mahagoni*), giant wild pine (*Tillandsia utriculata*), cowhorn orchid (*Cyrtopodium punctatum*), and leafy vanilla orchid (*Vanilla phaeantha*). Notable animals include the Halloween pennant dragonfly (*Celithemis eponina*), zebra longwing (*Heliconius charitonius*), eastern indigo snake (*Drymarchon corais*), great egret (*Ardea alba*), wood stork (*Mycteria americana*), roseate spoonbill (*Ajaia ajaja*), snail kite (see Fig. 7-35), white ibis (*Eudocimus albus*), and Everglades mink (*Mustela vison*). Alligators (*Alligator mississippiensis*) are common in the freshwater marshes, but the relatively rare American crocodile (*Crocodylus acutus*) remains in brackish to marine environments (see Fig. 7-17).

Everglades National Park is one of the crown jewels in the U.S. national park system; yet, its position at the downstream end of the drainage basin makes it quite vulnerable to upstream changes in terms of both the quantity and quality of water supply. Human impact on the Everglades is substantial. Significant modification of drainage began in 1909 with completion of the Miami Canal, which connected Lake Okeechobee to the sea (see Fig. 15-13B). Hurricane San Felipe in 1928 caused a storm surge from Lake Okeechobee that inundated the southern shore and killed more than 3000 people (Bryan, Scott and Means 2008). Only the Galveston hurricane in 1900 killed more people. As a result of Hurricane San Felipe, the federal government built Herbert Hoover Dike, which completely encircles Lake Okeechobee. Renewed flooding in the 1940s led to the creation of the Central and South Florida Control project, in which many more canals and levees were constructed. Water flow in southern Florida is presently constrained by over 3000 km of canals and levees as well as 150 major water-control structures (Dugan 2005).

Rapid population growth in southern Florida requires massive supplies of fresh water, and agriculture consumes still more water. Water for human uses is derived from pumping ground water and diversion of surface water. The net result is diminished flow of surface water throughout the Everglades drainage system with many environmental consequences. Periodic fires, for example, are ignited by lightning and help to maintain the grassland habitat by limiting invasion of woody brush and trees. Prior to human intervention, these fires swept lightly over the marshes, and the sawgrass sprouted more vigorously following fires. However, draining the Everglades and fire suppression have led to a new burning regime, in which fires burn deeply, destroy roots and underlying peat, and even penetrate into hammocks (Niering 1985).

Surface water in the Everglades is naturally nutrient-poor. However, upstream agricultural runoff delivers large quantities of fertilizer, namely nitrogen and phosphorus, to the Everglades. Where this has occurred, sawgrass prairie has converted into cattail marsh, with deleterious effects on organisms and water quality. One result of higher phosphorus concentration is increased methanogenesis (Castro, Ogram and Reddy 2004). Recognizing these negative impacts, the State of Florida initiated a multimillion dollar *Save our Everglades* program in 1983. However, the state and federal governments came into conflict in 1988, when the federal government sued the State of Florida and the South Florida Water Management District because of their failure to protect ecosystems in the Loxahatchee National Wildlife Refuge. The state approved the Everglades Forever Act in 1994, and the federal Comprehensive Everglades Restoration Plan in 2000 is the largest environmental recovery project ever in the United States (Dugan 2005). Full implementation will require billions of dollars and decades of work, much of which remains to be done.

15.5.2 Mississippi River delta

The delta of the Mississippi River is the end point of one of the world's largest terrestrial drainage systems, which includes the Mississippi, Ohio, Tennessee, Missouri, Platte, Arkansas, and other major rivers draining the continental interior of the United States and a small portion of Canada. All these river systems are heavily modified and managed from their headwater sources to the Gulf of Mexico for navigation, flood control, hydropower, water supply, and recreation. The term "delta" is commonly applied to Mississippi bottomland from Memphis, Tennessee southward; however, this discussion focuses on the coastal region of southern Louisiana wherein freshwater drainage systems grade into marine conditions of the Gulf of Mexico (Fig. 15-16). Louisiana contains 40 percent of all wetlands in the contiguous United States and contributes 30 percent of the nation's seafood (Whigham et al. 2010), which emphasizes the extreme importance of the Mississippi Delta and adjacent coastal environments.

The Mississippi Delta was built during the past several millennia as postglacial sea level rose and sediment was washed down from the continental interior. Many times, delta lobes have shifted as the river meandered back and forth across the low-lying coastal region between the Atchafalaya Bay and Mississippi Sound (see Color Plate 8-13). The modern delta is a birdfoot delta, so-called because of the way the distributary channels branch out at Head of Passes (Fig. 15-17). Between the early 1800s and 1985, numerous changes took place in the dynamic delta complex (Fig. 15-18). Among the major developments are infilling of Bay Ronde and Garden Island Bay, the appearance of Main Pass, extension of Pass á Loutre, and reductions of Southeast, South and Southwest passes. During the past two decades, however, the lower delta complex has lost considerable area (Fig. 15-19). Among the major losses are the Breton and Chandeleur islands, which have nearly disappeared, and reductions of Pass á Loutre and Southeast Pass, whereas South Pass has gained land.

The current paths of the Mississippi River and its distributary channels across the delta are maintained by an extensive network of levees and channel dredging, and the river is confined upstream for more than 1000 km of its course (Dugan 2005). Left to its own devices, the Mississippi would have breached its channel upstream and followed a shorter, steeper route

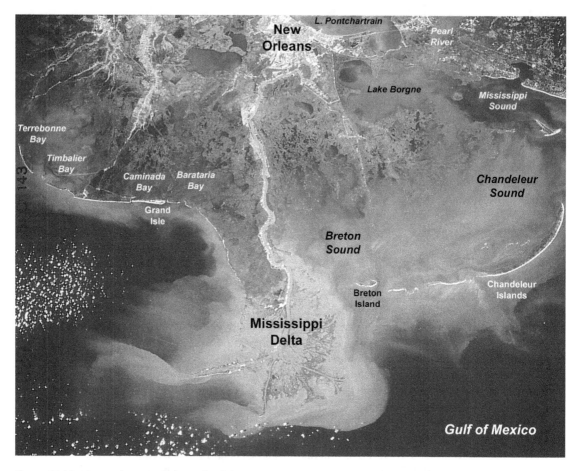

Figure 15-16. Manned space photograph of the New Orleans and Mississippi River delta vicinity. Note the nearly symmetrical arrangement of barrier islands, bays, sounds, and lakes on either side of the delta. This image was acquired in January 1985, before a series of hurricanes all but destroyed the Chandeleur Islands and parts of the delta complex. Image courtesy of K. Lulla; STS-51C-143-027.

to the sea via Atchafalaya Bay. Dredging to deepen the navigation channel leads to increased flow velocities, so that sediment is transported far offshore into deep water. For many decades, dredged spoil from the main channel was dumped at the head of Pass á Loutre, which has effectively clogged the pass and cut off the flow of fresh water and sediment into this distributary system. According to Nyman (2010, p. 12), Pass á Loutre is dying as a distributary channel, and the Pass A Loutre Wildlife Management Area is dying as a wetland habitat. This situation is symptomatic of the entire delta complex.

Coastal Louisiana is rapidly losing marshes and swamps. Since the 1930s, an area roughly the size of Delaware (~5000 km^2) has been converted from emergent wetlands to open water (Nyman 2010). The rate of loss was greatest during the 1960s and 70s, averaging one hectare every 55 minutes; the rate slowed to approximately one hectare every 95 minutes during the 1980s and 90s (Nyman 2010). Many natural and human causes contribute to the current loss of fresh-water wetlands and intrusion of the sea:

- Eustatic rise in sea level – Current sea level is rising approximately 3 mm per year globally due to gradual melting of land-based glaciers and thermal expansion of ocean water (see chapter 8.3.4). This process is

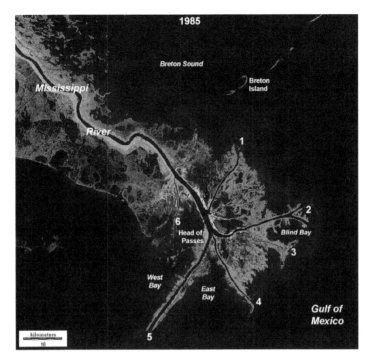

Figure 15-17. Satellite image of the Mississippi River delta vicinity, 11 October, 1985. Major distributary channels: 1. Main Pass, 2. Pass á Loutre, 3. Southeast Pass, 4. South Pass, 5. Southwest Pass, and 6. Grand Pass. Based on Landsat TM band 4 (near-infrared), which is particularly sensitive to emergent vegetation (light gray) and water bodies (black). Image from NASA; processing by J.S. Aber.

slow but inexorable, and predictions are that it will continue throughout the twenty-first century for a total of 1–2 m increase in sea level worldwide.

- Compaction of deltaic sediment – Clay and silt deposits, which dominate the marshes, bays and sounds of the delta region are initially quite porous with high water content. As the sediment becomes buried, the water is gradually expelled by increasing pressure, the sediment compacts, and the land surface subsides. This phenomenon happens in all deltas; continuous deposition of new sediment is necessary to maintain the delta surface.
- Hurricanes – Storm surges and strong waves may erode shores rapidly, breach barrier islands and shoals, and redistribute sediment into lagoons and bays. This region has frequent hurricanes. Grand Isle, Louisiana (Fig. 15-20), for example, has been struck by hurricanes 31 times since 1851, a recurrence interval of one hurricane about every five years (Doyle 2009). The popular concept that hurricanes are rare or unusual events is simply wrong; hurricanes are part of the natural regime of coastal Louisiana.
- Reduction in sediment supply – From their headwaters in the Rocky Mountains, northern Plains, and Appalachians all major rivers and many lesser streams of the Mississippi drainage basin are controlled by various human structures (Fig. 15-21). This impact extends to tens of thousands of smaller artificial ponds and retention basins on minor tributaries (Fig. 15-22). The net result is that much sediment is retained in upstream reservoirs; the Mississippi Delta is starved of sediment.
- Conversion of wetlands for agriculture – Federal agricultural subsidies led to clearing and draining of wetlands for conversion to cropland, mainly for soybean and cotton production, particularly from the 1950s to

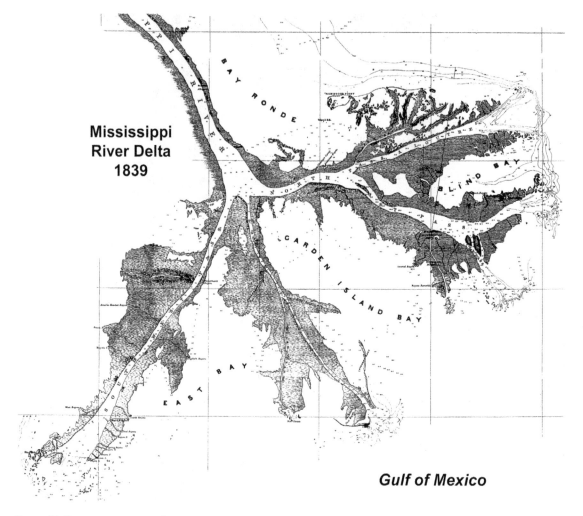

Figure 15-18. Lower portion of Mississippi River delta as surveyed in 1839 by the Bureau of Topographical Engineers. Note prominent passes and bays; Bay Ronde and Garden Island Bay have been largely infilled. Compare with previous figure; 5-minute lat/long grid; adapted from Morgan and Andersen (1961, pl. 2).

70s (Dugan 2005). Substantial wetland losses took place. Since the 1980s, however, new policies attempt to balance multiple, sustainable uses of wetlands for timber, fish and shellfish, wildlife, and recreation. Wetland losses have slowed, but still continue.

- Oil and gas production – Since World War II, Louisiana has become a major source for oil and gas production both on- and offshore. In order to accomplish inland production, many canals have been cut through marsh and swamp to service drilling and production rigs, which has allowed rapid water exchanges. Extraction of subsurface fluids leads eventually to surficial subsidence. Many scientists believe this is the primary cause for rapid increases in relative sea level (10–15 mm per year) along the Louisiana coast (Morton, Bernier and Barras 2006; Milliken, Anderson and Rodriguez 2008).

Among these factors, most are widespread and intractable in terms of effecting any significant changes – eustatic rise in sea level, sediment compaction and starvation, and hurricane

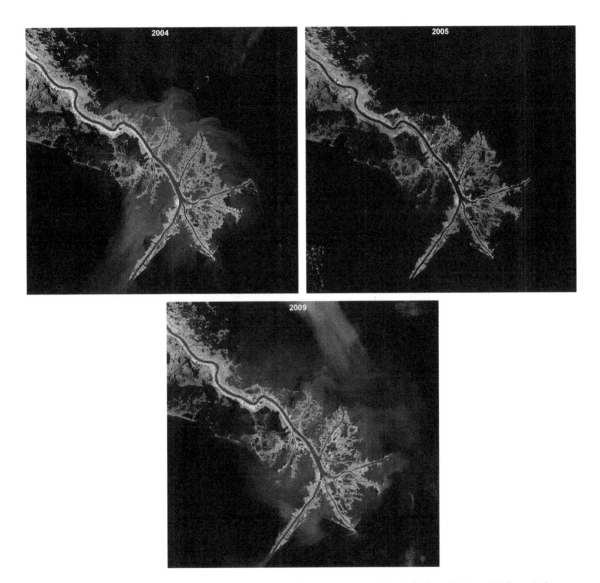

Figure 15-19. Satellite images of the Mississippi River delta vicinity immediately before (15 Oct. 2004) and after (16 Sept. 2005) Hurricane Katrina and four years later (29 Oct. 2009). Compare with image from 1985 (Fig. 15-17). Note changes particularly evident for Breton Island, Pass á Loutre, Southeast Pass, South Pass, and the area on the southwestern side of the Mississippi River. Based on Landsat TM band 4 (near-infrared), which is particularly sensitive to emergent vegetation (light gray) and water bodies (black). Image from NASA; processing by J.S. Aber.

frequency. Hydrocarbon production is most likely to continue for decades, considering the U.S. dependence on fossil fuels (Sever 2010). Given these trends, it seems inevitable that the Mississippi Delta and Louisiana Gulf coast would continue losing substantial ground to the sea; another 700 km² of wetlands are forecast to disappear within the next half century (Dugan 2005). Already many larger mammals, such as the jaguar (*Panthera onca*) and black bear (*Ursus americanus*), are in jeopardy.

Hurricanes Katrina and Rita in 2005 and the BP Deepwater Horizon oil spill in 2010 severely damaged coastal wetlands stretching from Louisiana to western Florida. Human population and economic activities endured great losses.

Figure 15-20. Grand Isle, Louisiana in 2004 prior to the major hurricanes of 2005. A. View westward with a gated recreational community in right foreground, behind which is an oil terminal. The Gulf of Mexico is visible in the left background, and Caminada Bay appears in the right background. B. Eastern bay side of the island with the Grand Isle Coast Guard station to right and recreational boat harbor to left. Kite aerial photos by S.W. Aber and J.S. Aber.

Figure 15-21. Dam, reservoir, and recreational park at Fall River Lake, part of the Arkansas drainage basin, in southeastern Kansas, United States. The reservoir is operated by the U.S. Army Corps of Engineers for flood control, recreation, and water supply. Photo by J.S. Aber.

Figure 15-22. Overview looking inland on the Mississippi Sound coast at Buccaneer State Park, near Waveland, Mississippi in 2004 prior to the major hurricanes of 2005. Notice the dam and reservoir in right background at Jackson Marsh, which block most inland sediment from reaching the tidal marsh in the left foreground. Kite aerial photo by S.W. Aber, J.S. Aber and M. Giardino.

Mangroves, in particular, are extremely vulnerable to oil spills and are quite slow to recover following oil pollution; whereas grass marshes recover relatively quickly (Whigham et al. 2010). Barrier islands suffered especially. The Chandeleur and Breton islands are nearly gone, as noted above (see Figs. 15-17 and 19). These islands, neighboring islands and shoals, and adjacent shallow waters, comprise the Breton National Wildlife Refuge, which was established in 1904 and is the second oldest national wildlife refuge in the United States. The brown pelican population suffered especially during the 2005 storm season (U.S. Fish and Wildlife Service Breton 2011). The refuge was closed beginning in May 2010 because of the BP oil spill, but was reopened in 2011 for routine visitation.

Dauphin Island, Alabama is another barrier island on the Mississippi Sound near Mobile Bay (see Color Plate 8-13). The western spit of the island is 18 km long, <2 m above sea level, and only 200-300 m wide. This portion of the island has been developed for recreational properties. Hurricane Katrina completely destroyed ~300 houses and severely damaged another ~150 (Crozier 2009). This part of the island has been damaged repeatedly during many previous storms. However, infrastructure is repaired and new houses are erected following each storm with federal funding from the National Flood Insurance Program (1968) and Stafford Act (1988), which provide financial support to rebuild. According to Crozier (2009), this failed public policy contributes to irresponsible fiscal commitment to constant redevelopment of high-risk coastal locales.

15.5.3 Padre Island and Laguna Madre

Padre Island is among the most famous barrier islands in the world. Stretching 185 km (115 miles) from Port Isabella to Corpus Christi, it is the longest barrier island in the United States (Weise and White 1980). A segment some 80 miles (129 km) long is designated as Padre Island National Seashore, which was established in 1962. Padre Island is separated from the mainland by Laguna Madre (Fig. 15-23). In addition to their environmental and ecological importance, Padre Island and Laguna Madre are extremely popular for recreational purposes (Fig. 15-24).

Padre Island is constructed primarily of sand deposited along the beach and in wind-blown dunes behind the beach. The island formed during the past several thousand years, as the last great ice sheets melted, sea level rose, and the shoreline retreated inland. Based on various lines of evidence, sea level stood in the range 3–10 m below present from 8000 to 4000 calendar years ago along the northern Gulf coast (Milliken, Anderson and Rodriguez 2008), at which time the South Texas shoreline stabilized with drowned river valleys, such as Baffin Bay, and submerged bars offshore (Weise and White 1980). For the past four millennia, sea level rose gradually at roughly 5 cm per century, until it began accelerating rapidly in the past century. As sea level approached modern, submerged bars built up initially as short, segmented barrier islands, which were situated mainly in the divides between drowned river valleys. Long-shore sediment drift caused the barriers to extend parallel to the coast, and eventually they grew together into Padre Island.

The modern island contains several distinct environmental zones passing from the Gulf side to the lagoon side (Fig. 15-25). Erosion, transportation and the deposition of sand by wind and waves determine the morphology of the island and may lead to changes in its size or position. In general, the northern portion of Padre Island is in stable equilibrium today; whereas, the southern section is undergoing loss, as erosion is predominant (Weise and White 1980; Brown and Huey 1991). Long-shore drift moves sand generally southward or northward along the upper shoreface and beach environments, and beach sand is blown inland by prevailing onshore wind (Fig. 15-26). The fore-island dune ridge is a fragile feature that protects the island from storm waves and surges (Weise and White 1980); the dunes are stabilized by marshhay cordgrass (*Spartina patens*), morning-glory (*Ipomoea* sp.), sea purslane (*Sesuvium portulacastrum*), bitter panicum

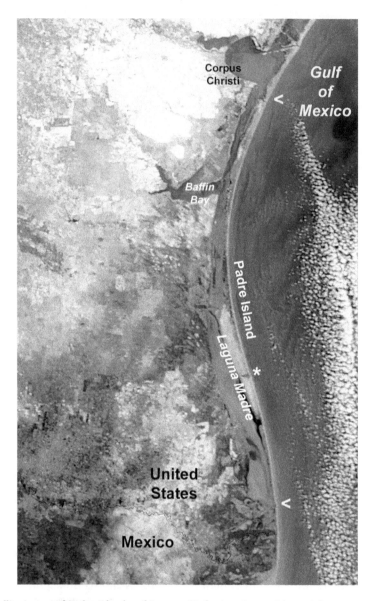

Figure 15-23. Satellite image of Padre Island and Laguna Madre in winter with partial snow cover on the mainland. The island is ~115 miles (185 km) long. Arrows (<) mark the approximate northern and southern ends of the island; the Mansfield Channel is indicated by the asterisk (*). Panchromatic version of original natural-color Terra/MODIS image; date of acquisition 27 December 2004. Adapted from NASA <http://visibleearth.nasa.gov/>.

(*Panicum amarum*), gulf croton (*Croton punctatus*), and sea oats (Fig. 15-27).

Behind the fore-island dune ridge, the vegetated barrier flat contains a complex of grasses, shrubs, marshes, and small pools (Fig. 15-28). These are typically fresh to brackish wetlands supported mainly by precipitation; cattail (*Typha domingensis*) grows in moist habitats (Richardson 2002). During hurricanes and tropical storms, the dune ridge may be breached, and washover fans and blowout dunes are deposited across the interior of the island. The lagoon side of the island is marked by a back-island dune field.

Laguna Madre is a shallow lagoon that receives little runoff from the mainland and has few connections to the open sea. Owing to minimal circulation and high evaporation rates, lagoon water is hypersaline. Water depth in much of the lagoon is around 1 m; the deepest point is only 2.5 m (Weise and White 1980). The lagoon is virtually tideless; water fluctuations result from wind-driven waves (Brown and Huey 1991). Sediment of the lagoon consists of mud, sand, and shell berms, much of which is stablized by marine grass. Shoalgrass (*Halodule wrightii*) is the most abundant marine grass, which prefers shallow water (<1 m deep) and tolerates high salinity, as its Latin name suggests. Grassflats are highly productive environments that support abundant invertebrates (snails and clams) and are spawning grounds or nurseries for many fish and crustaceans, such as shrimp and crabs (Weise and White 1980).

Laguna Madre contains, in addition, patch reefs composed of calcium carbonate and constructed by serpulid (annelid) worm tubes in the Baffin Bay vicinity (Weise and White 1980). Reefs tops are commonly exposed at the water surface or during low tides. The reefs are currently degrading, and none are living today; they may represent a wetter climatic past (Tunnell 2002). These serpulid reefs are well known for excellent fishing. The Laguna Atascosa National Wildlife Refuge includes the southwestern margin of Laguna Madre and adjacent mainland (Fig. 15-29). This region has a complex mixture of marine and terrestrial ecosystems at the interface of temperate, subtropical and desert environments (Fig. 15-30).

Both Padre Island and Laguna Madre are modified by human activities. The Mansfield Channel cuts across the lagoon and island, connecting Port Mansfield on the mainland directly to the Gulf of Mexico (see Fig. 15-23). This channel has altered the flow and exchange of water between the lagoon and open sea. Salinity in Laguna Madre has moderated, and fish can migrate between the lagoon and ocean (Weise and White 1980). Another channel is the Intracoastal Waterway, a canal running the entire

Figure 15-24. Preparation for the annual autumn fishing tournament at Padre Island National Seashore, Texas. A. Temporary fishing camp on the beach; vertical kite aerial photograph by S.W. Aber and J.S. Aber. B. Ground view of surf fishing from the beach. Photo by J.S. Aber.

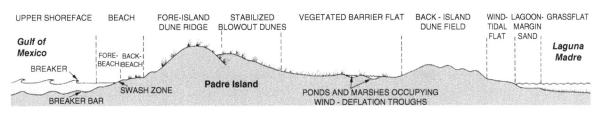

Figure 15-25. Generalized section across Padre Island showing environmental zones. Adapted from Weise and White (1980, Fig. 2).

Figure 15-26. Overview of the upper shoreface, beach, and fore-island dune ridge (right) in the northern portion of Padre Island, looking toward the south. Note people and vehicles on beach for scale. Kite airphoto by S.W. Aber and J.S. Aber.

Figure 15-27. Sea oats (*Uniola paniculata*), the quintessential dune plant of the U.S. Gulf coast, caps a small dune in southern Padre Island, Texas. Photo by J.S. Aber.

length of Laguna Madre. Sediment dredged from the channel was placed in spoil piles, which have been modified by erosion and vegetation growth into a chain of small islands and shoals along the canal. Some of these spoil islands have become important bird nesting sites, and many are used for sport fishing. The southern tip of Padre Island has been built up as a tourist resort (Fig. 15-31) and protected as a major shipping avenue (Fig. 15-32).

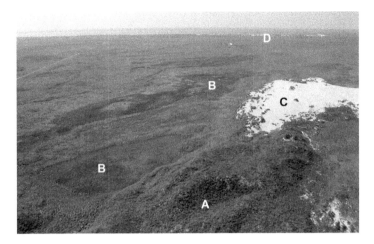

Figure 15-28. View across the interior of the northern portion of Padre Island. A. Fore-island dune ridge, B. Marshes, C. Blowout dune, and D. Back-island dune field. Laguna Madre is visible in the distance. Kite airphoto by S.W. Aber and J.S. Aber.

Figure 15-29. Overview looking southward at Laguna Atascosa National Wildlife Refuge, southern Texas, United States. Laguna Madre to left; mainland to right. Note the intricate shore complex of mud flats, tiny barrier ridges, peninsula, and enclosed pools. Kite airphoto by S.W. Aber and J.S. Aber; see also Fig. 7-18.

The pressure to develop barrier islands and their lagoons along the U.S. Gulf coast is intense. In spite of the high level of protection given by the national seashore, national wildlife refuges, state parks, and other nearby private reserves, Padre Island and Laguna Madre still face threats from beyond. Damming of inland rivers has reduced sediment supply to the coast. Pollution is derived from offshore oil production, shipping, and toxic compounds transported from diverse sources. Continued dredging of the Intracoastal Waterway is another point of concern (Brown and Huey 1991). Padre Island is a prime example of the delicate interplay of natural forces, which are always in a state of flux. Human impacts add more factors to the mix; only through careful management may the dynamic balance be maintained, as the following example illustrates.

The Kemp's Ridley sea turtle (*Lepidochelys kempii*) is the smallest of five species of sea

Figure 15-30. Plant and animal inhabitants on the mainland of Laguna Atascosa National Wildlife Refuge. A. Spanish dagger, palma pita (*Yucca treculeana*) and Texas prickly pear (*Opuntia engelmannii*) are indicative of the subtropical scrub vegetation (Richardson 2002). Laguna Madre appears in the background. B. Texas gopher tortoise (*Gopherus berlandieri*). The male (shown here) has a distinctive extension of the gular scutes (lower shell below the neck). Mainly a herbivore, it feeds on succulent plants and cacti as well as insects and snails (Ferri 2002). Photos by J.S. Aber.

Figure 15-31. Northward overview of South Padre Island, Texas. U.S. Coast Guard station (left center), Queen Isabella Causeway, which connects to Port Isabel on the mainland (upper left), and central South Padre Island (upper right). Kite airphoto by S.W. Aber and J.S. Aber.

Figure 15-32. Brazos Santiago Pass, Texas. A. Towboat cruises through the pass between southern Padre Island (lower left) and Brazos Island (upper right). Large stone blocks protect the pass from erosion. Kite airphoto by S.W. Aber and J.S. Aber. B. Offshore drilling rig is maneuvered through the pass by towboats. Photo by J.S. Aber.

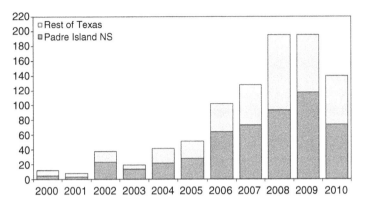

Figure 15-33. Number of Kemp's Ridley sea turtle nests found on Texas beaches, 2000–2010. Padre Island National Seashore accounts for more than half of the nests, which reached totals of nearly 200 in 2008 and 2009. Adapted from Padre Island (2010a).

turtles in the Gulf of Mexico (Padre Island 2010a). Historically these turtles laid eggs on beaches stretching from Mustang Island, Texas to Veracruz, Mexico. The primary nesting beach was near the village of Rancho Nuevo in Tamaulipas, Mexico. As recently as 1947 about 40,000 females laid eggs there, and Mexico began protecting this site in the 1960s from egg and meat poachers (Brown and Huey 1991). However, turtle numbers had plummeted to about 200 by 1977. The United States and Mexico joined efforts in 1978 to revive the Kemp's Ridley sea turtle. A designated beach in the Padre Island National Seashore was established for transplanting turtle eggs from Mexico each year as a safeguard against extinction. Hatchlings are collected and taken to an incubation facility at the national seashore, and then released from the beach when they reach appropriate behavioral development (Padre Island 2010b). After more than three decades of effort, the number of nests began to increase markedly in the first decade of the twenty-first century (Fig. 15-33). This international cooperation has led to nascent recovery for the Kemp's Ridley sea turtle in both Mexico and Texas.

15.6 Summary

This chapter provides wetland case studies from across the low latitudes. Climatic conditions are tropical or subtropical with high average summer temperatures and distinct dry and wet seasons. Rainfall and annual flood patterns play a significant and influential role in molding the ecosystem characteristics of these wetlands. Each of the wetlands considered in this chapter is of global consequence in terms of its biodiversity, functional and resource contributions. Moreover, three of the four regions considered include wetlands that span across country borders – complicating issues of conservation and protection.

The Sundarbans contains the world's largest contiguous mangrove swamps, providing essential flood and storm protection services for the densely populated Ganges–Brahmaputra delta region of India and Bangladesh. The mangroves fulfill important carbon sequestration and nutrient cycling functions, as well as providing a habitat for a diversity of flora and fauna. Moreover, they also provide a home to the endangered Royal Bengal tiger. The Okavango Delta provides a life-sustaining resource to what would otherwise be a dry subtropical desert. Dependent on precipitation in the Angolan highlands more than 1000 km from the delta, the Okavango River transports water to the inland delta during the driest and most critical time of year. This flood pulse supports an immense diversity of life resident to the region as well as migratory species that depend on its waters for survival. The high species richness and density observed across the delta is paradoxical to the surrounding Kalahari Desert, one of the driest places on Earth. Angola, Namibia and Botswana share the waters of the Okavango and use OKACOM, a multilateral institution, to provide coordinated conservation and management strategies and mediate resource conflicts.

The Pantanal, which is nourished by precipitation and the floodwaters of the Paraguay River, is shared by Bolivia, Brazil and Paraguay. The wetlands comprise a diverse mosaic of habitat types based on soil characteristics and inundation patterns. The flood-pulse system, dependent on rainfall in the upper Paraguay Basin and highlands and plateaus of Brazil, is a key driver of life across the Pantanal. Flora and fauna life cycles are synchronized with the flood and have adapted to distinct aquatic and terrestrial phases. Human activity across the Pantanal accelerated with the introduction of cattle ranching during the nineteenth century. Today, commercial-scale agricultural development, expanding urbanization, and related activities threaten the integrity of this ecosystem.

On the United States Gulf of Mexico coast, the Florida Everglades is a vast expanse of wet sawgrass prairie that extends from Lake Okeechobee in south-central Florida to the mangroves along the coast of southern Florida. It was once among the world's largest fresh-water marsh systems and still contains the biggest contiguous tract of mangrove swamp in the subtropics. The delta of the Mississippi River is the end point of one of the world's largest terrestrial drainage basins, where inland fresh-water systems intergrade with the marine environment. Coastal Louisiana and the delta are losing ground to the sea due to a combination of factors, including eustatic rise in sea level, compaction of delta sediment, frequent hurricanes, retention of sediment in upstream reservoirs, conversion of wetlands for agriculture, and production of oil and gas. Padre Island and Laguna Madre are the longest such barrier island and lagoon system in the United States. Substantial portions of these U.S. Gulf sites are protected in a national park, seashore, and several wildlife refuges as well as state parks and private nature reserves. However, many other portions of the coast have been heavily developed and altered by human activities often with deleterious impacts on coastal processes and wetland environments.

16 Middle-latitude wetland case studies

16.1 Introduction

The middle latitudes span the world north and south of the equator generally between 30° and 60°, ranging from subtropical deserts to regions with permafrost. For most of the middle latitudes, climate is distinctly seasonal with markedly hot and cold, wet and dry intervals driven principally by changes in solar radiation as the Earth orbits the Sun. Climatic conditions are modified regionally and locally by proximity to large water bodies, prevailing winds, mountain ranges, ocean currents, and other geographic factors, which may influence precipitation, cloud cover, evapotranspiration and, thus, available water for wetlands. The influence of high altitude on wetland environments is considered in chapter 17.

In this chapter, wetland case studies are presented for the Great Plains region of North America, including the Arkansas River valley in Colorado and Kansas, the Nebraska Sand Hills, and the Missouri Coteau region of southern Saskatchewan. These regions are sparsely populated for the most part with agriculture and resource production as the main economic activities. The Atlantic coastal region is reviewed for Maine and Massachusetts in the northeastern United States, where denser human population has a longer history of landscape modification and impacts on wetlands. The final case study comes from central Europe, specifically Estonia and its neighbors bordering the eastern end of the Baltic Sea. While subject to a long history and prehistory of human occupation and land modification, this region still preserves substantial wetlands, some of which are heavily impacted while others remain relatively pristine.

16.2 Great Plains of North America

The North American Great Plains extend from Texas and New Mexico northward into Saskatchewan and Alberta, with forested lowlands to the east and the Rocky Mountains to the west. The geological basis is flat-lying to slightly inclined sedimentary strata ranging from Paleozoic to Holocene in age. Rivers draining from the Rocky Mountains toward the east and south have sculpted the current landscape, glaciation molded the northern plains, and wind has played a conspicuous role in shaping many parts of the Great Plains. Vegetation consists mostly of grasslands, prairie and savanna with sparse trees following stream valleys.

The plains exhibit strong north–south and east–west gradients. From the Rocky Mountains, the High Plains slope from >2000 m altitude downward toward the east into lower plains and hills, eventually descending below 300 m elevation. Precipitation also follows an east–west pattern with more than 1 m per year at lower elevations to the east, decreasing westward to less than 30 cm in the rain shadow of the Rocky

Mountains (High Plains Regional Climate Center 2011). This precipitation gradient is reflected in the transition from tallgrass prairie in the east, to mixed-grass prairie, to shortgrass prairie in the west. Great Plains climate is continental and highly seasonal. In the southern plains, summers are long and hot, and winters are short and cool. The northern plains are the opposite; the greatest temperature contrast occurs in the northern plains where summer maximum may exceed 40 °C and winter minimum may dip below −40 °C.

A tremendous variety of wetland types exists within this vast territory. Toward the east, water is abundant and wetlands are mostly perennial and fresh. In contrast, precipitation declines westward and alkali water chemistry predominates, except for rivers draining the Rocky Mountains. Western wetlands are more subject to droughts and periodic dry conditions. Great Plains wetlands occur in four main geologic settings:

- Glaciated terrain – Recently glaciated regions have abundant undrained or poorly drained depressions, channels and potholes of all sizes and shapes. This region includes parts of the Dakotas, Iowa, Minnesota and Montana, and most of the Canadian Prairies. Older glaciated regions to the south are more eroded and do not support so many wetlands.
- Sand hills terrain – High water tables and good ground-water recharge make sand hills prime candidates for wetlands in the swales and basins between dunes. Good examples include the Nebraska Sand Hills and the Great Sand Hills of southwestern Saskatchewan.
- River floodplains – Large and small river valleys have in many regions broad floodplains that contain wetland habitats. The Missouri River and its many tributaries drain most of the northern Plains. Other major drainages include the Red River, Saskatchewan River, and Arkansas River.
- Playas – Shallow wind-eroded depressions are found primarily in the southern High Plains, particularly western Texas and Oklahoma, eastern New Mexico, southwestern Kansas, and southeastern Colorado. They are dry much of the time, but turn into rich marsh and ephemeral lake habitats during wet years.

16.2.1 Upper Arkansas River valley, Colorado and Kansas

The Arkansas River begins in central Colorado on the Continental Divide surrounded by peaks that exceed 14,000 feet (>4260 m) elevation. It descends through the Royal Gorge (Fig. 16-1), flows across the High Plains of southeastern Colorado, and continues into southwestern Kansas. Along the way, every drop of water derived from mountain snowmelt and summer thunderstorms is extracted for human use (see Fig. 1-8). Human diversion of water begins in the mountains with irrigation for cattle grazing and hay production in wet meadows (Fig. 16-2).

As the Arkansas River leaves the mountains, it is contained in Lake Pueblo (Fig. 16-3). Lake Pueblo and the Pueblo Dam were constructed by the U.S. Bureau of Reclamation as part of the Fryingpan–Arkansas Project, in which waters are diverted from the western slope of the Continental Divide and transferred down the Arkansas River to the southeast. Lake Pueblo is the principal storage reservoir in the project. Its waters are used for irrigation, drinking and recreation, and it provides flood control to the Pueblo community.

Across southeastern Colorado, water is extracted from the Arkansas River to support irrigated agriculture (Fig. 16-4). Water is diverted via canals directly to irrigated fields and into reservoirs in tributary valleys, from whence the water is further distributed to agricultural fields. These reservoirs have become wetlands with hydric vegetation, abundant fish, and migrating waterfowl (Fig. 16-5). As the Arkansas River passes into Kansas, only a small stream remains as required by the Kansas-Colorado Arkansas River Compact and affirmed by the U.S. Supreme Court (Kansas Department of Agriculture 2010). The vicinity of Lakin–Garden City is most instructive (Fig. 16-6). Water extraction takes place via canals into fields and holding reservoirs (Fig. 16-7). Irrigation between Lakin and Garden City north of the Arkansas River is conducted mainly with surface water diverted from the river; however, ground water supports irrigation south of the river valley and

Figure 16-1. Bridge over Royal Gorge just west of Cañon City, Colorado. The Arkansas River and a railroad line are visible at the bottom of the deep canyon. Photo by J.S. Aber.

elsewhere. The combination of surface- and ground-water extraction consumes essentially all the available water. As shown in the satellite image, the Arkansas River channel dries up between Lakin and Garden City.

The region immediately north of Garden City contains numerous playa basins of various types. Many are shallow, elongated depressions that occur in clusters with parallel alignment (see Fig. 3-24). These depressions were presumably formed by wind erosion of the unconsolidated loess (silty loam) sediment. Some of the deeper ones contain small perennial lakes, but many are dry most years and have been converted for agriculture. Across the southern High Plains, "playas are islands in the vast sea of crop monoculture that is virtually devoid of any diversity of habitat" (Steiert and Meinzer 1995, p. 2). In addition to providing wetland habitats, playas are important points for ground-water recharge to the High Plains aquifer.

Playa basins of western Kansas bear a remarkable similarity to pocosins, also known as

Figure 16-2. Cuchara River valley draining the Sangre de Cristo Mountains near La Veta in south-central Colorado. A. Small irrigation ditch and water-flow structure. B. Harvested hay bales in a wet meadow with the Culebra Range in the background. Photos by J.S. Aber.

Figure 16-3. Overview of Lake Pueblo and Pueblo Dam on the Arkansas River, just west of Pueblo, Colorado. Part of a fish hatchery is visible to right. Canal across the bottom of scene is supplied by Lake Pueblo and is operated by the Bessemer Irrigating Ditch Co. Kite aerial photograph by S.W. Aber, J.S. Aber and D. Eberts.

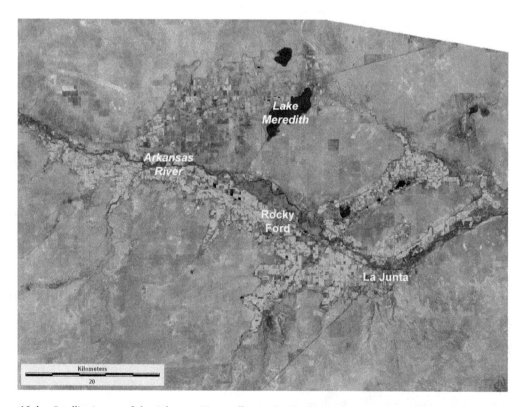

Figure 16-4. Satellite image of the Arkansas River valley in the Rocky Ford–La Junta vicinity, southeastern Colorado, United States (see Color Plate 16-4). Bright green indicates irrigated crops; maroon-pink shows dry upland areas. Lake Meredith is a major storage point for water diverted from the Arkansas River. Landsat TM bands 2, 4 and 5 color coded as blue, green and red; 7 August 2009. Image from NASA; processing by J.S. Aber.

MIDDLE-LATITUDE WETLAND CASE STUDIES 303

Figure 16-5. View over the eastern portion of Lake Meredith with Sugar City in the left background. Note the well-developed wetland vegetation in the foreground; the bushy trees are saltcedar. Colorado, United States; helium blimp airphoto by S.W. Aber and J.S. Aber.

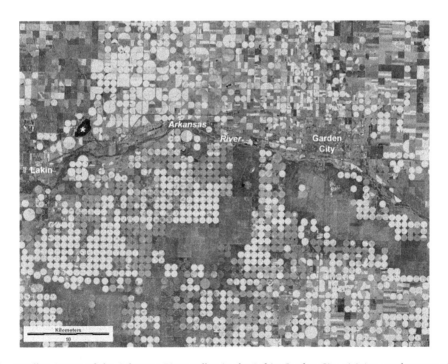

Figure 16-6. Satellite image of the Arkansas River valley in the Lakin-Garden City vicinity, southwestern Kansas, United States (see Color Plate 16-6). Bright green indicates irrigated crops; maroon-pink shows dry-fallow areas. Small irrigation circles are one-half mile (0.8 km) in diameter; large circles are one mile (1.6 km) in diameter. Asterisk (*) indicates Lake McKinney, a holding point for water diverted from the Arkansas River via the Amazon Ditch (next figure). Landsat TM bands 2, 4 and 5 color coded as blue, green and red; 4 August 2007. Image from NASA; processing by J.S. Aber.

Figure 16-7. Irrigation water in vicinity of Lakin, Kansas. A. Amazon Ditch and water-flow structure. B. Lake McKinney during a drought year with low water level. Road on left runs along the dam. The band of dark vegetation (*) is curly dock (*Rumex crispus*), a facultative wetland plant. Photos by J.S. Aber and S.W. Aber.

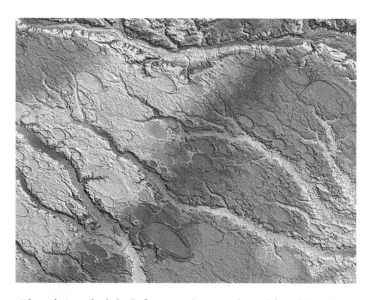

Figure 16-8. High-spatial-resolution, shaded-relief topographic map derived from lidar elevation data showing pocosin basins in vicinity of Rex, southeastern North Carolina. Elongated basins trend NW–SE. Multiple ages of basins are shown by overlapping patterns, and small streams have eroded valleys into some basins. Highways and a railroad run diagonally across the scene. Map covers ~600 km^2; adapted from original color image by M. Davias; obtained from Wikipedia Commons <http://commons.wikimedia.org/>.

Carolina Bays, which are found on the Atlantic Coastal Plain from Virginia to northern Florida (Cleveland 2008). Pocosins are elongated, oval basins that occur in parallel sets or trains and are located on uplands between stream valleys (Fig. 16-8). As for playas, wind is considered to have played a major role in the formation of pocosin basins (Blair 1986).

Several internal drainage basins are situated immediately north of the Arkansas River valley in western and central Kansas. Various tectonic and sedimentary processes may be responsible for isolating these basins from the Arkansas drainage system, including subsurface solution, infiltration of fine sediment, animal activity (buffalo wallows), wind action, meteorite impact,

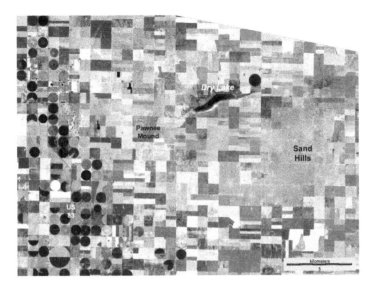

Figure 16-9. Satellite image of Dry Lake vicinity. Dry Lake occupies an enclosed drainage basin between Pawnee Mound and a tract of sand hills. Local relief ranges from >910 m elevation at Pawnee Mound to <870 m around Dry Lake. Dark circles represent irrigated crops during a drought year. Landsat TM band 7 (mid-infrared); 15 August 2011. Image from NASA; processing by J.S. Aber.

or a combination of these processes (Merriam 2011). Dry Lake, near Garden City, is one such example (Fig. 16-9). Dry Lake occupies a former stream valley that trended from west to east. The western drainage was cut off, presumably by movement of buried faults that uplifted the Pawnee Mound ridge. These faults may be associated with the Bear Creek and Crooked Creek fault zones of southwestern Kansas, which are collapse structures formed by subsurface salt solution (Miller and Appel 1997). The Dry Lake basin is separated from another unnamed basin to the east by a low barrier of sand dunes blown northward from nearby sand hills.

Long-term average annual precipitation in the vicinity of Dry Lake is approximately 50 cm (~20 inches), but individual years range from more than 90 cm to less than 25 cm precipitation (High Plains Regional Climate Center 2011). Given the semi-arid climate, highly variable precipitation, and relatively small size of its drainage basin, Dry Lake undergoes conspicuous changes annually and seasonally from full of water, which is a rare event, to completely dry, which is common (Fig. 16-10). When dry, the salt flat grades from low-salinity calcium-carbonate and sodium-halide salts around the margin to high-salinity bitter (potassium-halide) salts toward the center of the basin. Even when dry, saline ground water remains just below the surface (see Fig. 5-8). Dry Lake, thus, represents a salina or salada wetland habitat. When it contains water, migrating shorebirds and waterfowl may be seen in abundance.

16.2.2 Biocontrol of saltcedar along the upper Arkansas River valley

Saltcedar is a shrub or small tree that is native to Asia, the southern Mediterranean, and northern Africa. It comprises several species within the genus *Tamarix* that were introduced in the United States beginning in 1823 (DeLoach 2004). During the Dust Bowl of the 1930s, saltcedar was planted widely for windbreaks and to control stream erosion in the Great Plains and western United States. Since then, however, saltcedar has spread rapidly, becoming an invasive plant throughout arid and semiarid portions of the western United States (see Fig. 6-7) and northern Mexico, where it is responsible for a major ecological disaster (DeLoach et al. 2007).

Figure 16-10. Overviews from the middle of Dry Lake, Kansas looking toward its eastern end (see Color Plate 16-10). A. Lake completely full of water following heavy snowmelt and spring rains, May 2007. B. Pool of rust-colored water surrounded by mud/salt flat, May 2010. Water color is presumably a result of suspended sediment as well as saline invertebrates. C. Salt flat completely covers the dry lake basin, October 2010. Kite and blimp aerial photos by J.S. Aber, S.W. Aber, G. Corley, D. Leiker, C. Unruh and B. Zabriskie.

Saltcedar forms dense thickets along waterways, including the upper Arkansas River valley. These thickets crowd out native vegetation, degrade wildlife habitat, harm some 50 endangered or threatened native species, cause increases in wildfires and soil salinity, inhibit grazing and agricultural land use, and impede recreational use of parks and natural areas.

As a facultative phreatophyte and halophyte, saltcedar has significant advantages over native, obligate phreatophytes (Zouhar 2003). Following fire or cutting of the above-ground plant, new growth takes place readily from the root crown. It grows in soil with up to 50‰ salinity. Excess salt is collected in special glands in the leaves and then excreted onto the leaf surface. When these leaves fall to the ground, they contribute to soil salinity. For controlled rivers, lacking regular flooding and scouring, increased soil salinity inhibits seed germination and growth of native plants. In the Arkansas River valley, for instance, damming and water extractions dramatically reduce the intensity and recurrence interval of downstream flooding and may give *Tamarix* a competitive advantage over such native species as cottonwood (*Populus deltoides*) and willow (*Salix exigua*) (Lovell, Gibson and Heschel 2009).

Once established, saltcedar is extremely tenacious and difficult to eradicate through mechanical (cutting) or chemical (herbicide) means, or with fire (Zouhar 2003). Integrated (multiple) methods are more successful – for example, application of selected herbicides to freshly cut stumps prior to the growing season (e.g. Fick and Geyer 2010), which is labor-intensive and expensive. Another approach involves periodic dam releases to create pulse floods and overbank flooding, which simulate natural river behavior and favor native species (Zouhar 2003). Because of the numerous problems created by the *Tamarix* invasion, the Saltcedar Biological Control Consortium (SBCC) was established with numerous federal and state agencies, private organizations, and international partners (DeLoach and Gould 1998; SBCC 2011).

A leaf-eating beetle, *Diorhabda "elongata"* (saltcedar leaf beetle), now named *D. carinulata*, is a natural enemy of saltcedar (Fig. 16-11).

Figure 16-11. *Diorhabda carinulata*, imported from northwestern China, on saltcedar. Third instar larva (~8 mm long). Photo courtesy of D. Eberts.

In their larval and adult stages, *D. carinulata* eat the leaves and outer layers of stem tissue of saltcedar (and no other plants), and may defoliate the plant to such an extent that it eventually dies. These beetles were imported from Fukang, northwestern China and Chilik, eastern Kazakhstan (Dudley et al. 2010). The carefully managed international biocontrol program proceeded in several steps (DeLoach et al. 2007):

- Ten years testing *D. carinulata* imported from central Asia under quarantine conditions in Texas and California, starting in 1986.
- Release of *D. carinulata* into secure field cages at authorized research sites in 1997.
- Field release into the open within circumscribed areas at ten approved sites in six western states in 2001. One of these sites was located on the Arkansas River floodplain near Pueblo, Colorado (Fig. 16-12).
- Additional field releases of *Diorhabda* sp. imported from Crete, Greece in two southwestern states in 2003.
- Release of Greek *Diorhabda* sp. at study sites in northeastern Mexico along the Rio Grande in 2007.

Field releases of saltcedar leaf beetles had rapid and dramatic impacts on saltcedar with

Figure 16-12. Overview of the Pueblo, Colorado, saltcedar study site on the Arkansas River floodplain just below Lake Pueblo. Ground cover is a patchwork of saltcedar thickets, salt sacaton grass, rabbitbrush and assorted weeds, and bare sandy soil. Part of the dam is visible across the top of the scene. This site is supervised by the U.S. Bureau of Reclamation. Kite aerial photo by S.W. Aber, J.S. Aber and D. Eberts (lower left).

immediate success, particularly in arid plateau canyons and intermontane valleys of Nevada, Utah and western Colorado. Large tracts of saltcedar were cleared from river valleys and willow are beginning to recover (Bean, Ortega and Jandreau 2009; Dudley et al. 2010).

At the Pueblo study site, ongoing research focused on the biology and behavior of the beetles and the effects of their release on wildlife, saltcedar, and non-target vegetation (Eberts et al. 2003). Initial release of *D. carinulata* was successful in defoliating saltcedar quite well (Fig. 16-13), but the beetles did not migrate away from the Pueblo site in large numbers (Bean, Ortega and Jandreau 2009). Other releases were made subsequently downstream along the Arkansas River in southeastern Colorado. Saltcedar was defoliated in some situations (Fig. 16-14), but in general beetles did not become well established in many locations. The reason for limited success of saltcedar leaf beetles along the Arkansas River may be attributed to a couple of factors in particular (Bean, Ortega and Jandreau 2009):

- Latitude of ~38°N has a shorter summer day compared with the home ranges of the beetles imported from central Asia (~44°N).

Figure 16-13. Saltcedar largely defoliated by saltcedar leaf beetles at Pueblo, Colorado study site. This ground picture was taken near the end of the growing season in October 2003, two years after field release of the beetles. Saltcedar bush stands about 4 m high. Photo courtesy of D. Eberts.

Beetle reproduction is tied closely to length of summer daylight; at 38°N beetles achieve only 1–2 generations per summer. At the home-range latitude, longer summer days allow three generations per year.

- Predators – birds, other insects, and spiders – are much more numerous on the semi-arid plains than in the arid plateau canyons of

Figure 16-14. Largely defoliated saltcedar thicket beside the Arkansas River at Holly, Colorado near the border with Kansas. Mid-summer view taken in 2010 by J.S. Aber.

western Colorado. Beetles from western Colorado at 38° N have adapted for two full generations per summer; transplanting them to southeastern Colorado might improve chances for success there.

The long-running, highly successful saltcedar biocontrol program was brought to an abrupt halt in July 2010 with a moratorium issued by the U.S. Department of Agriculture (Dowdy 2010). This moratorium terminated the biological control program in 13 western states, prohibited any new permits for field-cage or greenhouse studies, and put an immediate end to interstate transport and environmental release of the beetles. The only permitted activity is continuation of existing containment studies of *Diorhabda* sp. This action came in response to a lawsuit filed by the Center for Biological Diversity and Maricopa Audubon Society in 2009 regarding the southwestern willow flycatcher (*Empidonax traillii extimus*). This bird was listed as an endangered species in 1995, and the cause for its listing was loss of cottonwood and willow vegetation across the southwestern United States (Dudley et al. 2010). Its primary range is Arizona, New Mexico, and southern California. Ironically, habitat change brought about by the *Tamarix* invasion was cited as a major factor in the flycatcher decline.

Public and scientific response was swift. The decision to cancel the saltcedar biocontrol program was widely condemned as regulatory dysfunction or bureaucratic disharmony. It was said to be based on the desire to avoid a costly and protracted lawsuit rather than on any scientific basis for the relationship between saltcedar and the flycatcher (Wildlife Management Institute 2010). In fact, restoration of native habitats, specifically willow stands on floodplains, is a primary long-term goal of the Saltcedar Biological Control Consortium (Dudley et al. 2010). The moratorium may in fact have little impact for those arid regions in which the beetles are released and thriving already. The fate of the upper Arkansas River valley, however, is less certain. This area is far from the range of the endangered flycatcher, but without additional releases of beetles better adapted for the region, saltcedar may recover and continue to flourish there.

16.2.3 Cheyenne Bottoms, Kansas

Another well-known enclosed basin, Cheyenne Bottoms, is situated in central Kansas adjacent to the Arkansas River valley (Fig. 16-15). It is considered to be among the most significant sites for shorebird and waterfowl migration in the United States (Zimmerman 1990). The "bottoms" is famous for great flocks of migrating waterfowl and shorebirds. More than 330 bird species have been spotted at Cheyenne Bottoms, some of which are threatened or endangered – whooping crane, peregrine falcon, piping plover and least tern (Penner 2010). The site is an important point for rest and nourishment for hundreds of thousands of birds in their annual migrations between the northern plains and Arctic summer breeding grounds and southern winter ranges along the Gulf Coast, Caribbean and South America. The birds are attracted by a feast of bloodworms (midge larvae) that live in the marsh muck (Zimmerman 1990). Cheyenne Bottoms is designated as a Ramsar Wetland of International Importance and also is recognized as a site of hemispheric importance by the Western Hemisphere Shorebird Reserve Network (Kostecke, Smith and Hands 2004).

Cheyenne Bottoms has existed as a closed depression for more than 100,000 years

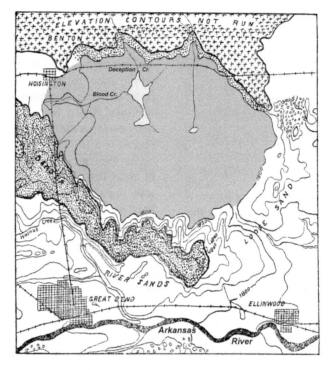

Figure 16-15. Early sketch map of Cheyenne Bottoms near Great Bend, central Kansas, United States. The oval shape of the bottoms is shown in gray. Deception and Blood creeks are the natural tributaries, which have built a delta complex in the northwestern portion of the bottoms near Hoisington. Elevation contours in feet; 1800 feet is ~550 m. Modified from Haworth (1897, Fig. 1).

(Zimmerman 1990). The geological origin of the basin may be connected with deep-seated tectonic movements associated with an ancient meteorite impact structure in the basement rock (Merriam 2011), subsurface salt solution and flowage, as well as surficial wind and stream action (Aber and Aber 2009). Mean annual precipitation for the Cheyenne Bottoms vicinity is about 65 cm per year, but yearly precipitation fluctuates greatly from less than 40 cm to more than 1 m (High Plains Regional Climate Center 2011), which leads to highly variable water conditions. Several major floods have occurred in the past century; some of these were severe enough to transform the bottoms into a large lake covering more than 80 km². At other times, during droughts, the bottoms has dried up completely.

These extreme fluctuations have made the area subject to water management endeavors since 1899, when the first water diverted from the Arkansas River was channeled into the bottoms. The current water-supply scheme was enacted in the 1950s and substantially enhanced in the 1990s for the state wildlife area, which encompasses nearly 20,000 acres (8000 ha). It involves dams, canals, dikes, and high-capacity pumps to regulate water levels in several artificial pools (Fig. 16-16). The state wildlife area is managed primarily for migratory waterfowl and shorebirds (Penner 2010).

The Nature Conservancy (TNC), on the other hand, makes no attempt to manipulate water supplies on its land covering almost 8000 acres (3200 ha). TNC land is managed for waterfowl and shorebirds as well as grassland birds (Penner 2010). Since the 1990s, in fact, TNC has removed barriers or artificial controls, where possible, in order to restore and maintain natural wetland habitats (see Fig. 13-2). Overflow from TNC land spills into the state wildlife area pools.

Cheyenne Bottoms has experienced cattail expansion since the 1970s, which was quite

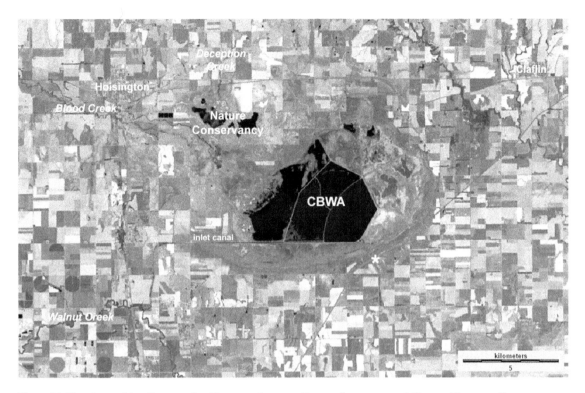

Figure 16-16. General locality map for Cheyenne Bottoms, Barton County, central Kansas. The state Cheyenne Bottoms Wildlife Area (CBWA) occupies the downstream or sump portion of the bottoms; its water supply is diverted from Walnut Creek and delivered via an inlet canal. Nature Conservancy land is located in the upstream deltaic portion of the Bottoms; its water is derived from natural overflows of Blood and Deception creeks. Location of the Kansas Wetland Education Center is indicated by the asterisk (*). Panchromatic image based on Landsat TM bands 2, 5 and 7; 3 September 2009. Image from NASA; processing by J.S. Aber.

dramatic during the 1990s, as documented by multitemporal Landsat imagery (Pavri and Aber 2004). This was the result of relatively cool and wet climate, water management schemes, and general expansion of cattails across the Great Plains (Zimmerman 1990). This invasion led to a decline of open-water and mudflat habitats, which subsequently caused a reduction in shorebird use (Kostecke, Smith and Hands 2004). Various methods for cattail management are practiced, including disking, mowing, controlled burning, flooding, drying, grazing, and combinations thereof. In TNC marshes, cattail infestation reached a crisis point at the beginning of the twenty-first century (Fig. 16-17A). At that time, TNC decided to control cattail using natural drought-and-flood cycles.

Beginning in the summer of 2002, a drought took place in which water levels dropped and portions of the marsh dried up. In the spring of 2003, much of the cattail had died, although small patches and narrow zones of cattails survived, mainly around the margins of thickets (Fig. 16-17B). Drought conditions continued through 2003 and early 2004. After the demise of the cattails, TNC began an attempt at marsh restoration in early spring 2004. Three treatments were applied to selected sections of dead cattail thatch. Some areas were mowed, some were burned, and some were left untouched. The result of this experiment was an unsystematic pattern in the removal of cattail thatch. Drought conditions ended rather abruptly in June 2004, heavy rain fell and the marshes were flooded. Wet conditions continued into 2005, when the marsh contained a mosaic of shallow pools and mudflats, with a mixture of emergent wetland vegetation.

Figure 16-17. Cattails at Nature Conservancy marsh, Cheyenne Bottoms, Kansas (see Fig. 13-2 for location). A. Healthy cattail thickets (dark gray) cover much of the marsh prior to drought. This phase represented the culmination of cattail expansion during the previous decade of relatively wet, cool climatic conditions. View westward with Hoisington in the background. Image date May 2002. B. View toward northwest over partially dead cattail thickets (light gray) following drought. Small patches and narrow zones of cattails survived (dark gray), particularly around the margins of thickets. Image date June 2003. Kite aerial photographs by J.S. Aber and S.W. Aber.

Drought conditions soon returned, and by the autumn of 2006 the marshes were completely dry. TNC now conducted extensive disking of dry mud flats and mowing of cattail thatch in an attempt to mimic heavy bison grazing (Fig. 16-18). Wet conditions quickly returned in 2007 with flooding of historic magnitude at Cheyenne Bottoms (Fig. 16-19). Much of the bottoms was inundated for several months. A multitemporal Landsat image illustrates the dramatic changes in water bodies, vegetation, and other land cover between 2006 and 2009 (Fig. 16-20). This image also highlights the deltas built by Deception and Blood creeks respectively in the northern and western portions of the bottoms. These deltas remained above most of the 2007 flooding; whereas nearly all the state wildlife area was submerged, and TNC marshes were inundated in the embayment between the two deltas.

As the impact of flooding receded, a bloom of mosquito fern (*Azolla* sp.) took place in TNC marshes in 2009 (Fig. 16-21A). This was remarkable because such a bloom had not been observed before in TNC marshes (see Color Plate 6-11). Widespread flooding in 2007 is considered the most likely means for transporting *Azolla* spores or plant fragments into TNC marshes. Several other dispersal mechanisms are possible, however, including waterfowl, boats and fishing equipment, or sewage discharge. Regardless of how *Azolla* entered TNC marshes, it had nearly disappeared by the following year (Fig. 16-21B), and cattail became dominant again the next year (Fig. 16-22). This discussion demonstrates the dynamic and highly variable wetland conditions of TNC marshes at Cheyenne Bottoms.

In addition to its environmental importance, Cheyenne Bottoms has a significant impact on the local and state economy, primarily through hunting and bird watching, which bring in state hunting fees as well as money spent locally for lodging, food, and supplies. In recognition of the economic impact of tourism, the Kansas Wetland Education Center was constructed in 2007–08 and opened to the public in 2009 (see Fig. 16-16). The center is a cooperative venture of the Kansas Department of Wildlife and Parks, which owns the land, and Fort Hays State University, which operates the center. Other partners include Ducks Unlimited and the City of Great Bend. In addition, the Wetlands and Wildlife National Scenic Byway links Cheyenne Bottoms to Quivira National Wildlife Refuge, another Ramsar Wetland of International Importance south of the Arkansas River (see Fig. 11-19).

Figure 16-18. Overview of TNC marsh in October 2006. Dead thatch was mowed (A), and completely dry mud flat was disked (B). Inset image shows tractor mowing dead thatch. Kite aerial photos by S.W. Aber and J.S. Aber.

Figure 16-19. Waxing flood in TNC marsh in May 2007. Flood waters continued to rise well into the summer. View westward with Hoisington in the background. Photo by J.S. Aber and S.W. Aber.

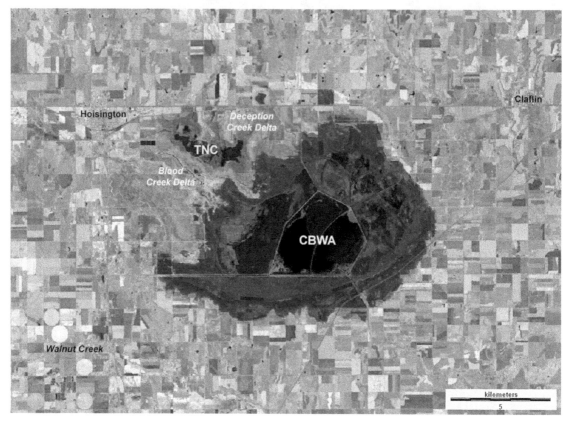

Figure 16-20. Multitemporal Landsat image based on TM band 4 (near-infrared) for 2006, 2007 and 2009, color coded respectively as blue, green and red (see Color Plate 16-20). Bright colors represent significant changes in land cover from year to year; dull-gray colors indicate little change in land cover. The broad maroon-purple zone shows the extent of high water in 2007; black and dark blue show perennial water bodies; compare with Fig. 16-16. CBWA – Cheyenne Bottoms Wildlife Area; TCN – The Nature Conservancy. Image from NASA; processing following the method of Pavri and Aber (2004), central Kansas, United States.

16.2.4 Nebraska Sand Hills

Extensive tracts of sand dunes are common throughout the central Great Plains and Rocky Mountains. The greatest among these is the Nebraska Sand Hills, which cover nearly 20,000 square miles (52,000 km²) in north-central Nebraska and southernmost South Dakota (Fig. 16-23). The Nebraska Sand Hills are the biggest expanse of sand dunes in the western hemisphere, more than three times larger than the state of Massachusetts, and one of the largest areas of native prairie in the Great Plains (Labedz 1990). The sand hills are located along the principal route of the North American central or Great Plains flyway for migrating waterfowl and shorebirds (Birds & Nature 2009), which find food, shelter, and rest in the myriad wetlands of the region. The main end points for this migration path are central Canada and the Gulf of Mexico, but some species travel from the Arctic to Patagonia.

The Nebraska Sand Hills are now largely stabilized by prairie vegetation, which has preserved the dune forms. The dunes resemble giant waves on the ocean (Fig. 16-24) and are comparable to other great sand seas in Africa, Arabia, Asia, and Australia. Individual large transverse dunes are typically several kilometers long, one kilometer wide, and up to 100 m tall (Keech and Bentall 1978). The dunes appear to have been active several times during the past

Figure 16-21. *Azolla* bloom in TNC marshes, Cheyenne Bottoms, Kansas, United States (see Color Plate 16-21). Comparable views both taken in mid-October. A. 2009. *Azolla* indicated by maroon color; cattle in lower right corner. B. 2010. Cattail (green) is prominent; *Azolla* is not evident in this aerial view, but small patches were observed on the ground at the edges of the marsh. Blimp and kite airphotos by J.S. Aber, S.W. Aber, G. Corley and B. Zabriskie.

10,000 years, most importantly during the mid-Holocene climatic optimum, some 8000 to 5000 years ago, and again from about 3000 to 1500 years ago (Swinehart 1990).

Thick dune sand rests on unconsolidated alluvial sand and gravel that cover poorly consolidated Ogallala strata, all of which are parts of the High Plains aquifer system. Throughout nearly all of the Nebraska Sand Hills, the saturated thickness of the aquifer is at least 100 m, and it exceeds 150 m in the central portion (Bleed 1990). This represents the greatest volume of water storage in any portion of the High Plains aquifer and is a key for understanding wetland hydrology in the Sand Hills. Essentially wetlands occur everywhere the water table intersects the land surface in swales and troughs between the dunes. Throughout the sand hills, cattle ranching is the primary land use, based mainly on wet meadows (Fig. 16-25).

This immense ground-water reserve is sustained by precipitation that averages about 0.5 m per year through the central portion of the sand hills (Wilhite and Hubbard 1990). Toward the west precipitation falls to less than 0.4 m per year; to the east it rises to about 0.6 m per year. This

316 MIDDLE-LATITUDE WETLAND CASE STUDIES

Figure 16-22. View over TNC marsh toward the northeast showing extensive cattail thickets (dark gray). May 2011; kite aerial photo by S.W. Aber and J.S. Aber.

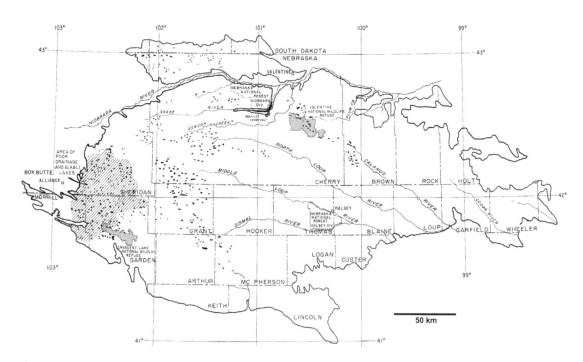

Figure 16-23. Nebraska Sand Hills showing lakes (black), rivers, and drainage features. National wildlife refuges are indicated by gray shading. Adapted from Keech and Bentall (1978, Fig. 2).

transition in available water gives rise to different wetland zones (see Fig. 16-23). In the central and eastern portions, perennial streams display remarkably constant discharge fed by groundwater baseflow; these streams rarely flood or dry up. In addition, sand hills streams have slightly alkaline pH (7.7–8.1), low content of dissolved solids (100–200 ppm), and low hardness (Bentall 1990). To the north and west, countless shallow lakes occupy enclosed basins, as at Valentine National Wildlife Refuge. Most of these are slightly alkaline, fresh-water lakes associated

MIDDLE-LATITUDE WETLAND CASE STUDIES 317

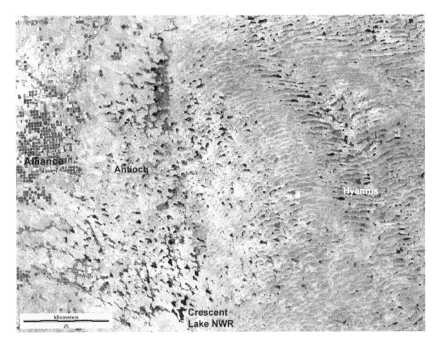

Figure 16-24. Satellite image of the western Nebraska Sand Hills, United States (see Color Plate 16-24). The eastern side of this scene displays massive transverse dunes, and smaller barchan dunes are found to the west. The most active vegetation appears dark green; less active vegetation is pale green. Numerous lakes (blue to black) occupy the troughs and swales between dunes. Landsat TM bands 2, 5 and 7 color coded as blue, green and red; image date 10 August 2010. Image from NASA; processing by J.S. Aber.

Figure 16-25. Predominant land use in the Nebraska Sand Hills. A. Cattle grazing in a wet meadow. B. Harvesting hay from a wet meadow. Photos by J.S. Aber.

with extensive marshes and wet meadows that follow elongated swales between large transverse dunes (Fig. 16-26).

The westernmost portion of the Nebraska Sand Hills consists of lower barchan-style dunes with lakes of irregular size and shape. These lakes are hyperalkaline; in some cases, alkalinity exceeds 90,000 mg/L, among the highest alkalinity values ever measured in natural lakes (Bleed and Ginsberg 1990). Such high salinity restricts most wetland vegetation and wildlife, but does support brine shrimp, brine flies, and rotifers. These invertebrates attract large flocks of avocets and phalaropes during their fall migrations. In more alkaline lakes, some invertebrates display bright red and orange colors (see Color Plate 4-6). This presumably is a result of carotenoid pigments that may protect

Figure 16-26. Steverson Lake with marsh and wet meadow in the foreground. Part of the Cottonwood-Steverson Wildlife Management Area of the Nebraska Game and Parks Commission. Located in western Cherry County, Nebraska, United States; kite aerial photo by S.W. Aber and J.S. Aber.

Figure 16-27. Remains of potash-processing plant at Antioch, Sheridan County, Nebraska. Photo by J.S. Aber.

the organisms from photo-oxidation (Bleed and Ginsberg 1990).

In the second decade of the twentieth century a potash industry flourished in the westernmost Nebraska Sand Hills based on the hypersaline lakes. Brine was pumped from the lakes to processing plants near railroads, and the chemicals were used for manufacturing fertilizer, epsom salt, soda, and other products. When World War I broke out, European sources for potash were cut off, and western Nebraska became the most important source in the United States. At least ten companies operated potash-extraction plants, each capable of producing 100 tons per day, most in Sheridan County (Miller 1990). With the end of the war, however, the companies were unable to compete with cheaper German potash, and the plants were shut down in 1920. Few remains of this industry are preserved today (Fig. 16-27).

16.2.5 Missouri Coteau, southern Saskatchewan

Southern Saskatchewan, Canada, is part of the prairie pothole region of recently glaciated terrain that stretches from central Iowa to southern Alberta covering some 300,000 square

Figure 16-28. Shallow pond and marsh occupy a typical pothole in hummocky terrain of the Missouri Coteau southwest of Moose Jaw, Saskatchewan, Canada. Photo by J.S. Aber.

miles (~780,000 km^2) and including millions of wetlands (Larsen 1991). Potholes of all sizes and shapes may contain shallow ponds, saline lakes, marshes, wet meadows and fens, and their density may reach 60 per km^2 (Dugan 2005). Known as the "duck factory" of North America, this region plays a vital role for waterfowl breeding. Other pothole functions include recharging ground water, storing flood water, absorbing nutrients, forming habitats for diverse flora and fauna, and providing water and forage for domestic livestock.

Throughout much of this vast region, many wetlands have been drained and converted for crop agriculture. The Missouri Coteau region of southern Saskatchewan, however, has retained a substantial proportion of its natural habitats (Dugan 2005). Several reasons explain the restriction of agriculture and preservation of coteau wetlands, including semi-arid climate that limits soil moisture, general lack of surface and ground water sufficient for irrigation, and locally hilly landscape that inhibits mechanized cropping (Fig. 16-28). The Missouri Coteau is in the mixed-grassland ecoregion with mean July temperature of 19 °C, mean January temperature about −13 °C, and annual water deficit greater than 0.5 m (Padbury and Acton 1994).

The Missouri Coteau is a prominent northeast-facing escarpment that extends from North Dakota across southern Saskatchewan. The coteau is a bedrock feature resulting from preglacial erosion and subsequently modified by repeated glaciations (Aber and Ber 2007). In particular, large ice-shoved ridges were thrust up along the edge of the coteau by glacier advances, which created steep, hilly landscapes (Fig. 16-29). Other conspicuous glacial landforms include hummocky moraine, melt-water spillway valleys, outwash plains, and proglacial lake basins. Wetlands are found in several geomorphic settings:

- Small potholes of irregular size and shape situated between ice-shoved ridges and in swales of hummocky moraine. These are particularly numerous, for example in the Dirt Hills, and vary from shallow ponds to overgrown marshes and wet meadows (Fig. 16-30).
- Kettle holes in outwash plains. Kettle holes mark places where blocks of stagnant glacier ice were buried by sand and gravel deposited from meltwater rivers. When the ice later melted, deep basins were created with steep sides. A good example is Lake Oro, located in the outwash plain south of the Dirt Hills (Fig. 16-31).
- Long, narrow, shallow lakes, known as ribbon lakes, that occupy spillway valleys such as Lake of the Rivers. This spillway carried glacial melt water from the Dirt Hills and Cactus Hills as well as overflow from the proglacial lake that occupied Old Wives basin. Nowadays the valley contains a series of shallow lakes and marshes during wet years (Fig. 16-32).
- Broad, shallow lakes that occupy former proglacial lake basins. Proglacial lakes were dammed between the ice sheet to the north

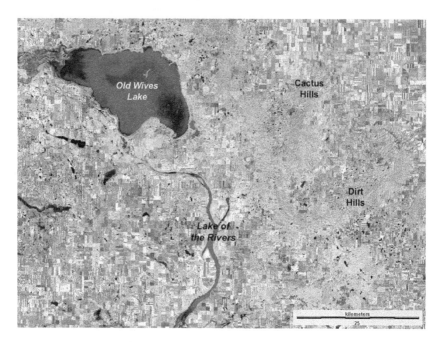

Figure 16-29. Satellite image of the Missouri Coteau, southern Saskatchewan, Canada (see Color Plate 16-29). Dirt Hills and Cactus Hills are large ice-shoved ridge complexes. Old Wives Lake and other water bodies are full during a wet episode. Active vegetation appears in green and yellow-green colors. Landsat TM bands 1, 5 and 7 color coded as blue, green, and red; image date 9 September, 2002. Image from NASA; processing by J.S. Aber.

and higher ground to the south. During deglaciation, the size, shape and volume of such proglacial lakes changed frequently as outlets were eroded and the ice margin retreated northward. Old Wives Lake illustrates this type of wetland setting (see Color Plate 6-29).

The Missouri Coteau is part of the Palliser Triangle region of southern Saskatchewan and southeastern Alberta, which represents the driest portion of Canada (Sauchyn 1997). This region is characterized by brown soils, temperature extremes, and a water deficit; yet it is agriculturally one of the most important regions in Canada for production of wheat, canola, flax, and other grain crops. The continental climate of the Palliser Triangle is subject to wet and dry cycles over periods of decades and longer time intervals. Droughts are common events historically. Thus lakes and wetlands may fill and overflow at times and completely dry up at other times (Fig. 16-33).

Water quality tends to be quite hard (dissolved calcium and magnesium) due to abundant limestone and dolostone in glacial sediments. Many shallow lakes and wetlands become eutrophic during the summer and support large algal blooms, which are enhanced by runoff of fertilizer from agricultural fields. Alkali chemistry is found in many lakes of the Palliser Triangle and adjacent United States (Fig. 16-34). Sodium sulfate is present both in the brine and in salt deposits, mainly as the mineral mirabilite (Glauber's salt). Sodium sulfate has been produced commercially from some of these lakes for various industrial applications – manufacturing pulp and paper, replacement for phosphate in detergent, pollution control at coal-fired power plants, and other purposes (*Encyclopedia of Saskatchewan* 2006a). For example, Frederick Lake was the site of a salt-processing plant using saline water diverted from Old Wives Lake, especially in the 1950s and 60s (*Encyclopedia of Saskatchewan* 2006b), but the plant is now abandoned. Salt extraction

MIDDLE-LATITUDE WETLAND CASE STUDIES   321

Figure 16-30. Pothole wetlands in the Dirt Hills, southwest of Regina, Saskatchewan, Canada. A. Abandoned farmstead (left) and a small marsh (right). The marsh is almost completely infilled with a mosaic of cattail (a) and bulrush (b); only a small area of open water remains on the far side of the township road. B. Another pothole in hummocky moraine. Note distinctive vegetation zones; the pond is largely covered by algae. Kite airphotos by S.W. Aber and J.S. Aber.

continues at five other plants, making Saskatchewan a world leader in naturally produced sodium sulfate. Few large commercial deposits of such salt exist outside this region of North America (Slezak and Last 1985).

Severe loss of wetlands and declining waterfowl were recognized as serious environmental issues, when Canada and the United States agreed to the innovative North American Waterfowl Management Plan (NAWMP) in 1986, and Mexico joined the plan in 1994 (Dugan 2005). The plan involves many partners and funding

Figure 16-31. Lake Oro occupies a deep kettle hole. This depression formed when glacial melt-water floods buried a stagnant block of ice in sand and gravel. Later, when the ice melted, a pronounced depression was left in the outwash plain. Lake Oro is today a small regional park near Crane Valley, Saskatchewan, Canada. Kite airphoto overview by S.W. Aber and J.S. Aber.

Figure 16-32. View across shallow lake and cattail-marsh in Lake of the Rivers spillway at Ardill, Saskatchewan, Canada. This is a favored nesting site for many duck species. Photo by J.S. Aber.

Figure 16-33. Satellite images of Old Wives Lake in wet and dry phases (see Color Plate 16-33). A. Wet phase with lakes full of water, September 2002. Old Wives Lake displays swirling patterns of suspended sediment driven by wind and waves as well as an algal mat (green) in the southeastern portion. B. Old Wives Lake is completely dry with exposed mud/salt flats, July 1988. Some water remains in Wood River and Frederick Lake. Landsat ETM/TM bands 3, 4 and 5 color coded as blue, green and red. Active vegetation appears in green and yellow-green colors; fallow/bare ground is pale purple to red; water bodies are light blue to black. Images from NASA; processing by J.S. Aber.

Figure 16-34. Partly dry alkali lake with salt flat, Cactus Hills, southern Saskatchewan, Canada. Photo by J.S. Aber.

from national, provincial, and local agencies as well as non-governmental organizations, such as Ducks Unlimited Canada and the Nature Conservancy of Canada. This plan has proven highly successful; through 2009, US$4.5 billion had been spent to protect more than 625,000 ha of wetland habitat (NAWMP 2009). The prairie pothole region was recognized as a high-priority transnational area in which conservation efforts were much needed (Larsen 1991). As part of the NAWMP, the Saskatchewan Watershed Authority has converted 7000 ha of cropland to pasture or permanent ground cover in the Missouri Coteau region (Saskatchewan Water Authority 2011).

The Chaplin-Old Wives-Reed Lakes complex west and southwest of Moose Jaw has been designated as a site of hemispheric importance by the Western Hemispheric Shorebird Reserve Network (WHSRN 2009). In this case, recognition was based on >100,000 shorebirds visiting annually and >30 percent of all sanderlings (*Calidris alba*) using the site as a spring migration stopover (Fig. 16-35). In addition, several wetland sites in the Missouri Coteau of Saskatchewan are designated as Important Bird Areas, including Old Wives-Frederick Lakes, Chaplin Lake, Big Muddy Lake, Willow Bunch Lake, and others (Nature Saskatchewan 2010). These non-governmental efforts involve a combination of public and private lands, which may or may not overlap with legally protected areas. Such multiple international and local efforts have led to significant wetland restoration and preservation in the Missouri Coteau region, which have greatly benefitted migrating

Figure 16-35. Sanderling (*Calidris alba*) at Farallon, Panama in winter. This individual might well migrate on the Great Plains flyway via Old Wives Lake, Saskatchewan, Canada. Adapted from original color photo by Mdf; obtained from Wikipedia Commons <http://commons.wikimedia.org/>.

waterfowl and shorebirds as well as other flora and fauna.

16.3 Coastal wetlands of Maine and Massachusetts, United States

Tidal salt, brackish and fresh-water marshes occur across coastal regions in the upper latitudes beyond the mangrove-dominated subtropics. Sheltered estuarine and delta regions along the New England coastline provide the right conditions for marsh development. These highly productive systems are dominated by grasses in low-lying tidally inundated zones near estuaries, sheltered bays and coves, inlets, creeks and channel edges. In addition to accretion by marine processes, alluvium deposited near the mouths of rivers builds layers of marsh substrate of sediment and organic matter (Mann 2000). Over long periods this accumulation allows marsh elevation to maintain itself in areas experiencing relative sea-level rise (Dionne, Dalton and Wilhelm 2006). Sea-level anomalies allow estuarine marshes to expand landward in the case of rising sea levels or seaward in the case of the opposite (Dame, Childers and Koepfler 1992). However, questions remain regarding marsh integrity in the face of rapidly rising sea levels during this century. Studies have suggested several possible scenarios, including the contraction of estuarine marshes that are bounded by higher elevation developed land on one side and rising sea level on the other. Other instances may see high marsh areas becoming low marsh or being completely drowned out.

Intricate patterns of channels and creeks allow sea water to reach far inland during high tides and flood vast sections of New England coastal marshes (see Fig. 4-4). Daily tidal action not only provides an important flushing function, transporting materials in and out of the marsh, but also introduces and deposits vital nutrients to the low and high marsh zones. Away from the daily influence of tides and with greater influxes of fresh water, both brackish and fresh-water marsh systems dominate. These inland zones may support a greater diversity of plant species. Species competition may influence the location of halophytes across a marsh. In addition, studies have indicated that salinity tolerance, distance from water, and tidal influence are also important variables in determining species zonation (Bertness and Ellison 1987; Sanderson, Foin and Usten 2001). However, considerable variability may be observed from site to site based on local conditions and latitudinal influences (Mann 2000).

Special morphological, root, and cellular adaptations are common in marsh plants. Across northern New England, salt marshes are dominated by tall forms of *Spartina alterniflora* (smooth cordgrass), an obligate halophyte, bordering creek and drainage channel banks within what is considered the low marsh. The short

Figure 16-36. An almost dry panne feature with blue-green algae and surrounded by short form *S. alterniflora*. Photo by Firooza Pavri.

form of *S. alterniflora* (generally <0.4 m tall) frequently occurs adjacent to this zone. *Spartina patens* (saltmeadow cordgrass), by contrast, may be found in extensive stands across higher sections of the marsh and may at times be interspersed with *Distichlis spicata* (saltgrass). *Juncus gerardii* (saltmeadow rush) is generally found toward the upper edges of the high marsh.

Pools and panne (or pan) systems are found dispersed across the high marsh landscape and provide unique microhabitats. These areas retain water from tidal flooding for varying periods of time. Pools are deeper depressions and vary in species composition depending on local factors. They provide habitat for fish stranded during flooding, and forage sites for bird populations. Pannes generally dry out between periods of inundation and are varied in their composition. Due to evaporation, salinity levels may be quite high. Some pannes are bordered by short form *S. alterniflora*, others are dominated by mud, blue-green algae, or forbs along their edges (Fig. 16-36).

The upland edges of a marsh are important transition zones. These are areas only occasionally flooded by exceptionally high tides. In New England, *Typha angustifolia* (narrow-leaf cattail) and *Phragmites australis* (common reed) in monotypic stands are often found across terrestrial upland edges. Some have attributed these invasive communities of *Phragmites* to increased nitrogen loading (Bertness, Ewanchuk and Silliman 2002).

16.3.1 Wells Reserve, southeastern Maine

Along southern coastal Maine, the Wells National Estuarine Research Reserve (NERR) is an example of a small yet productive salt-marsh system supporting wide floral and faunal diversity. The NERR program is administered by the National Oceanographic and Atmospheric Administration (NOAA) and includes a system of monitored coastal ecosystems across the United States. The Wells Reserve is the culmination of several conservation efforts, starting with the establishment of the Coastal Maine National Wildlife Refuge, renamed the Rachel Carson National Wildlife Refuge (NWR) in the 1960s (see Preface, Fig. 1). In 1984, the Laudholm Trust, formed by concerned citizens and residents, along with NOAA purchased farmland surrounding the Rachel Carson NWR. This entire conservation area was eventually listed a NERR site in 1986 (Dionne, Dalton and Wilhelm 2006).

Today the Wells Reserve serves a mix of educational, scientific and conservation functions (Wells Reserve 2011). It includes approximately 910 hectares of salt marsh across the Webhannet and Little River estuaries and is managed by various agencies (Fig. 16-37). The marsh was formed over the past 3000 to 4000 years behind a barrier spit which shelters it from the Gulf of Maine's strong tidal and wave action (Dionne, Dalton and Wilhelm 2006). The reserve is bordered by agricultural fields, tree cover, beaches, high-value ocean-front real estate, and intensive urban development (Fig. 16-38).

During the nineteenth and twentieth centuries, tidal restrictions such as roads, culverts and dikes reduced tidal exchange and fragmented and changed the configuration of these marshes (Dionne, Dalton and Wilhelm 2006). However, recent conservation efforts have sought to restore marsh functions (see Color Plate 2-16). Periodically conducted surveys report a diverse estuarine system with more than 260 recorded bird species, 93 species of flowering plants, 57 fish species and 288 invertebrates (Dionne, Dalton and Wilhelm 2006; Wells Reserve 2011).

The reserve's wetlands comprise a complex mix of species that loosely follow the typical species gradient pattern observed across New England coastal marshes. Tall form *Spartina alterniflora* is dominant along creek edges. Away from creek edges and across sections of the high marsh, *S. patens* is dominant or

Figure 16-37. Panoramic view over Laudholm beach and the mouth of the Little River at the Wells National Estuarine Research Reserve on the Atlantic coast of southeastern Maine, United States. Aside from the beach, a path through the forest (lower left), and a small observation platform (*), this wetland environment is closed to the public. Superwide-angle view looking northward; blimp airphoto by J.S. Aber, S.W. Aber and V. Valentine.

Figure 16-38. The Wells NERR marshes with *Spartina patens* in the foreground, taller *S. alterniflora* along creek edges in the mid-ground, and high-value ocean-front real estate in the background. Photo by Firooza Pavri.

Figure 16-39. A patch of *Phragmites australis* (common reed) over 1.6 m tall along an upland edge in the Wells NERR marshes, Maine, United States. Photo by Firooza Pavri.

co-dominant with *D. spicata* or short form *S. alterniflora*. Deeper pools with dependent fish communities and forb pannes are commonly observed across the high marsh. Fairly large patches of *P. australis* and *T. angustifolia* are found along upland edges (Fig. 16-39). *P. australis* is of particular concern in the Wells Reserve and elsewhere across the Atlantic coast. It grows well and out-competes such native species as *Juncus gerardii* along upland zones where anthropogenic modifications have dropped salinity levels. Successful control methods include the removal of tidal restrictions to restore natural tidal flushing, or burning and mechanical removal (Blossey 2003).

16.3.2 Plum Island Ecosystem, northeastern Massachusetts

Approximately 110 km south of the Wells NERR is the coastal marsh system of Plum Island Sound of northeastern Massachusetts (see Fig. 2-2B). The Plum Island Ecosystem (PIE) is the largest estuarine system within New England and receives runoff from the combined Parker and Ipswich watersheds (PIE 2011). It was designated as part of the National Science Foundation's (NSF) Long Term Ecological Research (LTER) program in 1998. This designation enables research institutions to conduct long-term ecological experiments and set up monitoring sites across the ecosystem.

Land use and development activities have a long history here. Paleo–Indian artifacts from 10,000 years ago indicate early human settlement of the region (Department of Conservation and Recreation 2011). Over the past 300-plus years, the region hosted productive farmland, pasture for cattle grazing, valuable commercial fisheries, and other associated activities. In the 1800s and early 1900s, extensive networks of drainage ditches were built into the marshes to permit further settlement and development (see Fig. 13-6). These ditches are still clearly observed in a recent near-infrared image of West Creek within the Parker River basin (Fig. 16-40). With the decline of agriculture across New England by the second half of the twentieth century, some farms and pastures were recolonized by forests. Today the PIE is just an hour drive north from Boston, New England's largest city and, as such, faces all the residential, commercial and industrial development pressures that accompany its location. Even so, parts of the region are protected from development by state and local ordinances as well as the Parker River National Wildlife Refuge.

PIE encompasses a vast system of beaches, dunes, mudflats, marsh complexes, vegetated platforms, river channels and upland areas, and has more than 490 species of vascular plants (Department of Conservation and Recreation 2011). *Spartina alterniflora* dominates creek edges, while mixes of *S. patens* and *J. gerardii* dominate high-salt-marsh zones closer to the coastline. Pools, pannes, algal mats and mudflats are found across the high marsh. Brackish and fresh-water communities including *Typha* extend farther inland along the Parker and Ipswich rivers, where vegetated platforms are more typical. Greater species mixing is found in the brackish areas when compared to saline areas (Pavri and Valentine 2008). The Parker River NWR is a well-known stopover point for migratory birds along the Atlantic flyway and has recorded more than 300 bird species, of which about 60 species breed within the marshes. Spring and fall migrations bring in large concentrations of waterfowl including up to 25,000 ducks and 6000 Canada Geese (Department of Conservation and Recreation 2011).

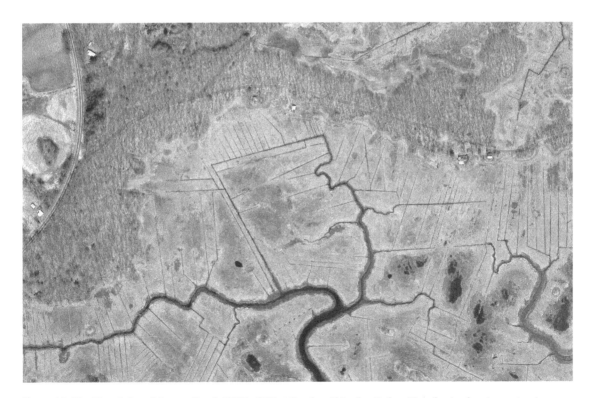

Figure 16-40. Near-infrared image (April 2005) of West Creek within the Parker River basin showing extensive drainage ditches across the salt marsh (see Color Plate 16-40). Data Source: 1:5000 digital orthoimagery from Massachusetts Office of Geographic Information (MassGIS 2011). Adapted from Pavri and Valentine (2008).

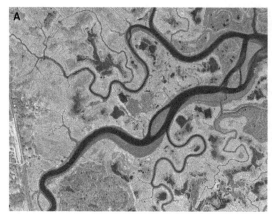 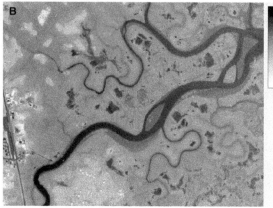

Figure 16-41. High-resolution imagery of Rowley River study site (see Color Plate 16-41A). A. Near-infrared image (April 2005) of the Rowley River salt-marsh system with a number of pools and pannes in black, active vegetation in red across upland areas and marsh areas yet to green-up in pale shades of pink and gray. B. Adjusted Normalized Difference Water Index (ANDWI) image. Wet pixels including pools, pannes, rivers, creeks and wet mud are indicated by negative ANDWI values, while active vegetation is represented by positive ANDWI values. Data source 1:5000 digital orthoimagery from Massachusetts Office of Geographic Information (MassGIS 2011). Adapted from Pavri and Valentine (2008).

The vicinity of Rowley River, a tributary of Parker River, has been investigated in some detail. High-spatial-resolution digital orthoimagery from 2005 provides a detailed view of the salt-marsh landscape (Pavri and Valentine 2008). In this early spring season image (Fig. 16-41A), a larger number of pools and pannes in black are quite distinctive. Active vegetation appears in red across the upland areas, while the marsh itself appears in pale shades of pink or gray. Green-up is yet to occur within the marsh and hence the image still indicates the previous year's dormant vegetation and dead matter. Water is an essential component of a wetland, and the spatial configuration of marsh features are often of some interest to ecologists and wetland managers. An Adjusted Normalized Difference Wetness Index (ANDWI) provides one way to easily distinguish pools, pannes and saturated soils across a marsh from drier upland zones (Hurd et al. 2005; Pavri and Valentine 2008). ANDWI values range between −1 and +1. Pixel values closer to −1 are water-based pixels, whereas those on the positive end of the spectrum are typically terrestrial based (Fig. 16-41B). The ANDWI image easily distinguishes wet areas from higher ground and provides some indication of water levels in the creeks, channels and drainage ditches.

Globally, coastal marshes are under threat from development and the ramifications of climate and sea-level change. Some of the highest population densities are found along continental coastlines, threatening the productive capacity of estuarine zones. Even so, strong economic arguments could be made for more stringent protection of salt marsh. These systems provide habitats necessary for commercially viable fisheries, offer recreational opportunities, and sustain tourism economies. These are industries of vital importance to both Maine and Massachusetts. Maintaining the integrity of coastal marsh systems, thus, not only serves ecological purposes but provides equally important economic benefits.

16.4 Estonia, eastern Baltic region

The country of Estonia is situated at the eastern end of the Baltic Sea and is surrounded by water on four sides – Gulf of Finland to the north, Baltic Sea to the west, Gulf of Riga to the south, and Lakes Peipsi-Pihkva to the east (Fig. 16-42).

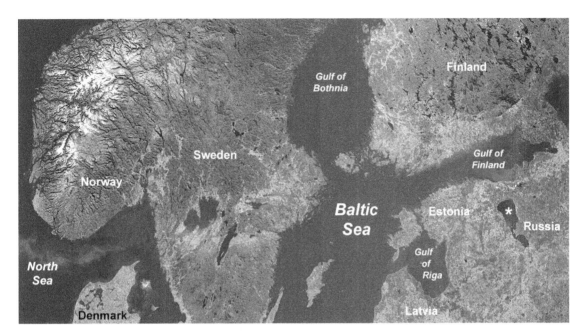

Figure 16-42. Satellite image showing the central Baltic region of northern Europe. Mosaic assembled from nearly cloud-free Multi-angle Imaging SpectroRadiometer (MISR) vertical imagery draped over a shaded-relief topographic map, equidistant conic projection. Asterisk (*) indicates Lakes Peipsi-Pihkva. Adapted from NASA Visible Earth <http://visibleearth.nasa.gov/>.

The Baltic Sea is brackish with salinity of 6–8‰ in the central portion and only 1–2‰ in the easternmost Gulf of Finland (Helsinki Commission 2011), which has a major influence on coastal wetland conditions. Given its location and proximity to large water bodies, Estonia has a temperate climate transitional between continental and maritime (Arold 2005). Summers are cool, winters are long with snow cover, and annual precipitation averages 50–70 cm (Ilomets 1997). Much of the country consists of low-relief terrain under 100 m above sea level, but the southeastern portion has hilly relief exceeding 200 m in elevation. The whole of the country is underlain by Paleozoic sedimentary bedrock mantled by Pleistocene glacial sediments of variable thickness and composition (Raukas and Kajak 1997).

Estonia is rich in lakes and wetlands. Some 1200 lakes exceed one hectare in area, and about 20,000 smaller bog pools exist (Arold 2005). More than one-fifth of the territory is covered by swamps, marshes, fens and bogs (Orru, Širokova and Veldre 1993; Ilomets 1997), and more than 40 of these are protected as national parks, nature reserves, or mire conservation areas. A dozen of these are designated as Ramsar wetlands of international importance (Ramsar 2010h), ranging from coastal estuaries, to raised bogs, to the delta of the Emajõgi River draining into Lake Peipsi (Fig. 16-43). The peatlands of Estonia may be divided in three general types (Valk 1988):

- Minerotrophic – Fens in depressions that are supplied by ground water or surface runoff rich in nutrients. Fens display great variety with diverse flora including >30 species of *Carex* (sedges) and are highly productive for *Vaccinium oxycoccos* (cranberries). This category accounts for more than half of Estonian mires.
- Mixotrophic – Transitional bogs generally receive less ground water and more surface runoff with fewer nutrients than fens. They are intermediate in character between fens and bogs and represent about one-eighth of Estonian peatlands.

Figure 16-43. Estonian postage stamps depicting wetland habitats. Matsula Looduskaitseala (nature protection area; later designated as a national park), an estuary system on the western mainland coast. Tolkuse raba, a bog in the southwestern mainland. Lakes Peipsi-Pihkva on the eastern margin, which together are Europe's fourth largest lake. Original stamps printed in multiple colors.

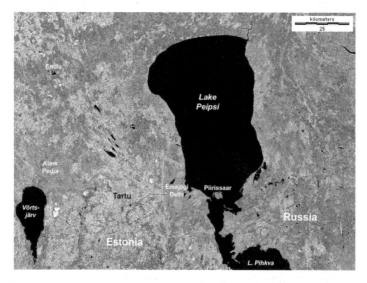

Figure 16-44. Satellite image of Lake Peipsi and surroundings, Estonia and Russia. Endla, Alam Pedja, and Emajõgi-Piirissaar are Ramsar wetlands of international importance. Landsat ETM+ band 5 (mid-infrared). Date of acquisition 10 July 1999. Image from NASA; processing by J.S. Aber.

- Ombrotrophic – Raised bogs that are fed entirely by atmospheric precipitation. Bogs may include pines (*Pinus sylvestris*) or be treeless, depending on the amount of nitrogen and phosphorus available for tree growth. *Sphagnum fuscum* is the principal peat-forming plant along with several other *Sphagnum* species. Rate of peat accumulation averages 0.5 to 1.0 mm per year, depending mainly on precipitation. Bogs amount to about one-third of Estonian mires.

Since the last ice sheet retreated, isostatic crustal adjustment has caused northern Estonia to uplift, while the southern portion of the country has subsided. This phenomenon continues today and is illustrated by Lake Peipsi (Fig. 16-44); the northern end is rising at 0.2–0.4 mm per year, while the southern part is sinking at 0.8 mm per year (Hang and Miidel 1999). Thus, the lake is transgressing toward the south, which has led to extensive shore erosion, submergence of medieval villages and churches, and paludification of wetlands around the southern margin. For example, the island of Piirissaar, a Ramsar site near the southern end of Lake Peipsi, decreased from ~20 km² area in 1796 to only 7.5 km² today (Hang and Miidel

Figure 16-45. Rumpo habitat/species management area of the Vormsi Maastikukaitseala (landscape reserve). Southern coast of the island of Vormsi plus the adjacent Väinameri shallow seafloor and islets, western Estonia. Kite aerial photo by S.W. Aber and J.S. Aber.

1999). Other small historical islands have disappeared completely.

The western mainland, islands, and northern margin of Estonia are parts of the Baltic Coast Bog Province (Fig. 16-45), which corresponds roughly to the region that was submerged in the Baltic Ice Lake following deglaciation in the latest Pleistocene (Valk 1988; Ilomets 1997). Eastern Estonia, on the other hand, remained above the Baltic Ice Lake and is part of the East Baltic Bog Province (Fig. 16-46). Bogs in these two regions have distinct flora; for example, *Trichophorum cespitosum* (tufted bulrush) and *Myrica gale* (sweetgale) grow only in the western region, whereas *Chamaedaphne calyculata* (leatherleaf) is found only in the eastern region (Valk 1988).

The oldest known peat in Estonia is dated nearly 10,000 radiocarbon years ago (Ilomets 1997). Paludification (rising ground water) leading to substantial peat accumulation began around 8500 years ago, and the next phase took place about 7100 to 6500 years ago, which corresponded to transgression of the Littorina Sea in the Baltic basin. The first significant phase of peat initiation by terrestrialization (lake infilling) happened some 6500–4500 years ago during the first half of the Atlantic period, and paludification reached a peak between 5100 and 4100 years ago. For the most part, basal peats accumulated in fens and are rich in *Bryales* (mosses) and *Carex* species.

The initiation of raised bogs, characterized by *Sphagnum* peat, was somewhat later. Three minor phases of bog formation occurred around 7000, 6000 and 5000 years ago, each lasting about half a millennium. The major time for bog development was between 4000 and 2000 years ago, however, when more than 40 percent of modern Estonian bogs came into existence as a result of hydroseral succession (Ilomets 1997). A few younger bogs formed between 1500 and 500 years ago. Given approximately 12,000 years since deglaciation of Estonia, several millennia were required for the processes of paludification and terrestrialization to initiate fens in which peat could begin to accumulate, which took place generally between 7000 and 4000 years ago during the mid-Holocene climatic optimum (Fig. 16-47). At least one or two more millennia were then necessary for raised bogs to appear via hydroseral succession. Within these general time frames, bogs and fens displayed distinct differences in the timing of their Holocene development between eastern and western Estonia (Ilomets 1997).

Human activities have influenced bog development during the past few centuries (see

Figure 16-46. Overview of Männikjärve Bog in the Endla mire complex of east-central Estonia. Raised bog displays pools and hummocks with dwarf pines in foreground; conifer forest surrounds the bog in background. Photograph taken from an observation tower by J.S. Aber.

chapter 14.3.4), and Estonia has a long history of mire investigations. The Baltic Peatlands Improvement Society began in 1908 in Tartu, and the Tooma Experimental Station was established in the Endla mire complex in 1910 (Valk 1988). This has continued as a research station into modern times, and Endla has been elevated to the status of a Ramsar Wetland of International Importance (Ramsar 2010h). The Endla mire system encompasses approximately 25,000 ha, of which more than 10,000 ha are preserved as the Endla Nature Reserve (Endla Looduskaitseala Administratsioon 2004). The present nature reserve was created in 1985 as an expansion of the previous smaller Endla–Oostriku mire reserve. It is located immediately south of the Pandivere Upland in east-central Estonia (see Fig. 16-44). The reserve is biologically diverse with at least 180 bird species and greater than 460 vascular plants including many protected species (Endla Looduskaitseala Administratsioon 2004).

The Endla mire complex grew up in the depression of former Great Endla Lake (Allikvee and Masing 1988). Several remnants of this lake still survive, notably Endla Lake and Sinijärv (Blue Lake). These lakes were subjected to several episodes of draining (1872, 1949, 1950) and were reflooded in 1968. The Endla mire complex contains several bogs separated by narrow rivers, and over 30 large springs rise in the western part of the complex (Fig. 16-48). The lakes, bogs, and springs are important sources of recharge for the Põltsamaa River.

Among the bogs, Männikjärve Bog has been studied intensively since the early 1900s, and a small meteorological station is located in the bog. An elevated, wooden walkway allows visitors to travel across the bog without disturbing the surface and without sinking into the peat and mud (Fig. 16-49; Aaviksoo, Kadarik and Masing 1997). The Endla mire complex has attracted a great deal of interest from international scientists for all types of investigations,

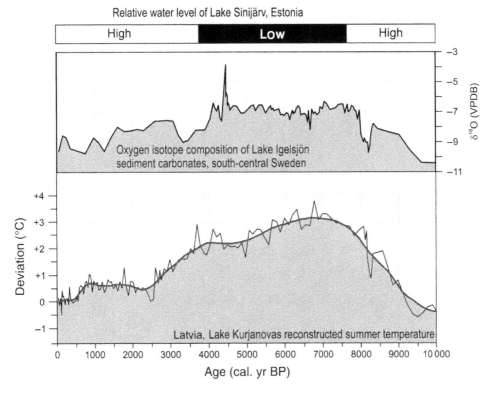

Figure 16-47. Records of Holocene climate from Estonia, southern Sweden, and Latvia. A warm, dry period took place across the central Baltic region in the middle Holocene between ~8000 and 4000 calendar years ago. Adapted from Heikkilä, M. and Seppä, H. 2010. Holocene climate dynamics in Latvia, eastern Baltic region: A pollen-based summer temperature reconstruction and regional comparison. *Boreas* 39, p. 705–719, Fig. 6.

because of its good access, long observation record, and nearly pristine conditions (Fig. 16-50; see also Fig. 2-21).

16.5 Summary

The middle latitudes span the world north and south of the equator generally between 30° and 60°. Temperate climate is distinctly seasonal with marked hot and cold, wet and dry intervals. Consequently wetland habitats vary greatly within these conditions, ranging from subtropical deserts to regions with permafrost. The Great Plains of North America display strong gradients in precipitation and temperature from Texas to the Canadian Prairie region. Toward the east, wetlands and water bodies are perennial and fresh; whereas, to the west water bodies are ephemeral and saline in the rain shadow of the Rocky Mountains. Western wetlands are subject to frequent droughts, and flooding may take place throughout the Great Plains. Millions of waterfowl and shorebirds utilize the Great Plains flyway in their twice yearly migrations. They exploit wetland habitats in sand hills, river valleys, glacial potholes, and playas for breeding, nourishment, protection, and rest during their annual life cycles. The Great Plains are an essential part of the North American Waterfowl Management Plan, which involves Canada, Mexico, and the United States.

Coastal salt-water marshes, brackish and fresh-water marshes are highly productive ecosystems dominated by grasses in low-lying, tidally inundated zones near estuaries, sheltered bays and inlets. Across northern New England, these marshes exhibit species zonation based on

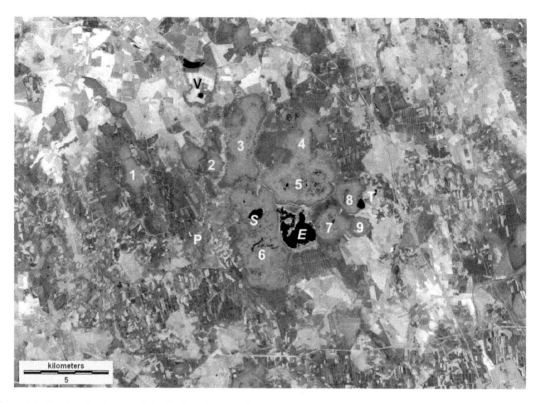

Figure 16-48. Satellite image of the Endla mire complex in east-central Estonia. Bogs: 1 – Punaraba, 2 – Rummallika raba, 3 – Kanamatsi raba, 4 – Mardimäe raba, 5 – Linnusaare raba, 6 – Endla raba, 7 – Kaasikjärve raba, 8 – Männikjärve raba and 9 – Teosaare raba. Other features: P – Põltsamaa River, S – Sinijärv (Blue Lake), E – Endla Lake, T – Tooma, and V – Väinjärve raba, a peat mine. Lateral drainage ditches can be seen in conifer forest surrounding the bogs. Landsat TM band 5 (mid-infrared). Date of acquisition 28 May 2007. Image from NASA; processing by J.S. Aber.

Figure 16-49. Elevated boardwalk and 8-m-tall observation tower near the center of Männikjärve Bog, east-central Estonia. Kite aerial photo by S.W. Aber, J.S. Aber, K. Aaviksoo and E. Karofeld.

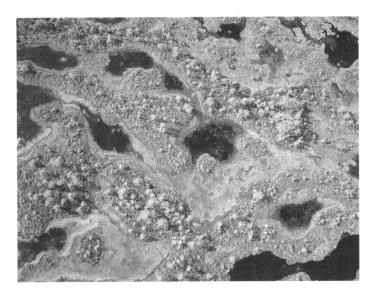

Figure 16-50. Vertical view over central portion of Männikjärve Bog, east-central Estonia (see Color Plate 16-50). Striking color zonation and intricate spatial patterns of pools, moss, and other vegetation are displayed in this autumn scene, which is ~100 m across. Kite aerial photo (Aber and Aber 2005).

distance from a water channel and salinity levels. The long history of human settlement across the North Atlantic coast has influenced these ecosystems. Anthropogenic changes through the building of tidal restrictions have altered salt-marsh systems and made them less productive and more susceptible to invasive species like *Phragmites australis*. Conservation and restoration efforts are supported by the NERR program at the Wells Reserve in Maine and research from LTER sites like Plum Island Estuary in Massachusetts. Such efforts are fundamental to securing the future of these ecosystems and in turn supporting economically vital industries for both Maine and Massachusetts.

Estonia is situated at the northern margin of the middle-latitude zone. The country is surrounded on three sides by brackish water of the Baltic Sea, and Lakes Peipsi-Pihkva bound the eastern side. More than one-fifth of the country is covered by mires, and coastal wetlands are extensive in the western islands and shallow seafloor. The western mainland, islands, and northern margin of Estonia are parts of the Baltic Coast Bog Province; whereas, eastern Estonia is in the East Baltic Bog Province. Given approximately 12,000 years since deglaciation of Estonia, several millennia were required for the processes of paludification and terrestrialization to initiate fens in which peat could begin to accumulate, which took place generally between 7000 and 4000 years ago during the mid-Holocene climatic optimum. The major time for raised bog development was between 4000 and 2000 years ago. Estonia has a long history of mire investigations, in which the Endla mire complex has played a leading role for more than a century.

High-latitude and high-altitude wetland case studies 17

17.1 Introduction

High latitudes (>60°) and high altitudes (>2500 m) include regions with mild to severe climates characterized by cold temperatures. Glaciers, permafrost and periglacial phenomena are common features in these realms. The ground and surface water bodies may remain frozen much of the year, and the growing season is quite short. Vegetation is restricted to those hardy species that tolerate extreme cold, strong winds, and prolonged snow cover. Included in these zones are vast expanses of boreal forests across northern Eurasia and North America as well as tundra regions still farther north. At lower latitudes, similar conditions may exist on high mountain peaks and plateaus, even in glaciated tropical locations such as the Andes Mountains, East Africa, the Himalayas and Tibet Plateau, and Irian Jaya.

Mountains display distinct climatic differences from surrounding terrain, which include colder temperature, enhanced cloud cover (Fig. 17-1), and more precipitation than for adjacent lowlands. Such mountains were also frequently sites for alpine glaciation. Just as with ice-sheet glaciation of lowlands (see chapter 16), alpine glaciers created numerous landscape settings for potential wetlands in cirques, valleys, and moraines (Fig. 17-2). High-altitude environments have received relatively less scientific study than lower-lying regions of the Earth's surface. More than half of all humans live at elevations below 300 m, and nearly 90% live below 1000 m altitude (Cohen and Small 1998). Thus, it is no surprise that most wetland science has been undertaken at relatively low altitudes. Similar circumstances have limited scientific work done at high latitudes. Nonetheless, high-latitude and high-altitude wetlands play important environmental roles for global biogeochemical cycles and habitats.

Case studies presented here include the Andes Mountains of Venezuela, Rocky Mountains and San Luis Valley of southern Colorado, United States, the North Slope and Yukon Delta of Alaska, United States, and the Lena River delta of Russia. These sites span low, middle and high latitudes with altitudes decreasing northward from 5000 m in Venezuela to sea level on the Yukon Delta and with climatic conditions ranging from relatively mild in the tropics to quite harsh in the Arctic.

17.2 Andes Mountains, Venezuela

The Andes Mountains are the second highest mountain system in the world and extend from Tierra del Fuego at the southern tip of South America to Colombia in the north, thus spanning several climatic and environmental zones. One branch of the Andes, the Cordillera de Mérida, trends from southwest to northeast across the western portion of Venezuela at latitudes 7–9°N (Fig. 17-3). Between deep valleys

---

Wetland Environments: A Global Perspective, First Edition. James Sandusky Aber, Firooza Pavri, and Susan Ward Aber.
© 2012 James Sandusky Aber, Firooza Pavri, and Susan Ward Aber. Published 2012 by Blackwell Publishing Ltd.

HIGH-LATITUDE AND HIGH-ALTITUDE WETLAND CASE STUDIES   337

Figure 17-1. Cumulus clouds building over the High Tatra Mountains, while the adjacent foreland remains cloud-free in this typical, midday, summer view near Strané pod Tatrami, Slovakia. Similar conditions prevail in most other temperate and tropical mountains during the summer monsoon season. Adapted from Aber et al. (2008, fig. 3).

Figure 17-2. Glacial landforms in the High Tatra Mountains of Slovakia include jagged horns and ice-carved valleys with small lakes. The ponds are dammed behind rock thresholds and moraine ridges. Note zig-zag hiking trail on lower left side. Picture taken by J.S. Aber from observation station at top of Lomnicky Peak (2634 m).

338 HIGH-LATITUDE AND HIGH-ALTITUDE WETLAND CASE STUDIES

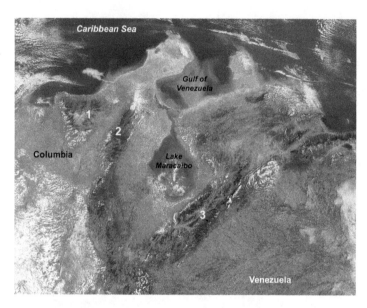

Figure 17-3. MODIS image of western Venezuela and northeastern Colombia. Segments of the Andes Mountains: 1. Sierra Nevada de Santa Marta, 2. Sierra de Perijá, 3. Cordillera de Mérida. Acquired 7 January 2003 during the dry winter season; adapted from visible-color image of NASA <http://visibleearth.nasa.gov/>.

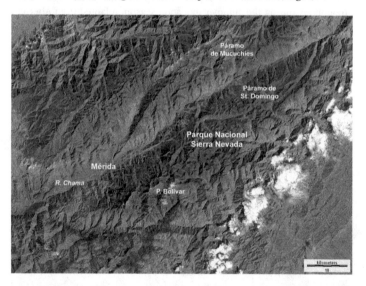

Figure 17-4. Rare, nearly cloud-free satellite image of Andes region in western Venezuela (see Color Plate 17-4). Bright green indicates evergreen and deciduous forest; dark olive-green is alpine shrub of the páramo. Landsat MSS bands 1, 4 and 2 color coded as blue, green and red. Acquired 13 January 1979. Image from NASA; processing by J.S. Aber.

and high peaks, topographic relief ranges from less than 1000 m elevation along the Río Chama valley below the city of Mérida to above 5000 m on Pico Bolívar (Fig. 17-4). This great range of elevations over short distances gives rise to dramatic variations in local climate, vegetation, wildlife, geomorphic processes, and human land use (Fig. 17-5; Ruiz 1992).

Distribution of precipitation is controlled strongly by elevation and prevailing wind. Most precipitation falls on the southeastern side of the major mountain ridge located to the southeast of Mérida. Annual precipitation exceeds 2.8 m in this zone; whereas the Río Chama valley generally has less than 1 m annual precipitation (Ruiz 1992). Summer is the rainy season, and

Figure 17-5. Climatic and environmental contrasts in the Venezuelan Andes Mountains. A. Desert zone with cactus vegetation in the Río Chama valley southwest of Mérida, below 1000 m elevation. B. Alpine zone at Pico Águila (4118 m), northeast of Mérida. P. Águila is a horn shaped by glacial erosion of the mountain sides. Photos by J.S. Aber.

winter is relatively dry. Temperature for a given site and elevation varies little on average during the year, although daily temperature may fluctuate considerably. Semi-deciduous forest predominates at lower elevations on wet slopes, and evergreen forest (bosque siempreverdes), including cloud forest (bosque nublados), is found from about 2000 m to 3000–3200 m altitude (Ruiz 1992). Higher up is the alpine shrub zone, known as páramo, an area described as summer in the daytime and winter at night.

The term páramo is derived from Spanish, generally referring to high, desolate, cold, wet terrain (Luteyn 2011). Far from being desolate, however, the páramo is now considered to be the most diverse high-altitude ecosystem in the world (Diazgranados 2009). It is situated in the interval between the upper forest limit and perennial snow cover in Colombia, Ecuador and Venezuela. In the latter, páramo is found from about 3500 m to 4500 m elevation mostly within the Parque Nacional Sierra Nevada. This zone is almost constantly in clouds. Rain falls most days during the cool, wet summer season, and snow falls during the slightly colder, dry winter months. Wetland conditions are widespread in formerly glaciated valleys and mesas, including mires and wet meadows with histosols (Fig. 17-6).

The most characteristic plants in the páramo are members of the subtribe (taxonomic

Figure 17-6. Meandering stream courses through wet meadow in a fog-shrouded valley, Mesa del Caballos, Parque Nacional Sierra Nevada, Venezuela. Photo by J.S. Aber.

subdivision between family and genus) Espeletiinae (family Asteraceae), commonly known as frailejóns, which includes eight genera and nearly 140 named species (Diazgranados 2009). They are highly endemic; many species are found only in a single páramo locale along the Andes in Colombia, Venezuela and, to a lesser extent, Ecuador. Frailejón grande (*Coespeletia moritziana*) and frailejón pequeño (*Espeletia schultzii*) are typical in the Venezuelan páramo (Ruiz 1992). Frailejóns are peculiar plants with thick, woody stems and leaf rosettes at the top. They display diverse growth habits, and in some locations they comprise more than 40% of the vegetation cover (Fig. 17-7). Frailejóns have great ecological importance for the páramo, as they regulate the hydrologic cycle, produce the most biomass, prevent soil erosion, and are associated with more than 125 animal species (Diazgranados 2009).

Climatic fluctuations associated with Pleistocene glacial–interglacial cycles resulted in vertical shifts of vegetation zones, particularly marked by the forest–páramo boundary. For example, on the high plain of Bogotá, Colombia (~2600 m), forest and páramo alternated some 15–20 times during the past one million years (Luteyn 2011). During the last glacial maximum (late Pleistocene), alpine glaciers advanced downslope even into forested zones, and then between 21,000 and 14,000 years ago glaciers receded and climate became quite cold and dry with mean annual temperature 6–7 °C lower than today (Luteyn 2011). During this interval the páramo was continuous over long distances above 2000 m. The end of the Pleistocene brought warming and further recession of glaciers, continuing into the middle Holocene, when the forest–páramo boundary rose about 300–400 m higher than today. Late Holocene cooling caused the forest–páramo boundary to descend to its current position, resulting in many disjunct (isolated) páramo locales.

Much of the modern páramo is heavily influenced by human activity, particularly during the past three centuries (Luteyn 2011). The greatest alteration is at the forest–páramo boundary, where logging has reduced the forest. The páramo is used primarily for cattle and is maintained by cutting, periodic burning, and grazing (see Fig. 1-5). The alpine water resources are diverted from streams and carried in pipelines to lower valleys (Fig. 17-8), where the water is used for human consumption, hydropower and irrigation. The tropical climate allows continuous grazing and multiple yearly harvests of potatoes, carrots, and other crops.

17.3 Southern Colorado, United States

The Rocky Mountain region in southern Colorado includes several distinct mountain ranges and intervening valleys (Fig. 17-9), in which elevations range from around 7500 feet (2300 m) to >14,000 feet (4260 m). Following long-lasting tectonic activity during the early and middle Cenozoic, the present elevation and relief of the region was brought about by vertical uplift of

Figure 17-7. Frailejón alpine vegetation in the Venezuelan Andes Mountains. A. Frailejón plants cover the steep slope in the foreground, vicinity of Pico Águila. Photo by J.S. Aber. B. Close-up view of frailejón plants, Páramo de Piedras Blancas. Modified from original photo by P.M. Vásconez; obtained from Wikimedia Commons <http://commons.wikimedia.org/>.

Figure 17-8. Water-supply systems in the Venezuelan Andes Mountains northeast of Mérida. A. Small dam and water pipe. B. Holding tank and pipelines on valley side. Photos by J.S. Aber.

Figure 17-9. MODIS image of southern Colorado and northernmost New Mexico, United States. Mountain ranges: 1. San Juan Mountains, 2. Sangre de Cristo Mountains, 3. Wet Mountains and 4. Culebra Range. Acquired 26 October 2001; adapted from visible-color image of NASA <http://visibleearth.nasa.gov/>.

at least one mile (1.6km) during the past few million years (Sahagian, Proussevitch and Carlson 2002; McMillan, Heller and Wing 2006; Pelletier, 2009). Mountains were deeply eroded as a consequence (see Fig. 16-1), and valleys were infilled with thick sediment accumulation. Erosion was further enhanced during the Pleistocene by a combination of increased precipitation, runoff, and glaciation (Dethier 2001).

As with the Andes Mountains, climatic conditions and vegetation cover vary greatly over short distances, according to elevation and prevailing wind, from sagebrush desert, to subalpine conifer forest, and, at higher elevations, to alpine tundra (Fig. 17-10). The following case studies focus on the Culebra Range, which forms the front range of the southern Rocky Mountains, and the San Luis Valley, which is part of the Rio Grande rift system.

17.3.1 Culebra Range

The Culebra Range is the southern segment of the Sangre de Cristo Mountains, which extend from south-central Colorado into northern New Mexico (Fig. 17-11). The Culebra Range comprises mainly sandstone, siltstone and shale of the Sangre de Cristo Formation (see Color Plate

Figure 17-10. Climatic and environmental contrasts in the Rocky Mountain region, southern Colorado, United States. A. sagebrush desert of the San Luis Valley with Blanca Peak (14,345 feet, 4372 m) of the Sangre de Cristo Mountains in the right background. Blimp airphoto by S.W. Aber and J.S. Aber. B. Northern portion of the Culebra Range seen from the east, looking up the Cucharas Creek valley. Mixed conifer and aspen forest covers the mountain slopes. Upper forest limit is about 3500–3600 m. Two glacial cirques (*) are visible as basins at the heads of valleys. C. Crest of the Culebra Range looking south toward Trinchera Peak (13,517 feet, 4120 m). Sparse turf partly covers quartzose sandstone bedrock in the alpine tundra. B and C photos by J.S. Aber.

2-10) as well as other sedimentary strata and crystalline basement rocks. The crest of the range is supported by thick quartzose sandstone, which is most resistant to erosion. Elevations along the crest generally exceed 3750 m up to peaks higher than 4000 m. The lower slopes of the range are covered by mixed conifer-aspen forest, and above the treeline the crest has alpine-tundra vegetation (see Fig. 17-10).

The Culebra Range was glaciated repeatedly during the Pleistocene (Richmond 1965, 1986; Armour, Fawcett and Geissman 2002), most recently during the Pinedale Glaciation, about 23,000 to 12,000 years ago. Glaciers descended from mountain sides, carved deep valleys with rock basins, and deposited rubbly moraines in those valleys, which created numerous closed depressions. These basins are today sites for lakes, fens, marshes and wet meadows (Fig. 17-12), and beaver are quite active in downstream creeks (Fig. 17-13). The Blue-Bear Lakes moraine complex in the Cucharas Creek valley possesses a variety of pothole habitats ranging from deep lakes to mires with significant peat accumulation (Fig. 17-14).

Two mires have been investigated in some detail in the Blue-Bear Lakes moraine complex. They are nearly identical in their overall physical settings in terms of climate, geology, and surrounding spruce-aspen forest vegetation. The substratum is composed of till rich in sandstone. Both contain peat more than 2 m deep. Yet the two mires are strikingly different in their morphology, vegetation, and surface water.

Cottongrass fen – This informal name indicates the presence of cottongrass growing around the margin of the fen. The fen is elliptical, about 40 m × 50 m in extent, and lies at an altitude of approximately 10,380 feet (3165 m). The surface is largely open, and shallow water extends across the entire fen without any discernible flow. This fen has a nearly ideal arrangement of concentric vegetation zones and water depths around its central point; the site displays no evidence for beaver activity. The center of the fen has somewhat taller emergent vegetation and deeper water, and small spruce trees occupy the western margin (Fig. 17-15). Most of the fen

Figure 17-11. Satellite image of Culebra Range and vicinity, south-central Colorado, United States. Selected peaks along the mountain crest: 1. Trinchera Peak (13,517 feet, 4120 m), 2. Mt. Maxwell (13,335 feet, 4065 m) and 3. Culebra Peak (14,069 feet, 4288 m). Based on Landsat TM bands 2, 5 and 7. Active vegetation is dark gray. Acquired 10 August 2010. Image from NASA; data processing by J.S. Aber.

is dominated by *Carex limosa* (mud sedge), and the center is characterized by *Carex vesicaria* (blister sedge) (E. Volkova, pers. com. 2006). The cottongrass is thought to be *Eriophorum angustifolium* (tall cottongrass, Fig. 17-16A). Cottongrass is relatively scarce in montane wetlands of Colorado, and this is the first report of it in Huerfano County (U.S. Department of Agriculture 2011). Green mosses are found on microrelief bumps, barely above normal water level (Fig. 17-16B).

Several shallow and one long peat cores were collected from the cottongrass fen (see Figure 3-31A). Sedge is the primary constituent of the shallow peat along with green mosses, cottongrass, and various herbs with a low degree of decomposition (E. Volkova, pers. com 2006); the upper meter of peat is especially loose and watery near the center of the fen. Toward the margin of the fen, the peat is denser and made up mostly of *Picea* (spruce) wood and bark. Toward the center, the peat is up to 3.5 m thick and rests on a silt bottom; basal peat is radiocarbon dated at 1530 to 1340 calendar years old (Beta 262070). This fen is relatively young in terms of conditions suitable for peat accumulation, and it is not yet close to transforming into a raised bog.

Beaver mire – Named for the obvious evidence of past beaver activity, this mire has highly varied vegetation cover, along with numerous water channels, and several generations of beaver dams in different parts of the mire complex (see Fig. 7-42). The overall shape is elliptical, roughly 55 m × 75 m in size, at about 10,430 feet (3180 m) elevation. The beaver dams are in a state of disrepair, and water level is currently around half a meter below the crest of the main dam. The mire is dominated by *Calamagrostis stricta* (slimstem reedgrass; E. Volkova, pers. com. 2006). Bushy willows occupy the western margin of the mire, and dead spruce remain standing toward the eastern side (Fig. 17-17).

A long-term soil-temperature record has been collected from near the center of the mire at a depth of about 0.5 m (see Fig. 1-20). Peak temperature of around 10 °C is achieved at the end of summer (late September) (Fig. 17-18). Throughout the autumn and winter, soil temperature declines gradually, reaching a low just above 0 °C in spring (March–May). Temperature

Figure 17-12. Alpine wetlands at the head of Trinchero Creek valley on the northwestern side of Trinchera Peak; elevation of valley floor ~3600 m. A. Overview of meandering stream and wet meadow (right) and small basin with pond (left). B. Close-up view of meandering Trinchero Creek and wet meadow. The stream is fed by a spring that emerges along a fault zone (see Fig. 4-8). C. Basin formed by ridge of bedrock mantled by moraine. Photos by J.S. Aber.

then begins to increase again in June and July. This region is currently included in plant hardiness zone 5, in which minimum winter air temperature is typically in the range −10 to −20 °F (−23 to −29 °C) (Arbor Day 2006). However, the

Figure 17-13. Newly built beaver lodge beside Cucharas Creek, downstream from the Blue-Bear Lakes moraine complex. Photo by J.S. Aber.

insulating effect of peat and snow cover prevents freezing at shallow depth in the mire throughout the long, cold winter, which has important implications for year-round biological activity and organic conditions in the shallow peat.

Several shallow and one long peat cores were acquired also from the beaver mire. The shallow peat is dominated by slimstem reedgrass and displays a high degree of decomposition of plant remains (Volkova, pers. com. 2006). Near the center of the mire, peat is at least 2.5 m deep; basal peat yielded a corrected calendar age of 3700 to 3630 years (Beta 232587). Given a maximum age of 3700 years and thickness of 2.5 m, the average rate of peat accumulation would be about two-thirds of a mm per year. However, this site has been disturbed repeatedly by beaver, which undoubtedly altered water levels and, thus, vegetation and peat accumulation.

Both mires began accumulating peat during the Neoglacial period of cooler, wetter climate in the late Holocene. This suggests that moraine lake infilling and hydroseral succession required several millennia after the late Pinedale glaciers disappeared in order to reach the status wherein plant remains could form peat. The southwestern United States was markedly warmer and drier during the mid-Holocene altithermal episode (Polyak et al. 2001), when most lakes in this vicinity may have dried up. This condition happens even nowadays during short-term

Figure 17-14. Blue-Bear Lakes vicinity seen from the crest of the Culebra Range looking toward the northeast. The location of pothole mires within the moraine complex is indicated. For opposite view, see Figure 17-10B. Photo by J.S. Aber.

Figure 17-15. Panoramic view of cottongrass mire looking southward; east to right, west to left. *Carex* sp. (sedge) covers most of the open surface; spruce and aspen surround the fen. E. Volkova stands in water to left side; photograph by J.S. Aber from a tree on the northern edge of the mire.

droughts (Fig. 17-19). Thus, mire development and peat accumulation may have been interrupted many times, especially during the middle Holocene.

17.3.2 San Luis Valley

The San Luis Valley is part of the Rio Grande rift system, which was active tectonically in particular during the middle Cenozoic (see Fig. 17-9 and Fig. 17-11). The valley dropped down along deep faults, while volcanic eruptions and igneous intrusions took place along the valley margins. As the valley basement sank, thick sediment was washed in from adjacent mountains; the sedimentary fill in the San Luis Valley exceeds 30,000 feet (9000 m) in depth, and large alluvial fans are built against the mountain flanks of the valley.

The San Luis Valley of southern Colorado has been called the highest, largest mountain desert in North America (Trimble 2001). It is approximately 160 km (100 miles) long, north–south, and 80 km wide; elevations span from about 2300 m (7500 feet) toward the central portion to more than 2500 m (8200 feet) around the margins. Winter temperature is quite cold (plant hardiness zone 5; Arbor Day 2006). Because of surrounding mountain ranges with peaks exceeding 14,000 feet (4265 m), an extreme rain

Figure 17-16. Vegetation of the cottongrass mire. A. Tall cottongrass, *Eriophorum angustifolium*. The cottongrass is scattered among *Carex limosa* (mud sedge) around the margin of the fen. B. Green moss growing on a microbump just above water level. Photos by J.S. Aber.

Figure 17-17. Central portion of beaver mire looking toward the northwest. Dead spruce trees to right side grew on an old beaver dam. Location of soil-temperature data logger marked by asterisk (*). Photo by J.S. Aber from a tree on the southeastern edge of the mire.

shadow exists. Average annual precipitation is in the range 6–10 inches (~15–25 cm) at most reporting stations (High Plains Regional Climate Center 2011). Nonetheless, water is abundant in the San Luis Valley. Surface and ground water are derived from spring snow melt and summer monsoon thunderstorms in the adjacent mountains.

The Rio Grande drains the southern part of the valley, entering from the west and exiting southward through a gorge in volcanic rocks along the Colorado–New Mexico border. The northern portion of the San Luis Valley is a closed depression, however, with no surface outlet for drainage (Fig. 17-20). Surface runoff from the Sangre de Cristos and San Juans soaks

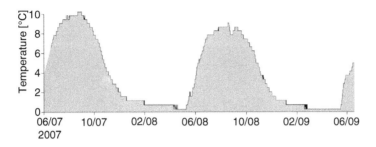

Figure 17-18. Two-year continuous record of soil temperature in the beaver mire near Bear Lake, Culebra Range, south-central Colorado, United States. The record begins in June 2007 and continues until June of 2009. Note that freezing never occurs at shallow depth in the mire.

Figure 17-19. Lower Blue Lake in the Blue-Bear Lakes moraine complex, Cucharas Creek valley, Culebra Range, south-central Colorado, United States. A. June 2002, a drought year, in which the lake was completely dry and many forest fires occurred nearby. B. June 2003 with the lake full of water. Note beaver-cut aspen in foreground. These trees were cut ~2 m above the ground, presumably when snow pack was present. Photos by J.S. Aber.

into alluvial fans and sand dunes, and ground water migrates toward the low point at San Luis Lake (Fig. 17-21). Abundant ground water gives rise to many lakes, springs and flowing wells, and supports considerable irrigation in the valley. Throughout most of the valley, ground water is less than 12 feet (3.6 m) below the land surface (Emery 1971). Irrigation is also carried out with water diverted from the Rio Grande via canals (Fig. 17-22), Culebra Creek (see Fig. 11-20), and other streams. Meanwhile, the state of Colorado is obligated to supply water southward under terms of the Rio Grande Compact with New Mexico and Texas (Emery 1971). In order to do so, ground water is salvaged via deep wells in the northern portion of the valley and transported via canal southward to New Mexico (Fig. 17-23).

The abundance of ground and surface water gives rise to many wetland habitats, as evidenced by Great Sand Dunes National Park and Preserve, as well as Monte Vista, Alamosa, and Baca national wildlife refuges, and several state wildlife areas. Access to a state wildlife area requires a Colorado Wildlife Habitat Stamp, funds from which are used to improve and maintain the wildlife areas. Spring and fall are migration seasons for vast flocks of waterfowl, including some 25,000 sandhill cranes (U.S. Fish and Wildlife Service 2010b). In fact, a Crane Festival is held each March in Monte Vista (Fig. 17-24). Summer is time for shorebird breeding and nesting, including American avocet, Wilson's phalarope, white-faced ibis, and many others (see Fig. 7-22). The modern abundance and diversity of waterfowl and

Figure 17-20. Satellite image of the central portion of San Luis Valley (see Color Plate 17-20). The region north and east of the Rio Grande is an internal basin draining into several lakes. GSD – Great Sand Dunes, RL – Russell Lakes, SLL – San Luis Lake. Landsat TM bands 2, 5 and 7 color coded as blue, green and red. Active vegetation appears green. Acquired 21 August 2009. Image from NASA; processing by J.S. Aber.

Figure 17-21. Water in vicinity of the Great Sand Dunes, Colorado, United States. A. Medano Creek drains from the Sangre de Cristo Range in the background, flows beside the southern edge of the Great Sand Dunes, and soaks into the sandy sediment. Note people for scale. B. San Luis Lake in foreground with the Great Sand Dunes in the distance against the Sangre de Cristo Range. Ground water recharged from Medano Creek at the Great Sand Dunes emerges downslope to replenish the lake. Great Sand Dunes exceed 2500 m elevation; San Luis Lake is <2300 m altitude. Kite aerial photos by S.W. Aber and J.S. Aber.

shorebirds in the San Luis Valley are similar to pre-European accounts (U.S. Fish and Wildlife Service 2010c).

Russell Lakes State Wildlife Area is an excellent example of relatively natural wetland habitats in the northwestern portion of San Luis Valley. Russell Lakes SWA consists of 2159 acres (~874 ha), just under 7600 feet (2340 m) elevation, including numerous marshes, pools and intervening subtle dunes (Fig. 17-25). The

Figure 17-22. Irrigation near Del Norte on the western side of the San Luis Valley. A. Rio Grande Canal with the San Juan Mountains in the background. Photo by J.S. Aber. B. Irrigated crops are supplied by the Rio Grande Canal and are marked by dark circles in this view looking east toward the center of the valley. Kite aerial photo by S.W. Aber and J.S. Aber.

complex is fed by Russell Creek that heads at Russell Springs a few kilometers to the west. To this natural flow, several artesian wells provide additional water to the complex (see Fig. 4-11). Emergent hydrophytes include bulrush, cattail, various sedges, and pink smartweed.

Surface water is saline, as evidenced by crusts of salt and bitter salt on dry mudflats. This promotes the growth of algae and salt-loving invertebrates (Fig. 17-26). The pools and marshes are separated by low silty dunes that rise 1–2 m above the general surface. Other than scattered mesquite bushes, these dunes are largely devoid of vegetation, which attests to the arid climate. The site offers hunting in season and wildlife viewing, and a raised boardwalk and gravel path facilitate access (Fig. 17-27). The area is closed

Figure 17-23. High-capacity pumps aerate ground water into a canal in the northern San Luis Valley for delivery southward to New Mexico. Sangre de Cristo Range in the background. Photo by J.S. Aber.

Figure 17-24. Mural painted on the brick side of the Crane Building in downtown Monte Vista, Colorado. Original mural in full color; photo by J.S. Aber.

Figure 17-25. Panorama of Russell Lakes State Wildlife Area looking toward the northeast with the Sangre de Cristo Range in the far background; east to right, north to left. Saline lakes and marshes occupy hollows between low, mesquite-covered dunes. Blimp aerial photo by S.W. Aber and J.S. Aber.

Figure 17-26. Close-up view of a pool at Russell Lakes State Wildlife Area (see Color Plate 17-26). Dark green emergent vegetation is mainly bulrush, and algae forms a mat on water surface. Blood-red patches presumably reflect carotenoid pigments of invertebrates in saline water. Blimp aerial photo by S.W. Aber and J.S. Aber.

Figure 17-27. Elevated wooden boardwalk (lower left) and gravel path (center) for wildlife viewing at Russell Lakes State Wildlife Area, Colorado. The position of an observation platform is marked (*). Blimp airphoto by S.W. Aber and J.S. Aber.

February 15 through July 15 during the waterfowl breeding season.

17.4 The Arctic

Anticipated impacts of rapid climate change during the next century are expected to have a substantial influence on the polar regions of the world. The most recent report of the Intergovernmental Panel on Climate Change (IPCC) suggested that the changes witnessed in these regions would be far reaching and drive feedback loops of global consequence (Anisimov et al. 2007). The average surface temperatures in the Arctic over the past 20 to 40 years have increased at a greater rate than the global average (McBean et al. 2005). This has been accompanied by a reduction in the extent of sea ice and permafrost (Anisimov et al. 2007). Such processes could potentially have regional and global-scale influences. Changes in ice-albedo feedback and the release of stored greenhouse gases, such as CO_2 and CH_4, from melting and disintegrating permafrost may further feed into global climate change (Callaghan et al. 2005). Moreover, such processes would impact wetlands directly across the Arctic.

17.4.1 Arctic Coastal Plain, Alaska

The North Slope of Alaska extends above 68° N latitude and covers a vast area almost to the Canadian border to the east, the Chukchi Sea to the west, the Beaufort Sea on the north, and the Brooks Range to the south. The northernmost section of the North Slope includes the Arctic Coastal Plain, which stretches across 50,000 km² of flat tundra landscape with a short two-to-three month summer, when July temperatures average around 10 °C (Martin et al. 2009). Ice plays a significant role in molding this Arctic landscape. It gives rise to open polygonal permafrost features roughly 5 to 10 m across caused by subsurface ice wedges. Pingos, or elevated mounds of ice cores covered by earth, are common as are other thermokarst features.

The summer transforms this area from a frozen snow- and ice-covered plain into an extensive wetland drained by several rivers, with a shallow layer of thawed soil, numerous lakes, and abundant wildlife and plant communities. At the surface, the soil layer varies in thickness between about 35 cm and 4 m and freezes and thaws seasonally (Martin et al. 2009). Below this active layer, permafrost may extend to a depth below 600 m. Permafrost typically prevents water percolation into the ground. Spring floods and overland river flows inundate vast sections of the coastal plain, feeding a landscape of thousands of lakes, troughs, shallow depressions and polygonal features (Fig. 17-28). These floods turn the plain into a highly productive system supporting enormous species densities and hosting hundreds of species of plants, mammals, insects, birds, and fish.

The vegetation of the Arctic Coastal Plain takes advantage of the long daylight hours during the brief summer to leaf-out within a

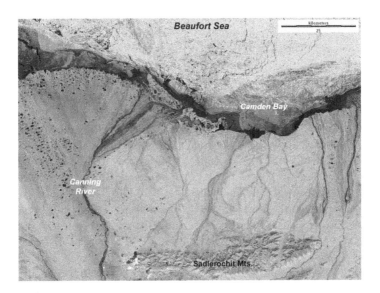

Figure 17-28. Satellite image of a portion of the North Slope region, including the Canning River and its delta, north of the Sadlerochit Mountains, northeastern corner of Alaska, United States (see Color Plate 17-28). Land vegetation is just beginning to green up (yellow-green), and sea ice (cyan) has pulled away from the shore in this early summer scene. Small lakes and streams are ice-free (black), but larger lakes and rivers still have partial ice cover. Most of this area is part of the Arctic National Wildlife Refuge, including the coastal plain 1002 Area, as well as Kaktovik Inupiat Corporate Lands. Landsat ETM+ bands 1, 4 and 5 color coded as blue, green and red. Image date 6 July 2000. Image from NASA; processing by J.S. Aber.

Figure 17-29. Sketch of musk ox (*Ovibos moschatus*) by P.S. Foresman. Obtained from Wikimedia Commons <http://commons.wikimedia.org/> March 2011.

span of a few weeks after spring thaws begin. Sedges, mosses, dwarf shrubs and tussock sedge dominate the low coastal and lacustrine landscape. Depending on standing water depth, sedge marsh, such as *Carex aquatilis* (water sedge) and *Eriophorum angustifolium* (tall cottongrass), along with forbs and mosses are common (Martin et al. 2009). Sedge and dwarf shrub groupings are more common on higher terrain toward the interior. The Arctic Coastal Plain is perhaps best known for its abundant bird life and large mammals, such as the reindeer or caribou (*Rangifer tarandus*), polar bear (see Fig. 7-39), Arctic fox (*Alopex lagopus*), and musk ox (Fig. 17-29). Recent surveys, however, suggest declines in the musk ox population to just over 200 individuals across the region (USFWS 2009b).

With the exception of a few species, like the rock ptarmigan and snowy owl, most of the Arctic Coastal Plain's 200-plus bird species comprise migratory waterfowl, shorebirds, terns and loons and use this region from May through September for nesting and breeding (U.S. Fish and Wildlife Service 2010d). Some of the more remarkable birds include the Arctic tern (*Sterna paradisaea*), which undertakes the longest annual bird migration recorded. It covers a distance of about 40,000 km in a round trip between its summer breeding grounds in the Arctic and Antarctica where it flies to escape the northern hemisphere winter. Another example is the Alaskan bar-tailed godwit (*Limosa lapponica baueri*), which flies an astonishing 11,000-km, non-stop journey between the Alaskan Arctic and New Zealand (Hedenström 2010).

Over the years, studies have tracked natural fluctuations within caribou populations across the Arctic (Douglas, Reynolds and Rhode 2002). Recent counts suggest declines across the circumpolar region, including within the Porcupine and Western Arctic caribou herds of northern coastal Alaska. These declines have been attributed to climate-induced changes to plant and insect phenology and anthropogenic influences (Vors and Boyce 2009). The caribou are a particularly important cultural symbol and socio-economic contributor to many indigenous communities including the Gwich'in, Yup'ik and Iñupiat in northern Alaska and Canada, as well as the Saami in Fennoscandia, among others. Their persistent decline would undoubtedly influence the ability of dependent indigenous communities to sustain long-standing cultural traditions and their economic livelihoods.

Since the production of oil from Prudhoe Bay began in the 1970s, Alaska's North Slope has seen significant development from oil exploration, production and transportation. Development has spread outward from Deadhorse/Prudhoe Bay and included offshore drilling as well. While stricter environmental regulation and improved drilling, extraction and transportation technology have helped to reduce the footprint of oil production across the North Slope, the fragility of this ecosystem and its vulnerability to change require special attention and careful decision making. Four major entities are responsible for land management and resource stewardship in the region. These include the Bureau of Land Management, State of Alaska, U.S. Fish and Wildlife Service, and Native lands, which collectively manage approximately 96% of the land area (Martin et al. 2009). The Arctic National Wildlife Refuge in the easternmost part of the North Slope is the country's largest and most northerly refuge managed by the U.S. Fish and Wildlife Service (see Fig. 17-28). The refuge covers approximately 75,000 km^2, almost the size of the state of South Carolina.

The U.S. Census Bureau lists the 2010 human population of the North Slope Borough of Alaska at 9,430 spread out across a handful of small communities that mostly dot the northern coast. Barrow, with a population of more than 4,000, is the largest. These communities are either directly related to the oil economy or encompass native settlements (U.S. Census Bureau 2011). Past climate oscillations played an influential role in shaping the relationship between indigenous communities across the Arctic, the land and its resources. Human and ecological systems adapted to changes then. The vulnerability of contemporary human and ecological systems to climate change may depend in part on the magnitude and timing of such changes. Undoubtedly, traditional ways of life will face increased pressures, ecosystems may witness shifts, and the adaptive capacities of both will be tested (Anisimov et al. 2007).

17.4.2 Yukon Delta, Alaska

The Yukon River drains an area of more than 850,000 km² as it travels a distance of nearly 2,900 km from its headwaters in the Yukon Territory of the Canadian Rockies before entering the Bering Sea (Brabets and Schuster 2008). Its basin extends between approximately 59°N to 69°N latitude and 139°W to 166°W longitude. Several tributaries drain into the Yukon River, including the glacially fed Tanana River. The meandering Yukon and its channels form an extensive low-lying wetland delta dotted by lakes before entering into the Bering Sea (Fig. 17-30). Most of the delta lies below an elevation of 100 m. The river discharges fresh water at ~6,500 m³/s into the Bering Sea (Brabets, Wang and Meade 2000). This large influx of fresh water provides essential nutrients and sediment runoff to the adjacent marine ecosystem (Brabets and Schuster 2008). Across the delta, shallow permafrost provides the conditions for the development of saturated soils with a mix of peatlands, heath meadows, sedges, rushes and other marsh communities that are suited to seasonally wet and flooded conditions (U.S. Fish and Wildlife Service 2011a).

Native communities inhabited this region in numerous villages scattered across the delta and have long since recognized the importance of the Yukon as a resource-rich system. During the

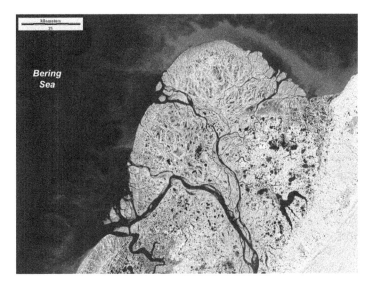

Figure 17-30. Satellite image of the Yukon River and Yukon Delta National Wildlife Refuge, Alaska, United States (see Color Plate 17-30). Land vegetation is just beginning to green up. Lakes, rivers and the sea are ice free in this late spring image; suspended sediment is highlighted in light blue colors. Intricate patterns are displayed in the delta and coastal environments. Landsat ETM+ bands 1, 4, and 5 color coded as blue, green and red. Image date 18 June 2002. Image from NASA; processing by J.S. Aber.

nineteenth century, the Yukon's role as a transportation link between the coast and the interior of Alaska was established with the fur trade and gold rushes. Today, the larger Yukon basin's ecological importance is perhaps best illustrated by the large number of established conservation areas. The basin contains wholly or partly seven national wildlife refuges (NWR), including the Yukon Delta NWR (U.S. Fish and Wildlife Service 2011b). It also contains sections of four national parks, including parts of Denali and Wrangell–St. Elias, and areas managed and protected by the Bureau of Land Management (National Park Service 2011). The Yukon Delta NWR is one of the largest of its kind in the United States and includes the deltas of both the Yukon and Kuskokwim rivers. It is particularly well known for supporting millions of migratory waterfowl, shorebirds, cranes, loons, grebes and other species. The delta also provides essential fish habitat and is vital for commercial Pacific salmon fisheries as well as local salmon and other fish economies.

Salmon has been a long-standing part of the culture and economy of resident native communities like the Yup'ik. It is considered an anadromous fish species, which spends part of its life cycle, including hatching, in fresh water, maturing in ocean, returning eventually to fresh-water systems to spawn and ultimately die. Chinook (*Oncorhynchus tshawytscha*), Chum (*O. keta*) and Coho (*O. kisutch*) salmon use upstream and downstream areas of the Yukon basin to spawn, feed, and also partially mature before entering the ocean. During the spawning season, salmon travel hundreds of kilometers inland from the ocean to reach spawning beds in the upper Yukon. Alongside food obtained by hunting and other subsistence activities, salmon is considered a staple and is preserved, dried or smoked to provide a year-long source of food and economic security for resident populations.

Salmon is also a commercially significant species and growing demand has led to increased harvesting pressures on the fishery. The fish is an important link in the Arctic trophic food chain and studies suggest that changing salmon populations could have far-reaching consequences on other dependent fish and mammal populations (Loring and Gerlach 2010). The Alaska Department of Fish and Game, through management strategies, attempts to strike a fine balance between supporting the subsistence needs of upstream and downstream local populations, the demands of commercial fisheries, and maintaining healthy biologically viable salmon stocks (Loring and Gerlach 2010). Yet meeting such competing needs is a difficult task. Compounding these challenges, climate change impacts including the thawing of permafrost, drying of wetlands, warmer water temperatures, earlier ice-out conditions, increased storm activity and runoff may all influence salmon runs as well (Anisimov et al. 2007; Loring and Gerlach 2010).

17.4.3 Lena River delta, Russia

Moving to the Arctic coast of Eurasia, the delta of the Lena River is among the most spectacular wetland complexes in the world. The Lena River is noteworthy for both its length and the volume of fresh water it brings to the Laptev Sea in the Arctic Ocean. The Lena traverses nearly 4,300 km in length from its source in the Baikal highlands region of central Russia to its mouth. Freshwater inputs influence the temperature and local circulation of the Laptev Sea, and the significant sediment nutrient load supports a vibrant terrestrial and marine ecosystem (Costard and Gautier 2007). From its source, the river flows northeastward till it reaches the city of Yakutsk, which is the largest in the region and the capital of Russia's Sakha Republic (also known as Yakutia). The Lena then turns and flows northward to its mouth in the Laptev Sea. The greater Lena basin, along with its four primary tributaries, extends over several different biomes including steppe grasslands, a wide belt of taiga forests, and permafrost tundra landscapes in its far northern reaches.

The river breaks into numerous branches and channels in a wide fan-shaped delta extending over roughly 28,000 km^2 before entering the Laptev Sea (Costard and Gautier 2007). The Lena Delta stretches between approximately 72° N to 74° N latitude and 122° E to 129.5° E longitude. It comprises an entirely tundra landscape, frozen for up to seven months of the year, with a permafrost layer >500 m deep and a mean annual temperature of −13 °C (Schneider, Grosse and Wagner 2009). Yet, as the active layer, extending just 10–100 cm below the surface, melts during the brief summer, the region turns into a biologically rich wetland oasis for tundra vegetation, birds and mammals (Center for Russian Nature Conservation 2011).

Data from several studies combined have reported 122 total bird species in the delta, which is one of the highest numbers of any region this far north of the Arctic Circle (Gilg et al. 2000). Among other factors, studies attribute this high species diversity and terrestrial and marine productivity to the mixing of the Lena's warmer fresh water and sediments into the cold Laptev Sea (Gilg et al. 2000). Mammals found in the delta include voles, Siberian lemmings (*Lemmus sibiricus*) and collared lemmings (*Dicrostonyx torquatus*), which provide an important source of food for other predators and birds such as the snowy owl (*Nyctea scandiaca*) (Gilg et al. 2000). The Arctic fox (*Alopex lagopus*) and reindeer (*Rangifer tarandus*), are among the other species observed in the delta (Center for Russian Nature Conservation 2011).

An image from Landsat's Thematic Mapper sensor shows a summer landscape that is typified by thousands of channels, islands, sand bars, patterned tundra and water-filled depressions, along with grasses, sedges, lichen and moss species (Fig. 17-31). The combination of visible and infrared bands used in this false-color composite highlights terrestrial and water boundaries, narrow water channels, soil moisture differences and vegetation type in vivid detail. As in other wetland sites, the presence of water influences vegetation configurations across the Lena Delta (Schneider, Grosse and Wagner 2009). Water-saturated areas are dominated by sedges (*Carex aquatilis*). Grasses and mosses are found on poorly drained moist soils, whereas dwarf shrubs and water-intolerant grasses are found in drier areas. This geographic landscape provides ideal waterfowl and shorebird breeding, nesting and foraging habitats.

Given its remote location and forbidding climate, the Lena Delta has largely been spared from intensive human activity. The greater Lena basin, however, is rich in natural resources including gold, other metals and coal that have been exploited commercially. While there are no settlements within the delta itself, hunting and fishing are common. The closest town to the delta is Tiksi, an important coastal port city on the Laptev Sea. In 1985, the Lena Delta Nature Reserve, also known as the Ust-Lensky Zapovednik, was established to protect this tundra landscape and habitat (Center for Russian Nature Conservation 2011). The greatest short-term threat to this delta is its vulnerability to the impacts of climate change. These impacts may include higher average winter and summer temperatures, thawing of the permafrost, and the effect of more frequent and catastrophic flood events upstream on the Lena.

With warmer temperatures recorded across the Arctic in the past 20–40 years (McBean et al. 2005), greenhouse gas emissions from sequestered sources in northern permafrost regions have been a growing concern. Northern latitude wetlands already contribute significant methane emissions. It is expected that thawing permafrost would speed up the decomposition of organic matter, enhancing microbial activity and methane releases. Across the Russian Arctic including the Lena basin, the thawing of yedoma, an organically rich permafrost with a high carbon content, is also a matter of concern (Sachs et al. 2008). The natural release of methane and other greenhouse gases from degraded thermokarst landscapes in northern latitudes over the coming decades and centuries is expected to amplify feedback on the climate system.

Studies suggest that there has already been some permafrost degradation of Arctic coastal lowland regions around the Lena basin (Grosse, Shirrmeister and Malthus 2006). Thawing permafrost and warming temperatures could have significant impacts on the Lena Delta. Enhanced erosion from wave action along the delta front may occur due to the loss of permafrost, while increased temperatures and further thawing could enhance methane releases or drain the thermokarst landscape (Anisimov et al. 2007). Such changes across the Arctic would

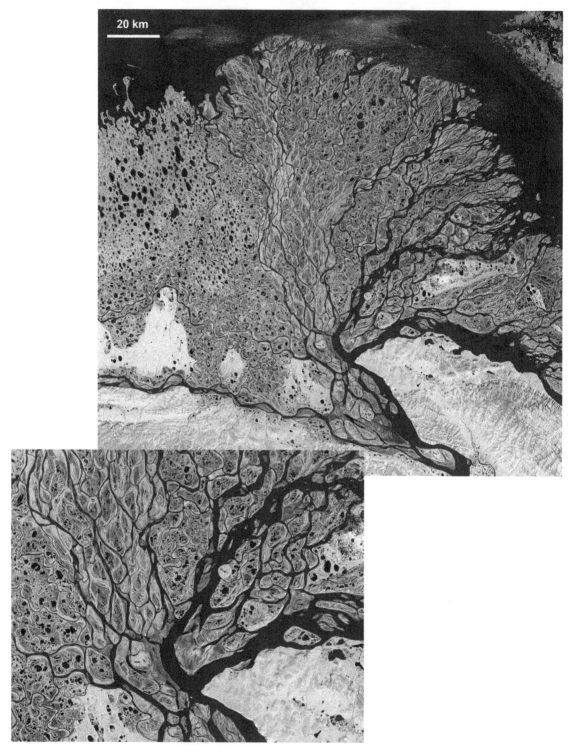

Figure 17-31. The Lena River delta on the Arctic coast of eastern Siberia, Russia (see Color Plate 17-31). This is one the most beautiful false-color Landsat images ever produced; it shows the intricate network of meandering waterways in the classic fan-shaped delta. The inset image depicts details of the distributary system. The delta is a highly protected wetland wilderness in a permafrost environment and is an important refuge and breeding ground for many wildlife species. Adapted from NASA; obtained from Wikimedia Commons <http://commons.wikimedia.org/> October 2011.

necessitate local ecosystem adaptations and they would also pose serious implications for the regional and global climate system.

17.5 Summary

High latitudes and altitudes include regions with mild to severe climates characterized by cold temperatures. Glaciers, permafrost and periglacial phenomena are common features in these realms. Included in these zones are vast expanses of boreal forests across northern Eurasia and North America as well as tundra regions still farther north. At lower latitudes, similar conditions may exist on high mountain peaks and plateaus. Mountains display distinct climatic differences from surrounding terrain, which include colder temperature, enhanced cloud cover, and more precipitation. High-altitude and -latitude environments have received relatively less scientific study than have more hospitable regions of the Earth's surface. Nonetheless, high-altitude and -latitude wetlands play important environmental roles for habitats and global biogeochemical cycles.

The páramo habitat consists of alpine shrub and grassland situated in the interval between the upper forest limit and perennial snow cover in Colombia, Ecuador and Venezuela. This zone is almost constantly in clouds; rain falls most days during the summer, and snow falls during the winter. Wetland conditions are widespread in formerly glaciated valleys and mesas including mires and wet meadows with histosols. The most characteristic plants are Espeletiinae, commonly known as frailejóns, which are endemic to this environment and include nearly 140 named species. Pleistocene and Holocene climatic fluctuations resulted in substantial vertical shifts of vegetation zones. Much of the modern páramo is heavily influenced by human activity.

The Rocky Mountain region in southern Colorado includes several distinct mountain ranges and intervening valleys, in which elevations range from around 2300 m to over 4000 m. Climatic conditions and vegetation cover vary greatly over short distances. Mires and fens are found in moraine potholes of glaciated valleys above 3000 m in the Culebra Range. These montane wetlands display diverse histories depending on when peat began to accumulate, drought episodes, and beaver activity. The San Luis Valley is a desert environment that surprisingly enjoys considerable surface and ground water derived from surrounding mountain ranges. Many wetland habitats are protected in a national park, three national wildlife refuges, and several state wildlife areas. The modern abundance and diversity of waterfowl and shorebirds in the San Luis Valley are similar to pre-European accounts.

The impacts of climate change may have far-reaching consequences for the polar regions of the world. Wetlands and dependent populations in these fragile high-latitude environments face unique challenges. Ice and permafrost play significant roles in molding these landscapes, while short summer months allow for surface melting and support a tremendous diversity and density of species. The Arctic Coastal Plain, the Yukon Delta in Alaska and the Lena Delta in Russia have landscapes with thousands of lakes, water-filled troughs and shallow depressions, which host migratory nesting birds by the millions during the summer. Large mammals like the caribou and important fish species, such as the salmon, play vital roles in the cultural traditions and economic livelihoods of resident native communities. Changing climate patterns and warming trends may influence ecosystem shifts across the Arctic. As in the past, the adaptive capacities of resident communities will be tested in the decades to come.

Sustainability for wetlands 18

18.1 Introduction

The term "sustainability" suggests using resources and ecosystems in such a way that ensures their long-term viability. Sustainable development built on this notion refers to a redirected approach toward future economic development and growth (World Commission on Environment and Development 1987). Most critically, it links long-term human well-being and economic stability to environmental health and maintaining the Earth's biogeochemical systems (Clark and Levin 2009).

In 1987, the Brundtland Report by the World Commission on Environment and Development (WCED) defined sustainable development as that which "meets the needs of the present without compromising the ability of future generations to meet their own needs" (WCED 1987, p. 45). Although there is still much disagreement about how this goal may be met, this succinct definition of sustainable development is generally accepted. The ideas put forth in the Brundtland Report suggested a path to development, and one that resonated with politicians and policy makers across the globe. So much so that today the aim of encouraging sustainable societies is discussed in mainstream public debate and is considered a realistic and reachable goal.

The last five decades have witnessed an improved and expanded understanding of wetland ecosystem structure, process, function, and management. Today, wetlands are recognized for the important contributions they make to the Earth's biophysical and ecological cycles and the valuable services they provide to human socio-economic systems. Many key wetlands are now protected in various ways and to differing degrees (Fig. 18-1). Nonetheless, the pattern of resource use, materials consumption and population growth since the Industrial Age have placed severe strains on the Earth's environment, wetlands included.

Despite the considerable adaptive capacity of Earth systems to natural or anthropogenic stressors, the rapid rate and large scale of environmental modifications have resulted in greater ecosystem vulnerability to changes and a reduced ability to mitigate impacts. In the face of such challenges, how do societies foster the application of sustainability principles to environmental and wetland use? Clearly much remains to be done. Yet, improved understanding of how biogeochemical and socio-economic systems function and interact provides clues that help navigate the path of sustainability.

In 2000, the then Secretary-General of the United Nations, Kofi Annan commissioned the Millennium Ecosystem Assessment to determine the current state of knowledge regarding the Earth's ecosystems, document the services they provide, and outline policies to more effectively and sustainably manage ecosystems and resources (Millennium Ecosystem Assessment 2011). Contributions from more than 1300

Wetland Environments: A Global Perspective, First Edition. James Sandusky Aber, Firooza Pavri, and Susan Ward Aber.
© 2012 James Sandusky Aber, Firooza Pavri, and Susan Ward Aber. Published 2012 by Blackwell Publishing Ltd.

Figure 18-1. Superwide-angle view looking southward over the Atlantic coast and salt marsh at Moody Division, Rachel Carson National Wildlife Refuge, Maine, United States. Note the dense recreational and residential development along both margins of the marsh; however, access to the marsh itself is strictly controlled. Blimp aerial photo by J.S. Aber, S.W. Aber and V. Valentine.

scientists and experts working in subgroups from different countries across the world arrived at a comprehensive documentation of current trends in ecosystem health. The assessment identified key problem areas that required urgent attention and offered policy responses and management strategies for adoption. Wetlands were one of the focus areas of the Millennium Ecosystem Assessment (Millennium Ecosystem Assessment 2005). Key findings of this report endorsed a focus on integrated ecosystem-scale approaches to wetland conservation. They also emphasized addressing the drivers of wetland change while at the same time encouraging Ramsar-supported wise-use approaches to management (see also chapter 12).

18.2 Key risks to wetlands

Despite the global focus on the environment and the renewed attention to sustainable development, wetlands face urgent threats and increasing pressures in the twenty-first century. These threats pose a greater problem across developing nations where increasing demand and limited resources heighten the vulnerability of already stressed ecosystems. They require immediate attention from scientists, land managers, stakeholders, policy makers, and relevant

Figure 18-2. Schematic illustration of key risks for wetlands.

institutions. The integration of collective knowledge and collaborative action could make a difference to the long-term viability of this ecosystem. Some of the risks to wetlands are specified below (Fig. 18-2).

Population increase – The world's population is expected to exceed nine billion people by 2050 from the current total of seven billion (United Nations Population Fund 2011). It is, however, in the developing world where the greatest expansion will take place. By 2050, more than 80% of the world's population will reside in developing countries. Agricultural efficiencies and advances in related technologies will be needed to feed expanding populations. Recent history indicates, however, that population size is not the sole driver of environmental change and degradation. If anything, it was the prosperity and wealth of industrialized nations and their drive for economic growth that led to significant regional and global environmental damage throughout most of the twentieth century. However, more stringent environmental protection policies across the developed world have helped ameliorate some of the past errors (Fig. 18-3).

Figure 18-3. Overview of landfill in southern Poland. Former landfill (top), recycling station (center), and active landfill (left). Such careful management of human trash and waste reduces pollution of wetlands in the vicinity. Kite airphoto by J.S. Aber, S.W. Aber, R. Zabielski and M. Górska-Zabielska.

As the developing world similarly pursues the race for industrialization and economic growth, it remains to be seen whether past mistakes may be avoided. Through global efforts such as the U.N. Millennium Project, which focuses on poverty reduction and education, governments may be urged to adopt a longer-term vision and to embrace the concept of sustainable development (Millennium Project 2011). Likewise, advances in technology and technology transfers may allow countries to leap-frog older polluting technologies for greener alternatives. The fundamental linkage of environmental health to human well-being and a country's development aspirations may aid in reorienting policy to incorporate sustainable development principles.

Resource demands – The demand for fresh water and land for agricultural and urban development remain key threats to wetland environments across the globe. In the past, deliberate conversions to other land uses were driven by a lack of understanding of the role wetlands play within the Earth's system. Despite the current state of scientific knowledge about wetlands, the demand for agricultural land, urban expansion and industrial development often supersede conservation priorities. Mire drainage is still being carried out in many places to enhance forestry, expand agriculture, and produce peat. As examples from chapter 11 illustrate, increased resource extraction would further stress wetland sites and make them acutely vulnerable to change.

Poverty – Poverty can be an influential driver in wetland habitat degradation, often working in tandem with other local socio-economic factors including inequality, landlessness, and the social and economic marginalization of rural populations. Under such circumstances, human vulnerability to even small disturbances within the natural system may have acute consequences. Extreme poverty drives the overuse of resources and limits the ability to reinvest in habitat protection or sustainable harvesting practices. Such practices may perpetuate and undermine local economies as natural capital is systematically depleted. While poverty alleviation is a key focus of the U.N. Millennium Project, it remains to be seen how much progress will be made toward such a goal (Millennium Project 2011).

Climate change – Climate oscillations have impacted and shall continue to influence the structure and function of wetlands (Fig. 18-4; see chapter 8). In many cases, wetlands may need to adapt more quickly to rapidly changing conditions brought on by anthropogenic influences in climate change. Changing climatic patterns driven by rising global temperature could influence freeze–thaw cycles, affect variability in precipitation, increase the risk of inundation, trigger seasonal increases in sea-surface temperature, induce latitudinal or altitudinal shifts in plant and animal species, and increase the frequency of extreme storm events – to name just a few. The influence of each of these factors on wetlands would vary geographically and in magnitude. Moreover, the adaptive capacity of wetlands in the face of such changing conditions is also influenced by local human patterns of use.

Institutional and governance failures – As chapter 11 elaborates, property rights institutions direct the use of resources, including wetlands. Those habitats with clearly defined

Figure 18-4. During the Little Ice Age in Norway, Jostedalsbreen (ice cap on far horizon) expanded and outlet glaciers descended into its surrounding valleys. Many farms, such as the one to the right, had to be abandoned because the climate turned so cold, and the local population suffered great poverty (Grove 1988). Glacier advances reached a maximum c. AD 1750; rapid glacier retreat took place after about 1850 (Aarseth et al. 1980). Vanndalstølen in foreground; Jostedal (valley) in middle distance. Photo by J.S. Aber.

and enforceable property and tenure rights, whether based on private, state or common property systems, have demonstrated their ability to balance ecosystem and human uses and remain viable systems. Integrating principles of sustainability into resource management practices could prove effective in supporting the long-term health of an ecosystem.

In the face of extreme pressures and competing demands for a wetland site, the lack of clearly defined institutional rules of resource use, access, and extraction, as seen in open-access systems, may result in overuse and degradation. Likewise, the lack of adequate environmental governance at the local, state or federal levels may also lead to "wild west" exploitation and resource degradation. Corrupt governance systems lacking transparency and accountability provide few incentives for sustainable resource management.

Shared wetland challenges – Many of the world's great wetlands, including the Pantanal, Okavango, Sundarban and prairie pothole region straddle country borders (see chapters 14 and 15). In addition, upstream uses of a river and water diversions within one country can influence the health of downstream floodplain and deltaic habitats in another. Such instances of shared wetlands are numerous around the globe. As the demand for fresh water becomes more acute, conflicts over access and use are likely to become more common. Conservation and management in such instances is confounded by regional politics. Equitably formulated joint agreements may help in the shared governance of such resources. In other cases, the involvement of international or non-governmental organizations might become necessary to ensure equity of access and use.

Wetland degradation – Pollution and habitat fragmentation heighten the vulnerability of wetlands to further degradation (see also chapter 10). Such stressors reduce the adaptive capacity of ecosystems and make them particularly susceptible to even small threats. Stricter pollution control mechanisms could help in such instances, but excessive nutrient loading from non-point source pollutants is more difficult to control and legislate. Habitat contiguity and intactness are necessary for the preservation of biodiversity and the long-term viability of resident species. Yet, wetland habitat fragmentation due to land-use conversions, the demand for resources, and encroachment are still commonplace across the globe.

18.3 Key opportunities in wetland conservation

Since the signing of the Ramsar Convention on Wetlands four decades ago, wetland conservation has matured into a fully developed approach championed by powerful international and national interests. Some of the key scientific developments of the past few decades could help propel wetland management and conservation forward. Here, we highlight some of the key opportunities and challenges in wetland conservation and management (Fig. 18-5).

Integrated human-natural systems approach – Recognizing the fundamental connection between human well-being and environmental health suggests that elements of these systems cannot be studied in isolation from each other. A systems approach allows for an integrated and holistic understanding of the biogeochemical and socio-economic aspects of wetlands, their availability, use and allocation. Such approaches may be complex and require multi-disciplinary teams of academics, practitioners and policy makers. Yet they are also critical to understanding the drivers of wetland change, identifying interactions between different components of the system at different spatial and temporal scales, and recognizing key vulnerabilities within systems.

Knowledge–action linkages – Linking scientific and practitioner *knowledge* of environmental systems and human–environment exchanges with *action* to conserve and manage ecosystems needs better coordination. Work within the sustainability sciences recognizes the importance of such linkages (Clark and Levin 2009). Knowledge–action linkages may be iterative, responsive, and creative to reflect changing conditions and new information gleaned from field experiences or data monitoring efforts.

Flexible institutions – Under many anticipated climate-change scenarios, set patterns of human–environment interaction within wetland sites are bound to face challenges. Institutions of environmental governance must be flexible enough to respond to such

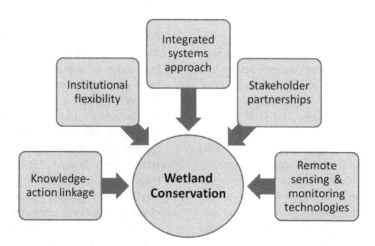

Figure 18-5. Schematic illustration of key opportunities and challenges in wetland conservation and management.

Figure 18-6. During a recent 5-year period, Dry Lake has fluctuated between a full body of water to completely dry salt- and mudflats (see also Color Plate 16-10). Seen here with moist, salty mudflats; highly saline ground water is just below the surface. Overview from eastern end looking westward; kite airphoto by J.S. Aber, G. Corley and B. Zabriskie, May 2011, western Kansas, United States.

unexpected perturbations. The adaptive capacity of institutions must be strengthened in anticipation of such changing conditions.

Stakeholder partnerships – Engagement with stakeholder groups and their participation in planning and implementation are critical to long-term successes in wetland management (see also chapter 12). Their participation ensures that wetland management and conservation will be relevant to local needs. Democratizing and devolving the process of conservation to the local level would provide legitimacy to a plan and give a greater chance of success.

Remote sensing technologies – Throughout this book we have used aerial photographs and satellite imagery to demonstrate the value of this important source of information for wetland investigations. Visual information is holistic and offers the means to make temporal and spatial connections. These datasets help monitor changing wetland conditions (Fig. 18-6). They also provide information on wetland function and health (see also chapter 3). Advances in this technology would improve our understanding of these unique and complex ecosystems and provide clues for their more efficient future management.

18.4 Future directions

This chapter outlines some threats, challenges and opportunities that wetland environments face in the next half century. It is important to recognize that at no time in the past have different facets of wetland environments been so well studied and understood. Today, wetlands are recognized globally as ecosystems vital to Earth-system functions and for supplying critical services. Furthermore, the renewed focus on wise use and sustainability translates into more institutional resources and greater public attention and awareness of wetlands. In other words, there has never been a more opportune time to engage in wetland conservation and sustainable management.

The history of environmental conservation has shown that in most cases, societies have been *re-active* in their approach. In other words, it takes serious environmental damage before public sentiment and governments are motivated to act. In this age of sustainability, public opinion is perhaps finally changing to adopt a *pro-active* approach or one where we act to protect ecosystems before serious damage occurs. Water is essential to life, and wetlands are essential for water.

Glossary of wetland types and terms

This glossary includes common names as well as some specialized or local terms for wetlands and their physical, chemical, biological, or cultural conditions. Names of individual plant and animal species are not included, however. Neither are names for various governmental agencies and most private organizations. The definitions are based mainly on Neuendorf, Mehl and Jackson (2005) and Mitsch and Gosselink (2007), as well as many other sources. See index for additional names, terms, and geographic locations.

Aapa – Mire in high latitude or altitude in which gently sloping (<1°) land forms narrow ridges of peat that are separated by deep pools (flarks). Ridges and pools are elongated perpendicular to slope. Also known as ribbed fens, patterned fens, and string bogs. Typically found in boreal or subarctic settings, south of palsa peatland. Etymol: Sami.

Acequias – Spanish term for a ditch-and-canal system used for irrigation; specifically a commonly owned and operated irrigation system in the San Luis Valley, Colorado, United States.

Acid rain – Precipitation with pH <5.

Acrotelm – Upper portion of the soil, in which oxygen is present and decay is relatively rapid. Zone generally above the long-term average water table.

Adaptive management – Structured process, approach or plan that allows flexibility and change based on new information or conditions.

Aerenchyma – Porous tissue with large, air-filled spaces between cells of stems and leaves that allows oxygen and methane to diffuse between submerged aquatic plant roots and the air and provides buoyancy.

Aerobic – Referring to the presence of oxygen or an oxidizing chemical environment.

Albedo – Reflectivity of the Earth's surface and atmosphere for visible light and invisible radiation.

Allogenic – Refers to wetland changes induced by external forcing factors, which are of sufficient magnitude to influence or alter internal processes.

Altithermal – Climatic optimum of warmer-than-present conditions that existed in the middle Holocene, ~8000 to 4000 calendar years ago. Also known as the hypsithermal or Holocene thermal maximum (HTM).

Amber – Fossil resin derived primarily from extinct conifer and evergreen trees that may contain extinct animal and plant remains, the most common being winged insects.

Amphibious plants – Those plants that are able to live both on land and in water and may grow as either submerged or emergent types. The emergent morphology may be quite different from the submerged plant form.

Anadromous fish – Species that spend part of their lives in fresh water, usually to spawn and hatch, and part of their lives in the open ocean.

Anaerobic – Referring to the absence of oxygen or a reducing chemical environment.

Anaerobic glycolysis – Biochemical processes that allow some plants to convert food into energy without oxygen.

Aquifer – Rock, sediment or soil that stores and transmits ground water in economically useful quantity.

Aquaculture – Farming of aquatic organisms such as fish, crustaceans and mollusks in confined or controlled environments for food, shells, pearls, or other products.

Artesian – Flowing well in which water rises to the surface under its own pressure; applies to human-made wells drilled into the ground, not natural springs.

Atlantic period – Middle Holocene interval of warm, mild climate and climax forest development in northern Europe representing the same interval as the Altithermal or Holocene thermal maximum (HTM).

Atmosphere – Gaseous outer portion of the Earth composed mainly of nitrogen and oxygen with many other trace gases.

Autogenic – Refers to changes that take place within a wetland habitat caused by natural growth and evolution due to internal forcing factors.

Barrier island – Long, narrow, low island constructed mainly of sand fronting the open ocean on one side and separated from the mainland by a shallow lagoon or bay on the other side.

Bayou – Term in lower Mississippi River basin that generally applies to a small stream or secondary tributary to a larger water body, especially a sluggish or stagnant stream that follows a winding course through alluvial lowland, coastal swamp or marsh, river delta or tidal estuary. Etymol: French, *boyau* – passageway; Choctaw, *bayuk* – small stream.

Beel – Expansive low-lying seasonally flooded area adjacent to major floodplains in eastern India and Bangladesh.

Billabong – Australian term for a riparian wetland that is periodically flooded by the adjacent stream or river.

Biocontrol – Using one species to control or limit another species.

Biogeochemical – Refers to the combined relationships and interactions between biological, geological and chemical components of the Earth's environmental system.

Biosphere – Living organisms on or near the Earth's surface; all microbes, plants and animals of the air, land, and sea.

Bitter salt – Various potassium chloride and other extremely soluble salts that have a bitter taste.

Blanket bog – Growing moss and peat layer that drapes over the landscape.

Blytt-Sernander – Sequence of late glacial and Holocene pollen/climatic intervals, particularly in northern Europe, including the Boreal, Atlantic, Sub-boreal and Sub-atlantic periods.

Bofedales – Peruvian term for **puna** peatland (see below). Characterized by vascular plants forming large cushions and utilized for livestock grazing. Also known as **oconales** (Salvador, Monerris and Rochefort 2010).

Bog – Waterlogged, spongy ground, consisting primarily of mosses and decaying vegetation that may develop into peat. Receives only direct precipitation; characterized by acidic water, low alkalinity, and low nutrients.

Bog body – Human mummy preserved by burial and natural tanning processes in a bog.

Bolson – Deserts of southwestern U.S. Extensive, flat, alluvial basin or depression with internal drainage from surrounding mountains. Playa or temporary lake at center, usually saline.

Boreal – High northern latitudes, specifically those regions nominally covered by conifer forest, particularly spruce forest. Also the name for an early Holocene interval in which climatic amelioration took place following retreat of the last great ice sheets, particularly in Europe.

Bottomland – Lowland along streams and rivers, usually alluvial land, that is periodically flooded.

Brackish – Water salinity intermediate between normal marine and fresh water; typical of coastal and tidal wetlands as well as inland saline lakes.

Carr – Term used in Europe for a forested wetland characterized by alder (*Alnus*) and willow (*Salix*).

Catotelm – Lower portion of soil and underlying sediment in which oxygen is lacking and decay is relatively slow. Zone generally below the long-term average water table.

Chinampa – Method of building artificial islands in shallow lakes in ancient Mesoamerica; used for growing crops and flowers. Eytmol: Nahuatl word chinamitl – square made of canes (*The Free Dictionary* 2011).

Coal – Sedimentary rock and fossil remains consisting primarily of carbon and formed by compaction and chemical alteration from vegetation remains. Varieties anthracite, bituminous coal, and lignite differ in density, carbon percentage, and energy content.

Cost-benefit analysis – Means of evaluating the economic costs and benefits of a development project.

Coteau – Hilly upland forming a drainage divide often with steeply sloping sides. Applied in the Great Plains of North America; for example, Coteau du Missouri and Coteau des Prairies. Eytmol: French – slope (*The Free Dictionary* 2011).

Crustal depression/rebound – Subsidence and uplift of the crust (lithosphere) as a consequence of loading and unloading during ice-sheet glaciation, which may amount to several 100 m.

Cryosphere – Frozen portions of the Earth's surface, including glaciers, ice sheets, permafrost, pack ice, perennial snowfields, etc.

Cumbungi swamp – Cattail (*Typha*) marsh in Australia.

Dambo – Seasonally waterlogged and grass-covered linear depression located in headwater zone of rivers with no marked stream channel or woodland vegetation. Eytmol: ChiChewa (central Africa) – meadow grazing.

Delta – Wetland-river-upland complex located where a river forms distributaries as it merges into a sea or lake. The delta is built of sediment delivered by the river; sediment supplied by river is greater than can be redistributed along the coast or offshore. Also inland deltas such as the Peace–Athabasca Delta in Canada and Okavango Delta in Botswana.

Dhap – Floating cultivation on mats of dead aquatic vegetation in wetlands of Bangladesh.

Drainage tile – Buried drainage pipes designed to lower water table through gravity flow typically installed in agricultural fields to promote crop growth.

Duck stamp – Popular name for the U.S. Migratory Bird Hunting and Conservation Stamp, which is required for hunting migrating waterfowl in the United States and is also required for entering certain U.S. National Wildlife Refuges.

Ecotourism – Travel for the purpose of viewing natural environments and wildlife, bird and whale watching, etc.

Eemian – Refers to the high stand of sea level during the last interglacial period, ~130,000 to 120,000 years ago. European term equivalent to Sangamon in North America.

Estuary – Where a river flows into the sea in a deep or wide valley without sufficient sediment to build a delta. Fresh and salt water mix with intermediate salinity, and tides ebb and flow within the estuary system.

Eustasy – Refers to worldwide sea level and global changes in sea level.

Eutrophic – Refers to wetland or water with a high level of nutrients (phosphate, nitrate) with high primary productivity.

Evaporite – Deposits resulting from the evaporation of saline water. Highly varied mineral types depending upon chemical composition and concentration of evaporating water body.

Evapotranspiration – Combined water loss due to evaporation from water bodies and exposed soils as well as transpiration through plant bodies.

Externality – Positive or negative byproducts of production or exchange that are not part of the economic market supply and demand.

Facultative – Refers to hydrophytic vegetation that often grows in wetland conditions (34–66%). Also may refer to the transition interval in soils between aerobic and anaerobic zones.

Fadama – Floodplains underlain by shallow aquifers along Nigerian river systems; often developed for irrigated agriculture.

Feedback – Relationships between phenomena, such that a change in one leads to a change in another, which may be positive (reinforcing) or negative (damping).

Fen – Waterlogged, spongy ground, containing alkaline decaying vegetation, characterized by reeds, sedge, shrub or forest that may form peat, sometimes occurs in sinkholes of karst regions. Receives some surface runoff and/or ground water, which has neutral pH and moderate to high nutrients.

Floodplain – Overbank portion of alluvial or tidal environments that may be flooded frequently or rarely when water flows exceed channel capacity.

Fossil fuel – Ancient (buried) carbon- and hydrocarbon-rich sediment (coal, oil shale) or fluid (oil, natural gas) that may be burned to provide energy.

Geobotanical – Refers to mapping and analysis of landscapes based on vegetation distribution as a primary indicator of topography, soils, landforms, sediment, and underlying bedrock.

Glossopteris – Generic name given to temperate flora common across Gondwana during the late Paleozoic.

Gondwana – Ancient megacontinent consisting of modern South America, Africa, Arabia, India, Antarctica and Australia during the Paleozoic. It made up the southern portion of Pangaea.

Greenhouse gases – Gases (CO_2 and CH_4) that absorb thermal radiation emitted from the Earth, thus warming the lower atmosphere. The atmospheric content of greenhouse gases is one factor influencing the Earth's climate.

Halophyte – Plant that tolerates high salt levels in saline soils or hypersaline water.

Histosol – Soil order composed largely of organic materials derived from plant tissues, including peat and muck soils.

Hoar – Inland wetland system in Bangladesh and eastern India that is characterized by deep, bowl-shaped depressions that contain water most of the year.

Holocene – Last 10,000 years of Earth history. The youngest epoch of the geological time scale.

Holocene thermal maximum (HTM) – (see Altithermal).

Hydraulic conductivity – One factor that determines the rate at which ground water is able to move depending upon the size, shape, and connections of pore spaces within the sediment or rock body. Permeability (k) of the sediment or rock.

Hydric soil – Refers to wetland soils that were formed under conditions of saturation, flooding or ponding long enough during the growing season to develop anaerobic conditions in the upper part (NRCS 2010a).

Hydrograph – Graphical display of stream discharge or stage versus time. Primary tool for analyzing stream flow and flooding.

Hydroperiod – Seasonal pattern of water level in a wetland, as determined by climate and human management.

Hydrophyte – Plants that grow only in standing water or saturated soils, such as moss, sedges, reeds, cattail and horsetail, mangroves, cypress, and rice.

Hydroseral succession – Gradual transition involving both water and vegetation beginning with open-water habitat and ending with a raised bog or forest cover.

Hydrosphere – Liquid water at the Earth's surface, including the ocean, lakes, rivers, soil moisture, and shallow ground water.

Hypoxic – Water bodies that contain <2 mg/L dissolved oxygen, which results in fish kills and dead zones, especially in coastal marine waters.

Hypsithermal – (see Altithermal)

Invasive species – Exotic or foreign species whose introduction may cause environmental or economic harm.

Irrigation – Diversion or extraction of surface or ground water for agricultural purposes, namely watering dry land for growing annual or perennial crops.

Isostasy – Density-mass balance within the crust and lithosphere that determine the general level (elevation) of the Earth's surface.

Karst – Landforms created by bedrock solution, including caves, sinkholes and related features.

Kavir – Term for a salt desert, specifically the Great Kavir of central Iran. Series of closed basins noted for marshy conditions and high salinity.

Lacustrine – Refers to lakes including the water body, sediment, plants and animals.

Lagoon – Deepwater enclosed or partially opened aquatic system, especially in coastal delta regions.

Lake Agassiz – Great proglacial lake in north-central North America that developed as the last ice sheet retreated from the region. Considered to be one of the largest fresh-water lakes that every existed, it covered substantial parts of the northern Great Plains in Canada and the United States.

Laurasia – Ancient megacontinent consisting of modern North America and Eurasia during the Paleozoic. It made up the northern portion of Pangaea.

Lignite – Soft, brown coal with density and carbon content between peat and bituminous coal.

Lithosphere – Solid, outer portion of the Earth, including the upper mantle and crust, averaging about 100 km thick, thinner under oceans and thicker beneath continents.

Littorina Sea – Stage in development of the Baltic Sea when water level rose and the sea transgressed over surrounding coastal areas during the middle Holocene.

Little Ice Age – Prolonged period of cold temperature, c. AD 1200 to 1900, during which the Earth's climate was 1–2 °C lower than today (Grove 1988).

Mangrove – Subtropical and tropical marine coastal ecosystem dominated by halophytic trees and shrubs growing in brackish to saline water. Also called mangal.

Marsh – Water-saturated, poorly drained area, intermittently or permanently covered by water, having aquatic and grasslike vegetation. In Europe, must have a mineral substrate and lack peat accumulation.

Maunder minimum – Time interval of reduced solar output during which sun spots all but disappeared, AD 1645–1715, that corresponded with the Little Ice Age.

Medieval climatic optimum – Time period of anomalously warm conditions, ca. AD 700–1200, known from historical records in Europe and geological evidence elsewhere.

Mere – Lake, pond or pool. A marsh or fen. A levee lake behind a barrier of sediment carried upstream by the tide. Mainly British regional usage.

Mesotrophic – Refers to wetland or water having an intermediate level of nutrients, in between eutrophic and oligotrophic.

Minerotrophic – Peatland whose water originated from mineral soils or rocks, but sometimes via lakes, streams or ground water. It may be eutropic, mesotrophic or oligotrophic.

Mire – Marshy, swampy or boggy ground. Collective term for all peat-forming ecosystems, including bogs, fens, and swamps of ombrotropic and minerotropic types. Used mainly in Europe.

Mitigation – Balancing wetland loss due to development or conversion by restoring, enhancing or creating another similar wetland elsewhere. Related to the "no net loss" wetland policy in the United States.

Monsoon – Climatic regime of certain continental areas characterized by heavy summer precipitation and relatively dry winter conditions, as in South Asia.

Moor – German term meaning peatland with similar usage in England.

Moraine – Glacial landforms created by deposition of till and other sediment.

Muck – Organic-rich soil consisting mainly of humus; organic matter has decayed beyond the point of recognition.

Munsell Color – Quantitative means to describe color based on hue, value and chroma (or saturation); widely used in soil science and other disciplines.

Muskeg – Large expanse of peatland or bogs, a term used particularly in Canada and Alaska.

Nacre – Secretion of aragonite and conchiolin layers that form pearls and mother of pearl in mollusk shells.

Neoglaciation – Last four millennia in which global climate cooled and glaciers expanded compared with the preceding mid-Holocene thermal maximum.

Obligate – Refers to hydrophytic vegetation that grows only in wetland conditions (>99%).

Oligotrophic – Refers to wetland or water that has low level or deficiency of nutrients available for plant growth.

Ombrotrophic – Peatland that receives all of its nutrients exclusively from rain (including snow and atmospheric fallout), which makes nutrients extremely oligotrophic. Also named ombrogenous.

Oxbow – Abandoned river channel, usually curved or loop shaped, that may contain a lake, swamp or marsh.

Pakihi – Peatland in southwestern New Zealand dominated by sedges, rushes, ferns, and scattered shrubs. Usually found on terraces or plains of glacial or fluvial outwash; acidic and exceedingly infertile.

Paleosol – Ancient, fossil or buried soil that formed under past climatic and biologic conditions that may be quite different from the modern environment.

Palsa – Peatland found near the southern limit of tundra region, consisting of large peat plateaus that are 20 to 100 m in width and length and up to 3 m high, generally underlain by permafrost. Also spelled paalsa. Etymol: Sami.

Paludification – Creation of wetlands by rising ground water which leads to saturation of the soil and ponding at the surface.

Palustrine – Non-tidal wetlands that are substantially covered with emergent vegetation including bogs, swamps, floodplains and marshes.

Palynology – Scientific study of pollen, its classification and environmental significance.

Panne (pan) – Bare, muddy spots within saltmarsh complex, normally above tidal range, but may be flooded occasionally.

Pangaea – Ancient supercontinent that included all modern continents during the late Paleozoic and early Mesozoic.

Pântano – Swamp or bog in South America. Etymol: Portuguese.

Páramo – Term specific to the Andes Mountains of Venezuela and Columbia referring to the wet alpine zone characterized by perennial cloud cover in which annual precipitation exceeds 2.5 m.

Pearl – Organic concretion formed within certain mollusks and often valued as gemstones.

Peat – Organic-rich soil characterized by anaerobic and acidic conditions and consisting primarily of partly decayed plant materials that are still recognizable. Varieties are fibric, hemic, and sapric, which are respectively least, partly, and most decomposed.

Peatland – All types of wetlands, including bogs, fens and swamps, in which peat has built up and continues to accumulate from partly decayed plant materials. Equivalent to mire.

Permafrost – Soil and ground that remains permanently frozen at depth; a shallow active layer may thaw during the summer. Typical of high-latitude regions.

Phreatophyte – Vegetation with deep roots that reach the water table; typical in semi-arid and arid riparian environments.

Plant hardiness zone – Geographic range of climatic conditions, primarily minimum winter temperature, that limit plant survival.

Playa (coast) – a) Small, generally sandy, land area at the mouth of a stream or along the shore of a bay. b) Flat, alluvial coastland, as distinguished from a beach. Etymol: Spanish – beach, shore, strand or coast.

Playa (geomorph) – Dry, vegetation-free flat area at the lowest part of an undrained depression underlain by stratified clay, silt or sand and commonly by soluble salts. May be marked by an ephemeral lake. See also salina, salar, salada. Syn: dry lake, vloer, sabkha, kavir, takir. Playa basin = bolson.

Pleistocene – Next to last epoch of the geologic time scale, ~2.6 million to 10,000 years ago. Popularly associated with the Ice Age, during which ice sheets, sea level, and climate underwent multiple cycles of glaciation and interglaciation.

Pneumatophore – Plant breathing roots that grow from saturated soil up into the air to allow for gas exchange.

Pocosin – Peat-accumulating, non-riparian, fresh-water wetland, generally dominated by evergreen shrubs and trees. Coastal plain of southeastern U.S. Etymol: Algonquin – swamp on a hill.

Pothole – Shallow, marshlike pond, particularly in the U.S. Dakotas and Canadian Prairie provinces; prairie pothole region.

Prop (stilt) roots – Root system that holds plant body above the ground or water surface, as in many mangroves.

Property regime – Types of property ownership, rights, access, and rules of use.

Puna – High-altitude zone above forest and below permanent snow in the Andes Mountains of Peru. Relatively dry grasslands that contain areas of peatland (see bofedales).

Raised bog – Growing moss and peat layer that stand above the surrounding terrain, such that no surface runoff or ground water enters the bog from outside.

Ramsar Treaty – Convention on Wetlands of International Importance, first adopted in Ramsar, Iran in 1971, and now with 160 member countries protecting >1900 sites worldwide (Ramsar Home 2011).

Raupo swamp – Cattail (*Typha*) marsh in New Zealand.

Reedmace swamp – Cattail (*Typha*) marsh in the United Kingdom.

Reedswamp – Marsh dominated by common reed (*Phragmites*), term used particularly in Europe.

Redoximorphic – Soil processes that cause alternating oxidation and reduction, particularly of iron, resulting in rusty mottles and streaks.

Restoration – Returning a degraded wetland as close as possible to pre-existing conditions.

Riparian – Refers to a zone or ecosystem with a high water table in close proximity to an aquatic environment, usually a stream or river.

Sabkha – Supratidal environment of deposition formed under arid and semiarid conditions on restricted coastal plains just above normal high-tide level. A saline marine marsh. Characterized by tidal flooding, evaporite-saline minerals, including dolomite, and aeolian deposits. Common around the Persian Gulf and the Sea of Cortez. Etymol: Arabic, also spelled sabkhah and sebkha.

Salada – Term in southwestern U.S. for a salt-covered plain where a lake has evaporated.

Salar – Term in southwestern U.S. and South America for a salt flat or salt-encrusted depression that may represent the basin of a salt lake.

Salina – Salt-encrusted playa or wet playa. Also refers to a body of saline water, such as pond, lake, spring, well or playa.

Salt marsh – Halophytic grassland on alluvial sediment bordering saline water bodies where water level may fluctuate either tidally or nontidally.

Sand sea – Geographically extensive tract of large and well-preserved or active sand dunes.

Sangamon – Refers to the last interglacial period, ~130,000 to 120,000 years ago, when sea level stood a few meters higher than today. North American term equivalent to Eemian in Europe.

Sedge meadow – Shallow wetland dominated by several species of sedges (e.g. *Carex, Scirpus, Cyperus*).

Shorebirds – Birds that typically walk on beaches, mudflats and wet meadows or wade in shallow water, such as sandpipers, herons, cranes, flamingos, etc.

Slough – Elongated swamp or shallow lake system, often adjacent to a river or stream. Slowly flowing shallow swamp or marsh in the southeastern U.S. (see bayou). Etymol: Old English, sloh – watercourse in a hollow.

Stromatolite – Reef built by cyanobacteria (blue-green algae) in shallow marine environments. Oldest known fossils; organisms are still extant in the world today.

Swamp – Low, waterlogged ground having shrubs or trees (U.S.), with or without formation of peat. In Europe, forested fens and reed (*Phragmites*) wetlands are also called swamps.

Takir – Russian term for clay-silt playa. Also spelled takyr.

Terrestrialization – Gradual sedimentary infilling of lakes and other small water bodies leading eventually to wetland habitats.

Tidal flat – Mud or sand flats along sea coasts that are alternately inundated and exposed by the flood and ebb tides.

Treatment wetland – Artificial wetland designed to filter, decontaminate, or clean sewage or polluted water.

Tundra – Treeless, level or undulating plain characteristic of Arctic and subarctic regions. Usually has a marshy surface that supports mosses, lichens, and low shrubs and is underlain by mucky soil and permafrost. Etymol: Sami.

Turlough – Areas seasonally flooded by karst ground water with sufficient frequency and duration to produce wetland characteristics. Generally flooded in winter and dry in summer, filling and emptying via underground passages. Specific to western Ireland.

Uniformitarianism – Principle that modern geologic processes are analogous to past processes that shaped the Earth.

Várzea – Portuguese term for nutrient-rich forests growing on floodplains adjacent to Amazonia rivers. The Várzea River is a tributary of the Uruguay River in southern Brazil.

Vegetation index – Ratio of red light to near-infrared reflectivity, which is an indication of the amount of photosynthetically active, emergent vegetation present in a scene, such as an aerial photograph or satellite image.

Vernal pond – Shallow, intermittently flooded wet meadow and small pools. Typically of Mediterranean climate; wetlands temporarily flooded during spring and dry in summer and fall.

Vlei(s) – Seasonal wetland(s) similar to dambo. Term used in Zimbabwe and other parts of southern Africa.

Vloer – South African term for flat surface with caked mud and high salt content, generally devoid of vegetation; resembles playa. Etymol: Afrikaans – floor.

Wad (pl. Wadden) – Unvegetated tidal flat on the North Sea coast in the northern Netherlands and northwestern Germany. The term is now used for coastal areas worldwide.

Water table – Level to which water rises in a well within an unconfined aquifer or soil.

Waterfowl – Aquatic birds that habitually swim or dive, such as ducks, geese, pelicans, etc.

Wet meadow – Grassland with waterlogged soil near the surface but without standing water for most of the year.

Wet prairie – Similar to a marsh, but with water level intermediate between a marsh and wet meadow.

Wetland – Area that is regularly wet or flooded and has a water table that stands at or above the land surface for at least part of the year during the growing season.

Wetlands Reserve Program – U.S. program authorized by the 1990 Farm Bill that pays farmers who restore and protect once-farmed wetlands in areas that are subject to periodic flooding.

Wrack – Dead vegetation, thatch, specifically material washed up and deposited during storm or high tide.

Zakaznik – Russian nature sanctuary, many of which were traditionally managed as game reserves.

Zapovednik – Russian nature preserve given the highest level of environmental protection to maintain wild conditions.

References

Aapala, K. and Aapala, K. 1997. Pohjois-Suomen murteiden suosanastosta (Mire words in the dialects of northern Finland). *Suo* 48/1, pp. 1–8.

Aarseth, I., Orheim, O., Holtedahl, H., Kjeldsen, O., Liestøl, O., Østrem, G. and Sollid, J.L. 1980. Glaciation and deglaciation in central Norway. Norsk Polarinstitutt, Field Guide to Excursion, 31 August– 3 September, 1980.

Aaviksoo, K., Kadarik, H. and Masing, V. 1997. Kaug- ja lähivõtteid 30 Eesti soost: Esimene raamat telmatograafiast. [Aerial views and close-up pictures of 30 Estonian mires: The first book on telmatography]. Tallinna Raamatutrükikoda, Tallinn, Estonia, 96 pp.

Aber, J.S. 1991. The glaciation of northeastern Kansas. *Boreas* 20, pp. 297–314.

Aber, J.S., Aaviksoo, K., Karofeld, E. and Aber, S.W. 2002. Patterns in Estonian bogs depicted in color kite aerial photographs. *Suo* 53/1, pp. 1–15.

Aber, J.S. and Aber, S.W. 2001. Potential of kite aerial photography for peatland investigations with examples from Estonia. *Suo* 52/2, pp. 45–56.

Aber, J.S. and Aber, S.W. 2005. 2005 visualization challenge: Photography. American Association for the Advancement of Science. Accessed online <http://www.sciencemag.org/content/309/5743/1991.2.full> January 2011.

Aber, J.S. and Aber, S.W. 2009. *Kansas physiographic regions: Bird's-eye views*. Kansas Geological Survey, Educational Series 17, 76 pp.

Aber, J.S., Aber, S.W., Buster, L., Jensen, W.E. and Sleezer, R.L. 2009. Challenge of infrared kite aerial photography: A digital update. *Kansas Academy of Science, Transactions* 112, pp. 31–39.

Aber, J.S., Aber, S.W., Pavri, F., Volkova, E. and Penner, R.L. 2006. Small-format aerial photography for assessing change in wetland vegetation, Cheyenne Bottoms, Kansas. *Kansas Academy of Science, Transactions* 109, pp. 47–57.

Aber, J.S., Aber, S.W., Janočko, J., Zabielski, R. and Górska-Zabielska, M. 2008. High-altitude kite aerial photography. *Kansas Academy of Science, Transactions* 111, pp. 49–60.

Aber, J.S., Aber, S.W., Manders, G. and Nairn, R.W. 2010. *Route 66 – Geology and legacy of mining in the Tri-state district of Missouri, Kansas and Oklahoma*. Geological Society of America, Field Guide 17, pp. 1–22.

Aber, J.S. and Ber, A. 2007. *Glaciotectonism. Developments in Quaternary Science* 6, Elsevier, Amsterdam, 246 pp.

Aber, J.S., Eberts, D. and Aber, S.W. 2005. Applications of kite aerial photography: Biocontrol of salt cedar (*Tamarix*) in the western United States. *Kansas Academy of Science, Transactions* 108, pp. 63–66.

Aber, J.S., Marzolff, I. and Ries, J.B. 2010. *Small-format aerial photography: Principles, techniques and geoscience applications*. Elsevier, Amsterdam, 266 pp.

Aber, J.S., Owens, L.C., Aber, S.W., Eddy, T., Schulenberg, J.H., Sundberg, M. and Penner, R.L. 2010. Recent *Azolla* bloom at Cheyenne Bottoms, Kansas. *Kansas Academy of Science, Transactions* 113, pp. 56–63.

Aber, J.S., Spellman, E.E., Webster, M.P. and Rand, L.L. 1997. Applications of Landsat imagery in the Great Plains. *Kansas Academy of Science, Transactions* 100/1–2, pp. 47–60.

Aber, J.S., Zupancic, J. and Aber, S.W. 2003. Applications of kite aerial photography: Golf course management. *Kansas Academy of Science, Transactions* 106, pp. 211–214.

Aber, S.W. and Kosmowska-Ceranowicz, B. 2001a. Bursztyn i inne żywice kopalne świata. Kredowe

żywice kopalne Ameryki Północnej: cedaryt (czemawinit), jelinit. *Polski Jubiler* 2/13, pp. 22–24.
Aber, S.W. and Kosmowska-Ceranowicz, B. 2001b. Kansas amber: Historic review and new description, in A. Butrimas (ed.), *Baltic Amber*, pp. 27–33. *Proceedings of the International Interdisciplinary Conference: Baltic Amber in Natural Sciences, Archaeology, and Applied Arts*. Vilnius Academy of Fine Arts Press, Vilnius, Lithuania.
Abraham, K. and Keddy, C. 2005. The Hudson Bay lowland, in Fraser, L.H. and Keddy, P.A. (eds), *The world's largest wetlands: Ecology and conservation*, pp. 118–148. Cambridge University Press, Cambridge.
Acheson, J.M. and Brewer, J. 2003. Changes in the territorial system of the Maine lobster industry, in Dolšak, N. and Ostrom, E. (eds), *The Commons in the New Millennium*, pp. 37–60. MIT Press, Cambridge, MA.
Adams, J. 1999. An inventory of data, for reconstructing "natural steady state" carbon storage in terrestrial ecosystems. Accessed online <http://www.esd.ornl.gov/projects/qen/carbon1.html> December 2010.
Alho, C. 2005. The Pantanal, in Fraser, L.H. and Keddy, P.A. (eds), *The world's largest wetlands: Ecology and conservation*, pp. 203–271. Cambridge University Press, Cambridge.
Alho, C.J.R., Lacher Jr., T.E., and Gonçalves, H.C. 1988. Environmental degradation in the Pantanal ecosystem of Brazil. *BioScience* 38, pp. 164–171.
Allikvee, H. & Masing, V. 1988. Põhja-Eesti kõrgustiku suurte mosaiiksoode valdkond, in Valk, K.U. (ed.), *Eesti sood* (Estonian peatlands), pp. 264–270. Valgus, Tallinn, 344 pp.
Allison, M.A. 1998. Geologic framework and environmental status of the Ganges–Brahmaputra delta. *Journal of Coastal Research* 14/3, p. 826.
Alongi, D.M. 2002. Present state and future of the world's mangrove forests. *Environmental Conservation* 29/3, pp. 331–349.
American Association for the Advancement of Science (AAAS) 1991. Background: Malaria in sub-Saharan Africa. Malaria and development in Africa: A cross-sectional approach. AAAS accessed online <http://www.aaas.org/international/africa/malaria91/background.html> September 2010.
American Museum of Natural History (AMNH) 2002. Marine pearls: Fossil pearls. American Museum of Natural History. Accessed online <http://www.amnh.org/exhibitions/pearls/marine/fossil.html> June 2011.
Anderson, J.B. and Rodriguez, A.B. 2008. Preface: Response of upper Gulf Coast estuaries to Holocene climate change and sea-level rise. *Geological Society of America, Special Paper* 443, pp. v–vi.
Anderson, M.K. 2006. Plant fact sheet: Narrowleaf cattail, *Typha angustifolia* L. U.S. Department of Agriculture, Natural Resources Conservation Service. Accessed online <http://plants.usda.gov/plantguide/pdf/cs_tyan.pdf> September 2010.
Anderson, R. 2009. *Castor canadensis* (Kuhl, 1820): American beaver. Encyclopedia of Life. Accessed online <http://www.eol.org/pages/328025> October 2010.
Anderson, R.C. 1988. Reconstruction of preglacial drainage and its diversion by earliest forebulge in the upper Mississippi Valley region. *Geology* 16/3, pp. 254–257.
Andrews, J.T. 1970. A geomorphological study of postglacial uplift with particular reference to Arctic Canada. Institute of British Geographers, Special Publication 2, 156 pp.
Animal and Plant Health Inspection Service (APHIS) 2005. Nutria. U.S. Department of Agriculture, Animal and Plant Health Inspection Service, Factsheet, October 2005. Accessed online <http://www.aphis.usda.gov/publications/wildlife_damage/content/printable_version/fs_nutria10.pdf> October 2010.
Anisimov, O.A., Vaughan, D.G., Callaghan, T.V., Furgal, C., Marchant, H., Prowse, T.D., Vilhjálmsson, H. and Walsh, J.E. 2007. Polar regions (Arctic and Antarctic), in Parry, M.L., Canziani, O.F., Palutikof, J.P., van der Linden, P.J. and Hanson, C.E. (eds), *Climate Change 2007: Impacts, Adaptation and Vulnerability. Contribution of Working Group II to the Fourth Assessment Report of the Intergovernmental Panel on Climate Change*, pp. 653–85. Cambridge University Press, Cambridge.
Anonymous 1997. Tidal wetlands in New York state. *New York State Conservationist* 51/6, p. 4.
Anundsen, K. 1985. Changes in shore-level and ice-front position in Late Weichsel and Holocene, southern Norway. *Norsk Geografisk Tidsskrift* 39, pp. 205–225.
Applegate, R.D., Flock, B.E. and Finck, E.J. 2003. Changes in land use in eastern Kansas, 1984–2000. *Kansas Academy of Science, Transactions* 106/3–4, pp. 192–197.
Arbor Day Foundation 2006. 2006 arborday.org Hardiness Zone Map. Accessed online <http://www.arborday.org/media/zones.cfm> September 2010.
Armour, J., Fawcett, P.J. and Geissman, J.W. 2002. 15 k.y. paleoclimatic and glacial record from northern New Mexico. *Geology* 30/8, pp. 723–726.
Arold, I. 2005. Eesti maastikud. Tartu Ülikooli Kirjastus, Tartu, Estonia, 453 pp.

Arruda, J.A. 1992. *Water in southeast Kansas*. Kansas Academy of Science, Guidebook 5, pp. 81–91. Kansas Geological Survey, Open-file Report 92-22.

Assine, M.L. and Silva, A. 2009. Contrasting fluvial styles of the Paraguay River in the northwestern border of the Pantanal wetland, Brazil. *Geomorphology* 113/3, pp. 189–99.

Australian Bryophytes 2008. Ecology: *Sphganum*. Australian National Botanic Gardens. Accessed online <http://www.anbg.gov.au/bryophyte/ecology-sphagnum.html> August 2011.

Australian Government 2010a. Wetlands. Accessed online <http://www.environment.gov.au/water/topics/wetlands/index.html> August 2010.

Australian Government 2010b. Murray-Darling Basin. Accessed online <http://www.environment.gov.au/water/locations/murray-darling-basin/index.html> August 2010.

Australia Report 2008. National report on the implementation of the Ramsar Convention on Wetlands: Australia. Accessed online <http://www.ramsar.org/pdf/cop10/cop10_nr_australia.pdf> August 2010.

Avery, T.E. and Berlin, G.L. 1992. *Fundamentals of remote sensing and airphoto interpretation* (5th edn), Macmillan, New York, 474 pp.

Avila-Serrano, G.E., Téllez-Duarte, M.A. and Flessa, K.W. 2009. Ecological changes on the Colorado River delta: The shelly fauna evidence, in Johnson, M.E. and Ledesma-Vázquez, J. (eds), *Atlas of coastal ecosystems in the western Gulf of California: Tracking limestone deposits on the margin of a young sea*, pp. 95–103. University of Arizona Press, Tucson, Arizona.

Azar, D., Gèze, R., El-Samrani, A., Maalouly, J. and Nel, A. 2010. Jurassic amber in Lebanon. *ACTA Geologica Sinica – English Edition* 84/4, pp. 977–983.

Azar, D., Prokop, J. and Nel, A. 2010. The first damsel fly from the Early Cretaceous Lebanese amber (Odonata, Zygoptera, Lestomorpha). *Alavesia: The Journal of the International Palaeoentomological Society* 3, pp. 73–79. Museo de Ciencias Naturales de Álava, Spain.

BackyardGardener 2010. Plant hardiness zone map. Accessed online <http://www.backyardgardener.com/zone/index.html> September 2010.

Banks, V. 1991. *The Pantanal: Brazil's forgotten wilderness*. Sierra Club Books, San Francisco, 254 pp.

Baroni, C. and Orombelli, G. 1994. Abandoned penguin rookeries as Holocene paleoclimatic indicators in Antarctica. *Geology* 22/1, pp. 23–26.

Bass, K. (ed.) 2009. *Nature's great events*. University of Chicago Press, Chicago, 320 pp.

Bean, D., Ortega, S. and Jandreau, C. 2009. Arkansas River tamarisk biocontrol project report 2009. Accessed online <http://arkwipp.org/pdf/Ark-River-Tamarisk-Biocontrol-Project-2009.pdf> June 2011.

Beckemeyer, R.J. and Huggins, D.G. 1997. Checklist of Kansas dragonflies. *The Kansas School Naturalist* 43/2. Accessed online <http://www.emporia.edu/ksn/v43n2-february1997/index.html> October 2010.

Beckemeyer, R.J. and Huggins, D.G. 1998. Checklist of Kansas damselflies. *The Kansas School Naturalist* 44/1. Accessed online <http://www.emporia.edu/ksn/v44n1-march1998/index.html> October 2010.

Bentall, R. 1990. Streams, in Bleed, A.S. and Flowerday, C. (eds), *An atlas of the Sand Hills. Conservation and Survey Division, Resource Atlas No. 5a* (2nd edn), pp. 93–114. Institute of Agriculture and Natural Resources, University of Nebraska-Lincoln.

Bertness, M.D. and Ellison, A.M. 1987. Determinants of pattern in a New England salt marsh plant community. *Ecological Monographs* 57/2, pp. 129–47.

Bertness, M.D., Ewanchuk, P.J. and Silliman, B.R. 2002. Anthropogenic modification of New England salt marsh landscapes. *Proceedings of the National Academy of Sciences* 99/3, pp. 1395–98.

Bickel, D.J. 2009. The first species described from Cape York Amber, Australia: *Chaetogonopteron bethnorrisae* n. sp. (Diptera: Dolichopodidae), in Berning, B. and Podenas, S. (eds), *Amber: Archive of deep time*, pp. 35–39. Verlag Bibliothek der Provinz, Weitra, Austria, 294 pp.

Birds & Nature 2009. North American migration flyways. Birds & Nature. Accessed online <http://www.birdnature.com/flyways.html> January 2011.

BirdLife International (BLI) 2010a. North Indian wetlands. Accessed online <http://www.birdlife.org/action/science/species/asia_strategy/pdf_downloads/wetlandsW12.pdf> October 2010.

BirdLife International (BLI) 2010b. BirdLife's flyway program. Accessed online <http://www.birdlife.org/flyways/index.html> October 2010.

Birkes, F., George, P., Preston, R., Hughes, A., Turner, J. and Cummins, B. 1994. Wildlife harvesting and sustainable regional native economy in the Hudson and James Bay Lowland, Ontario. *Arctic* 48, pp. 81–93.

Birks, H.J.B. and Seppä, H. 2010. Late-Quaternary palaeoclimatic research in Fennoscandia – A historical review. *Boreas* 39/4, pp. 655–673.

Blair, R.W. Jr. 1986. Karst landforms and lakes, in Short, N.M. Sr. and Blair, R.W. Jr. (eds), Geomorphology from space. *NASA Special Publication* 486, pp. 402–446.

Blatt, H., Middleton, G. and Murray, R. 1972. *Origin of sedimentary rocks*. Prentice-Hall, Englewood Cliffs, New Jersey, 634 pp.

Bleed, A. 1990. Groundwater, in Bleed, A.S. and Flowerday, C. (eds), *An atlas of the Sand Hills. Conservation and Survey Division, Resource Atlas No. 5a* (2nd edn), pp. 67–92. Institute of Agriculture and Natural Resources, University of Nebraska-Lincoln.

Bleed, A. and Ginsberg, M. 1990. Lakes and wetlands, in Bleed, A.S. and Flowerday, C. (eds), *An atlas of the Sand Hills. Conservation and Survey Division, Resource Atlas No. 5a* (2nd edn), pp. 115–122. Institute of Agriculture and Natural Resources, University of Nebraska-Lincoln.

Blossey, B. 2003. A framework for evaluating potential ecological effects of implementing biological control of *Phragmites australis*. *Estuaries* 26/2B, pp. 607–17.

Blum, M.D., Tomkin, J.H., Purcell, A. and Lancaster, R.R. 2008. Ups and downs of the Mississippi Delta. *Geology* 36/9, pp. 675–678.

Bonsteel, J.A. 1912a. Muck and Peat. U.S. Department of Agriculture, Bureau of Soils, Circular No. 65.

Bonsteel, J.A. 1912b. Marsh and Swamp. U.S. Department of Agriculture, Bureau of Soils, Circular No. 69.

Boojh, R., Patry, M. and Smart, M. 2008. Report on the UNESCO–IUCS mission to Keoladeo National Park, India. UNESCO World Heritage Report Mission Report 340. Accessed online <http://whc.unesco.org/en/list/340/documents/> October 2010.

Botkin, D.B. 1995. *Our natural history: The lessons of Lewis and Clark*. Grosset/Putnam Book, Putnam, New York, 300 pp.

Boule, M.E. 1994. An early history of wetland ecology, in Mitsch, W.J. (ed.), *Global Wetlands: Old World and New*, pp. 57–74. Elsevier, Amsterdam.

Brabets, T.P. and Schuster, P.F. 2008. Transport of water, carbon and sediment through the Yukon River Basin. U.S. Geological Survey, Factsheet 2008–3005, Anchorage, Alaska, 4 pp. Accessed online <http://pubs.usgs.gov/fs/2008/3005/pdf/fs20083005.pdf> March 2011.

Brabets, T.P., Wang, B. and Meade, R.H. 2000. Environmental and hydrologic overview of the Yukon River Basin, Alaska and Canada. U.S. Geological Survey, Water-Resources Investigations Report, 99-4204, 106 pp. Accessed online <http://ak.water.usgs.gov/Publications/pdf.reps/wrir99.4204.pdf> March 2011.

Bradsher, K. and Barboza, D. 2006. Pollution from Chinese coal casts a global shadow. *New York Times*. Accessed online <http://www.nytimes.com/2006/06/11/business/worldbusiness/11chinacoal.html> December 2010.

Brady, L.L. and Dutcher, L.F. 1974. Kansas coal: A future energy resource. *Kansas Geological Survey Journal*, 28 pp.

Bray, P. and Anderson, K.B. 2009. Identification of Carboniferous (320 million years old) class 1c amber. *Science* 326/5949, pp. 132–134. Accessed online <http://www.sciencemag.org/content/326/5949/132.abstract> June 2011.

Briggler, J.T. and Johnson, T.R. 2008. *Missouri's toads and frogs*. Missouri Conservation Commission, Jefferson City, Missouri, 32 pp.

Brinson, M.M. and Malvárez, A.I. 2002. Temperate freshwater wetlands: Types, status, and threats. *Environmental Conservation* 29/2, pp. 115–133

British Trust for Ornithology (BTO) 2011a. BTO History. Accessed online <http://www.bto.org/about-bto/history/> January 2011.

British Trust for Ornithology (BTO) 2011b. Breeding Bird Survey Methodology and Survey Design. Accessed online <http://www.bto.org/volunteer-surveys/bbs/research-conservation/methodology> January 2011.

Britton, A. 2009a. *Alligator mississippiensis* (DAUDIN, 1801). Crocodilians: Natural history & conservation, Florida Museum of Natural History. Accessed online <http://www.flmnh.ufl.edu/cnhc/csp_amis.htm> October 2010.

Britton, A. 2009b. *Crocodylus acutus* (CUVIER, 1807). Crocodilians: Natural history & conservation, Florida Museum of Natural History. Accessed online <http://www.flmnh.ufl.edu/cnhc/csp_cacu.htm> October 2010.

Bromley, D. 1991. *Environment and economy: Property rights and public policy*. Basil Blackwell, Cambridge, 247 pp.

Brown, J.E. and Huey, G.H.H. 1991. *Padre Island: The national seashore*. Southwest Parks and Monuments Association, Tucson, Arizona, 62 pp.

Bryan, J.R., Scott, T.M. and Means, G.H. 2008. *Roadside geology of Florida*. Mountain Press Publishing Co., Missoula, Montana, 376 pp.

Buol, S.W., Hole, F.D., McCracken, R.J. and Southard, R.J. 1997. *Soil genesis and classification* (4th edn), Iowa State University Press, Ames, Iowa.

Burch, B.L. 1995. Pearly shells: A pearling family in North America – Foragers, farmers, wholesale jewelers. *Hawaiian Shell News* 3, pp. 6–7. Accessed online <http://internethawaiishellnews.org/HSN/1995/9508.pdf> June 2011.

Burridge, M. and Mandrak, N. 2010. Freshwater Ecoregions of the World: Upper Yukon. Accessed online <http://www.feow.org/ecoregion_details.php?eco=102> October 2010.

Burrough, P.A. 1988. Principles of geographical information systems for land resources assessment (reprinted with corrections). *Monographs on soil and resources survey* 12. Oxford University Press, Oxford, 194 pp.

Burt, J.E., Barber, G., and Rigby, D.L. 2009. *Elementary Statistics for Geographers* (3rd edn), Guilford Press, New York, 654 pp.

Busman, L. and Sands, G. 2002. Agricultural drainage. University of Minnesota Extension Service report, BU-07740-S. Accessed online <http://www.extension.umn.edu/distribution/cropsystems/components/07740.pdf> November 2010.

Cairns, J. Jr. 1988. Restoration ecology: The new frontier, in Cairns, J. Jr. (ed.), *Rehabilitating damaged ecosystems*, Vol. 1, pp. 2–11. CRC Press, Boca Raton, Florida.

Callaghan, T.V., Björn, L.O., Chernov, Y.I., Chapin III, F.S., Christensen, T.R., Huntley, B., Ims, R., Johansson, M., Riedlinger, D.J., Jonasson, S., Matveyeva, N.V., Oechel, W., Panikov, N., Shaver, G., et al. 2005. Arctic tundra and polar desert ecosystems, in Symon, C., Arris, L. and Heal, O.W. (eds), *Arctic Climate Impacts Assessment*, pp. 243–351. Cambridge University Press, Cambridge. Accessed online <http://www.acia.uaf.edu/PDFs/ACIA_Science_Chapters_Final/ACIA_Ch07_Final.pdf> June 2011.

Camill, P., Barry, A., Williams, E., Andreassi, C., Limmer, J. and Solick, D. 2009. Climate-vegetation-fire interactions and their impact on long-term carbon dynamics in a boreal peatland landscape in northern Manitoba, Canada. *Journal of Geophysical Research* 114/4, G04017, 10 pp.

Campbell, I.D., Campbell, C., Apps, M.J., Rutter, N.W. and Bush, A.B.G. 1998. Late Holocene ~1500 yr climatic periodicities and their implications. *Geology* 26/5, pp. 471–473.

Campbell, R.W. (ed.) 1999. Iraq and Kuwait: 1972, 1990, 1991, 1997. Earthshots: Satellite images of environmental change. U.S. Geological Survey. Accessed online <http://earthshots.usgs.gov> December 2010.

Caputo, M.V. and Crowell, J.C. 1985. Migration of glacial centers across Gondwana during Paleozoic Era. *Geological Society America, Bulletin* 96/8, pp. 1020–1036.

Carson, R. 1962. *Silent Spring*. Houghton Mifflin, Boston.

Carter, V. 1997. Technical aspects of wetlands: Wetland hydrology, water quality, and associated functions. National Water Summary on Wetland Resources, U.S. Geological Survey, Water Supply Paper 2425. Accessed online <http://water.usgs.gov/nwsum/WSP2425/hydrology.html> October 2010.

Carter, V., McGinness, J.W. Jr. and Anderson, R.R. 1976. Wetland mapping in a large tidal brackish-water marsh in Chesapeake Bay, in Williams, R.S. and Carter, W.D. (eds), *ERTS-1 a new window on our planet*, pp. 286–289. U.S. Geological Survey, Professional Paper 929.

Casco Bay Estuary Project (CBEP) 2003. Presumpscot River: A plan for the future. Accessed online <http://www.cascobay.usm.maine.edu/pdfs/prmp_introduction.pdf> June 2011.

Castro, H., Ogram, A. and Reddy, K.R. 2004. Phylogenetic characterization of methanogenic assemblages in eutrophic and oligotrophic areas of the Florida Everglades. *Applied and Environmental Microbiology* 70/11, pp. 6559–6568.

Cater, E. 1994. Introduction, in Cater, E. and Lowman, G. (eds), *Ecotourism: A sustainable option?* pp. 3–17. Wiley, New York.

Center for Aquatic and Invasive Plants (CAIP) 2008. Mosquito fern. University of Florida, IFAS Extension. Accessed online <http://aquat1.ifas.ufl.edu/node/59> Oct. 2009.

Center for Russian Nature Conservation (CRNC) 2011. Ust-Lensky. Accessed online <http://www.wild-russia.org/bioregion1/Ust-Lensky/1_ustlen.htm> October 2011.

Centers for Disease Control (CDC) 2009. West Nile virus: Entomology. Centers for Disease Control and Prevention. Accessed online <http://www.cdc.gov/ncidod/dvbid/westnile/mosquitospecies.htm> September 2010.

Chakrabarti, R. 2009. Local people and the global tiger: An environmental history of the Sundarbans. *Global Environment: A Journal of History and Natural and Social Sciences* 3, pp. 73–95.

Chang, K.-T. 2008. *Introduction to geographic information systems*. McGraw-Hill, Boston, 450 pp.

Chapman, A.D. 2009. *Numbers of living species in Australia and the world*, 2nd edn, Australian Biological Resources Study, Canberra, Australia. Accessed online <http://www.environment.gov.au/biodiversity/abrs/publications/other/species-numbers/2009/pubs/nlsaw-2nd-complete.pdf> June 2011.

Charman, D. 2002. *Peatlands and environmental change*. J. Wiley & Sons, London and New York, 301 pp.

Chattopadhyay, S. 2010. Ecosystem in peril. CEERA, National Law School of India. Accessed online <http://www.ceeraindia.org/documents/sunderbans.htm> June 2011.

Chertok, M.A. and Sinding, K. 2005. Federal jurisdiction over wetlands: "Waters of the United States", in Connolly, K.D., Johnson, S.M., and Williams, D.R. (eds). *Wetlands Law and Policy: Understanding Section 404*, pp. 59–104. American Bar Association, Chicago.

Church, J.A. and White, N.J. 2006. A 20th century acceleration in global sea-level rise. *Geophysical Research Letters* 33, L01602, doi: 10.1029/2005GL 024826.

Cintron, G. and Novelli, Y.S. 1984. Methods for studying mangrove structure, in Snedaker, S.C. and Snedaker, J.G. (eds), *The Mangrove Ecosystem: Research Methods*, pp. 91–113. UNESCO, Paris.

Clague, J.J. 1989. Sea levels on Canada's Pacific coast: Past and future trends. *Episodes* 12/1, pp. 29–33.

Clark, W.C. and Levin, S.A. 2009. Toward a Science of Sustainability: Executive Summary of the 2009 Airlie Center Workshop, in Clark, C. and Levin, S.A. (eds), *Toward a science of sustainability*, pp. 7–10. Report from Toward a Science of Sustainability Conference. Princeton University, New Jersey.

Clements, F.E. 1916. *Plant succession: An analysis of the development of vegetation*. Carnegie Institute of Washington, Pub. 242. Washington, D.C.

Cleveland, C. 2008. Pocosins. *Encyclopedia of the Earth*. Accessed online <http://www.eoearth.org/article/Pocosins> January 2010.

Clewell, A., Rieger, J. and Munro, J. 2005. *Guidelines for developing and managing ecological restoration projects*, 2nd edn, Society for Ecological Restoration International, Tucson, Arizona, 16 pp. Accessed online <http://www.ser.org/pdf/SER_International_Guidelines.pdf> June 2011.

Clymo, R.S. 1984. The limits to peat bog growth. *Philosophical Transactions of the Royal Society of London B303*, pp. 605–654.

Cohen, J.E. and Small, C. 1998. Hypsographic demography: The distribution of human population by altitude. *Proceedings of the National Academy of Sciences, Applied Physical Sciences, Social Sciences* 95 pp. 14009–14014. Accessed online <http://www.pnas.org/content/95/24/14009.full.pdf> June 2011.

Coles, B. and Coles, J. 1989. *People of the wetlands: Bogs, bodies and lake-dwellers*. Thames and Hudson, New York, 215 pp.

Colin, M. 1996. Glacial rebound. *Harvard Magazine*, May–June, 1996, pp. 14–18.

Collins, J.T. and Collins, S.L. 2006. *Amphibians, turtles and reptiles of Cheyenne Bottoms* (2nd (revised) edn), U.S. Fish & Wildlife Service, 76 pp.

Collins, J.T., Collins, S.L. and Gress, B. 1994. *Kansas wetlands: A wildlife treasury*. University Press of Kansas, Lawrence, Kansas, United States, 128 pp.

Colona, R., Farrar, R., Kendrot, S., McKnight, J., Mollett, T., Murphy, D., Olsen, L. and Sullivan, K. 2003. Nutria (*Myocastor coypus*) in the Chesapeake Bay: A draft bay-wide management plan. The Chesapeake Bay Nutria Working Group. Accessed online <http://archive.chesapeakebay.net/pubs/calendar/marp_03-31-05_Handout_2_6079.pdf> October 2010.

Comer, R.P., Kinn, G., Light, D. and Mondello, C. 1998. Talking digital. *Photogrammetric Engineering & Remote Sensing* 64, pp. 1139–1142.

Cook, E., Bird, T., Peterson, M., Barbetti, M., Buckley, B., D'Arrigo, R., Francey, R. and Tans, P. 1991. Climatic change in Tasmania inferred from a 1089-year tree-ring chronology of Huon Pine. *Science* 253/5025, pp. 1266–1268.

Corday, A. and Dittrich, H. 2009. Amber – The Caribbean approach. *InColor, International Colored Gemstone Association*, Fall/Winter, pp. 1–6.

Costanza, R., d'Arge, R., de Groot, R., Farberk, S., Grasso, M., Hannon, B., Limburg, K., Naeem, S., O'Neill, R.V., Paruelo, J., Raskin, R.G., Suttonkk, P. and van den Belt, M. 1997. The value of the world's ecosystem services and natural capital. *Nature* 387, pp. 253–260.

Costard, F. and Gautier, E. 2007. The Lena River: Hydromorphodynamic features in a deep permafrost zone, in Gupta, A. (ed.), *Large rivers: Geomorphology and management*, pp. 225–233. John Wiley and Sons, Sussex, England.

Cowardin, L.M., Carter, V., Golet, F.C. and LaRoe, E.T. 1979. Classification of wetlands and deepwater habitats in the United States. U.S. Department of the Interior, U.S. Fish & Wildlife Service, Publication FWS/OBS-79/31.

Crean, K. 2000. Contrasting approaches to the management of common property resources: an institutional analysis of fisheries development strategies in Shetland and the Solomon Islands. *Australian Geographer*, 31/3, pp. 367–382.

Creighton, J.L. 1992. *Involving citizens in community decision making: A guidebook*. Program for Community Problem Solving, Washington, D.C., 223 pp.

Crozier, G.F. 2009. Continued redevelopment of the west end of Dauphin Island, Alabama – A policy review. *Geological Society of America, Special Paper* 460, pp. 121–126.

Curtis, F. 2007. Control of malaria vectors in Africa and Asia. Radcliffe's IPM World Textbook, University of Minnesota. Accessed online <http://ipmworld.umn.edu/chapters/curtiscf.htm> September 2010.

Cutter, S. and Renwick, W. 2004. *Exploitation, conservation and preservation: A geographic perspective on natural resource use*, 4th edn, Wiley and Sons, New York, 390 pp.

Dahl, T.E. 1990. *Wetland losses in the United States, 1780s to 1980s*. U.S. Department of the Interior, Fish and Wildlife Service, Washington, D.C., 21 pp.

Dahl, T. 2000. *Status and trends of wetlands in the conterminous US 1986–1997*. Fish and Wildlife Service, U.S. Dept. of the Interior, Washington D.C., 82 pp.

Dahl, T.E. 2006. *Status and trends of wetlands in the conterminous United States 1998 to 2004*. U.S. Department of the Interior; Fish and Wildlife Service, Washington, D.C. 112 pp.

Dahms, D. 1999. *Rocky Mountain wildflowers: Pocket guide*. Paragon Press, Fort Collins, Colorado, 123 pp.

Dame, R., Childers, D. and Koepfler, E. 1992. A geohydrological continuum theory for the spatial and temporal evolution of marsh-estuarine ecosystems. *Netherlands Journal of Sea Research* 30, pp. 63–72.

Davis, S.N. and DeWiest, R.J.M. 1966. *Hydrogeology*. John Wiley & Sons, New York, 463 pp.

Davit, J. and Cebrian, C. 2007. Victoria: The reigning queen of waterlilies. Fairchild Tropical Botanic Garden, Virtual Herbarium. Accessed online <http://www.virtualherbarium.org/GardenViews/victoriaamazonica.html> September 2010.

De Bord, D. 2009. *Alces alces* (Linnaeus, 1758): Moose. *Encyclopedia of Life*. Accessed online <http://www.eol.org/pages/328654> October 2010.

DeBusk, T. and DeBusk, W. 2001. Wetlands for water treatment, in Kent, D. (ed.), *Applied wetlands science and technology*, pp. 241–280 (2nd edn), Lewis Publishers, Boca Raton, Florida.

De Groot, R.S., Stuip, M.A., Finlayson, C.M. and Davidson, N. 2006. Valuing wetlands: Guidance for valuing the benefits derived from wetland ecosystem services. *Ramsar Technical Report No. 3/CBD Technical Series No. 27*. Ramsar Convention Secretariat, Gland, Switzerland & Secretariat of the Convention on Biological Diversity, Montreal, Canada, 54 pp.

Deil, U. 2005. A review on habitats, plant traits and vegetation of ephemeral wetlands – a global perspective. *Phytocoenologia*, 35/2–3, pp. 533–706.

DeLoach, C.J. Jr. 2004. *Tamarix ramosissima*. Crop Protection Compendium Database [CD-ROM]. CAB International, Wallingford, United Kingdom.

DeLoach, C.J. Jr. and Gould, J. 1998. Biological control of exotic, invading saltcedar (*Tamarix* spp.) by the introduction of *Tamarix*-specific control insects from Eurasia. Research proposal to U.S. Fish and Wildlife Service, 28 August 1998. 45 pp.

DeLoach, C.J., Jr., Knutson, A.E., Moran, P.J., Michels, G.J., Thompson, D.C., Carruthers, R.I., Nibling, F., Fain, T.G. 2007. Biological control of Saltcedar (Cedro salado) (*Tamarix* spp.) in the United States, with implications for Mexico, in Lira-Saldivar, R.H. (ed.), *Bioplaguicidas y Control Biologico. International Symposium of Sustainable Agriculture*, 24–26 October 2007, Saltillo, Coahula, Mexico, pp. 142–172.

Denman, K.L., Brasseur, G., Chidthaisong, A., Ciais, P., Cox, P.M., Dickinson, R.E., Hauglustaine, D., Heinze, C., Holland, E., Jacob, D., Lohmann, U., Ramachandran, S., da Silva Dias, P.L., Wofsy, S.C., and Zhang, X. 2007. Couplings between changes in the climate system and biogeochemistry, in Solomon, S., D. Qin, M. Manning, Z. Chen, M. Marquis, K.B. Averyt, M. Tignor and H.L. Miller (eds), *Climate change 2007: The physical science basis. Contribution of Working Group I to the Fourth Assessment Report of the Intergovernmental Panel on Climate Change*. Cambridge University Press, Cambridge and New York.

Denton, G.H. and Hendy, C.H. 1994. Younger Dryas Age advance of Franz Josef Glacier in the Southern Alps of New Zealand. *Science* 3, 264/5164, pp. 1434–1437.

Department of Conservation and Recreation (DCR) 2011. Areas of critical environmental concern: The Great Marsh. Accessed online <http://www.mass.gov/dcr/stewardship/acec/acecs/l-parriv.htm> March 2011.

Dethier, D.P. 2001. Pleistocene incision rates in the western United States calibrated using Lava Creek B tephra. *Geology* 29/9, pp. 783–786.

DeVantier, L., Alcala, A. and Wilkinson, C. 2004. The Sulu-Sulawesi Sea: Environmental and socioeconomic status, future prognosis and ameliorative policy options. *Ambio* 33/1–2. Accessed online <http://www.unep.org/dewa/giwa/publications/articles/ambio/article_11.pdf> June 2011.

Diazgranados, M. 2009. The Espeletia project. Accessed online <http://espeletia.org/Espeletia/Home.html> January 2011.

Dionne, M., Dalton, C. and Wilhelm, H. (eds) 2006. Site profile of the Wells National Estuarine Research

Reserve. Wells National Estuarine Research Reserve, Wells, Maine, 312 pp. Accessed online <http://nerrs.noaa.gov/Doc/PDF/Reserve/WEL_SiteProfile.pdf> June 2011.

Dise, N. 2009. The critical role of carbon in wetlands, in Maltby, E. and Barker (eds), *The wetland handbook*, pp. 249–265. Blackwell Publishing, Oxford.

Douglas, D.C., Reynolds, P.E. and Rhode, E.B. (eds) 2002. Arctic Refuge Coastal Plain terrestrial wildlife research summaries. U.S. Geological Survey, Biological Resources Division, Biological Science Report USGS/BRD/BSR-2002-0001. Accessed online <http://alaska.usgs.gov/BSR-2002/index.html> March 2011.

Dowdy, A.K. 2010. USDA APHIS PPQ moratorium for biological control of saltcedar (*Tamarix* species) using the biological control agent *Diorhabda* species (Coleoptera: Chrysomelidae). Invertebrate and Biological Control Programs, Plant Protection and Quarantine, Animal and Plant Health Inspection Service, U.S. Department of Agriculture. Accessed online <http://www.usbr.gov/uc/albuq/library/eaba/saltcedar/pdfs/2010/BeetleMemoUSDA.pdf> June 2011.

Doyle, T.W. 2009. Hurricane frequency and landfall distribution for coastal wetlands of the Gulf coast, USA. *Wetlands* 29/1, 35–43.

Doyle, T.W., Krauss, K.W. and Wells, C.J. 2009. Landscape analysis and pattern of hurricane impact and circulation on mangrove forests of the Everglades. *Wetlands* 29/1, pp. 44–53.

Drumm, A. and Moore, A. 2005. *Ecotourism development: A manual for conservation planners and managers, Volume 1: An introduction to ecotourism planning*, 2nd edn, The Nature Conservancy, Arlington, VA.

Drury, S.A. 1987. *Image interpretation in geology*. Allen & Unwin, London, 243 pp.

Dudley, T., Bean, D., DeLoach, C.J. Jr., Tracy, J., Longland, W., Kazmer, D., Eberts, D.,Thompson, D. and Brooks, M. 2010. *Tamarix* biocontrol, an endangered bird and regulatory dysfunction: Can restoration provide resolution? Biological Control for Nature, University of California, Riverside. Accessed online <http://biocontrolfornature.ucr.edu/pdf/dudley,tom-tamarix_biocontrol.pdf> January 2011.

Dugan, P. (ed.) 2005. *Guide to wetlands: An illustrated guide to the ecology and conservation of the world's wetlands*. Firefly Books, Buffalo, New York, 304 pp.

Earth Island Institute (EII) 2010. Mangrove Action Project. Accessed online <http://mangroveactionproject.org/> June 2011.

Eberts, D., White, L., Broderick, S., Nelson, S.M., Wynn, S. and Wydoski, R. 2003. Biological control of saltcedar at Pueblo, Colorado: Summary of research and insect, vegetation and wildlife monitoring – 1997–2002. Technical Memorandum No. 8220-03-06. U.S. Bureau of Reclamation, Technical Service Center, Denver, CO.

Eddy, J.A. 1977. The case of the missing sunspots. *Scientific American* 236/5, pp. 80–92.

Eertman, R.H., Kornman, B.A., Stikvoort, E. and Verbeek, H. 2002. Restoration of the Sieperda tidal marsh in the Scheldt estuary, The Netherlands. *Restoration Ecology*, 10/3, pp. 438–449.

Elgin, R. 2004. Crocodylomorpha. Department of Earth Sciences, University of Bristol. Accessed online <http://palaeo.gly.bris.ac.uk/palaeofiles/fossilgroups/crocodylomorpha/> October 2010.

Ellery, W.N. and McCarthy, T.S. 1998. Environmental change over two decades since dredging and excavation of the lower Boro River, Okavango Delta, Botswana. *Journal of Biogeography* 25/2, pp. 361–78.

Elliot, L., Gerhardt, C. and Davidson, C. 2009. *The frogs and toads of North America: A comprehensive guide to their identification, behavior, and calls*. Houghton Mifflin Harcourt, Boston, 343 pp.

Emery, P.A. 1971. Water resources of the San Luis Valley, Colorado, in James, J.L. (ed.), *Guidebook of the San Luis Basin, Colorado*, pp. 129–132. New Mexico Geological Society.

Encyclopedia of Saskatchewan (EoS) 2006a. Sodium sulphate. Canadian Plains Research Center, University of Regina. Accessed online <http://esask.uregina.ca/entry/sodium_sulphate.html> January 2011.

Encyclopedia of Saskatchewan (EoS) 2006b. Old Wives Lake. Canadian Plains Research Center, University of Regina. Accessed online <http://esask.uregina.ca/entry/old_wives_lake.html> January 2011.

Endla Looduskaitseala Administratsioon (ELA) 2004. Endla Looduskaitseala. Accessed online <http://www.endlakaitseala.ee/?id=824> January 2011.

Engel, M. 2001. A monograph of the Baltic amber bees and evolution of the Apidea (Hymenoptera). *Bulletin of the American Museum of Natural History*, No. 259, New York, 192 pp.

Engel, M. and Grimaldi, D.A. 2005. Primitive new ants in Cretaceous amber form Myanmar, New Jersey, and Canada (Hymenoptera: Formicidae). *American Museum Novitates* 3485, pp. 1–23.

Engels, S., Helmens, K.F., Väliranta, M., Brooks, S.J. and Birks, J.B. 2010. Early Weichselian (MIS 5d and 5c) temperatures and environmental changes in northern Fennoscandia as recorded by chironomids and macroremains at Soki, northeast Finland. *Boreas* 39/4, pp. 689–704.

Enos, P. and Perkins, R.D. 1977. *Quaternary sedimentation in South Florida*. Geological Society America, Memoir 147.

Evans, C.S. 2010. Playas in Kansas and the High Plains. Kansas Geological Survey, Public Information Circular 30, 6 p. Accessed online <http://www.kgs.ku.edu/Publications/PIC/pic30.html> August 2010.

Evans, J.M. 2011. The water cycle (illustration). U.S. Geological Survey. Accessed online <http://ga.water.usgs.gov/edu/watercycle.html> May 2011.

Fellows, S. and Gress, B. 2006. *A pocket guide to Great Plains shorebirds*. Great Plains Nature Center, Wichita, Kansas, United States, 69 pp.

Fennessy, M.S., Jacobs, A.D. and Kentula, M.E. 2004. Review of rapid methods for assessing wetland condition. EPA/620/R-04/009. U.S. Environmental Protection Agency, Washington, D.C. Accessed online <http://water.epa.gov/type/wetlands/assessment/upload/2004_04_21_monitor_RapidMethodReview.pdf> June 2011.

Ferri, V. 2002. *Turtles & tortoises*. Firefly Books, Willowdale, Ontario, 255 pp.

Fick, W.H. and Geyer, W.A. 2010. Cut-stump treatment of saltcedar (*Tamarix ramosissima*) on the Cimarron National Grasslands. *Kansas Academy of Science, Transactions* 113/3–4, pp. 223–226.

Field, B.C. and Field, M.K. 2006. *Environmental economics: An introduction*, 4th edn, McGraw Hill Irwin, New York, 503 pp.

Finlayson, C.M., Davidson, N.C., Spiers, A.G. and Stevenson, N.J. 1999. Global wetland inventory – Current status and future priorities. *Marine and Freshwater Research* 50/8, pp. 717–727.

Foley, J. 2010. Boundaries for a healthy planet. *Scientific American* 302/4, pp. 54–57.

Food and Agriculture Organization (FAO) 2007. The world's mangroves: 1980–2005. FAO Forestry Paper 153. FAO of the United Nations, Rome, 89 pp.

Food and Agriculture Organization (FAO) 2009. *Yearbook of fishery statistics 2007*. Food and Agriculture Organization of the United Nations, Rome. Accessed online <ftp://ftp.fao.org/docrep/fao/012/i1013t/i1013t.pdf>October 2010.

Franco, M.S. and Pinheiro, R. 1982. Geomorfologia, in Brasil (ed.), Ministério das Minas e Energia. Departamento Nacional de Produção Mineral. Projeto RADAMBRAZIL. Folha SE. 21 Corumbá and SE. 20. *Levantamento de Recursos Naturais* 27, pp. 161–224.

Franzén, L.G. 1994. Are wetlands the key to the ice-age cycle enigma? *Ambio* 23, pp. 300–308.

Franzén, L.G. and Ljung, T.L. 2009. A carbon fibre composite (CFC) Byelorussian peat corer. *Mires and Peat* 5, article 2.

Frenzel, P. and Karofeld, E. 2000. CH_4 emission from a hollow-ridge complex in a raised bog: The role of CH_4 production and oxidation. *Biogeochemistry* 51/1, pp. 91–112.

Fricke, H.C., O'Neil, J.R. and Lynnerup, N. 1995. Oxygen isotope composition of human tooth enamel from medieval Greenland: Linking climate and society. *Geology* 23/10, pp. 869–872.

Galkina, E.A. 1946. Bolotnye landshafty i principy ikh klassifikatsii [Mire landscapes and principles of their classification], in Sbornik nauchnykh rabot Botan. Instituta imemi V.L. Komarova AN SSSR, pp. 139–156 (in Russian).

Gaston, K.J. 2007. Latitudinal gradient in species richness. *Current Biology* 17/15, p. R574.

Gauci, V., Dise, N. and Fowler, D. 2002. Controls on suppression of methane flux from a peat bog subjected to simulated acid rain sulfate deposition. *Global biogeochemical cycles* 16, pp. 1004–1016.

Gemological Institute of America (GIA) 2000. *Pearl grading: Pearl description system manual*. Gemological Institute of America, Carlsbad, California, pp. 1–3.

Geological Society of America (GSA) 2009. 2009 Geological time scale. Geological Society of America. Accessed online <http://www.geosociety.org/science/timescale/timescl.pdf> November 2010.

Germany Report 2005. National planning tool for the implementation of the Ramsar Convention on Wetlands: Germany. Accessed online <http://www.ramsar.org/pdf/cop9/cop9_nr_germany.pdf> June 2011.

Germany Report 2008. National report on the implementation of the Ramsar Convention on Wetlands: Germany. Accessed online <http://www.ramsar.org/pdf/cop10/cop10_nr_germany.pdf> August 2010.

Gilg, O., Sane, R., Solovieva, D.V., Pozdnyakov, V.I., Sabard, B., Tsanos, D., Zockler, C., Lappo, E.G., Syroechkovski, E.E. and Eichhorn, G. 2000. Birds and mammals of the Lena Delta Nature Reserve, Siberia. *Arctic* 53/2, pp. 118–133.

Global Coral Reef Monitoring Network (GCRMN) 2010. Climate change and coral reefs: Consequences of inaction. Global Coral Reef Monitoring Network. Accessed online <http://www.gcrmn.org/pdf/P1 Status 2009 Brochure 7th draft.pdf> October 2010.

Gopal, B. and Chauhan, M. 2006. Biodiversity and its conservation in the Sundarban Mangrove Ecosystem. *Aquatic Science* 68/3, pp. 338–354.

Gorham, E., Lehman, C., Dyke, A., Janssens, J. and Dyke, L. 2007. Temporal and spatial aspects of peatland initiation following deglaciation of North America. *Quaternary Science Reviews* 26/3–4, pp. 300–311.

Gosselink, J.G., Conner, W.H., Day, J.W. Jr., and Turner, R.E. 1981. Classification of wetland resources: Land, timber, and ecology, in Jackson, B.D. and Chambers, J.L. (eds), *Timber harvesting in wetlands.* pp. 28–48. Louisiana State University, Baton Rouge, Louisiana.

Government of Canada 1991. The federal policy on wetland conservation, Environment Canada, Ottawa, Ontario, pp. 1–14. Accessed online <http://dsp-psd.pwgsc.gc.ca/Collection/CW66-116-1991E.pdf> May 2010

Grant, S.M. 2005. The applicability of international conservation instruments to the establishment of marine protected areas in Antarctica. *Ocean and Coastal Management* 48/9–10, pp. 782–812.

Greb, S.F. and DiMichele, W.A. (eds) 2006. Wetlands through time. Geological Society of America, Special Paper 399, Preface, pp. v–viii.

Greb, S.F., DiMichele, W.A. and Gastaldo, R.A. 2006. Evolution and importance of wetlands in Earth history, in Greb, S.F. and DiMichele, W.A. (eds), Wetlands through time. Geological Society of America, Special Paper 399, pp. 1–40.

Gregory, S.V., Swanson, F.J., McKee, W.A. and Cummins, K.W. 1991. An ecosystem perspective of riparian zones. *BioScience* 41/8, pp. 540–551.

Grimaldi, D.A. 1996. *Amber: window to the past.* Harry N. Abrams, New York.

Grimald, D.A., Lillegraven, J.A., Wampler, T.W., Bookwalter, D. and Shedrinsky, A. 2000. Amber from Upper Cretaceous through Paleocene strata of the Hanna Basin, Wyoming, with evidence for source and taphonomy of fossil resins. *Rocky Mountain Geology* 35/2, pp. 163–204.

Grimaldi, D., Nascimbene, P., Luzzi, K., and Case, G. 1998. North American ambers and an exceptionally diverse deposit from the Cretaceous of New Jersey, USA. World Congress on Amber Inclusions. *Museo de Ciencias Naturales de Álava, Abstracts.* Spain, 81 pp.

Grimaldi, D., Shedrinsky, A. and Wampler, T. 2000. A remarkable deposit of fossiliferous amber from the Upper Cretaceous (Turonian) of New Jersey, in Grimaldi, D. (ed.), *Studies on fossils in amber, with particular reference to the Cretaceous of New Jersey,* pp. 1–76. Backhuys Publishers, Leiden, 498 pp.

Grootjans, A., Iturraspe, R. Lanting, A. Fritz, C. and Joosten, H. 2010. Ecohydrological features of some contrasting mires in Tierra del Fuego, Argentina. *Mires and Peat* 6, article 1.

Grosse, G., Schirrmeister, L. and Malthus, T.J. 2006. Application of Landsat-7 satellite data and a DEM for the quantification of thermokarst-affected terrain types in the periglacial Lena-Anabar coastal lowland. *Polar Research* 25/1, pp. 51–67.

Grove, J.M. 1988. *The Little Ice Age.* Methuen, London, 498 pp.

GSA Rock-Color Chart 1991. Geological Society of America, 8th printing, 1995.

Guo, X. 2000. Aquaculture in China: Two decades of rapid growth. *Aquaculture Magazine* 26/3, p. 27.

Guthrie, R.L. 1985. Characterizing and classifying wetland soils in relation to food production, in *Wetland Soils: Characterization, classification and utilization, Workshop Proceedings,* pp. 11–22. International Rice Research Institute, Manila, Philippines.

Gwadz, B. 2001. Malaria in Africa: An overview. Report from a symposium on Malaria in Africa: Emerging prevention and control strategies. AAAS Annual Meeting, Feb. 17, 2001. Accessed online <http://www.aaas.org/international/africa/malaria/gwadz.html> June 2010.

Hackett, S.C. 2006. *Environmental and natural resources economics: Theory, policy and sustainable society,* 3rd edn, M.E. Sharpe, Armonk, New York, 524 pp.

Hait, A. and Behling, H. 2009. Holocene mangrove and coastal environmental changes in the western Ganga-Brahmaputra Delta, India. *Vegetation History and Archaeobotany* 18/2, pp. 159–169.

Håkanson, E. 1971. Stevns Klint, in *Geologi på Øerne* pp. 25–36. Varv, Ekskursionsfører Nr. 2.

Hall, R.C. 1997. Post war strategic reconnaissance and the genesis of project Corona, in R.A. McDonald (ed.), *Corona: Between the Sun & the Earth, The first NRO reconnaissance eye in space.* American Society of Photogrammetry and Remote Sensing, pp. 25–58.

Hambrey, M.J. and Harland, W.B. (eds) 1981. *Earth's pre-Pleistocene glacial record.* Cambridge University Press, United Kingdom, 1004 pp.

Hammer, D.A. 1997. *Creating freshwater wetlands,* 2nd edn, Lewis Publishers, Boca Raton, FL, 406 pp.

Hand, S., Archer, M., Bickel, D., Creaser, P., Dettmann, M., Godthelp, H., Jones, A., Norris, B. and Wicks, D. 2010. Australian Cape York amber, in Penney, D.

(ed.), *Biodiversity of fossils in amber from the major world deposits*, pp. 69–79. Siri Scientific Press: Manchester, United Kingdom.

Handy, F. 2001. Advocacy by environmental nonprofit organizations: An optimal strategy for addressing environmental problems? *International Journal of Social Economics* 28/8, pp. 648–666.

Hang, T. and Miidel, A. 1999. Late- and postglacial crustal movements, in Miidel, A. and Raukas, A. (eds), *Lake Peipsi geology*, pp. 52–56. Sulemees Publishers, Tallinn.

Hanley, N. and Spash, C.L. 1995. *Cost-benefit analysis and the environment*. Edward Elgar, Cheltenham/Aldershot, United Kingdom, 275 pp.

Hanna, S., Folke, C., Maler, C-G. 1995. Property rights, environmental resources, in Hanna, S. and Munasinghe, M. (eds), *Property rights and the environment: Social and ecological issues*, pp. 15–28. World Bank and Beijer International Institute of Ecological Economics, Washington D.C.

Hanna, S. and Munasinghe, M. 1995. *Property rights and the environment: Social and ecological issues*. World Bank and Beijer International Institute of Ecological Economics, Washington D.C., 165 pp.

Hardin, G. 1968. The tragedy of the commons, in Hardin, G. and Baden, J. (eds), *Managing the Commons*. W.H. Freeman & Co., San Francisco.

Harris, M.B., Tomas, W., Mourão, G., Da Silva, C.J., Guimarães, E., Sonoda, F. and Fachim, E. 2005. Safeguarding the Pantanal wetlands: Threats and conservation initiatives. *Conservation Biology* 19/3, pp. 714–720.

Harris, M.S. 2009. *Aix galericulata* (Linnaeus, 1758): Mandarin duck. *Encyclopedia of Life*. Accessed online <http://www.eol.org/pages/1048478> October 2010.

Hathaway, D.H. 2011. The sunspot cycle. NASA Solar Physics <http://solarscience.msfc.nasa.gov/SunspotCycle.shtml> Accessed September 2011.

Haworth, E. 1897. Physiography of western Kansas. University Geological Survey of Kansas, Vol. 2, pp. 11–49.

Hay, D.L. and Philippi, N.S. 1999. *A case for wetland restoration*. Wiley and Sons, New York, 215 pp.

Hearty, P.J., Kindler, P., Cheng, H. and Edwards, R.L. 1999. A +20 m middle Pleistocene sea-level highstand (Bermuda and the Bahamas) due to partial collapse of Antarctic ice. *Geology* 27/4, pp. 375–378.

Hedenström, A. 2010. Extreme endurance migration: what is the limit to non-stop flight? Public Library of Science: *Biology* 8/5: e1000362. Accessed online <http://www.plosbiology.org/article/info%3Adoi%2F10.1371%2Fjournal.pbio.1000362> March 2011.

Heikkilä, M. and Seppä, H. 2010. Holocene climate dynamics in Latvia, eastern Baltic region: A pollen-based summer temperature reconstruction and regional comparison. *Boreas* 39, pp. 705–719.

Heinl, M., Silva, J., Tacheba, B. and Murray-Hudson, M. 2007. The relevance of fire frequency for the floodplain vegetation of the Okavango Delta, Botswana. African *Journal of Ecology* 46, pp. 350–358.

Helama, S., Läänelaid, A., Tietäväinen, H., Fauria, M.M., Kukkonen, I.T., Holopainen, J., Nielsen, J.K. and Valovirta, I. 2010. Late Holocene climatic variability reconstructed from incremental data from pines and pearl mussels – A multi-proxy comparison of air and subsurface temperatures. *Boreas* 39/4, pp. 734–748.

Helama, S. and Valovirta, I. 2008. The oldest recorded animal in Finland: Ontogenetic age and growth of *Margaritifera margaritifera* (L. 1758) based on internal shell increments. Memoranda Soc. *Fauna Flora Fennica* 84, pp. 20–30. Accessed online <http://www.helsinki.fi/science/raakku/memoranda-2008.pdf> June 2011.

Helsinki Commission (HC) 2011. The brackish nature of the Baltic Sea. Helsinki Commission: Baltic Marine Environment Protection Commission. Accessed online <http://www.helcom.fi/environment2/nature/en_GB/salinity/> January 2011.

Heyns, P.S.v. 2007. Governance of a shared and contested resource: a case study of the Okavango River Basin. *Water Policy* 9/2, 149–167.

Higer, A.L., Cordes, E.H. and Coker, A.E. 1976. Water-management model of the Florida Everglades, in Williams, R.S. and Carter, W.D. (eds), *ERTS – 1 a new window on our planet*, pp. 159–161. U.S. Geological Survey, Professional Paper 929.

High Plains Regional Climate Center (HPRCC) 2011. Period of record general climate summary – Precipitation. Accessed online <http://www.hprcc.unl.edu/data/historical/index.php> January 2011.

Hillebrand, H. 2004. On the generality of the latitudinal diversity gradient. *American Naturalist* 163/2, pp. 192–211.

Hingston, R.W. 1931. Proposed British national parks for Africa. *The Geographical Journal* 77/5, pp. 401–422.

Holden, J. 2010. *Castor fiber* (Linnaeus, 1758): European beaver. *Encyclopedia of Life*. Accessed online <http://www.eol.org/pages/1036116> October 2010.

Hossain, S. 2001. Biological aspects of the coastal and marine environment of Bangladesh. *Ocean and Coastal Management* 44/3-4, pp. 261-282.

Howard, C. 2010. *Puma concolor coryi*: Florida panther. Encyclopedia of Life. Accessed online <http://www.eol.org/pages/10465959> October 2010.

Howes, N.C., FitzGerald, D.M., Hughes, Z.J., Georgiou, I.Y., Kulp, M.A., Miner, M.D., Smith, J.M. and Barras, J.A. 2010. Wetland loss during hurricanes: Failure of low salinity marshes. *National Academy of Sciences, Proceedings* 107/32, pp. 14014-19.

Huber, F.M. and Langenheim, J.H. 1986. Dominican amber tree had African ancestors. *Geotimes* 31, pp. 8-10.

Hughes, T.J., Denton, G.H., Andersen, B.G., Schilling, D.H., Fastook, J.L. and Lingle, C.S. 1981. The last great ice sheets: A global view, in Denton, G.H. and Hughes, T.J. (eds), *The last great ice sheets*, pp. 263-317. J. Wiley & Sons, New York, 484 pp.

Hurd, J. D., Civco, D. L., Gilmore, M. S., Prisloe, S., and Wilson, E.H. 2005. Coastal marsh characterization using satellite remote sensing and in-situ radiometry data: Preliminary results. *Proceedings of the ASPRS Annual Conference, Baltimore, Maryland*, March 7-11, 2005.

Ilomets, M. 1997. Holocene terrestrial processes: Genesis and development of mires, in Raukas, A. and Teedumäe, A. (eds), *Geology and mineral resources of Estonia*, pp. 293-298. Estonian Academy Publishers, Tallinn.

India Report 2008. National report on the implementation of the Ramsar Convention on Wetlands: India. Accessed online <http://www.ramsar.org/pdf/cop10/cop10_nr_india.pdf> July 2010.

Interagency Workgroup on Wetland Restoration (IWWR) 2003. An introduction and user's guide to wetland restoration, creation, and enhancement. Accessed online <http://www.epa.gov/owow/wetlands/pdf/restdocfinal.pdf> June 2011.

International Rivers Network (IRN) 2010. IRN's Okavango campaign. Accessed online <http://www.internationalrivers.org/en/node/2431/> June 2011.

International Union for Conservation of Nature (IUCN) 1997. Sundarban Wildlife Sanctuaries (Bangladesh). World Heritage Nomination, IUCN Technical Evaluation. Accessed online <http://whc.unesco.org/archive/advisory_body_evaluation/798.pdf> June 2011.

Iversen, J. 1973. The development of Denmark's nature since the Last Glacial. *Danmarks Geologiske Undersøgelse*, Series V, Nr. 7-C, 126 pp.

Jablonski, D., Roy, K. and Valentine, J.W. 2006. Out of the tropics: evolutionary dynamics of the latitudinal diversity gradient. *Science* 314/5796, pp. 102-106.

Jackson, C. 2009. *Cygnus atratus* (Latham, 1790): Black swan. *Encyclopedia of Life*. Accessed online <http://www.eol.org/pages/1047328> October 2010.

Jacobson, G.L. Jr., Webb III, T. and Grimm, E.C. 1987. Patterns and rates of vegetation change during the deglaciation of eastern North America. *Geological Society America*, Vol. K-3, pp. 277-288.

Jensen, J.R. 2007. *Remote sensing of the environment: An Earth resource perspective* (2nd edn), Prentice Hall Series in Geographic Information Science, Upper Saddle River, New Jersey, United States, 592 pp.

Jewellery News Asia (JNA) 2011a. Tahitian pearls mark 50th anniversary. Jewellery News Asia, 9 June 2011. Accessed online <http://www.jewellerynewsasia.com/en/Pearls/details.html?id=270> June 2011.

Jewellery News Asia (JNA) 2011b. Pearl farmer goes "green." Jewellery News Asia, 1 January 2011. Accessed online <http://www.jewellerynewsasia.com/en/Pearls/details.html?id=184> June 2011.

Johnson, G., Holmes, Jr. R. and Waite, L. 2010. The Great Flood of 1993 on the Upper Mississippi River – 10 years later. Accessed online <http://il.water.usgs.gov/hot/Great_Flood_of_1993.pdf> October 2010.

Johnson, W.M. 1975. Foreword, in Soil Survey Staff (1975). *Soil taxonomy: A basic system of soil classification for making and interpreting soil surveys*. U.S. Department of Agriculture, Agricultural Handbook No. 436.

Jørgensen, F., Sandersen, P.B.E. and Auken, E. 2003. Imaging buried Quaternary valleys using the transient electromagnetic method. *Journal Applied Geophysics* 53, pp. 199-213.

Jørgensen, F., Sandersen, P.B.E., Auken, E., Lykke-Andersen, H. and Sørensen, K. 2005. Contributions to the geological mapping of Mors, Denmark – A study based on a large-scale TEM survey. *Geological Society Denmark, Bulletin* 52, pp. 53-75.

Junk, W.J. and Cunha, C.N. de 2004. Pantanal: A large South American wetland at a crossroads. *Ecological Engineering* 24/4, pp. 391-401.

Junk, W., Da Silva, C., Wantzen, C.M., Nunes Da Cunha, C., and Nogueira, F. 2009. The Pantanal of Mato Grosso: Linking ecological research, actual use and management for sustainable development, pp. 908-943, in Maltby, E. and Barker, T. (eds), *The wetlands handbook*, Wiley-Blackwell Publishers, Chichester, West Sussex, United Kingdom, 1058 pp.

Junk, W. and Piedade, M. 2005. The Amazon River basin, in Fraser, L.H. and Keddy, P.A. (eds), *The world's largest wetlands: Ecology and conservation*, pp. 63–117, Cambridge University Press, Cambridge.

Kansas Department of Agriculture (KDA) 2010. Kansas-Colorado Arkansas River Compact update. Accessed online <http://www.ksda.gov/interstate_water_issues/content/143> January 2011.

Kansas Geospatial Community Commons (KGCC) 2010. Water resources: Kansas playa wetlands. Accessed online <http://www.kansasgis.org/> August 2010.

Karofeld, E. 1996. The effects of alkaline fly ash precipitation on the *Sphagnum* mosses in Niinsaare bog, NE Estonia. *Suo* 47/4, pp. 105–114.

Karofeld, E. 1998. The dynamics of the formation and development of hollows in raised bogs in Estonia. *The Holocene* 8/6, pp. 697–704.

Keddy, P.A. and Fraser, L.H. 2005. Introduction: Big is beautiful, in Fraser, L.H. and Keddy, P.A. (eds), *The world's largest wetlands: Ecology and conservation*, pp. 1–10. Cambridge: Cambridge University Press.

Keddy, P., Fraser, L., Solomeshch, A., Junk, W., Campbell, D., Arroyo, M. and Alho, C. 2009. Wet and wonderful: the world's largest wetlands are conservation priorities. *Bioscience* 59/1, pp. 39–51.

Keech, C.F. and Bentall, R. 1978. Dunes on the plains: The Sand Hills region of Nebraska. Conservation and Survey Division, Resource Report 4, 18 pp. Institute of Agriculture and Natural Resources, University of Nebraska-Lincoln.

Kenney, F.R. and McAtee, W.L. 1938. The problem: Drained areas and wildlife habitats, in United States Department of Agriculture, *1938 Yearbook of Agriculture, Soils and Man*, pp. 77–83. U.S. Government Printing Office, Washington D.C.

Kent, D.M. (ed.) 2000a. *Applied wetland science and technology*. Lewis Publishers, Boca Raton, 454 pp.

Kent, D.M. 2000b. Design and management of wetlands for wildlife, in Kent, D.M. (ed.), *Applied wetland science and technology*, pp. 281–321. Lewis Publishers, Boca Raton.

Knutsen, M.G., Sauer, J.R., Olsen, D.A., Mossman, M.J., Hemesath, L.M., and Lannoo, M.J. 1999. Effects of landscape composition and wetland fragmentation on frog and toad abundance and species richness in Iowa and Wisconsin, USA. *Conservation Biology* 13/6, pp. 1437–46.

Korholaa, A., Ruppela, M., Seppäb, H., Välirantaa, M., Virtanena, T. and Weckströma, J. 2010. The importance of northern peatland expansion to the late-Holocene rise of atmospheric methane. *Quaternary Science Reviews* 29/5–6, pp. 611–617.

Kosmowska-Ceranowicz, B. 1999. Succinite and some other fossil resins in Poland and Europe. *Estudios del Museo de Ciencias Naturales de Álava* 14/2, pp. 73–117. Vitoria-Gasteiz, Spain.

Kosmowska-Ceranowicz, B. 2009. Amber (succinite) deposits. Bursztynowy Portal. Accessed online <http://www.amber.com.pl/en/resources/amber/item/407-amber-succinite-deposits> June 2011.

Kosmowska-Ceranowicz, B., Giertych, M. and Miller, H. 2001. Cedarite from Wyoming: Infrared and radiocarbon data. *Prace Muzeum Ziemi* 46, pp. 77–80.

Kostecke, R.M. and Smith, L.M. and Hands, H.M. 2004. Vegetation response to cattail management at Cheyenne Bottoms, Kansas. *Journal of Aquatic Plant Management* 42, pp. 39–45.

Kramer, A. 2010. Past errors to blame for Russia'a peat fires. *The New York Times* – Europe, August 12. Accessed online <http://www.nytimes.com/2010/08/13/world/europe/13russia.html> June 2011.

Krauss, K.W., Doyle, T.W., Doyle, T.J., Swarzenski, C.M., From, A.S., Day, R.H. and Conner, W.H. 2009. Water level observations in mangrove swamps during two hurricanes in Florida. *Wetlands* 29/1, pp. 142–149.

Kremenetski, K.V., Velichko, A.A., Borisova, O.K., MacDonald, G.M., Smith, L.C., Frey, K.E., and Orlova, L.A. 2003. Peatlands of the western Siberian lowlands: Current knowledge on zonation, carbon content and late Quaternary history. XVI INQUA Congress, *Quaternary Science Reviews* 22/5–7, pp. 703–723.

Krielaars, M. 2010. Kaliningrad is trying to capitalize on its amber monopoly. NRCHandelsblad News Archive. Accessed online <http://vorige.nrc.nl/international/article2516018.ece/Kaliningrad_is_trying_to_capitalise_on_its_amber_monopoly?service=Print> June 2011.

Krinsley, D.B. 1976. Lake fluctuations in the Shiraz and Neriz playas of Iran, in Williams, R.S. and Carter, W.D. (eds), *ERTS-1 a new window on our planet*, pp. 143–149. U.S. Geological Survey, Professional Paper 929.

Kusler, J. 1999. Climate change in wetland areas part II: Carbon cycle implications. *Acclimations*, July–August 1999, Newsletter of the U.S. National Assessment of the Potential Consequences of Climate Variability and Change, pp. 2, 10–11. Accessed online <http://www.usgcrp.gov/usgcrp/Library/nationalassessment/newsletter/1999.08/issue7.pdf> June 2011.

Kusler, J. 2006. Common questions: Wetland restoration, creation and enhancement. Association of State Wetland Managers, Inc., Berne, NY. Accessed online <http://aswm.org/wetland-science/wetland-science/1088-common-questions-wetland-classification> June 2011.

Labedz, T.E. 1990. Birds, in Bleed, A.S. and Flowerday, C. (eds), *An atlas of the Sand Hills. Conservation and Survey Division, Resource Atlas No. 5a* (2nd edn), pp. 161–180. Institute of Agriculture and Natural Resources, University of Nebraska-Lincoln.

Lahring, H. 2003. *Water and wetland plants of the Prairie Provinces*. Canadian Plains Research Center, Regina, Canada, 326 pp.

Laine, J. and Vasander, H. 1996. Ecology and vegetation gradients of peatlands, in Vasander, H. (ed.), *Peatlands in Finland*, pp. 10–19. Finish Peatland Society, Helsinki, Finland.

Laing, G., Van de Moortel, A., Lesage, E., Tack, P. and Verloo, M.G. 2008. Factors affecting metal accumulation, mobility and availability intertidal wetlands of the Scheldt Estuary (Belgium), in Vymazal, J. (ed.), *Wastewater treatment, plant dynamics and management in constructed and natural wetlands*, pp. 121–134. Springer, Dordrecht, the Netherlands.

Landis, G.P. 2009. Air bubbles, amber, and dinosaurs. U.S. Geological Survey, Understanding our planet through chemistry. Accessed online <http://minerals.cr.usgs.gov/gips/na/amber.html> June 2011.

Langbein, W.B. 1976. Hydrology and environmental aspects of Erie Canal (1817–99). *U.S. Geological Survey, Water-Supply Paper* 2038, 92 pp.

Langenheim, J.H. 2003. *Plant resins*. Timber Press, Portland, Oregon, 586 pp.

Larsen, J.S. 1991. North America, in Finlayson, M. and Moser, M. (eds), *Wetlands*, pp. 57–84. International Waterfowl and Wetlands Research Bureau, Facts on File, Oxford and New York, 224 pp.

Larsen, L.W. 1996. The Great USA Flood of 1993. Hydrologic Research Laboratory, Office of Hydrology, NOAA/National Weather Service. Accessed online <http://www.nwrfc.noaa.gov/floods/papers/oh_2/great.htm> August 2010.

Larson, G.E. 2006. Aquatic and wetland vascular plants of the northern Great Plains. U.S. Geological Survey, Northern Prairie Wildlife Research Center. Accessed online <http://www.npwrc.usgs.gov/resource/plants/vascplnt/index.htm> September 2010.

Larsson, S.G. 1978. Baltic amber – *A palaeobiological study*. Entomonograph 1. Scandinavian Science Press, Klampenborg, Denmark, 192 pp.

Laubhan, M.K. and Roelle, J.E. 2001. Managing wetlands for waterbirds, in Rader, R., Batzer, D.P. and Wissinger, S.A. (eds), *Bioassessment and management of North American freshwater wetlands*, pp. 387–411. Wiley and Sons, New York.

Lauer, D.T., Morain, S.A. and Salomonson, V.V. 1997. The Landsat program: Its origins, evolution, and impacts. *Photogrammetric Engineering & Remote Sensing* 63, pp. 831–838.

Le Roy Ladurie, E. 1971. *Times of feast, times of famine: A history of climate since the year 1000*. Doubleday, Garden City, New York, 426 pp.

Le Roy Ladurie, E. 2004. *Histoire humaine et comparée du climat: Canicule et glaciers du XIIIe au XVIIIe siècles*. Librairie Arthème Fayard, Paris, 740 pp.

Leachtenauer, J.C., Daniel, K. and Vogl, T.P. 1997. Digitizing Corona imagery: Quality vs. cost, in R.A. McDonald (ed.), *Corona: Between the Sun & the Earth, The first NRO reconnaissance eye in space*. American Society Photogrammetry and Remote Sensing, pp. 189–203.

Leppälä, M., Laine, A.M. and Tuittila, E.-S. 2011. Winter carbon losses from a boreal mire succession sequence follow summertime patterns in carbon dynamics. *Suo* 62/1, pp. 1–11.

Lewis, R.R. III 1989. Wetland restoration/creation/enhancement terminology: Suggestions for standardization, in Wetland creation and restoration: The status of the science, v. II. EPA 600/3/89/038B. U.S. Environmental Protection Agency, Washington, D.C.

Lewis, W.M. Jr. 2001. *Wetlands explained: Wetland science, policy and politics in America*. Oxford University Press, New York, 147 pp.

Library of Congress (LOC) 2002. The evolution of the conservation movement, 1850–1920. Accessed online <http://memory.loc.gov/ammem/amrvhtml/conshome.html> November 2010.

Lillesand, T.M., Kiefer, R.W. and Chipman, J.W. 2008. *Remote sensing and image interpretation*. Wiley, Hoboken, New Jersey, United States, 756 pp.

Littell, J., Peterson, D. and McKenzie, D. 2010. Regional fire/climate relationships in the Pacific Northwest and beyond. Climate Impacts Group, University of Washington. Accessed online <http://cses.washington.edu/cig/res/fe/fireclimate.shtml> November 2010.

Lønne, I. and Lauritsen, T. 1996. The architecture of a modern push-moraine at Svalbard as inferred from ground-penetrating radar measurements. *Arctic and Alpine Research* 28, pp. 488–495.

Loring, P.A. and Gerlach, C. 2010. Food security and conservation of Yukon River salmon: Are we asking too much of the Yukon River? *Sustainability* 2/9, pp. 2965–87.

Lovell, J.T. Gibson, J. and Heschel, M.S. 2009. Disturbance regime mediates riparian forest dynamics and physiological performance, Arkansas River, CO. *The American Midland Naturalist* 162/2, pp. 289–304.

Luternauer, J.L., Clague, J.J., Conway, K.W., Barrie, J.V., Blaise, B. and Mathewes, R.W. 1989. Late Pleistocene terrestrial deposits on the continental shelf of western Canada: Evidence for rapid sea-level change at the end of the last glaciation. *Geology* 17/4, pp. 357–360.

Luteyn, J.L. 2011. Páramo ecosystem. Missouri Botanical Garden. Accessed online <http://www.mobot.org/mobot/research/paramo_ecosystem/introduction.shtml> January 2011.

Lutz, W., Sanderson, W. and Scherbov, S. 2001. The end of world population growth. *Nature* 412, pp. 543–545. Accessed online <http://www.nature.com/nature/journal/v412/n6846/full/412543a0.html> December 2010.

Lynch, D.K. and Livingston, W. 1995. *Color and light in nature*. Cambridge University Press, Cambridge, 254 pp.

Maas, P. 2010. *Neovision macrodon*. The Extinction Website. Accessed online <http://www.petermaas.nl/extinct/speciesinfo/seamink.htm> October 2010.

Maccarone, A.D. and Cope, C.H. 2004. Recent trends in the winter population of Canada geese (*Branta canadensis*) in Wichita, Kansas: 1998–2003. *Kansas Academy of Science, Transactions* 107, pp. 77–82.

Mahajan, M. 2010. The Mangroves. Soonabai Pirojsha Godrej Foundation, Mumbai, India, 28 pp.

Maine Department of Environmental Protection (MDEP) 2008. Integrated water quality monitoring and assessment report. Maine DEP 2008 305(b) Report and 303(d) List: DEPLW0895. Accessed online <http://www.maine.gov/dep/water/monitoring/305b/2008/report.pdf> June 2011.

Mann, K.H. 2000. *Ecology of coastal waters with implications for management*, 2nd edn, Blackwell Science, Malden, Massachusetts, 406 pp.

Mapes, G. and Mapes, R.H. 1988. *Regional geology and paleontology of Upper Paleozoic Hamilton quarry area in southeastern Kansas*, Kansas Geological Survey, Guidebook Series 6, 273 pp.

Marquardt, M. 2012. Neutralizing the rain. *Earth* 57/7, pp. 31–37.

Marsh, G.P. 1864. *Man and nature; or, Physical geography as modified by human action*. C. Scribner, New York, 560 pp.

Martin, P. D., Jenkins, J.L., Adams, F., Jorgenson, M.T., Matz, A.C., Payer, D.C., Reynolds, P.E., Tidwell, A.C. and Zelenak, J.R. 2009. Wildlife response to environmental Arctic change: Predicting future habitats of Arctic Alaska. *Report of the Wildlife Response to Environmental Arctic Change Alaska Workshop, 17–18 November 2008*. U.S. Fish and Wildlife Service, Fairbanks, Alaska, 138 pp.

Masing, V. 1984. Estonian bogs: Plant cover, succession and classification, in Moore, P.D. (ed.), *European mires*, pp. 120–148. Academic Press, London.

Masing, V. 1997. *Ancient mires as nature monuments*. Monumenta Estonica, Estonian Encyclopaedia Publ., Tallinn, Estonia, 96 pp.

Masing, V. 1998. Multilevel approach in mire mapping, research, and classification. Contribution to the IMCG Classification Workshop, March 25–29, 1998, Greifswald. Accessed online <http://www.imcg.net/docum/greifswa/masing.htm> August 2010.

MassGIS 2011. GIS datalayers: 1 : 5000 scale color ortho imagery, 2005. Accessed online <http://www.mass.gov/mgis/colororthos2005.htm> March 2011.

Matlins, A. 2006. *The pearl book: The definitive buying guide* (3rd edn), GemStone Press, Woodstock, Vermont.

Matthews, E. 1993. Global geographical databases for modelling trace gas fluxes. *International Journal of Geographical Information Systems* 7/2, pp. 125–142.

Matthews, E. and Fung, I. 1987. Methane emission from natural wetlands: Global distribution, area, and environmental characteristics of sources. *Global Biochemical Cycles* 1, pp. 61–86 (doi: 10.1029/GB001i001p00061).

Mausbach, M.J. 1994. Classification of wetland soils for wetland identification. *Soil Survey Horizons* 35, pp. 17–25.

McAlpine, J.F. and Martin, J.E.H. 1969. Canadian amber – a paleontological treasure-chest. *The Canadian Entomologist* 101/8, pp. 819–328.

McBean, G., Alekseev, G., Chen, D., Førland, E., Fyfe, J., Groisman, P.Y., King, R., Melling, H., Vose, R. and Whitfield, P.H. 2005. Arctic climate: past and present, in Symon, C., Arris, L. and Heal, O.W. (eds), *Arctic Climate Impacts Assessment*, pp. 21–60. Cambridge University Press, Cambridge.

McCarthy, T.S., Cooper, G.R.J., Tyson, P.D. and Ellery, W.N. 2000. Seasonal flooding in the Okavango Delta, Botswana – recent history and future prospects. *South African Journal of Science* 96/1, pp. 25–32.

McCay, G.A., Prave, A.R., Alsop, G.I. and Fallick, A.E. 2006. Glacial trinity: Neoproterozoic Earth history within the British-Irish Caledonides. *Geology* 34/11, pp. 909–912.

McIntyre, C. 2007. *Botswana: Okavango Delta, Chobe, northeastern Kalahari*. The Globe Pequot Press, Inc., USA, 502 pp.

McKellar, R.C., Wolfe, A.P., Tappert, R. and Muehlenbachs, K. 2008. Correlation of Grassy Lake and Cedar Lake ambers using infrared spectroscopy, stable isotopes, and palaeoentomology. *Canadian Journal of Earth Sciences* 45/9, pp. 1061–1082.

McKellar, R.C. and Wolfe, A.P. 2010. Canadian amber, in D. Penney (ed.) *Biodiversity of fossils in amber from the major world deposits*, pp. 149–166. Siri Scientific Press, Manchester, United Kingdom.

McMillan, M.E., Heller, P.L. and Wing, S.L. 2006. History and causes of post-Laramide relief in the Rocky Mountain orogenic plateau. *Geological Society America, Bulletin* 118/3, pp. 393–405.

McMurry, J., Castellion, M.E. and Ballantine, D.S. 2007. *Fundamentals of general, organic, and biological chemistry* (5th edn), Pearson Prentice Hall, Upper Saddle River, New Jersey, 889 pp.

Melcher, N.B. and Parrett, C. 1993. 1993 Upper Mississippi River floods. *Geotimes* 38/12, pp. 15–17.

Meltzer, D.J. 1999. Human responses to middle Holocene (Altithermal) climates on the North American Great Plains. *Quaternary Research* 52/3, pp. 404–416.

Mendelsohn, J. and el Obeid, S. 2004. *Okavango River: The flow of a lifeline*. Struik Publishers, Cape Town, South Africa, 176 pp.

Merriam, D.F. 1996. Kansas 19th century geologic maps. Kansas Academy of Science, *Transactions* 99, pp. 95–114.

Merriam, D.F. 2011. Topographic depressions on the High Plains of western Kansas. Kansas Academy of Science, *Transactions* 114/1–2, pp. 69–76.

Merron, G.S. and Bruton, M.N. 1995. Community ecology and conservation of the fishes of the Okavango Delta, Botswana. *Environmental Biology of Fishes* 43/2, pp. 109–119.

Miao, X., Mason, J.A., Swinehart, J.B., Loope, D.B., Hanson, P.R., Goble, R.J. and Liu, X. 2007. A 10,000 year record of dune activity, dust storms, and severe drought in the central Great Plains. *Geology* 35, pp. 119–122.

Middleton, B. 1999. *Wetland restoration, flood pulsing and disturbance dynamics*. John Wiley and Sons, Inc., New York, 389 pp.

Millennium Ecosystem Assessment (MEA) 2005. Ecosystems and human well-being: Synthesis. Island Press, Washington, D.C., 80 pp. Accessed online <http://www.millenniumassessment.org/documents/document.356.aspx.pdf> June 2011.

Millennium Ecosystem Assessment (MEA) 2011. Millennium Ecosystem Assessment. Accessed online <http://www.maweb.org/en/About.aspx> June 2011.

Millennium Project (MP) 2011. The Millennium Project. Accessed online <http://www.unmillenniumproject.org/goals/index.htm> June 2011.

Miller, C. 2002. Plant fact sheet: Pickerelweed, *Pontederia cordata* L. U.S. Department of Agriculture, Natural Resources Conservation Service. Accessed online <http://plants.usda.gov/factsheet/pdf/fs_poco14.pdf> September 2010.

Miller, J.A. and Appel, C.L. 1997. Ground water atlas of the United States: Kansas, Missouri, and Nebraska. U.S. Geological Survey, HA 730-D. Accessed online <http://pubs.usgs.gov/ha/ha730/ch_d/> May 2011.

Miller, S.M. 1990. Land development and use, in Bleed, A.S. and Flowerday, C. (eds), *An atlas of the Sand Hills. Conservation and Survey Division, Resource Atlas No. 5a* (2nd edn), pp. 207–226. Institute of Agriculture and Natural Resources, University of Nebraska-Lincoln.

Milliken, K.T., Anderson, J.B. and Rodriguez, A.B. 2008. A new composite Holocene sea-level curve for the northern Gulf of Mexico. *Geological Society of America, Special Paper* 443, pp. 1–11.

Mitchell, D.S. 1974. *Aquatic vegetation and its use and control*. United Nations Educational, Scientific, and Cultural Organization, Paris.

Mitsch, W.J. (ed.) 2006. *Wetland creation, restoration and conservation: The state of the science*. Elsevier, Amsterdam, p. 418.

Mitsch, W.J. and Gosselink, J.G. 2000. *Wetlands* (3rd edn), Wiley and Sons, New York, 920 pp.

Mitsch, W.J. and Gosselink, J.G. 2007. *Wetlands* (4th edn), John Wiley & Sons, Hoboken, NJ.

Mitsch, W.J. and Jørgensen, S.E. 2003. *Ecological engineering and ecosystem restoration*. Wiley and Sons, Hoboken, NJ, 411 pp.

Mittermeier, R.A., Harris, M.B., Mittermeier, C.G., Seligmann, P., Cardoso da Silva, J.M., Lourival, R. and da Fonseca, G.A.B. 2005. *Pantanal: South America's wetland jewel*. Firefly Books, Buffalo, New York, 176 pp.

Money, R.P., Wheeler, B.D., Baird, A.J. and Heathwaite, L. 2009. Replumbing wetlands – managing water for the restoration of bogs and fens, in Maltby, E. and Barker, T. (eds), *The wetland handbook*, pp. 755–799. Blackwell Publishing, Oxford.

Morgan, J.P. and Andersen, H.V. 1961. Genesis and paleontology of the Mississippi River mudlumps. Department of Conservation, Louisiana Geological Survey, *Geological Bulletin* 35.

Morton, R.A., Bernier, J.C. and Barras, J.A. 2006. Evidence of regional subsidence and associated wetland loss induced by hydrocarbon production,

Gulf Coast region, USA. *Environmental Geology* 50/2, pp. 261–274.

Mourao, G., Harris, M.B., Lourival, R., Mittermeier, R.A. and Mittermeier, C.G. 2005. Grasslands: Born of flood and drought, in Allofs, T., Mittermeier, R.A., Harris, M.B., Mittermeier, C.G., Cardoso da Silva, J.M., Lourival, R., da Fonseca, G. and Seligmann, P.A. (eds), *Pantanal: South America's wetland jewel*, pp. 108–131. Firefly Books, Buffalo, New York.

Muilenburg, G. (ed.) 1961. Stories of resource-full Kansas: Featuring the Kansas landscape. State Geological Survey, University of Kansas, Pamphlet 1, 42 pp.

Murphy, T. 2010. *Dreissena polymorpha* (Pallas, 1771): Wandering mussel. *Encyclopedia of Life*. Accessed online <http://www.eol.org/pages/493165> October 2010.

Murtha, P.A., Deering, D.W., Olson, C.E. Jr. and Bracher, G.A. 1997. Vegetation, in Philipson, W.R. (ed.), *Manual of photographic interpretation* (2nd edn), American Society for Photogrammetry and Remote Sensing, Bethesda, Maryland, United States, pp. 225–255.

Musgrove, P. 2010. *Wind power*. Cambridge University Press, 323 pp.

Nairn, R.W., Beisel, T., Thomas, R.C., LaBar, J.A., Strevett, K.A., Fuller, D., Strosnider, W.H., Andrews, W.J., Bays, J. and Knox, R.C. 2009. Challenges in design and construction of a large multi-cell passive treatment system for ferruginous lead-zinc mine waters, in Barnhisel, R.I. (ed.), *Revitalizing the environment: Proven solutions and innovative approaches*, pp. 871–892. *Joint conference of the 26th Annual Meetings of the American Society of Mining and Reclamation and 11th Billings Land Reclamation Symposium, May 30–June 5, 2009, Billings, Montana*.

National Aeronautical and Space Administration (NASA) 2010a. Hypoxia related data integration within COAST. NASA Stennis Space Center. Accessed online <http://www.coastal.ssc.nasa.gov/coast/COAST_hypoxia.aspx> December 2010.

National Aeronautical and Space Administration (NASA) 2010b. Reclaiming Mesopotamia's marshes. Accessed online <http://earthobservatory.nasa.gov/IOTD/view.php?id=38409> December 2010.

National Aeronautical and Space Administration (NASA) 2011a. LCDM: Landsat Data Continuity Mission. Accessed online <http://ldcm.nasa.gov/> May 2011.

National Aeronautical and Space Administration (NASA) 2011b. New solar cycle prediction. NASA Science News. Accessed online <http://science.nasa.gov/science-news/science-at-nasa/2009/29may_noaaprediction/> September 2011.

National Atlas 2009. Zebra mussels. U.S. National Atlas. Accessed online <http://www.nationalatlas.gov/articles/biology/a_zm.html#two> October 2010.

National Invasive Species Information Center (NISIC) 2008. Laws and Regulations. National Agricultural Laboratory, U.S. Department of Agriculture. Accessed online <http://www.invasivespeciesinfo.gov/laws/execorder.shtml> September 2010.

National Invasive Species Information Center (NISIC) 2010. Aquatic species. National Agricultural Laboratory, U.S. Department of Agriculture. Accessed online <http://www.invasivespeciesinfo.gov/aquatics/main.shtml> October 2010.

National Museum of Denmark (NMD) 2010a. The Sun Chariot. National Museum of Denmark. Accessed online <http://natmus.dk/historisk-viden/danmark/oldtid-indtil-aar-1050/bronzealderen-1700-fkr-500-fkr/solvognen/> November 2010.

National Museum of Denmark (NMD) 2010b. The chemistry of the bog bodies. National Museum of Denmark. Accessed online <http://natmus.dk/historisk-viden/danmark/oldtid-indtil-aar-1050/aeldre-jernalder-500-fkr-400-ekr/kvinden-fra-huldremose/moseligenes-kemi/> November 2010.

National Oceanographic Data Center (NODC) 2008. NODC Coastal Water Temperature Guide (CWTG). National Oceanographic and Atmospheric Administration. Accessed online <http://www.nodc.noaa.gov/dsdt/cwtg/> June 2011.

National Park Service (NPS) Cape Hatteras 2010. Moving the Cape Hatteras Lighthouse. U.S. National Park Service. Accessed online <http://www.nps.gov/caha/historyculture/movingthelighthouse.htm> November 2010.

National Park Service (NPS) 2011. National Park Service: Alaska. Accessed online <http://www.nps.gov/state/ak/index.htm?program=parks> March 2011.

National Technical Committee for Hydric Soils (NTCHS) 2010. NTCHS functions. Natural Resources Conservation Service, U.S. Department Agriculture. Accessed online <http://soils.usda.gov/use/hydric/ntchs/functions.html> August 2010.

National Trappers Association 2005. Mink. Accessed online <http://www.nationaltrappers.com/mink.html> May 2011.

National Wildlife Refuge System (NWRS) 2010. Welcome to the National Wildlife Refuge system. Accessed online <http://www.fws.gov/refuges/> June 2011.

NATO 2005. Water: A key security asset. Accessed online <http://www.nato.int/docu/water/html_en/water01.html> October 2010.

Natural Resources Conservation Service (NRCS) 1996. Wetlands programs and partnerships. USDA, RCA Issue Brief #8. Accessed online <http://www.nrcs.usda.gov/wps/portal/nrcs/detail/national/technical/nra/rca/?&cid=nrcs143_014214> November 2010.

Natural Resources Conservation Service (NRCS) 2010a. Hydric soils – introduction. Natural Resources Conservation Service, U.S. Department Agriculture. Accessed online <http://soils.usda.gov/use/hydric/intro.html> August 2010.

Natural Resources Conservation Service (NRCS) 2010b. Record enrollment in Wetlands Reserve Program benefiting farm operators, migratory birds, and recreational activities. Accessed online <http://www.mn.nrcs.usda.gov/news/news_release/2010/WRP_Enrollment_News_release.html> November 2010.

Natural Resources Conservation Service (NRCS) 2010c. Wetland restoration, enhancement, creation and construction. United States Department of Agriculture. Accessed online <http://www.nrcs.usda.gov/wps/portal/nrcs/detail/national/water/wetlands/restore/?&cid=nrcs143_010912> December 2010.

Natural Resources Conservation Service (NRCS) 2010d. Wetland Reserve Program: 2010 New Jersey Fact Sheet. U.S. Department of Agriculture, Natural Resources Conservation Service. Accessed online <http://www.nj.nrcs.usda.gov/programs/wrp/documents/2010WRPFactSheet.pdf> August 2010.

Nature Saskatchewan (NS) 2010. Important Bird Areas in Saskatchewan. Nature Saskatchewan. Accessed online <http://www.naturesask.ca/?s=stewardship&p=importantbirdareas> January 2011.

Nature of Shark Bay (NSB) 2010. Stromatolites of Shark Bay. Nature of Shark Bay. Accessed online <http://www.sharkbay.org/default.aspx?WebPageID=129> November 2010.

Nesje, A. and Dahl, S.O. 2000. *Glaciers and environmental change*. Arnold, London, 203 pp.

Nesje, A. and Kvamme, M. 1991. Holocene glacier and climate variations in western Norway: Evidence for early Holocene glacier demise and multiple Neoglacial events. *Geology* 19, pp. 610–612.

Neuendorf, K.K.E., Mehl, J.P. Jr. and Jackson, J.A. 2005. *Glossary of geology* (5th edn), American Geological Institute, Alexandria, Virgina, United States, 779 pp.

Neumann, A.C. and Hearty, P.J. 1996. Rapid sea-level changes at the close of the last interglacial (sub-stage 5e) recorded in Bahamian island geology. *Geology* 24/9, pp. 775–778.

New World Encyclopedia (NWE) 2008. Flamingo. New World Encyclopedia. Accessed online <http://www.newworldencyclopedia.org/entry/Flamingo> October 2010.

New Zealand Department of Conservation (NZDC) 1990. *The green print: The state of conservation in New Zealand in 1990*. Department of Conservation, Wellington, New Zealand.

New Zealand Report 2008. National report on the implementation of the Ramsar Convention on Wetlands: New Zealand. Accessed online <http://www.ramsar.org/pdf/cop10/cop10_nr_newzealand.pdf> July 2010.

Newell, T.L. 2010. *Ondatra zibethicus* (Linnaeus, 1766): Muskrat. Encyclopedia of Life. Accessed online <http://www.eol.org/pages/313678> October 2010.

Newfoundland Harbor Marine Institute (NHMI) 2010. Mangrove morphological and physiological adaptations. Accessed online <http://www.nhmi.org/mangroves/phy.htm> September 2010.

Niering, W.A. 1985. *Wetlands. The Audubon Society Nature Guides*. Alfred A. Knopf, New York, 638 pp.

North American Waterfowl Management Plan (NAWMP) 2009. North American Waterfowl Management Plan: History. U.S. Fish & Wildlife Service. Accessed online <http://www.fws.gov/birdhabitat/nawmp/index.shtm> January 2011.

North Carolina State University (NCSU) 2004. Mosquito control around the home and in communities. North Carolina State University, Insect Note ENT/rst-6. Accessed online <http://www.ces.ncsu.edu/depts/ent/notes/Urban/mosquito.htm> September 2010.

Nunn, P.D. 2003. Nature–society interactions in the Pacific Islands. *Geografiska Annaler, Series B: Human Geography* 85/4, pp. 219–229.

Nyerges, T.L. 1993. Understanding the scope of GIS: Its relationship to environmental modeling, in Goodchild, M.F., Parks, B.O. and Steyaert, L.T. (eds), *Environmental modeling with GIS*, pp. 75–93. Oxford University Press, 488 pp.

Nyman, A. 2010. A delta in the hand is worth two in a plan. *Wetland Science and Practice* 27/2, pp. 10–15.

Ohio Department of Natural Resources (ODNR) 2010. A to Z species guide – Fish – Longear Sunfish. Ohio Department of Natural Resources, Division of Wildlife. Accessed online <http://www.dnr.state.oh.us/Home/species_a_to_z/SpeciesGuideIndex/longearsunfish/tabid/6676/Default.aspx> October 2010.

Okavango Commission (OKACOM) 2010. The permanent Okavango River Basin Commission.

Accessed online <http://www.okacom.org/okacom-commission> July 2010.

Okavango Commission (OKACOM) 2011. Fact Sheet for the Cubango-Okavango Basin. Accessed online <http://www.okacom.org/okavango-fact-sheet/fact-sheet-for-the-cubango-okavango-basin> February 2011.

Olson, D.M, Dinerstein, E., Wikramanayake, E.D., Burgess, N.D., Powell, G.V.N., Underwood, E.C., D'amico, J.A., Itoua, I., Strand, H.E., Morrison, J.C., Loucks, C.J., Allnutt, T.F., Ricketts, T.H., Kura, Y., Lamoreux, J.F., Wettengel, W.W., Hedao, P. and Kassem, K.R. 2001. Terrestrial Ecoregions of the World: A New Map of Life on Earth. *BioScience* 51, pp. 933–938.

Orru, M., Širokova, M. and Veldre, M. 1993. Eesti sood (Estonian mires). Geological Survey of Estonia, map scale = 1:400,000.

Ostrom, E. 1990. *Governing the commons: The evolution of institutions for collective action.* Cambridge University Press, Cambridge, 280 pp.

Paal, J., Vellak, K., Liira, J. and Karofeld, E. 2009. Bog recovery in northeastern Estonia after the reduction of atmospheric pollutant input. *Restoration Ecology* 18, Issue Supplement s2, pp. 387–400.

Padbury, G.A. and Acton, D.F. 1994. *Ecoregions of Saskatchewan.* Saskatchewan Property Management Corporation, Regina, Canada.

Padre Island 2010a. The Kemp's Ridley sea turtle. U.S. National Park Service, Padre Island. Accessed online <http://www.nps.gov/pais/naturescience/kridley.htm> March 2011.

Padre Island 2010b. Sea turtle hatchling releases. U.S. National Park Service, Padre Island. Accessed online <http://www.nps.gov/pais/naturescience/releases.htm> March 2011.

Page, S.E., Siegert, F., Rieley, J.O., Boehm, H.-D.V., Jaya, A. and Limin, S. 2002. The amount of carbon released from peat and forest fires in Indonesia during 1997. *Nature* 420, pp. 61–65.

Papayannis, T. and Pritchard, D.E. 2008. Culture and wetlands – A Ramsar guidance document. Ramsar Convention, Gland, Switzerland, 78 pp. Accessed online <http://www.ramsar.org/pdf/cop10/cop10_culture_ group_e.pdf> January 2011.

Park, J.R. 2005. *Missouri mining heritage guide.* Stonehouse Pub. Co., South Miami, Florida, 279 pp.

Parrett, C., Melcher, N.B. and James, R.W. 1993. Flood discharges in the Upper Mississippi River Basin, in Floods in the Upper Mississippi River Basin, U.S. Geological Survey, Circular 1120-A. Accessed online <http://pubs.usgs.gov/circ/1993/circ1120-a/pdf/circ_1120-a_a.pdf> August 2010.

Pavri, F. and Aber, J.S. 2004. Characterizing wetland landscapes: Multitemporal and multispatial analysis of remotely sensed data at Cheyenne Bottoms, Kansas. *Physical Geography* 25/1, pp. 86–104.

Pavri, F. and Valentine, V. 2008. Image processing and analysis for salt marsh plant community and habitat mapping: Plum Island estuary, MA. Final Report for MA Office of Coastal Zone Management, MA, 65 pp.

Pearce, W. 2010. Quivira marsh short on water, but full of eagles. *Wichita Eagle*, 19 Dec. 2010.

Pearl-Guide 2011. Freshwater pearls. Pearl-Guide: The world's largest information source. Accessed online <http://www.pearl-guide.com/freshwater-pearls.shtml> June 2011.

Pearson, D.L. and Beletsky, L. 2005. *Brazil: Amazon and Pantanal.* Interlink Books, Northampton, Massachusetts, 492 pp.

Peat News (PN) 2010. VeriFlora certification for Canadian peat producers. *Peat News of the International Peat Society*, 10/2010, pp. 3–4.

Peatland Ecology Research Group (PERG) 2009. Bois-des-Bel: Plant diversity. Université Laval, Canada. Accessed online <http://www.gret-perg.ulaval.ca/12793.html> August 2011.

Peck, E. and Thieme, M. 2010. Freshwater ecoregions of the world: Upper Nile. Accessed online <http://www.feow.org/ecoregion_details.php?eco=522> October 2010.

Pelletier, J.D. 2009. The impact of snowmelt on the late Cenozoic landscape of the southern Rocky Mountains, USA. *GSA Today* 19/7, pp. 4–11.

Penner, R.L. II 2010. *The birds of Cheyenne Bottoms.* Kansas Department of Wildlife and Parks, Topeka, Kansas, 156 pp.

Perry, C.A. 2000. Significant floods in the United States during the 20th century – USGS measures a century of flood. U.S. Geological Survey, Factsheet 024-00. Accessed online <http://ks.water.usgs.gov/pubs/fact-sheets/fs.024-00.html> October 2010.

Persoz, F., Larsen, E. and Singer, K. 1972. Helium in the thermal springs of Unartoq, South Greenland. Grønlands Geologiske Undersøgelse, Rapport Nr. 44, 21 pp.

Philippe, M., Cuny, G., Suteethorn, V., Teerarungsigul, N., Barale, G., Thévenard, F., Le Loeuff, J., Buffetaut, E., Gaona, T., Košir, A., and Tong, H. 2005. A Jurassic amber deposit in southern Thailand. *Historical Biology: A Journal of Paleobiology* 17, pp. 1–6. Accessed online <http://dinosauria.academia.edu/JeanLeLoeuff/Papers/369779/A_Jurassic_amber_deposit_in_Southern_Thailand> June 2011.

Pilkey, O.H. and Young, R. 2009. *The rising sea.* Island Press/Shearwater Books, Washington, D.C., 203 pp.

Platt, R.H. 2004. *Land use and society: Geography, law and public policy*. Island Press, Washington, D.C., 455 pp.

Plum Island Ecosystem (PIE) 2011. Plum Island Ecosystem: Long Term Ecological Research. Accessed online <http://ecosystems.mbl.edu/PIE/> March 2011.

Poinar, G.O. 1992. *Life in amber*. University Press, Stanford, California, 349 pp.

Poinar, G.O. 1995. *Discovering the mysteries of amber*. Geofin, Udine, Italy, 67 pp.

Poinar, G.O. 1999. Cenozoic flora and fauna in amber. *Estudios del Museo de Ciencias Naturales de Álava* 14/2, pp. 151–154. Vitoria-Gasteiz, Spain.

Poinar, G.O. and Milki, R.K. 2001. *Lebanese amber: The oldest insect ecosystem in fossilized resin*. Oregon State University Press, Corvallis, Oregon, 96 pp.

Poinar, G.O. and Poinar, R. 1999. *The amber forest*. Princeton University Press, Princeton, New Jersey, 239 pp.

Pollowitz, G. 2010. Human intervention to blame for Russia's peat fires. *Climate Change Dispatch*. Accessed online <http://climatechangedispatch.com/climate-reports/7506-human-intervention-to-blame-for-russias-peat-fires> November 2010.

Polyak, V.J., Cokendolpher, J.C., Norton, R.A. and Asmerom, Y. 2001. Wetter and cooler late Holocene climate in the southwestern United States from mites preserved in stalagmites. *Geology* 29/7, pp. 643–646.

Poore, R.Z. and Dowsett, H.J. 2001. Pleistocene reduction of polar ice caps: Evidence from Cariaco Basin marine sediments. *Geology* 29/1, pp. 71–74.

Portney, P.R. 2008. Benefit–cost analysis, in *The concise encyclopedia of economics*. Library of Economics and Liberty. Accessed online <http://www.econlib.org/library/Enc/BenefitCostAnalysis.html> November 2010.

Post, L. von 1946. The prospect for pollen analysis in the study of the Earth's climatic history. *New Phytologist* 45, pp. 193–217.

Prasad, S.N., Ramachandra, T.V., Ahalya, N., Sengupta, T., Kumar, A., Tiwari, A.K., Vijayan, V.S. and Vijayan, L. 2002. Conservation of wetlands of India: A review. *Tropical Ecology* 43/1, pp. 173–186.

Project Tiger 2011. Sundarbans Tiger Reserve. Accessed online <http://projecttiger.nic.in/sundarbans.htm> June 2011.

Purdue 2008. Mosquito breeding sites in the community. Entomology Extension, University of Purdue. Accessed online <http://extension.entm.purdue.edu/mosquito_tool/> September 2010.

Quinty, F. and Rochefort, L. 2003. *Peatland restoration guide*, 2nd edn. Canadian Sphagnum Peat Moss Association and New Brunswick Department of Natural Resources and Energy. Accessed online <http://www.peatmoss.com/pdf/Englishbook.pdf> August 2011.

Ramberg, L., Hancock, P., Lindholm, M., Meyer, T., Ringrose, S. Silva, J., Van As, J. and VanderPost, C. 2006. Species diversity of the Okavango Delta, Botswana. *Aquatic Sciences* 68/3, pp. 310–337.

Ramsar 1971. Convention on Wetlands of International Importance especially as Waterfowl Habitat. Ramsar Convention Secretariat, Gland, Switzerland. Accessed online <http://www.ramsar.org/cda/en/ramsar-documents-texts/main/ramsar/1-31-38_4000_0__> June 2010.

Ramsar 1999. Wetland and climatic change – Background paper from IUCN. Ramsar Convention on Wetlands. Accessed online <http://www.ramsar.org/cda/en/ramsar-wetlands-and-climate/main/ramsar/1%5E21076_4000_0__#2> October 2010.

Ramsar 2003. Resolutions of the Conference of Contracting Parties: Principles and guidelines for wetland restoration. Accessed online <http://www.ramsar.org/cda/en/ramsar-documents-resol-resolution-viii-16/main/ramsar/1-31-107%5E21386_4000_0__> December 2010.

Ramsar 2010a. Contracting Parties to the Ramsar Convention on Wetlands. Accessed online <http://www.ramsar.org/cda/en/ramsar-about-parties-contracting-parties-to-23808/main/ramsar/1-36-123%5E23808_4000_0__> June 2010.

Ramsar 2010b. Launch of an animated 40th anniversary Ramsar logo. Accessed online <http://www.ramsar.org/cda/en/ramsar-activities-40ramsar/main/ramsar/1-63-443_4000_0__> June 2010.

Ramsar 2010c. The Ramsar sites information service. Accessed online <http://www.ramsar.org/cda/en/ramsar-activities-rsis-ramsar-sites/main/ramsar/1-63-97%5E7705_4000_0__> October 2010.

Ramsar 2010d. The Ramsar Convention and its mission. Accessed online <http://www.ramsar.org/cda/en/ramsar-about-mission/main/ramsar/1-36-53_4000_0__> July 2010.

Ramsar 2010e. What is the role of National Ramsar Committees? Accessed online <http://www.ramsar.org/cda/en/ramsar-about-faqs-what-is-role-of/main/ramsar/1-36-37%5E24077_4000_0__> July 2010.

Ramsar 2010f. The Ramsar Secretariat. Accessed online <http://www.ramsar.org/cda/en/ramsar-about-bodies-secr-ramsar-secretariat-7707/main/ramsar/1-36-71-77%5E7707_4000_0__> July 2010.

Ramsar 2010g. Criteria for identifying wetlands of international importance. Accessed online <http://www.ramsar.org/cda/en/ramsar-about-sites-criteria-for/main/ramsar/1-36-55%5E20740_4000_0__> July 2010.

Ramsar 2010h. The list of Wetlands of International Importance. Accessed online <http://www.ramsar.org/pdf/sitelist.pdf> January 2011.

Ramsar Convention Secretariat (RCS) 2007a. Managing wetlands: Frameworks for managing Wetlands of International Importance and other wetland sites, in *Ramsar Handbooks for the Wise Use of Wetlands*, 3rd edn, Vol. 16. Ramsar Convention Secretariat, Gland, Switzerland, 96, p. Accessed online <http://www.ramsar.org/pdf/lib/lib_handbooks 2006_e16.pdf> June 2011.

Ramsar Convention Secretariat (RCS) 2007b. *Wise use of wetlands: A conceptual framework for the wise use of wetlands*. Ramsar handbooks for the wise use of wetlands, 3rd edn, V.1. Ramsar Convention Secretariat, Gland, Switzerland, 27pp.

Ramsar Home 2011. Celebrating the Convention's 40th anniversary. Accessed online <http://www.ramsar.org/cda/en/ramsar-documents-notes-2010-2010-03-annex/main/ramsar/1-31-106-438^24441_4000_0__> May 2011.

Ramsar Strategic Plan 2006. Ramsar Strategic Plan, 2003–2008. Accessed online <http://www.ramsar.org/pdf/key_strat_plan_2003_e.pdf> June 2011.

Ramseier, D., Klotzli, F., Bollens, U. and Pfadenhauer, J. 2009. Restoring wetlands for wildlife habitat, in Maltby, E. and Barker, T. (eds), *The wetland handbook*, pp. 780–801. Blackwell Publishing, Oxford.

Randerson, J.T., Liu, H., Flanner, M.G., Chambers, S.D., Jin, Y., Hess, P.G., Pfister, G., Mack, M.C., Treseder, K.K., Welp, L.R., Chapin, F.S., Harden, J.W., Goulden, M.L., Lyons, E., Neff, J.C., Schuur, E.A.G. and Zender, C.S. 2006. The impact of boreal forest fire on climate warming. *Science* 17, 314/5802, pp. 1130–1132.

Raukas, A. and Kajak, K. 1997. Quaternary cover, in Raukas, A. and Teedumäe, A. (eds), *Geology and mineral resources of Estonia*, pp. 125–136. Estonian Academy Publishers, Tallinn.

Rebelo, L.M., Finlayson, C.M., and Nagabhatla, N. 2009. Remote sensing and GIS for wetland inventory, mapping and change analysis. *Journal of Environmental Management* 90, pp. 2144–53.

Reddy, K.R., Kadlec, R.H., Flaig, E. and Gale, P.M. 1999. Phosphorus retention in streams and wetlands: A review. *Critical Reviews in Environmental Science and Technology* 29, pp. 83–146.

Reed, P.B. 1988. National list of plant species that occur in wetlands: 1988 National Summary. U.S. Department of the Interior, Fish and Wildlife Service, *Biological Report* 88 (24), Washington, D.C., 244 pp.

Resh, V.H. 2001. Mosquito control and habitat modification: case history studies of San Francisco Bay wetlands, in Rader, R., Batzer, D.P. and Wissinger, S.A. (eds), *Bioassesment and management of North American freshwater wetlands*, pp. 413–449. Wiley and Sons, New York.

Richardson, A. 2002. *Wildflowers and other plants of Texas beaches and islands*. University of Texas Press, Austin, 247 pp.

Richardson, C. and Vaithiyanathan, P. 2009. Cycling and storage of phosphorus in wetlands, in Maltby, E. and Barker, T. (eds), *The wetland handbook*, pp. 229–248. Blackwell Publishing, Oxford.

Richmond, G.M. 1965. Glaciation of the Rocky Mountains, in Wright, H.E. Jr. and Frey, D.G. (eds), *The Quaternary of the United States*, pp. 217–230. Princeton University Press, Princeton, New Jersey.

Richmond, G.M. 1986. Stratigraphy and correlation of glacial deposits of the Rocky Mountains, the Colorado Plateau, and the ranges of the Great Basin, in Sibrava, V., Bowen, D.Q. and Richmond, G.M. (eds), *Quaternary glaciations in the northern hemisphere*, pp. 99–127. Pergamon Press, Oxford & New York.

Rockström, J., Steffen, W., Noone, K., Persson, Å, Chapin, F.S. III, Lambin, E., Lenton, T.M., Scheffer, M., Folke, C., Schellnhuber, H.J., Nykvist, B., de Wit, C.A., Hughes, T., van der Leeuw, S., Rodhe, H., Sörlin, S., Snyder, P.K., Costanza, R., Svedin, U., Falkenmark, M., Karlberg, L., Corell, R.W., Fabry, V.J., Hansen, J., Walker, B., Liverman, D., Richardson, K., Crutzen, P. and Foley, J. 2009a. Planetary boundaries: Exploring the safe operating space for humanity. *Ecology and Society* 14/2, article 32.

Rockström, J., Steffen, W., Noone, K., Persson, Å, Chapin, F.S. III, Lambin, E., Lenton, T.M., Scheffer, M., Folke, C., Schellnhuber, H.J., Nykvist, B., de Wit, C.A., Hughes, T., van der Leeuw, S., Rodhe, H., Sörlin, S., Snyder, P.K., Costanza, R., Svedin, U., Falkenmark, M., Karlberg, L., Corell, R.W., Fabry, V.J., Hansen, J., Walker, B., Liverman, D., Richardson, K., Crutzen, P. and Foley, J. 2009b. A safe operating space for humanity. *Nature* 461, pp. 472–475.

Rodriguez, R. and Lougheed, V.L. 2010. The potential to improve water quality in the middle Rio Grande through effective wetland restoration. *Water Science and Technology*, 62/3, pp. 501–509.

Roulet, N.T. 2000. Peatlands, carbon storage, greenhouse gases, and the Kyoto Protocol: Prospects and significance for Canada. *Wetlands* 20/4, pp. 605–615.

Rubec, C.D. 1994. Canada's federal policy on wetland conservation: A global model, in Mitsch, W.J. (ed.), *Global wetlands: Old World and New*, pp. 909–917. Elsevier, Amsterdam.

Rud, M. (ed.) 1979. *Arkæologisk håndbog*. Politikens Forlag, Copenhagen, 320 pp.

Ruddiman, W.F. 2005. *Plows, plagues, and petroleum: How humans took control of climate*. Princeton University Press, New Jersey, 202 pp.

Ruiz, G. (ed.) 1992. *Imagen de Venezuela: Una visión especial*. Petróleos de Venezuela, Caracas, 272 pp.

Rushforth, K. and Hollis, C. 2006. *Field guide to trees of North America*. National Geographic, Washington, D.C., 262 pp.

Russian Federation Report 2005. National report of the Russian Federation for the period 2003–2005. Accessed online <http://www.ramsar.org/pdf/cop9/cop9_nr_russia.pdf> June 2011.

Sachs, T., Wille, C., Boike, J. and Kutzbach, L. 2008. Environmental controls on ecosystem-scale CH_4 emission from polygonal tundra in the Lena River Delta, Siberia. *Journal of Geophysical Research* 113, G00A03, pp. 1–12.

Sahagian, D., Proussevitch, A. and Carlson, W. 2002. Timing of Colorado Plateau uplift: Initial constraints from vesicular basalt-derived paleoelevations. *Geology* 30/9, pp. 807–810.

Salsbury, G.A. and White, S.C. 2000. *Insects in Kansas*. Kansas Department of Agriculture, Topeka, Kansas, United States, 523 pp.

Saltcedar Biological Control Consortium (SBCC) 2011. Saltcedar Biological Control Consortium Grant. Accessed online <http://ucce.ucdavis.edu/universal/printedprogpageshow.cfm?pagenum=1732&progkey=1169&county=5362> January 2011.

Salvador, F., Monerris, J. and Rochefort, L. 2010. Peruvian peatlands (bofedales): From Andean traditional management to modern environmental impacts. *Peatlands International* 2/2010, pp. 42–48.

San Diego Museum of Natural History (SDMNH) 2010. After the dinosaurs: When crocodiles ruled. San Diego Museum of Natural History. Accessed online <http://www.sdnhm.org/exhibits/crocs/tguide/tgcrocs.html> October 2010.

Sanderson, E., Foin, T., and Ustin, S. 2001. A simple empirical model of salt marsh plant spatial distributions with respect to a tidal channel network. *Ecological Modeling* 139/2–3, pp. 293–307.

Saskatchewan Water Authority (SWA) 2011. North American Waterfowl Management Plan. Saskatchewan Water Authority. Accessed online <http://www.swa.ca/Stewardship/NorthAmericanWaterfowlManagementPlan/Default.asp?type=PrairieShores> January 2011.

Sauchyn, D.J. 1997. Practical and theoretical basis for mapping landscape sensitivity in the southern Canadian Interior Plains. *Kansas Academy of Science, Transactions* 100/1–2, pp. 61–72.

Savage, A. 1995. *The Anglo-Saxon chronicles*. Crescent Books, New York, 288 pp.

Scarborough, V. 2009. Beyond sustainability: Managed wetlands and water harvesting in ancient Mesoamerica, in Fisher, C., Hill, J.B., and Feinman, G. (eds), *The archaeology of environmental change: Socionatural legacies of degradation and resilience*, pp. 62–84. University of Arizona Press, Tucson.

Scheyvens, R. 1999. Ecotourism and the empowerment of local communities. *Tourism Management* 20/2, pp. 245–249.

Schneider, J., Grosse, G. and Wagner, D. 2009. Land cover classification of tundra environments in the Arctic Lena Delta based on Landsat 7 ETM+ data and its application for upscaling of methane emissions. *Remote Sensing of Environment* 113, pp. 380–391.

Scholz, M. and Lee, B.-H. 2005. Constructed wetlands: A review. *International Journal of Environmental Studies* 62/4, pp. 421–448.

Schot, P.P. 1999. Wetlands, in Nath, B. et al. (eds), *Environmental management in practice: Vol. 3*, pp. 62–85. Routledge, London & New York, 297 pp.

Schrope, M. 2006. The dead zones. *New Scientist*, 192/2581, pp. 38–42.

Schwarzbach, M. 1986. *Alfred Wegener the father of continental drift*. Science Tech, Madison, Wisconsin, 241 pp.

Scientific Committee on Antarctic Research (SCAR) 2010. The Antarctic Treaty System. Accessed online <http://www.scar.org/treaty/> August 2010.

Scott, D.A. (ed.) 1995. *A directory of wetlands in the Middle East*. IUCN, Gland, Switzerland and IWRB, Slimbridge, United Kingdom, 560 p. Accessed online <http://www.wetlands.org/RSIS/WKBASE/MiddleEastDir/IRAQ1.htm> December 2010.

Scott, D.A. and Jones, T.A. 1995. Classification and inventory of wetlands: a global review, in Finlayson, C.M. and van der Valk, A.G. (eds), *Classification and inventory of the world's wetlands*, pp. 3–16. Kluwer Academic Publishers, Dordrecht.

Scott, L. 2010. Freshwater ecoregions of the world: Okavango. Accessed online <http://www.feow.org/ecoregion_details.php?eco=569> October 2010.

Sea Cortez Pearl Blog (SCPB) 2011. The pearl farm tour. Accessed online <http://www.perlas.com.mx/blog/2011/03/30/the-pearl-farm-tour/> June 2011.

Secretariat of the Convention on Biological Diversity (SCBD) 2006. Guidelines for the rapid ecological assessment of biodiversity in inland water, coastal and marine areas. CBD Technical Series No. 22, Montreal, Canada, 57 pp. Accessed online <http://www.ramsar.org/pdf/lib/lib_rtr01.pdf> January 2011.

Seppä, H., Birks, J.B., Bjune, A.E. and Nesje, A. 2010. Current continental palaeoclimatic research in the Nordic region (100 years since Gunnar Andersson 1909) – Introduction. *Boreas* 39/4, pp. 649–654.

Seto, K.C., and Fragkias, M. 2007. Mangrove conversion and aquaculture development in Vietnam: A remote sensing-based approach for evaluating the Ramsar convention on Wetlands. *Global Environmental Change* 17, pp. 486–500.

Setoguchi, H., Osawa, T.A., Pintaud, J.C., Jaffré, T. and Veillon, J.M. 1998. Phylogenetic relationships within Araucariaceae based on *rbcL* gene sequences. *American Journal of Botany* 85/11, pp. 1507–1516. Accessed online <http://www.amjbot.org/content/85/11/1507.full.pdf> June 2011.

Sever, M. 2010. After the spill. *Earth* 55/12, p. 36.

Shaw, S.P. and Fredine, C.G. 1956. *Wetlands of the United States, Their extent and their value for waterfowl and other wildlife*. U.S. Department of the Interior, Fish and Wildlife Service, Circular 39, Washington, D.C., 67 pp.

Shier, C.J. and Boyce, M.S. 2009. Mink prey diversity correlates with mink-muskrat dynamics. *Journal of Mammalogy* 90/4, pp. 897–905.

Shine, C. and de Klemm, C. 1999. Wetlands, water and the law: Using law to advance wetland conservation and wise use. IUCN Environmental Policy and Law Paper 38, IUCN, Gland, Switzerland, 330 pp.

Sievert, G. and Sievert, L. 2006. *A field guide to Oklahoma's amphibians and reptiles*. Oklahoma Department of Wildlife Conservation, Oklahoma City, 205 pp.

Simpson, S. 2011. The blue food revolution. *Scientific American* 304/2, pp. 54–61.

Slezak, L.A. and Last, W.M. 1985. Geology of sodium sulfate deposits of the north Great Plains, in Glaser, J.D. and Edwards, J. (eds), *Twentieth forum on the geology of industrial minerals*. Maryland Geological Survey, Special Publication 12, pp. 105–115.

Smil, V. 2011. Global energy: The latest infatuations. *American Scientist* 99, pp. 212–219.

Smith, T.J., Anderson, G.H., Balentine, K., Tiling, G., Ward, G.A. and Whelan, K.R.T. 2009. Cumulative impacts of hurricanes on Florida mangrove ecosystems: Sediment deposition, storm surges, and vegetation. *Wetlands* 29/1, pp. 24–34.

Snodgrass, J.W. and Burger, J. 2001. Management of wetlands for fish populations: Population maintenance and control, in Rader, R., Batzer, D.P. and Wissinger, S.A. (eds), *Bioassesment and management of North American freshwater wetlands*, pp. 357–386. Wiley and Sons, New York.

Snowball, I., Muscheler, R., Zillén, L., Sandgren, P., Stanton, T. and Ljung, K. 2010. Radiocarbon wiggle matching of Swedish lake varves reveals asynchronous climate changes around the 8.2 kyr cold event. *Boreas* 39/4, pp. 720–733.

Society for Ecological Restoration International Science & Policy Working Group (SERI) 2004. The SERI primer on ecological restoration. Society for Ecological Restoration International, Tucson, Arizona, 15 pp. Accessed online <http://www.ser.org/pdf/primer3.pdf> June 2011.

Society for Ecological Restoration International and IUCN Commission on Ecosystem Management (SERI IUCN) 2004. Ecological restoration, a means of conserving biodiversity and sustaining livelihoods. Society for Ecological Restoration International, Tucson, Arizona, USA and IUCN, Gland, Switzerland, 6, p. Accessed online <https://www.ser.org/pdf/Global_Rationale.pdf> December 2010.

Society of Wetland Scientists (SWS) 2010. SWS Journal Wetlands. Accessed online <http://www.sws.org/wetlands/> June 2010.

Soil Classification Working Group (SCWG) 1998. *The Canadian system of soil classification* (3rd edn), Agriculture and Agri-Food Canada, Publication 1646, 187 pp. Accessed online <http://sis.agr.gc.ca/cansis/taxa/cssc3/> May 2011.

Soil Survey Staff 1975. *Soil taxonomy: A basic system of soil classification for making and interpreting soil surveys*. U.S. Department of Agriculture, Agricultural Handbook No. 436.

Solomeshch, A.I. 2005. The West Siberian lowland, in Fraser, L.H. and Keddy, P.H. (eds), *The world's largest wetlands: Ecology and conservation*, pp. 11–62. Cambridge University Press, Cambridge.

Southern Regional Climate Center (SRCC) 2011. Tropical cyclones. Southern Regional Climate Center. Accessed online <http://www.srcc.lsu.edu/tropdesk/> February 2011.

Srinivas, H. 2008. The Indian Ocean tsunami and its environmental impacts. Global Development Research Center. Accessed online <http://www.gdrc.org/uem/disasters/disenvi/tsunami.html> October 2010.

St. Amour, N.A., Hammarlund, D., Edwards, T.W.D. and Wolfe, B.B. 2010. New insights into Holocene atmospheric circulation dynamics in central Scandinavia inferred from oxygen-isotope records of lake-sediment cellulose. *Boreas* 39/4, pp. 770–782.

Steiert, J. and Meinzer, W. 1995. *Playas: Jewels of the plains*. Texas Tech University Press, Lubbock, Texas, United States, 134 pp.

Steiner, F. 2000. *The living landscape: An ecological approach to landscape planning* (2nd edn), McGraw Hill, New York, 477 pp.

Stevenson, G.B. 1969. *Trees of Everglades National Park and the Florida Keys* (2nd edn), Published by the author, 32 pp.

Strebeigh, F. 2002. Where nature reigns. *Sierra Magazine*, March–April 2002 Issue.

Swarts, F.A. (ed.) 2000. *The Pantanal of Brazil, Bolivia and Paraguay* (2nd edn), Waterland Research Institute, Hudson MacArthur Publishers, 289 pp.

Swinehart, J.B. 1990. Wind-blown deposits, in Bleed, A.S. and Flowerday, C. (eds), *An atlas of the Sand Hills. Conservation and Survey Division, Resource Atlas No. 5a* (2nd edn), pp. 43–56. Institute of Agriculture and Natural Resources, University of Nebraska-Lincoln.

Szwedo, J. and Szadziewski, R. 2009. Umbrella-pines – new contenders for the parent trees of Baltic amber. Bursztynowy Portal. Accessed online <http://www.amber.com.pl/en/resources/amber/item/1503-umbrella-pines-new-contenders-for-the-parent-trees-of-baltic-amber> June 2011.

Tans, P. 2010. Trends in atmospheric carbon dioxide: Mauna Loa, Hawaii. NOAA/ESRL. Accessed online <http://www.esrl.noaa.gov/gmd/ccgg/trends/> December 2010.

Tansley, A.G. 1939. *The British Islands and their vegetation*. Cambridge University Press, Cambridge.

Tatem, A.J., Goetz, S.J. and Hay, S.I. 2008. Fifty years of Earth-observation satellites. *American Scientist* 96, pp. 390–398.

Teller, J.T. 1987. Proglacial lakes and the southern margin of the Laurentide ice sheet. Geological Society America, *Geology of North America, vol. K-3, North America and adjacent oceans during the last deglaciation*, pp. 39–69.

Teng, W.L., Loew, E.R., Ross, D.I., Zsilinszky, V.G., Lo, C.P., Philipson, W.R., Philpot, W.D. and Morain, S.A. 1997. Fundamentals of photographic interpretation, in Philipson, W.R. (ed.), *Manual of photographic interpretation* (2nd edn), American Society for Photogrammetry and Remote Sensing, Bethesda, Maryland, United States, pp. 49–113.

TerraNature 2007. New Zealand ecology: Penguin in paradise. TerraNature Trust. Accessed online <http://terranature.org/penguin.htm> October 2010.

The Free Dictionary 2011. Accessed online <http://encyclopedia.thefreedictionary.com/> May 2011.

Thompson, L.G., Mosley-Thompson, E., Dansgaard, W. and Grootes, P.M. 1986. The Little Ice Age as recorded in the stratigraphy of the tropical Quelccaya Ice Cap. *Science* 234/4774, pp. 361–364.

Thompson, L.G., Mosley-Thompson, E., Davis, M.E., Henderson, K.A., Brecher, H.H. Zagorodnov, V.S., Mashiotta, T.A., Lin, P.-N., Mikhalenko, V.N., Hardy, D.R. and Beer, J. 2002. Kilimanjaro ice core records: Evidence of Holocene climate change in tropical Africa. *Science* 298, pp. 589–593.

Tilford, G.L. 1997. *Edible and medicinal plants of the West*. Mountain Press, Missoula, Montana, 239 pp.

Tiner, R.W. 1997. Wetlands, in Philipson, W.R. (ed.), *Manual of photographic interpretation* (2nd edn), American Society for Photogrammetry and Remote Sensing, Bethesda, Maryland, United States, pp. 475–494.

Tockner, K. and Stanford, J.A. 2002. Riverine flood plains: present state and future trends. *Environmental Conservation* 29, pp. 308–330.

Touré, Y. 2001. Malaria vector control in Africa: Strategies and challenges. Report from a symposium on Malaria in Africa: Emerging prevention and control strategies. AAAS Annual Meeting, Feb. 17, 2001. Accessed online <http://www.aaas.org/international/africa/malaria/toure.html> June 2010.

Tucker, C.J. 1979. Red and photographic infrared linear combinations for monitoring vegetation. *Remote Sensing Environment* 8, pp. 127–150.

Trimble, S. 2001. *Great Sand Dunes: The shape of the wind*. Southwest Parks and Monuments Association, 32 pp.

Tunnell, J.W. Jr. 2002. The environment, in Tunnell, J.W. Jr. and Judd, F.W. (eds), *The Laguna Madre of Texas and Tamaulipas*. Texas A&M University Press, College Station, Texas, 372 pp.

Turetsky, M.R. 2010. Peatlands, carbon, and climate: The role of drought, fire, and changing permafrost in northern feedbacks in climate change. *Proceedings of the 19th World Congress of Soil Science*. Accessed online <http://www.iuss.org/19th%20WCSS/Symposium/pdf/1929.pdf>June 2011.

Turner, B. 1984. Changing land use patterns in the fadamas of northern Nigeria, in Scott, E. (ed.), *Life before the drought*, pp. 149–170. Allen and Unwin, Boston.

Turner, K., Georgiou, S. and Fisher, B. 2008. *Valuing ecosystem service: The case of multi-functional wetlands*. Earthscan Publishing, London, 229 pp.

Turner, R.K., van den Bergh, J.C.J.M. and Brouwer, R. (eds) 2003. *Managing wetlands: An ecological *economics approach*. Edward Elgar, Cheltenham, United Kingdom, 318 pp.

Tushingham, A.M. 1992. Postglacial uplift predictions and historical water levels of the Great Lakes. *Journal Great Lakes Research* 18/3, pp. 440–455.

Tyre, A.J., Tenhumberg, B., Field, S.A., Niejalke, D., Parris, K. and Possingham, H.P. 2003. Improving precision and reducing bias in biological surveys: estimating false-negative error rates. *Ecological Applications* 13, pp. 1790–1801.

Tyson, K. 2009. *Rostrhamus sociabilis* (Vieillot, 1817): Everglade kite. Encyclopedia of Life. Accessed online <http://www.eol.org/pages/1049023> October 2010.

Tzoumis, K.A. 1998. Wetland policymaking in the U.S. Congress from 1789 to 1995. *Wetlands* 18/3, pp. 447–459.

Uekötter, F. 2004. The old conservation history – And the new: An argument for fresh perspectives on an established topic. *Historical Social Research* 29/3, pp. 171–191. Accessed online <http://hsr-trans.zhsf.uni-koeln.de/hsrretro/docs/artikel/hsr/hsr2004_626.pdf> June 2011.

United Kingdom (UK) Report 2005. National planning tool for the implementation of the Ramsar Convention on Wetlands: United Kingdom. Accessed online <http://www.ramsar.org/pdf/cop9/cop9_nr_uk.pdf> July 2010.

United Kingdom (UK) Report 2008. National report on the implementation of the Ramsar Convention on Wetlands: United Kingdom. Accessed online <http://www.ramsar.org/pdf/cop10/cop10_nr_uk.pdf> July 2010.

United Nations Environmental Programme (UNEP) 2004. Similarities and differences between cold-water and warm-water coral reefs. United Nations Environmental Programme. Accessed online <http://www.unep.org/cold_water_reefs/comparison.htm> October 2010.

United Nations Environmental Program (UNEP) 2010. Support for environmental management of the Iraqi Marshlands. Accessed online <http://marshlands.unep.or.jp/default.asp?site=marshlands&page_id=7B495B9E-13E0-4DFE-9D1D-C77580754FB4> December 2010.

United Nations Population Fund (UNPF) 2011. Linking population, poverty and development: Rapid growth in less developed regions. Accessed online <http://www.unfpa.org/pds/trends.htm> June 2011.

U.S. Army Corps of Engineers (USACE) 1987. Corps of Engineers wetlands delineation manual. Technical Report Y-87-1. U.S. Army Corps of Engineers Waterways Experiment Station, Vicksburg, Mississippi, 100 pp.

U.S. Army Corps of Engineers (USACE) 2002. Zebra mussel information system. U.S. Army Corps of Engineers. Accessed online <http://el.erdc.usace.army.mil/zebra/zmis/> October 2010.

U.S. Army Corps of Engineers (USACE) 2005. The results of mining at Tar Creek: Environmental case study by NRE 492 Group 5. University of Michigan. Accessed online <http://www.umich.edu/~snre492/cases_03-04/TarCreek/TarCreek_case_study.htm> July 2009.

U.S. Army Corps of Engineers (USACE) 2009. Kissimmee River restoration progress: Facts and information. Accessed online <http://www.saj.usace.army.mil/Divisions/ProgramProjectMgt/Branches/EcoSys/Everglades/KRR/DOCS/Kissimmee_FS_Nov 2009.pdf> December 2010.

U.S. Census Bureau (USCB) 2010. World POPClock projection. U.S. Census Bureau. Accessed online <http://www.census.gov/ipc/www/popclockworld.html> December 2010.

U.S. Census Bureau (USCB) 2011. American fact finder. Accessed online <http://factfinder2.census.gov/faces/nav/jsf/pages/index.xhtml> March 2011.

U.S. Department of Agriculture (USDA) 1996. Distribution of wetlands. U.S. Department of Agriculture, Natural Resources Conservation Service, Soil Survey Division, World Soil Resources. Accessed online <http://soils.usda.gov/use/worldsoils/mapindex/wetlands.html> April 2011.

U.S. Department of Agriculture (USDA) 2007. Indonesia: Palm oil production prospects continue to grow.

U.S. Department of Agriculture (USDA) 2010a. Plants profiles: *Nymphaea odorata* Aiton, American white waterlily. U.S. Department of Agriculture, Natural Resources Conservation Service. Accessed online <http://plants.usda.gov/java/profile?symbol=NYOD> October 2010.

U.S. Department of Agriculture (USDA) 2010b. Species profiles: Asian tiger mosquito. National Invasive Species Information Center, U.S. Department of Agriculture. Accessed online <http://www.invasivespeciesinfo.gov/animals/asiantigmos.shtml> October 2010.

U.S. Department of Agriculture (USDA) 2010c. Foreign agricultural service commodity intelligence report. Accessed online <http://www.pecad.fas.usda.gov/highlights/2007/12/Indonesia_palmoil/> October 2010.

U.S. Department of Agriculture (USDA) 2010d. U.S. beef and cattle industry: Background statistics and information. Accessed online <http://www.ers.usda.gov/topics/animal-products/cattle-beef/statistics-information.aspx> October 2010.

U.S. Department of Agriculture (USDA) 2011. Plants profile: *Eriophorum angustifolium* Honck. (tall cottongrass). Natural Resources Conservation Service. Accessed online <http://plants.usda.gov/java/profile?symbol=ERAN6> February 2011.

U.S. Environmental Protection Agency (USEPA) 2000. Principles for the ecological restoration of aquatic resources. EPA841-F-00-003. Office of Water (4501F), United States Environmental Protection Agency, Washington, D.C., 4 pp. Accessed online <http://www.epa.gov/owow/wetlands/restore/principles.html> December 2010

U.S. Environmental Protection Agency (USEPA) 2004. Constructed treatment wetlands. EPA 843-F-03-013, Office of Water, United States Environmental Protection Agency, Washington, DC., 2 pp. Accessed online <http://www.epa.gov/owow/wetlands/pdf/ConstructedW.pdf> June 2011.

U.S. Environmental Protection Agency (USEPA) 2007. Effects of acid rain. Accessed online <http://www.epa.gov/acidrain/effects/> June 2011.

U.S. Environmental Protection Agency (USEPA) 2010. Section 404 of the Clean Water Act: An overview. Accessed online <http://www.epa.gov/owow/wetlands/pdf/reg_authority_pr.pdf> June 2011.

U.S. Fish and Wildlife Service (USFWS) 1993. Wetlands of International Importance: United States participation in the "Ramsar" Convention. Department of the Interior, Washington D.C., 11 pp.

U.S. Fish and Wildlife Service (USFWS) 1998. National list of vascular plant species that occur in wetlands: 1998 summary of indicators. U.S. Fish and Wildlife Service. Accessed online <http://www.fws.gov/Pacific/ecoservices/habcon/pdf/1998%20National%20list.pdf> September 2010.

U.S. Fish and Wildlife Service (USFWS) 2006. National survey of fishing, hunting, and wildlife-associated recreation. U.S. Department of the Interior, U.S. Fish and Wildlife Service, and U.S. Department of Commerce, U.S. Census Bureau. Accessed online <http://wsfrprograms.fws.gov/subpages/nationalsurvey/nat_survey2006_final.pdf> October 2010.

U.S. Fish and Wildlife Service (USFWS) 2009a. Laguna Atascosa Refuge: Wildlife and habitat. U.S. Fish and Wildlife Service. Accessed online <http://www.fws.gov/refuges/profiles/index.cfm?id=21553> October 2010.

U.S. Fish and Wildlife Service (USFWS) 2009b. Arctic National Wildlife Refuge: Summary of 2008 survey activities. Accessed online <http://arctic.fws.gov/ct08summaries.htm> March 2011.

U.S. Fish and Wildlife Service (USFWS) 2010a. Migratory birds & habitat programs: Migratory Bird Treaty Act. Accessed online <http://www.fws.gov/pacific/migratorybirds/mbta.htm> June 2010.

U.S. Fish and Wildlife Service (USFWS) 2010b. Sandhill cranes: An ancient path. Alamosa/Monte Vista/Baca National Wildlife Refuge complex. Accessed online <http://www.fws.gov/alamosa/cranes.html> February 2011.

U.S. Fish and Wildlife Service (USFWS) 2010c. Wildlife: Changing seasons, changing wildlife. Alamosa/Monte Vista/Baca National Wildlife Refuge complex. Accessed online <http://www.fws.gov/alamosa/Wildlife.html> February 2011.

U.S. Fish and Wildlife Service (USFWS) 2010d. Arctic National Wildlife Refuge: Wildlife and wild landscapes. Accessed online <http://arctic.fws.gov/wildlife_habitat.htm> March 2011.

U.S. Fish and Wildlife Service (USFWS) 2011a. Yukon Delta National Wildlife Refuge. Accessed online <http://yukondelta.fws.gov/> March 2011.

U.S. Fish and Wildlife Service (USFWS) 2011b. National Wildlife Refuge: Alaska. Accessed online <http://alaska.fws.gov/nwr/map.htm> March 2011.

U.S. Fish and Wildlife Service (USFWS) Breton 2011. Breton National Wildlife Refuge. Southeast Louisiana National Wildlife Refuges. Accessed online <http://www.fws.gov/breton/index.html> March 2011.

U.S. Geological Survey (USGS) 1995. Historical Landsat data comparisons: Illustrations of the Earth's changing surface. EROS Data Center.

U.S. Geological Survey (USGS) 2001. Digital orthophoto quadrangles. U.S. Geological Survey, Fact Sheet 057-01. Accessed online <http://egsc.usgs.gov/isb/pubs/factsheets/fs05701.html> August 2010.

U.S. Geological Survey (USGS) 2007. Strategic plan for the North American Breeding Bird Survey: 2006–2010. *U.S. Geological Survey, Circular* 1307, p. 19.

U.S. Geological Survey (USGS) 2008. Zebra mussel. Great Lakes Science Center, U.S. Geological Survey. Access online <http://www.glsc.usgs.gov/main.php?content=research_invasive_zebramussel&title=Invasive%20Invertebrates0&menu=research_invasive_invertebrates> October 2010.

U.S. Geological Survey (USGS) Devils Lake 2010. Elevation of Devils Lake: Period of record graph for Devils Lake near Devils Lake, North Dakota. U.S. Geological Survey. Accessed online <http://nd.water.usgs.gov/devilslake/data/dlelevation.html> November 2010.

Valk, K.U. (ed.) 1988. *Eesti sood* (Estonian peatlands). Kirjastus Valgus, Tallinn, 344 pp.

van den Bergh, J.C.J.M., Barendregt, A., Gilbert, A.J. 2004. *Spatial ecological-economic analysis for wetland management: Modelling and scenario evaluation of land use.* Cambridge University Press, Cambridge, 239 pp.

van der Valk, A.G. 2009. Restoration of wetland environments: lessons and successes, in Maltby, E. and Barker, T. (eds), *The wetland handbook* pp. 729–754. Blackwell Publishing, Oxford.

Van Diver, B.B. 1990. *Roadside geology of Pennsylvania*. Mountain Press Pub. Co., Missoula, Montana, 352 pp.

van Kooten, G.C. and Schmitz, A. 1992. Preserving waterfowl habitat on the Canadian Prairies: Economic incentives versus moral suasion. *American Journal of Agricultural Economics* 74/1, pp. 79–89.

Vasilas, L.M., Hurt, G.W. and Noble, C.V. (eds) 2010. Field indicators of hydric soils in the United States, version 7.0. Natural Resources Conservation Service, U.S. Department Agriculture. Accessed online <ftp://ftp-fc.sc.egov.usda.gov/NSSC/Hydric_Soils/FieldIndicators_v7.pdf>August 2010.

Veatch, S.W. and Meyer, H.W. 2008. History and paleontology at the Florissant fossil beds, Colorado. *Geological Society of America, Special Paper* 435, pp. 1–18.

Verhoeven, J., Arheimer, B., Yin, C. and Hefting, M.M. 2006. Regional and global concerns over wetlands and water quality. *Trends in Ecology and Evolution* 21, pp. 96–103.

Viau, A.E., Gajewski, K., Fines, P., Atkinson, D.E. and Sawada, M.C. 2002. Widespread evidence of 1500 yr climatic variability in North America during the past 14,000 yr. *Geology* 30, pp. 455–458.

Viking Ship Museum (VSM) 2010. The five Skuldelev ships. Viking Ship Museum. Accessed online <http://vikingeskibsmuseet.dk/index.php?id=1404&L=1> November 2010.

Virtual Fossil Museum (VFM) 2010. Stromatolites: The oldest fossils. Virtual Fossil Museum. Accessed online <http://www.fossilmuseum.net/Tree_of_Life/Stromatolites.htm> November 2010.

Vors, L.S. and Boyce, M.S. 2009. Global declines of caribou and reindeer. *Global Change Biology* 5/11, pp. 2626–2633.

Vymazal, J. (ed.) 2008. *Wastewater treatment, plant dynamics and management in constructed and natural wetlands*. Springer, Dordrecht, the Netherlands, 350 pp.

Vymazal, J., Greenway, M., Tonderski, K., Brix, H. and Mander, U. 2006. Constructed wetlands for wastewater treatment, in Verhoeven, J.T., Beltman, B., Bobbink, R. and Whigham, D.F. (eds), *Wetlands and natural resource management*, pp. 69–91. Springer, Berlin.

Wais, I.R. and Roth-Nelson, W. 1994. Management strategy for a large South American floodplain wetlands system: The Parana–Paraguay basin, in Mitsch, W.J. (ed.), *Global wetlands: Old World and New*, pp. 713–723. Elsevier, Amsterdam.

Warburton, K. 2010. *Mirounga angustirostris* (Gill, 1866): Northern elephant seal. Encyclopedia of Life. Accessed online <http://www.eol.org/pages/328640> October 2010.

Ward, F. and Ward, C. 1998. *Pearls*. Gem Book Publishers, Bethesda, Maryland.

Washington Ecology 2010. Non-native invasive freshwater plants: Fragrant water lily (*Nymphaea odorata*). Department of Ecology, State of Washington. Accessed online <http://www.ecy.wa.gov/programs/wq/plants/weeds/lily.html> October 2010.

Watanabe, I. 1982. *Azolla-Anabaena* symbiosis – its physiology and use in tropical agriculture, in Dommergues, Y.R. and Diem, H.G. (eds), *Microbiology of tropical soils and plant productivity*, pp. 169–185. Martinus Nijhoff/W. Junk, The Hague.

Webber, M.E. 2010. A dirty secret, China's greatest imports: Carbon emissions. *Earth*. Accessed online <http://www.earthmagazine.org/earth/article/3cc-7da-b-16> December 2010.

Weckström, J., Seppä, H. and Korhola, A. 2010. Climatic influence on peatland formation and lateral expansion in sub-arctic Fennoscandia. *Boreas* 39/4, pp. 761–769.

Weersink, A. 2002. Policy options to account for the environmental costs and benefits of agriculture. *Canadian Journal of Plant Pathology* 24/3, pp. 265–273.

Weise, B.R. and White, W.A. 1980. *Padre Island National Seashore: A guide to the geology, natural environments, and history of a Texas barrier island*. Bureau of Economic Geology, Guidebook 17, 94 pp. University of Texas, Austin (second printing 1991).

Welcomme, R.L., Bene, C., Brown, C.A., Arthington, A., Dugan, P., King, J.M. and Sugunan, V. 2006a. Predicting the water requirements of river fisheries, in Verhoven, J.T., Beltman, B., Bobbink, R. and Whigham, D.F. (eds), *Wetlands and Natural Resource Management*, pp. 123–154. Springer-Verlag, Berlin, Heidelberg.

Welcomme, R.L., Brummet, R.E., Denny, P., Hasan, M.R., Kaggwa, R.C., Kipkemboi, J., Mattson, N.S., Sugunan, V. and Vass, K.K. 2006b. Water management and wise use of wetlands: Enhancing productivity, in Verhoven, J.T., Beltman, B., Bobbink, R., Whigham, D.F. (eds), *Wetlands and Natural*

Resource Management, pp. 155–182. Springer-Verlag, Berlin, Heidelberg.

Wells Reserve 2011. Wells Reserve Monitoring Program. Accessed online <http://www.wellsreserve.org/research/monitoring_program> March 2011.

Welsch, D.J., Smart, D.L., Boyer, J.N., Minkin, P., Smith, H.C. and McCandless, T.L. 1995. Forested wetlands. U.S. Department of Agriculture, Forest Service, Northeastern Area, NA-PR-01-95. Accessed online <http://na.fs.fed.us/spfo/pubs/n_resource/wetlands/index.htm> July 2010.

Western Hemispheric Shorebird Reserve Network (WHSRN) 2009. Chaplin Old Wives Reed Lakes. Western Hemispheric Shorebird Reserve Network. Accessed online <http://www.whsrn.org/site-profile/chaplin-old-wives-reed-lakes> January 2011.

Wheeler, B., Shaw, S.C., Fojt, W.J., and Robertson, R.A. (eds) 1995. *Restoration of temperate wetlands*, Wiley and Sons, Chichester, 562 pp.

Whelan, K.R.T., Smith III, T.J., Anderson, G.H. and Ouellette, M.L. 2009. Hurricane Wilma's impact on overall soil elevation and zones with the soil profile in a mangrove forest. *Wetland* 29/1, pp. 16–23.

Whigham, D., Dykyjova, D., and Hejny, S. 1993. *Wetlands of the world: Inventory, ecology and management*, Vol. 1, Kluwer, Dordrecht.

Whigham, D.F., Broome, S.W., Richardson, C.J., Simpson, R.L. and Smith, L.M. 2010. The Deepwater Horizon disaster and wetlands. *Wetland Science and Practice* 27/2, pp. 6–8.

White, J. and Reddy, K. 2009. Nitrogen cycling in wetlands, in Maltby, E. and Barker, T. (eds), *The wetland handbook*, pp. 213–227. Blackwell Publishing, Oxford.

Whitley, J.R., Bassett, B., Dillard, J.G. and Haefner, R.A. 1999. *Water plants for Missouri ponds*. Missouri Department Conservation, Jefferson City, Missouri, 151 pp.

Whitney, M. 1909. Soils of the United States. U.S. Department of Agriculture, Bureau of Soils, Bulletin No. 55.

Whitten, S., Bennett, J., Moss, W., Handley, M. and Phillips, W. 2002. Incentive measures for conserving freshwater ecosystems: Review and recommendations for Australian policy makers. Department of Environment and Heritage, Environment Australia, 145 pp. Accessed online <http://www.environment.gov.au/water/publications/environmental/ecosystems/incentive.html> November 2010.

Wiesnet, D.R., Wagner, C.R. and Philpot, W.D. 1997. Water, snow, and ice, in Philipson, W.R. (ed.), *Manual of photographic interpretation*, 2nd edn, American Society for Photogrammetry and Remote Sensing, Bethesda, Maryland, United States, pp. 257–267.

Wildflower and Wetlands Trust 2008. Saving wetlands for wildlife and people. Accessed online <http://www.wwt.org.uk/what-we-do/saving-wetlands/> June 2011.

Wildlife Management Institute (WMI) 2010. Saltcedar, flycatcher and saltcedar leaf beetle – Three part disharmony. Wildlife Management Institute, Outdoor News Bulletin. Accessed online <http://www.wildlifemanagementinstitute.org/index.php?option=com_content&view=article&id=462:saltcedar-flycatcher-and-saltcedar-leaf-beetlethree-part-disharmony&catid=34:ONB%20Articles&Itemid=54> June 2011.

Wilhite, D.A. and Hubbard, K.G. 1990. Climate, in Bleed, A.S. and Flowerday, C. (eds), *An atlas of the Sand Hills*. Conservation and Survey Division, Resource Atlas No. 5a (2nd edn), pp. 17–28. Institute of Agriculture and Natural Resources, University of Nebraska-Lincoln.

Williams, D.R. 2005. Agricultural programs, in Connolly, K.D., Johnson, S.M., and Williams, D.R. (eds), *Wetlands law and policy: Understanding Section 404*, pp. 463–488. American Bar Association, Chicago.

Williams, M. (ed.) 1990. *Wetlands: A threatened landscape*. Blackwell, Oxford, 400 pp.

Wilson, J.S. and Boles, R.J. 1967. Common aquatic weeds of Kansas ponds and lakes. *Emporia State Research Studies* 15/3, 36 pp.

Wisconsin Department of Natural Resources (WDNR) 2008. Eurasian water milfoil (*Myriophyllum spicatum*). Wisconsin Department of Natural Resources. Accessed online <http://dnr.wi.gov/invasives/fact/milfoil.htm> September 2010.

Wise, R.W. 2006. *Secrets of the gem trade: The connoisseur's guide to precious gemstones*. Brunswick House Press, Lennox, Massachusetts.

Wolfe-Simon, F., Switzer Blum, J., Kulp, T.R., Gordon, G.W., Hoeft, S.E., Pett-Ridge, J., Stolz, J.F., Webb, S.M., Weber, P.K., Davies, P.C.W., Anbar, A.D. and Oremland, R.S. 2010. A bacterium that can grow by using arsenic instead of phosphorus. *Science Express*, 2 December 2010. Accessed online <http://www.sciencemag.org/content/early/2010/12/01/science.1197258> December 2010.

Wood, A. and van Halsema, G. 2008. Scoping agriculture – Wetland interactions: Towards a sustainable multiple-response strategy. *FAO Water Reports* – 33, 178 pp. Accessed online <http://www.fao.org/docrep/011/i0314e/i0314e00.htm#Contents> October 2010.

World Bank Project Report 1998. Aral Sea Basin Program (Kazakhstan, Kyrgyz Republic, Tajikistan, Turkmenistan and Uzbekistan): Water and environmental management project. Report #17587-UZ. World Bank Global Environmental Division, Washington DC. Accessed online <http://www-wds.worldbank.org/external/default/WDSContentServer/WDSP/IB/1999/06/03/000009265_3980625101714/Rendered/PDF/multi_page.pdf> October 2010.

World Commission on Environment and Development (WCED) 1987. *Our common future*. Oxford University Press, Oxford, 400 pp.

World Resources Institute (WRI) 2011. Online global reefs map. World Resources Institute. Accessed online <http://www.wri.org/publication/reefs-at-risk-revisited/global-reefs-map> March 2011.

World Wildlife Federation (WWF) 2011. Coral triangle: The world's richest garden of corals and sea life. Accessed online <http://www.worldwildlife.org/what/wherewework/coraltriangle/> June 2011.

Yang, Y. 2007. A China environmental health project research brief: Coal mining and environmental health in China. Woodrow Wilson International Center for Scholars. Accessed online <http://www.wilsoncenter.org/index.cfm?topic_id=1421&fuseaction=topics.item&news_id=231749> December 2010.

Yin, C., Shan, B., and Mao, Z. 2006. Sustainable water management by using wetlands in catchments with intensive land use, in Verhoven, J.T., Beltman, B., Bobbink, R. and Whigham, D.F. (eds), *Wetlands and natural resource management*, Springer-Verlag, Berlin, Heidelberg, pp. 53–68.

Zedler, J.B. 2000. Progress in wetland restoration ecology. *Trends in Ecological Evolution* 15/10, pp. 402–407.

Zedler, J.B. (ed.) 2001. *Handbook for restoring tidal wetlands*. CRC Press, Boca Raton, FL, 439 pp.

Zedler, J. and Kercher, S. 2005. Wetland resources: status, trends, ecosystem services, and restorability. *Annual Review Environmental Resources* 30/1, pp. 39–74. Accessed online <http://www.salmontrout.org/files/issues/briefing_papers/zedler_and_kercher_2005.pdf> June 2011.

Zhang, L., Wang, M.-H., Huc, J., and Hob, Y.-S. 2010. A review of published wetland research, 1991–2008: Ecological Engineering and ecosystem restoration. *Ecological Engineering* 36/8, pp. 973–980. Accessed online <http://meed.whigg.cas.cn/ky/lunwen/201101/P020110114386426276606.pdf> June 2011.

Zherikhin, V.V. and Ekov, K.Y. 1999. Mesozoic and Lower Tertiary in former USSR. *Estudios del Museo de Ciencias Naturales de Álava* 14/2, pp. 119–131. Vitoria-Gasteiz, Spain.

Zhu, R., Liu, Y., Ma, J., Xu, H., Sun, L. 2008. Nitrous oxide flux to the atmosphere from two coastal tundra wetlands in eastern Antarctica. *Atmospheric Environment* 42/10, pp. 2437–2447.

Zimmerman, J.L. 1990. *Cheyenne Bottoms: Wetland in jeopardy*. University Press of Kansas, Lawrence, Kansas, 197 pp.

Ziuganov, V., Miguel, E.S., Neves, R.J., Longa, A., Fernandéz, C., Amaro, R., Beletsky, V., Popkovitch, E., Kaliuzhin, S. and Johnson, T. 2000. Life span variation of the freshwater pearl mussel: a model species for testing longevity mechanisms in animals. *Ambio: A Journal of the Human Environment* 29/2, pp. 102–105. Accessed online <http://www.bioone.org/doi/abs/10.1579/0044-7447-29.2.102> June 2011.

Zobel, A.M. 1999. Cedarite and other fossil resins in Canada, in Investigations into amber: *Proceedings of the international interdisciplinary symposium on Baltic amber and other fossil resins*, pp. 241–245. Gdansk, Poland.

Zouhar, K. 2003. *Tamarix* spp. U.S. Department of Agriculture, Forest Service, Fire Effects Information System. Rocky Mountain Research Station, Fire Sciences Laboratory Accessed online <http://www.fs.fed.us/database/feis/plants/tree/tamspp/all.html> July 2010.

Index

numbers in bold signify photographs, charts and other illustrations
numbers in italics signify tables

AAC
  see All American Canal
aapa 15, 364
Abernethy Pearl 204
Aborigines, Australian 213
acequia (irrigation system) 213, **214**, 364
acid rain 8, **177**, 183–184, 189, 364
acrotelm 180, 364
adaptive management 235, *236*, 237, 239–240, 250, 364
Adjusted Normalized Difference Wetness Index (ANDWI) 328
adventitious roots 86
Aedes aegypti (mosquito) **110**, 111, Plate 7-6
aerenchyma 86, 105, 364
aerial photography 30, 31, 32, 35, 41, 42, **43**
  airplane **33**
  balloon 42, **43**, 44
  blimp 42, **43**, 44
  color-infrared imagery 36, **37**, 41, 270, 271, Plates 3-8, 15-2
  digital orthophoto quadrangle (DOQ) 41
  kite 42, **43**, 44
  panchromatic aerial photography 39, *40*, 41, **41**, **182**, **234**, **292**, **311**

  paraglider 42
  small-format aerial photography (SFAP) 42, **43**, 44
  ultralights 42
aerobic (definition) 364
aerobic soil zone 78–79, 175, 178, 179, 180, 189
  see also acrotelm
Afghanistan 193
African buffalo (*Syncerus caffer*) 275
African crocodile (*Crocodylus niloticus*) **108**
African fish eagle (*Haliaeetus vocifer*) **108**, 124
African buffalo (*Syncerus caffer*) 275
African Humid Period 171
African lion (*Panthera leo*) **276**
agricultural tiling
  see tiling
agriculture, wetland conversion to 73, 75, 79, 172, 177, 190, 287
Akoya pearl (*Pinctada fucata*) 204, 205, *205*, Plate 11-11
Alamosa NWR, Colorado 347
Alam Pedja, Estonia **330**
Alaska 6, 102, **125**, 143, 154, 196, 336, 351–354, **351**, **353**, 357, Plates 17-28, 17-30

Alaska Department of Fish and Game 354
Alaskan bar-tailed godwit (*Limosa lapponica baueri*) 352
albedo 136, 148, **148**, 153, 154, 155, 181, 364
Alberta, Canada **20**, **48**, **62**, 299, 318, 320–321
alewife (*Alosa pseudoharengus*) 241
Alexandria, Egypt **3**
algarrobo forest 162
alkali bulrush (*Scirpus maritimus*) 95
All American Canal, United States 209, Plate 11-14
Alliance, Nebraska **182**
allogenic change 13, 30, 135, 154–155, 364
Altithermal 168, 173, 344–345, 364
  see also Holocene thermal maximum
aluminium 177, 185
Amazon Ditch, United States **304**
Amazon River, Brazil **11**, **210**
Amazonia, Brazil 11, **18**, 200, **210**, 210, 213
amber 157, 160–163, **161**, **162**, **163**, 173, 364

Wetland Environments: A Global Perspective, First Edition. James Sandusky Aber, Firooza Pavri, and Susan Ward Aber.
© 2012 James Sandusky Aber, Firooza Pavri, and Susan Ward Aber. Published 2012 by Blackwell Publishing Ltd.

American alligator (*Alligator mississippiensis*) **113**, 117, 284
American avocet (*Recurvirostra americana*) 118, **120**
American beaver (*Castor canadensis*) 126, 129
see also beaver dam; beaver lodge; beaver meadow; beaver mire
American bullfrog (*Rana catesbeiana*) **114**, 130, Plates 7-12, 7-13
American coot (*Fulica americana*) 121, **122**
American crocodile (*Crocodylus acutus*) 117, **117**, 284
American shad (*Alosa sapidissima*) 241
American sycamore (*Platanus occidentalis*) *101*
American white pelican (*Pelecanus erythrorhynchos*) 123, **124**
American white waterlily (*Nymphaea odorata*) 95, **97**
amphibians 53, 54, 113–115, **114**, 133, 193, 242
see also American bullfrog; frogs; toads; salamander
amphibious plants 98–99, 364
Amu Darya river 198, **199**, Plate 11-6B
anadromous (fish) 241, 354, 364–365
see also chinook salmon; chum salmon; coho salmon
anaerobic
definition 365
soil conditions 18–19, 24, 28–29, 72–73, 78–79, 82, 86, 138, 175, 178
anaerobic glycolysis 105, 365
anaerobiosis 78, 84
Anastasia Formation, Everglades 282
Andernach, Germany **3**
Andersson, Gunnar 164
Andes Mountains, Venezuela **3**, 336–340, **337**, **338**, **339**, **340**, **341**, Plate 17-4
Angola 265, 274, 276, 298

anhinga (*Anhinga anhinga*) 122–123, **124**, 133
Anopheles mosquito 110, 111
Antarctica 263–264
Antarctic Treaty 263
penguin nesting sites 171, *172*
Victoria Land *172*
Antioch, Nebraska **318**
Appalachian
Mountains 158, 184, 287
states 9
apple snail (*Pomacea paludosa*) 125
aquaculture 55, 193–194, 199, 203–204, *203*, *204*, 264, 273, 365
see also extractive industries; pearl farming
aquatic civilizations 9, 13
aquifer 61–62, 365
Arabia **148**, 316, Plate 8-20
Aral Sea 198, **199**, Plate 11-6B
araucarian trees (Araucariaceae) 162
Arbor Day Foundation 102
Arbuckle Mountains, Oklahoma **70**
Archaea bacteria 179
Archean Eon 156
Arctic 350–357, **351**, **353**, **356**
Arctic National Wildlife Refuge **351**, 352, Plate 17-28
Arctic Coastal Plain, Alaska 351–353, **351**, 352
Arctic fox (*Alopex lagopus*) 352, 355
Arctic tern (*Sterna paradisaea*) 352
Ardill, Saskatchewan **321**
Argentina 207, 279
Arkansas River **4**, **253**, 299, 300–312, **301**, **302**, **303**, Plates 16-4, 16-6
see also saltcedar, biocontrol of
Army Corps of Engineers (ACE) 11, 16, 56, 66, 132, 233, 241, 254
arrowhead (*Sagittaria* sp.) 93, **94**, 97
arsenic 177–178, 189
ash **27**, Plate 2-25B
Asia, South
see South Asia

Asian tiger mosquito (*Aedes albopictus*) 131
aspen 342, **342**, **345**, **347**
Atchafalaya, Louisiana 144, 196, 285, Plate 8-13
Atlantic period 166, 331, 365
see also Blytt-Sernander
Audubon Society 309
Australia 262–264
Aborigines 213
amber 162
amphibian extinction 115
coal deposits 158
mangrove coverage *270*
pearl production *205*
wetland losses 9
wetland rehabilitation 229
see also Cairns; Coral Sea; Great Barrier Reef; Murray-Darling Basin; New South Wales; Northern Territories; Queensland; Shark Bay; Victoria
autogenic
change 13, 30, 135–136, 152, 154–155
definition 13
Aztecs 190

Baca NWR, Colorado 347
Baffin Bay, Texas 291, 293
bagno 15
Bahamas **19**, **39**, 140, 156, Plate 2-5A
Baikal Highlands, Russia 354
Baker Creek, Colorado **82**
bald cypress (*Taxodium distichum*) 283
bald eagle (*Haliaeetus leucocephalus*) 125, **125**, 233
balloon (use in aerial photography) 42, **43**, 44
Baltic Coast Bog Province 331, **330**, 335
Baltic Ice Lake 331
Baltic Peatlands Improvement Society, Estonia 332
Baltic Sea **16**, 141, 163, 164, 197, **259**, 299, **329**
Bangkok, Bight of
see Bight of Bangkok

Bangladesh 8, 190, 198, 264, 270, **271**
see also Sundarbans
Bangladesh Wildlife Act 273
barnyardgrass (*Echinochloa muricata*) 99
barrier island 143, 195, **286**, 287, 291, 295, 298
Batchelder's Landing, Massachusetts **104**
Bay of Bengal 269, **269**, 270
see also Sundarbans
bayou 15, 24, 365
BBS
　see North American Breeding Bird Survey
Beaufort Sea 351
beaver
　see American beaver; Eurasian beaver
beaver dam 127, **128**, **129**, **130**, 343, 346
beaver lodge 128, **344**
beaver meadow 126, **129**
beaver mire 343–344, **344**, **346**, *347*, **347**
beel 198, 365
Bengal tiger (*Panthera tigris*) 4, 107, 210, 273
Bengal, Bay of
　see Bay of Bengal
Bering Sea 193, 353
Bermuda 140
Big Blue River, Kansas **68**
Big Brutus, Kansas **183**
Big Cypress National Preserve, Florida 281, 283
Big Muddy Lake, Saskatchewan, Canada 323
Big Pine Key, Florida **83**
Big Salt Marsh, Kansas 233, **235**
Bight of Bangkok **76**
billabong 15, 365
biocontrol
　case study: saltcedar, Arkansas River valley, USA 305–309, **306**, **307**, **308**
　definition 365
biodiversity loss xii, *xiii*
biofuels 203, 279
biogeography 135, 155
biosphere 12, 72, 82, 156, 175, **176**, 365

birch **27**, Plate 2-25B
bird migration
　see migration, birds
bird watching 6, 107, 211, 314
　see also ecosystem services; ecotourism
birds, wetland 117–125, **119**, **120**, **121**, **122**, **123**, **124**, **125**
　see also bird watching; shorebirds; waterfowl
bison 170, 312
bitter panicum (*Panicum amarum*) 291–292
bitter salt 349, 355
Bitterfeld amber 162–163
black bear (*Ursus americanus*) 289
black mangrove (*Avicennia germinans*) **87**, 92
black swan (*Cygnus atratus*) 122, **123**
black-lipped pearl (*Pinctada margaritifera*) 204, *205*, **206**, Plate 11-11
　see also pearl farming
black-necked stilt (*Himantopus mexicanus*) 118, **119**
bladderwort (*Utricularia* sp.) 96–97, **98**, 283
Blanca Peak, Colorado **342**
blanket bog
　see bog, blanket
blimp (use in aerial photography) 42, **43**, 44
blister sedge (*Carex vesicaria*) 343
BLM
　see Bureau of Land Management
Blood Creek, Kansas **310**, 312
Blue Lake, Utah **18**
Blue-Bear Lakes, Colorado **98**, 342, **344**, **345**, **347**
blunt spike rush (*Eleocharis obtusa*) **93**
Blytt-Sernander 164, 365
bofedales 365
bog 1, 5, 15, 25
　blanket 26, **26**, 29, 81, **81**, 89, 138, 365
　bodies 166, **167**, 365
　climate change and 150–152

　definition 365
　raised 329, 331, **332**, 335, 370, Plate 6-30
　study of 30–57
　transition to 136–139
　vegetation 89–105
　see also Baltic Coast Bog Province; Estonia; mire
Bogotá, Colombia 340
Bois-des-Bel, Quebec, Canada 242
Bolivia **27**, 265–267, **277**, **278**, 298, Plate 15-9
　see also Pantanal
boloto 15
bolson 365
Botswana
　Boteti River 274–275
　Makgadikgadi 265, **266**, 275
　Okavango River/Delta 1, **2**, 11, **12**, **21**, **108**, 128, 191, 193, **211**, 210, 252, 264–265, **265**
bottomland 285, 365
Boundary Glacier, Alberta, Canada **48**
BP Deepwater Horizon oil spill (2010) 9, 289, 291
Brachiopods 69, 82
Brahmaputra River 195, 269, **269**, 270, 273, 297
　see also Sundarbans
Brandis, Dietrich 220
Brazil
　Amazon River **11**, **210**
　Amazonia 11, **18**, 200, **210**, 210, 213
　Baía de Marajó, Belém **11**
　Cáceres 207
　Manaus **80**, **210**
　National Wetlands Institute 266
Brazos Santiago Pass, Texas **296**
Breton NWR, Louisiana 285, **289**, 291
brine flies **61**, 317, Plate 4-6
brine shrimp **61**, 120, Plate 4-6
British Columbia, Canada 141
British East Africa 220
British Trust for Ornithology (BTO) 55
Bronze Age
　bog finds, Denmark 166–167, **167**, **188**

bromine **200**
brook crest (*Cardamine cordifolia*) **61**
Brooks Range, Alaska 351
Brown coal
  see lignite
Bruguiera 272
Brundtland Commission, Report 221, 358
Bryales 331
Buccaneer State Park, Mississippi **290**
Buenos Aires 207
buffalo 193, 275
bulrush (*Scirpus* sp.) 94–95, **95**, 275, **321**, 331, 349, **350**, Plate 17-26
Bureau of Land Management (BLM) (USA) 56, 213, 352, 354
Bureau of the Census (USA) 56
buttonbush (*Cephalanthus occidentalis*) **21**, 91, **92**, **101**, Plate 6-14

cactus 339
Cactus Hills, Saskatchewan, Canada 319, **320**, **323**, Plate 16-29
caddisfly (*Limnephilus* sp.) 109, 162
cadmium 196, 246
Cairns, Australia 112, Plate 7-7
calcium carbonate 69
California **25**, **44**, **82**, 102, 117, **127**, **209**, Plate 11-14
Caminada Bay, Louisiana **290**
Canada goose (*Branta canadensis*) 122, **123**
Canadian Federal Policy on Wetland Conservation 255
canal-and-lock system 242
canals, effect on wetlands 8, 9, 60–61, 208
  see also All American Canal; Erie Canal; Forth and Clyde Canal; Miami Canal; Rio Grande Canal; Union Canal; Welland Canal
Cancún agreement 187
Canning River, Alaska 351, Plate 17-28
canoe **18**

Canon Beach, Oregon **24**
Cañon City, Colorado **301**
canvasback (*Aythya valisineria*) 121, **122**, Plate 7-26
Cape Fear Formation **162**
Cape Hatteras, North Carolina 144, 147, **147**
capybara (*Hydrochaeris hydrochaeris*) 278
carbamate insecticide 111
carbon (C)
  carbon dioxide (CO_2) (atmospheric) *xiii*, 3
  carbonic acid 69
  sequestration 194–195, 215, 261–262, 269, 297
  storage 155, 179, *179*, 181, 194, 215
carbon cycle *xii*, *xiii*, 3–4, 13, 150, 155, 179–181, 194
Carboniferous period 157, 173
Caribbean Sea 112, 113, 140, 279, 309
caribou/reindeer (*Rangifer tarandus*) 352, 355, 357
carnivorous plants **90**
carotenoid pigments 317, 350, Plates 4-6, 17-26
carp (*Cyprinus carpio*) 233
carr 366
Casamance River, Senegal **92**
Caspian Sea 131, 171, **252**, Plate 14-1
castoreum 126
catotelm 180, 366
  see also anaerobic soil conditions
cattail (*Typha domingensis*, sp.) 94, **96**, 293
cattail hybrid (*Typha glauca*) 94, 225
cattail infestation, control of **104**, 105, 127, 225, 311
cattle raising 199, 200, 213, **214**, 224, 226, 279, 298, 300, 315, **317**
Cenozoic era 157, 159–160, 163, 173
Center for Biological Diversity 309
Cereopsis goose (*Cereopsis novaehollandiae*) 122

Challenger II light sport aircraft **33**
Chandeleur Islands, Louisiana 285, **286**, 291
Chaplin Lake, Saskatchewan, Canada 323
Charadriidae 118
chemical elements
  see elements (chemical)
Chesapeake Bay 141, 197
Cheyenne Bottoms, Kansas **5**
  bird habitat 211, **226**, 309
  case study 309–312, **310**, **311**, **312**, **314**, **315**
  cattail control **104**, 154
  restoration of 233, **234** (*see also* case study)
  satellite views Plates 3-8, 16-20, 16-21
soil profile **73**
chikungunya 11
China 184, 187, 202–204, *204*, *205*, 224–225
chinampa 190, 366
chinook salmon (*Oncorhynchus tshawytscha*) 193, 354
Chukchi Sea 143, 351
chum salmon (*Oncorhynchus keta*) 354
chytrid fungus (*Batrachochytrium dendrobatidis*) 115, 133
  see also amphibians
cinnamon teal (*Anas cyanoptera*) 121, **121**, Plate 7-25
cirque 336, 343
clams 108, 293
classification, wetland
  see wetland classification
Clean Water Act (USA) 213, 233, 253, 254
Clear Lake, Nebraska **170**
coal 150, 157–160, 173, **182**, 187, 258, **260**, 366
  see also coal mining; fossil fuels; lignite
coal mining **158**, 160, 183–184, 187
coastal erosion **20**, 144, 195, 279
  see also shore migration

Coastal Maine National Wildlife Refuge
  see Rachel Carson National Wildlife Refuge
coho salmon (*Oncorhynchus kisutch*) 354
Cold War 260
collared lemming (*Dicrostonyx torquatus*) 355
colonialism 190, 220, 229
Colorado Wildlife Habitat Stamp 347
Colorado
  River 11–14, **40**, 63, **209**, Plate 3-14
  see also Arkansas River; Cuchara; Culebra; Russell Lakes; Sangre de Cristo Mountains; San Luis Valley; Southern Colorado case study
color-infrared imagery (in aerial photography) 36, **37**, 41, 270, 271, Plates 3-8, 15-2
  see also aerial photography
Commerce, Oklahoma 49, **246**, **247**, **249**, Plate 13-9
common bladderwort (*Utricularia vulgaris*)
common cattail (*T. latifolia*) 20, 94, *101*, **235**, **249**, 311, **316**, 320
common green darner dragonfly (*Anax junius*) **110**, Plate 7-5
common plantain (*Plantago major*) 99
common reed (*Phragmites australis*)101, 103, **104**, 324, **326**
Commonwealth of Independent States 257, 260–262
conifers 159, *160*, **258**, **332**
Conneaut Marsh, Pennsylvania **22**, **96**, **104**, **169**, Plates 6-21, 6-34
Connecticut 9, *9*
Conservation movement, thought 218–220
Continental Divide (North America) 300
continental shelf 22, 81

Convention on Wetlands of International Importance especially as Waterfowl Habitat
  see Ramsar Convention
coontail, hornwort (*Ceratophyllum demersum*) **22**, 97, **99**
copal 161
Copenhagen, Denmark **46**, **185**, 187
copper 105, 196, 245
  mining 184
coral reefs 4, 17, 22, **84**, 108, 111–113, **112**, 133, 150, 195, 207
Coral Sea, Australia 112, **112**, Plate 7-7
Coral Triangle 207
cordgrass (*Spartina* sp.) 191, **191**, 324, 325, **324**, **326**, 327
Cordillera de Mérida, Venezuela 336, **338**
Cordova Pass, Colorado 150
coring 51, **52**, **53**, 141
  see also Geoprobe; ice coring
cormorant 117, 122, 124, 133
corn 73, 207, 279
corn belt (U.S.), wetland losses 9, *9*
Corpus Christi, Texas 291
cost–benefit analysis 215, 222, 366
coteau 366
  see also Missouri Coteau
cotton 73, 198, 287
cottongrass (*Eriophorum angustifolium*) 343, **346**, 352
cottongrass mire **345**, **346**
cottonwood (*Populus deltoides*) **101**, 307, 309
Cottonwood River, Kansas **66**, 131
Cottonwood-Steverson Wildlife Management Area, Nebraska **318**
cowhorn orchid (*Cyrtopodium punctatum*) 284
cranberry (*Vaccinium oxycoccos*) *101*

crane 4, **5**, 193, 210, 233, 309, 347, **349**, 354
crabs 108, 197, 293
crayfish 108, 116, 116
Cree 200
Crestwynd, Saskatchewan, Canada 41
Cretaceous **150**, 157, 158–159, **159**, 173
crops
  see agriculture
cryosphere 366
Cuba *270*
Cuchara, Colorado 20, 98, Plate 2-10
Cuchara River **301**
Cucharas Creek, Colorado 128, 342, **342**, **344**, **347**
Cucharas Pass **101**
Culbertson, Montana **49**
Culebra
  Creek, (Colorado 213, **214**, 347
  Peak, Colorado **343**
  Range, Colorado **82**, **227**, **301**, 357, case study 341–345, **341**, **342**, **343**, **345**, **347**
Culex mosquito 110
cumbungi swamp 366
cypress 20

dabbling/dabbler duck 121, **121**, Plate 7-25
dambo 200, 366
damselflies (Zygoptera) 107, 109, **110**, Plate 7-5
Darcy's Law 62
Dartmoor, England **81**
Dauphin Island, Alabama 291
DDT 111
Dead Sea **200**
Deception Creek, Kansas **311**
Del Norte, Colorado **349**
deltas
  definition 366
  glaciomarine 143
  human use of 9, 195
  see also Amazonia; Amu Darya; Colorado River; Ganges River/Delta; Lena River; Mississippi River Delta; Okavango River/Delta; Sundarbans; Volga River Delta

Denali National Park, Alaska 354
Dené 200
dengue, dengue hemorrhagic
    fever 110, 111, Plate 7-6
denitrification 177
    see also nitrogen
Denmark
    Bovbjerg, Jylland 46
    Copenhagen 185, 187
    Fegge, Limfjord 59
    Kong Asger's høj, Møn 166
    Lindholm, Langerak 168
    Marmor Broen 185
    Mårup Kirke, Skallerup Strand
        145, 146, Plate 8-14
    Mosegård Museum, Århus 167
    Ramme Dige 188
    Ristinge Klint 71, Plate 4-23
    Roskilde 167, 168
    Roskilde Fjord, Sjælland, 167
    Silkeborg Museum, 167
    Skagen Peninsula, 144, 146
    Skaggerak Sea, 146
    Skærbæk, Jylland, 52
    Stevns Klint, 147, 159
    Trundholm Mose, Sjælland,
        Denmark 167
    Vensyssel, Jylland, Denmark
        145, Plate 8-14
    Viking Ship Museum, Roskilde,
        Denmark 168
Department of Agriculture, U.S.
    see U.S. Department of
        Agriculture
Des Moines River 66
Desulphovibrio bacteria 178
Deûle canalisée, Lambersart,
    France 132
Devils Lake, North Dakota 151,
    151, 152, 153, Plate 8-28
dhap 190, 366
diamond exploration 8, 184
diatom 82
Dirt Hills, Saskatchewan, Canada
    319, 320
Discovery space shuttle 33
diversion, water
    see water diversion
diving duck 121, 122, Plate 7-26
dolostone 69, 70, 79, 320
Dominican Republic 117, 162
dragonfly (Anisoptera) 109, 110,
    Plate 7-5

drumlins 27,137, Plates 2-25B,
    8-1
Dry Lake, Kansas 19, 60, 80,
    305, 305, 306, 363, Plates
    4-3, 5-11, 16-10
duck 107, 117, 120–121, 121
    see also dabbling duck; diving
        duck;
"duck factory" 319
"duck potato" 93
"duck stamps"
    see Migratory Bird Hunting
        and Conservation
        Stamp
Ducks Unlimited 10, 107, 222,
    312
Ducks Unlimited Canada 255,
    323
duckweed (Lemnaceae) 22,
    94–95, 97
Dust Bowl 305

eagle
    see bald eagle; fish eagle
East African rift system 274
East Baltic Bog Province 331,
    335
East Devils Lake, North Dakota
    151, 153, Plate 8-28
eastern forktail damselfly
    (*Ischnura verticalis*) Plate
    7-5
eastern indigo snake
    (*Drymarchon corais*) 284
eastern river cooter (*Pseudemys
    concinna*) 116
economic services, wetland 190,
    199–208
    see also ecosystem functions;
        ecosystem services;
        extractive industries;
        mining
ecosystem functions (wetlands)
    6, 194, 215, 216
ecosystem services (wetlands) 4,
    6, 27, 56, 191–195, 191,
    211, 215, 216, 230, 237
ecotourism 11, 12, 210, 366
Ecuador 339, 340, 357
    see also páramo
Eemian 140, 141, 366
    see also Sangamon
egret 117, 119, 278, 284

Egypt 3, 27, 190, 198
    see also Nile River
Egyptian lotus (*Nymphaea
    caerulea*) 21
elements (chemical), wetland
    69–71, 78, 84, 175, 187,
    189
Elbląg Canal, Poland 209
elephant (*Loxodonta africana*)
    265, 266, 275
elephant seal (*Mirounga
    angustirostris*) 44, 127
elm 27, Plate 2-25B
Emajõgi, Estonia 329, 330
Emerson, Ralph Waldo 218
Emporia, Kansas 55, 66, 87
Emporia State University, Kansas
    55
Endla mire complex, Estonia 25,
    27, 258, 332–335, 332,
    334, 335
Environment Canada 11
environmental processes,
    boundaries of *xiii*
Environmental Protection Agency
    see U.S. Environmental
        Protection Agency
Eocene Sea 162
Eocene Florissant Formation 157
EPA
    see U.S. Environmental
        Protection Agency
Erie Canal, United States 208
    see also New York State Barge
        Canal
Erie Lake
    see Lake Erie
Estonia
    Alam Pedja 330
    Emajõgi 329, 330
    Endla mire complex 25, 27,
        258, 332–335, 332, 334,
        335
    Kaasikjärve 334
    Kanamatsi 334
    Linnusaare 25, 334
    Männikjärve Bog 25, 27, 139,
        151, 332, 332, 334, 335,
        Plate 3-10
    Mardimäe 334
    Matsalu nature protection area
        259
    Meenikunno bog complex 140

Narva **186**
Nigula Bog **17**, **27**, **32**, **35**, Plate 2-2A
Piirissaar **331**, **330**
Põltsamaa River **332**, **334**
Punaraba **334**
Rannu Soo **202**
Rummallika **334**
Sinijärv lake **332**, **334**
Tartu **137**, **140**, **202**, **258**, **259**, **334**, Plate 8-1
Teosaare Bog **138**
Tolkuse Bog **330**
Tooma Experimental Station **332**
Ulila **258**
Väinameri Sea **36**, **331**
Valgesoo Bog **138**, **259**
Vooremaa drumlin field **138**, Plate 8-1
Vormsi **16**, **36**, **331**
Võrtsjärv **3**
Estonian oil shale 184–186, **186**, 189
estuarine wetlands 22, 24, *23*, **24**, 58, 59, 240, 328
estuary
  definition 366
  in hydroseral succession 136
Euphrates River 233, 243, **244**
Eurasian beaver (*Castor fiber*) 126
Eurasian elk
  see moose
Eurasian water milfoil (*Myriophyllum spicatum*) 96, 103–104
European mink (*Mustela lutreola*) **130**
European Union (EU)
  Birds Directive 256, 257
  Habitat Directive 256, 257
  Water Framework Directive 256, 257
eustatic sea-level change 141–143, 147
  see also eustasy, glacial; sea-level change
eustasy, glacial 139–141, **140**, **141**, 142–143
  see also sea-level change
evaporite 19, **19**, 366
evapotranspiration 366

Everglades, Florida
  case study 281–285, **282**, **283**, 298
  flamingos *121*
  human impact on 284–285
  mangroves in 92, 268
  map **281**
  Ramsar grading 252
  recreation 209
  restoration 233
  satellite images 280, **281**, Plate 15-10
  snail kite 125, **125**
  see also Everglades National Park; Florida Keys
Everglades mink (*Mustela vison*) 284
Everglades National Park, Florida 213, 282, 284
Everglades saw grass (*Cladium jamaicense*) 177
extractive industries 27, 199–204, *202*, *203*, **204**, 217
  see also ecosystem functions; ecosystem services; fisheries; hunting; mining; pearl production
extinction rates *xiii*, 268

fadama 200, 366
Fafard et Frères, Québec, Canada 229
Falkirk Wheel, Scotland **210**
Fall River Lake, Kansas **290**
Fallon, Nevada **89**
Federal Policy on Wetland Conservation
  see Canadian Federal Policy on Wetland Conservation
fen 1, 5, 15, 26, 28, 29, 367
Fennoscandia 164–167, **165**, **166**, **167**, **168**, 352
fern 158, 159
  see also mosquito fern
Ferruginous hawk (*Buteo regalis*) 114
fertilizer 79, 186
  runoffs from 134, 177, 194, 279, 285, 320
  tax (U.S.) 228, 320
Finland, Gulf of **16**, 328
Finney Wildlife Area, Kansas **64**, Plate 3-17

Finnish Peatland Society 12
First Dark Age 171
fisheries 1
  effects of pollution on 197
  in Central Asia 198
  land ownership and 213
  Plum Island, Massachusetts 326
  wetland management and 226
Fish and Wildlife Service (U.S.)
  see U.S. Fish and Wildlife Service
fish eagle
  see African fish eagle
fjords 24, **24**, 141, 142, **142**
flamingo (Phoenicopteridae, *Phoenicopterus chilensis*) 120, *121*
Flat Ridge Wind Farm, Kansas **188**
Flint Hills, Kansas **105**, 152, Plates 5-3, 6-37
flood cycle 9, 13, 17, 193
floodplains 1, 4, 24, **39**
  conversion 196, 198, 200, 224
  definition 367
  restoring 240–241
  see also individual case studies
floods
  biochemical adaptations to 88–89, **87**, 105
  control 6, 65, 66, 195, 196, 198
  damage caused by 64–65, **65**
  human contribution to 66–68, 70, 194, 195
  impact of bridges on **66**
  Mississippi (1993) 66–67, **67**, **68**, 196
  Poland (1997) 65, **65**
  wetlands role in mitigating 1, 6, 27, 66, 191, *191*, 196, 211, 215, 216, 242, 269
Florida Bay Plate 5-18
Florida Everglades
  see Everglades, Florida
Florida Keys **15**, **84**, **87**, 112, **112**, Plate 7-8
Florida panther (*Puma concolor coryi*) 129–130, **131**, 283

fly ash (effect on bogs) 184, 186, 189, 257
Food Security Act (U.S.) (swampbuster provision) 245
fool's gold 178–179
Fort Hays State University, Kansas
Fort Leavenworth, Kansas **42**, Plate 3-17
Forth and Clyde Canal, Scotland **210**
fossil(s)
  dinosaur 159
  macrofossils 25, **26**, 29, 156
  microfossils 25, 29, 156
  stromatolites 156, **157**, 173, 370
fossil fuels 1, 11, 177, **177**, 181–184, **183**, 185–189, 367
  see also carbon cycle; coal; Estonian oil shale; Industrial Revolution; lignite; natural gas; oil
foxtail barley (*Hordeum jubatum*) 99, **100**
frailejón (Espeletiinae) 340, **340**, 357
frailejón grande (*Coespeletia moritziana*) 340
frailejón pequeño (*Espeletia schultzii*) 340
Frederick Lake, Saskatchewan, Canada **84**, **89**, 322–323, **322**, Plate 16-33
French Polynesia (pearl production in) 205, *205*
  see also pearl farming
fresh-water pearl (*Hyriopsis cumingii, Cristaria plicata, Megalonaias nervosa Margaritifera margaritifera*) 204–205, **206**
  see also pearl farming
frogs 114, 115, 130, 158
  see also amphibians

gallium 246
Galveston, Texas 144, 284
Gammaproteobacteria 178
Gangamopteris 158

Ganges River/Delta 195, 264, 269, **269**, 270, 271, 272–273, 274, 297, Plate 15-2
Garden City, Kansas 300–301, **303**, Plate 16-6
gas, natural
  see natural gas
gastropods 82
gator holes 117
gaur (*Bos gaurus*) 273
geographic information system (GIS) 30, 46, 48, 57
Geoprobe 52, **55**
germanium 246
Germany
  Bitterfeld amber 162–163
  Jossa River, Mernes **129**
  National Federal Conservation Act 256–257
  National Strategy on Biological Diversity 257
  Rhine River **3**
GFAJ-1 bacterium 177–178
gharials (Gavialidae) 116–117
giant river otter (*Pteronura brasiliensis*) 278
giant waterlily (*Victoria amazonica*) 95, **98**
giant wild pine (*Tillandsia utriculata*) 284
Gillette, Wyoming **182**
giraffe (*Giraffa camelopardalis*) 275, **276**
GIS
  see geographic information system
glacial eustasy
  see eustasy, glacial
glaciation
  Greenland Ice Sheet 139, 140
  Laurentide Ice Sheet 167
  Lake Agassiz 168, 368
  Pinedale Glaciation 342
  proglacial lakes 167–168, **169**, 173, 319–320, **320**
  West Antarctic Ice Sheet 140
  Wisconsin Glaciation **169**, **181**
  see also moraine
Glauber's salt (mirabilite) 320
global positioning system (GPS) 46, 238
Glossopteris fauna 157, 158, 367
Gobi Desert 6

golden- and silver-lipped pearl (*Pinctada maxima*) 204, *205*, 205
golpatta (*Nypa fruticans*) 273
Gondwana 156, **158**, 367
goose 121–122, **259**
Gould Pond, Maine (pollen chart) **170**
GPR
  see ground-penetrating radar
GPS
  see global positioning system
Graham's crayfish snake (*Regina grahamii*) 116
Grand Isle, Louisiana 287, **290**
granite 72, **73**, 81,82
grass carp 105
Grauballe Man, Denmark **167**
gray mangrove (*Avicennia marina*) 88, **197**, **215**, 272, **272**
grazing, livestock 1, 8, 200, 224, 226, 233
Great Barrier Reef, Australia 112, **112**, Plate 7-7
Great Bend, Kansas **310**, 312
great blue heron (*Ardea herodias*) 123, 117, **119**
great bulrush (*Scirpus validus*) 95
great diving beetle (*Dytiscus marginalis*) 109, **109**
great egret (*Ardea alba*) 284
Great Grimpen Mire 1
Great Lakes, North America 9, 131–132, 141, 168
Great Plains, North America 120, 151, 170, **182**, 207, 299–323, 332
Great Sand Dunes National Park and Preserve, Colorado 347, **348**, Plate 17-20
Great Sand Hills, Saskatchewan, Canada 300
grebe 117, 354
green ash (*Fraxinus pennsylvanica*) 86, *101*
green moss **346**
greenhouse
  effect 152, 153, 155, 181, 183, 186–189
  see also carbon cycle, carbon dioxide, climate change, methane, nitrous oxide

gases **6**, 13, 27, 81, 136, 148, 150, 152, 154, 155, 172–173, 177, 180–181, 186–189
Greenland
 fjords **24**
 Hvalsey church **149**
 Ice Age **165**
 Ice Sheet 139, 140–141
 Inuit **18**, **149**
 Uunartoq springs **69**
 Viking settlement in 149, 166–167
greylag goose **259**
Grossmoor 30
ground-penetrating radar (GPR) 52–53
Guam 112
gulf croton (*Croton punctatus*) 292
Gulf of Mexico
 see Mexico, Gulf of
gulls 118
gymnosperms 158, 162
Gwich'in 352

Halloween pennant dragonfly (*Celithemis eponina*) 284
halophytes 88, 324, 367
 see also vegetation, wetland
Halstead, Kansas **68**
Hamilton Quarry, Kansas 157
hardiness, plant
 see plant hardiness
Hawaii 112
hay 200, **202**, **301**, **317**
Head of Passes, Louisiana 285, **287**
heavy metals *see* metals, heavy
helium 68, **69**
herbicides (and wetlands) 105, 134, 227, 279
heron 117–118, **119**, 120, 133, 278
herpetology 113–114
Hidrovia Project, South America 207, 279
High Plains, North America **4**, 46, **47**, **60**, 188, **254**, 299–300, 315, Plates 3-23, 4-3
High Plains aquifer 301, 315
Himalaya(s) 8, 336

hippopotamus (*Hippopotamus amphibius*) **2**
hoars 198, 367
hooded pitcher plant (*Sarracenia minor*) **90**
Hoisington, Kansas **310**, **312**, **313**, 234
Holly, Colorado **309**
Holmes, Sherlock 1
Holocene thermal maximum (HTM) 173, 367
 in Antarctica 171
 in Nordics 164, 167
 in North America 168–171, **169**, *169* (*see also* Altithermal)
 in Tropics 171
Holocene wetlands 25, 163–172, **165**, 173–174, 180–181
horn (glacial feature) **337**, **339**
horsetail (*Equisetum hyemale, E.* sp.) 20, 91, **91**
HTM
 see Holocene thermal maximum
Hudson Bay, Canada 141, 200
human activity, impact
 see ecosystem functions; ecosystem services; extractive industries; hunting; mining; wetland ecosystems, human use of
Hungary **225**
hunting, in wetlands 4, 5, 6, 11, 107, 134, 199, 200
 alligators 117
 colonial management of 220
 ducks 253
 fur-bearing animals 126, 128–129, 134
 revenues from 211
Huron, Lake
 see Lake Huron
hurricane(s) **280**, 281
 classification of 280, *280*
 Andrew 282
 Isabel 147
 Katrina 195, *280*, 282, 289, **290**, 291
 Rita 289, **290**
 San Felipe 284
 Wilma *280*, 282

hydraulic civilizations 9, 13–14, 198
hydraulic conductivity (k) 62–63, 367
hydraulic gradient 63, 70
hydrograph 367
hydrologic cycle 58, **59**, 70
hydroperiod 58, **59**–60, *61*, 63, 70, 175, 241, 367
hydrophytes 16, 20, 86, 367
 see also vegetation, wetland
hydroseral succession 136–139, **137**, **138**, **139**
hydrosphere 12, 72, 175, *176*, 367
Hymenaea protera, verrucosa, H. sp. 162
Hypsithermal *see* Altithermal ibis **17**, 117, 120, **120**, 278, 284, 347

Ice Age, Little
 see Little Ice Age
ice coring 139, 171, 173, 351
Iceland
 peat in **26**
 Tjórsa River **19**
 Vikings 167
Ikonos satellite 42, **42**, 234
Illinois 9, *9*, 66, 161
image interpretation
 ground sample distance (GSD) 32, *33*
 multispectral imagery 42
 multitemporal imagery 311–312, **313**, **314**, Plate 16-20
 National Imagery Interpretability Rating Scale (NIIRS) *34*, 35
 spatial resolution 32 (GSD), 37–38, 40, *40* (of thematic mapper) 41, 42, 52–53, 57, **234**, **304** (high spatial resolution)
IMCG
 see International Mire Conservation Group
impala (*Aepyceros melampus*) 275
Imperial Valley, California **209**, Plate 11-14
Independence Formation (Kansas) **163**

India
  agricultural output 203, *204*
  coal mining 184
  in Cretaceous
  Dal Lake, Srinagar 190
  fishing in 200, **201**, 228
  Forest Department 220
  Indian Forest Act (1878) 273
  Jharkali 265
  Keoladeo National Park, Rajasthan 193, 210, 213
  Kerala 195
  mangroves in **195**, **197**, **201**, **215**, 265, *270*, 272
  Ministry of Environment and Forests 264
  Mithi River 197
  Mughals 190
  Mumbai 111, **197**, **215**
  National Wetland Inventory 264
  Project Tiger 273
  Revenue Department 264
  Sundarban Tiger Reserve 4, 8, 265, 271
  Taj Mahal, Agra 190
Indian Javan rhinoceros (*Rhinoceros sondaicus inermis*) 273
Indiana 9, *9*
Indonesia
  aquaculture output *204*
  Bali agreement 187
  coral in 112
  Kalimantan 153, 203
  Mega-Rice Project 153
  pearl production *205*, 205–207
  peat fires 153, 203
Indus River 1, 27
Industrial Revolution 8–9, 149, 172, 174, 181
Intergovernmental Panel on Climate Change (IPCC) 350
International Mire Conservation Group (IMCG) 12
International Peat Congress 12
International Peat Society (IPS) 12
Intracoastal Waterway **8**, 293, 295
Iñupiat 351, **351**, Plate 17–28

invasive species
  animal 131–132, **132**, **133**, 193, 224, 368
  plant 102–105, **104**, **105**, 106, 224
Iowa 9, *9*, 300, 318
IPCC
  see Intergovernmental Panel on Climate Change
Iraq
  Ma'dan 243
  Mesopotamian Marshes 6–7, 243, *244*, **245**
  Saddam Hussein 243
Ireland **3**, *75*
Irian Jaya 336
iron (Fe)
  bog iron ore 70–71
  ferric **10**, 36, 69, 70, *71*, 73, 78, **80**
  ferrous 69, *70*, 78
  in soil 19–20
Iron Age 166, **188**
irrigation
  Alberta, Canada **62**
  Aral Sea region 198
  Arkansas River valley 300, **302**, **303**, **304**
  China 224–225
  Colorado River 63
  definition 367
  ecosystem service 6
  effects on mosquito 111
  effects on wetlands 1, 4, 63, 172, 195, 198, 208
  Mesopotamian Marshes 243
  páramo 340
  San Luis Valley 347, **349**
  Saskatchewan, Canada 258
  see also acequia
Israel **200**
Italy (Venice) 143

jabiru (*Jabiru mycteria*) 278
jaguar (*Panthera onca*) 4, 210, 289
Jamaica swamp saw grass (*Cladium jamaicense*) 283
James Bay, Canada 200
Japan
  Kyoto climate agreement 187
  pearl production 204, *205*, 205–206

jelinite **161**
Jordan **200**

Kahola Creek/Lake, Kansas **88**, **92**, **93**, **113**
Kaktovik Inupiat Corporate Lands, Alaska 35, Plate 17–28
Kalahari Desert 6, 265, 274, **275**, 298
Kamchatka Peninsula, Russia **148**
kame **143**
Kansas River 219
Kansas Wetland Education Center **311**, 312
Kansas-Colorado Arkansas River Compact 300
Karelia, Russia 80, **263**, Plate 5–12
karst **263**, 368
Kattegat Sea, Denmark **145**, **146**, Plate 8–14
Kauri pine (*Agathis australis*) 162
kavir 367
Kaw Point, Kansas **219**
kayak 17, **18**
Kazakhstan 198, 260
keelboat **219**
Kemp's Ridley sea turtle (*Lepidochelys kempii*) 295, 297, **297**
Key Largo Limestone **84**, **140**
Kilimanjaro 171
kingfisher 278
Kissimmee River, Florida 240–241
kite aerial photography **43**
Kivalina, Alaska 143
Kodiak, Alaska **125**
Kofi Annan 358
kudu antelope (*Tragelaphus strepsiceros*) 275
Kuskokwim River, Alaska 354
Kuwait **153**
Kyrgyzstan 198

lacustrine wetlands 23, 24, **25**
  see also individual lakes
Lago Enriquillo, Dominican Republic 117
lagoon 22, 35, 368

Laguna Atascosa NWR, Texas 117, **118**, 293, **295**, **296**, Plate 7–18A
Laguna Madre, Texas **118**, 268, 281, 291–297, **295**, **296**
La Junta, Colorado **302**, Plate 16–4
Lake Chad 171
Lake Diefenbaker, Saskatchewan, Canada 255
Lake Erie **5**, **83**, 131–132, **169**, 208, Plate 1–9
Lake Huron 132
Lake McKinney, Kansas **303**, **304**, Plate 16–4
Lake Meredith, Colorado **302**, **303**, Plate 16–4
Lake Michigan 141, 156
Lake Mono, California 177
Lake Okeechobee, Florida 125, 241, 281, **281**, 284, 298, Plate 15–13
  see also Florida Everglades
Lake Oro, Saskatchewan, Canada 319, **321**
Lake of the Rivers, Saskatchewan, Canada 319, **321**
Lake Pontchartrain, Louisiana **144**, Plate 8–13
Lake St. Clair 131
Lake Tahoe, California, Nevada **25**
Lake Titicaca, South America 171
lakes, proglacial
  see proglacial lakes
Lakeside, Nebraska **61**, Plate 4–6
Lakin, Kansas 300–301, **303**, **304**, Plate 16–6
landfill, management of 228, **360**
Landsat satellite 40, *40*, 42
Laptev Sea, Russia 354, 355
lapwing 118
Larix 161
Latendresse, John and Chessy 206
Latvia *333*
Laudholm, Massachusetts **325**
Laudholm Trust 325
Laurasia 158, *158*, 368
La Veta, Colorado **301**
lead **8**, *248*
  mining **8**, 184, 246, **246**, 248

leafy vanilla orchid (*Vanilla phaeantha*) 284
least bittern (*Ixobrychus exilis*) **119**, 309
least sandpiper (*Calidris minutilla*) 117
leatherleaf (*Chamaedaphne calyculata*) 331
lechwe (*Kobus leche*) 275
Lena Delta Nature Reserve, Russia 355
Lena River, Russia 336, 354–357, **356**, Plate 17–31
Leopold, Aldo 218
lesser duckweed (*Lemna minor*) 97
lesser snow goose (*Anser caerulescens*) 121–122, **123**
levee 8, 9, 66, 68, **68**, 223, **235**, 242, **256**, 284, 285
Lewis and Clark expedition **218**, 219
lichen **26**
lignite 19, **150**, 159, **160**
lily (*Nuphar* and *Nymphaes* sp.) 22, 283
lithosphere xii 175, *176*, 368
Little Arkansas River, Kansas **68**
Little Ice Age 149, **165**, 170, 171, *171*, 173, **361**, 368
Little River, Massachusetts 325, **325**
Littorina Sea 332, 368
live oak (*Quercus virginiana*) 101
livestock grazing
  see grazing, livestock
lodge, beaver
  see beaver lodge
longear sunfish (*Lepomis megalotis*) **113**, Plate 7–9
Long Term Ecological Research (LTER) 326
lotus (Nelumbonaceae) 159
Louisiana
  hurricanes in 280, *280*
  hypoxic zone 177, **178**
  marshes 195
  wetlands, extent of 285
  see also Mississippi River delta

Loxahatchee NWR, Florida 124, 285
Luck Lake, Saskatchewan, Canada **38**, **39**, 255, 256, Plates 3–9, 3–12

Macedonia 124
macrofossils 25, **26**, 29, 156
magnesium 177, **200**
magnesium carbonate 69
Maine
  case study: coastal wetlands 323–325, **325**
  lobster fishery 228
  tourism 328
  wetland losses 9, *9*
  see also Rachel Carson NWR; Presumpscot River; Wells National Estuarine Research Reserve
Makgadikgadi, Botswana 265, **266**, 275
malaria 11, 110–111, 133
Malaysia
  Badjao tribe 205
  pearl production *205*, 206–207
Maldive Islands **84**, 113, 143
mallard (*Anas platyrhynchos*) **10**, 121, **121**, Plate 7–25
Mandarin duck (*Aix galericulata*) 107, 121, **122**, Plate 17–27
mangal
  see mangrove
manganese 78, 84, 86, 187, 189
mangrove/mangal
  swamps 8, 22, 59, 283, 298
  trees 20, 86, 87, **87**, 88, 91–92, **92**, 159, 195, 195, **197**, **215**, 243, 264, *270*, **272**
  see also Everglades; Sundarbans
Manhattan, Kansas **68**
Manitoba, Canada 154, 163–164, 168
Männikjärve Bog, Estonia 25, 27, 139, 151, **332**, 332, **334**, **335**, Plate 3–10
Mansfield Channel, United States **292**, 293
maple (*Acer* sp.) **27**, Plate 2–25B
marine wetlands 5, **5**, *23*, 29, 58, 63, 81, 108, 150, 155

marsh deer (*Blastocerus dichotomus*) 278
Maryland 9, *9*
Massachusetts
see New England; Rowley River; Little River; Parker River;
Maunder minimum 149, 368
Mayan lowlands 190
Mekong River 1, 195
Middle Ages 144
see also Little Ice Age; medieval climatic optimum; Vikings
Medano Creek, Colorado **348**
Medieval climatic optimum **149**, 166–168, *171*, 368
mercury 128, 196, 245
mere 368
Mérida, Venezuela 338, **339**, 341
Mesa del Caballos, Venezuela 339
Mesopotamian Marshes, Iraq 6–7, 243, **244**, 245
mesotrophic conditions 79, 85, 368
mesquite 349, **350**
metal(s)
heavy 125, 134, 185, 196, 197
toxic 196, 245, 246, 247
meteorological measurement **49**
methane (CH$_4$) 3, 68, 81, 86, 148, 149–150, 151–152, 172, 179, 180, 195, 263, 255
see also greenhouse gases
Métis 200
Mexico 102, 190
amber in 162
mangrove coverage *270*
Sea of Cortez 207
turtle conservation 297
Yucatán Peninsula 279, **280**, Plate 15–10
Mexico, Gulf of 2, 9, 142, **144**, **178**, 197, 268, 279–297, **280**, 298, Plates 8–13, 10–14, 15–10
Miami Canal, United States **282**, 284, Plate 15–13
Miami, Florida **383**, Plate 15–13
Miami limestone (oolite) **140**, 281–282

Michigan, Lake
see Lake Michigan
microfossils 25, 29, 156
microbes
Archaea 179
Desulphovibrio bacteria 178
Gammaproteobacteria 178
GFAJ-1 177–178
Nitrobacter 177
Nitrosomonas 177
Thiobacillus 178
Micronesia 112
microorganisms
see microbes
Middle Loup River, Nebraska **61**, 97
Midwestern arrowhead (*Sagittaria brevirostra*) **94**
migration
birds 4, 11, 193, 309–310, 314–316, 317, 323, 327, 333, 347, 352
fish 278
plants 52, *53*, 102, 166
shore 144, 274, 139, 155
see also coastal erosion
Migratory Bird Hunting and Conservation Stamp (duck stamp) 10, **10**, 14, 253, **253**, 366
Migratory Bird Treaty Act 10, 253
Mikimoto, K. 205
mining
coral **113**
gold 184
lead and zinc **8**
peat **258**, 199, 202–203, **202**
pollution from **8**, 246–248, **246**, Plate 13-9
reclamation **248**
salt **68**
see also coal mining
mink 125, 127–129, **130**
Minnesota 9, 115, 151, 168, 300
mirabilite (Glauber's salt) 320
mire
beaver 343–344, **344**, 346, *347*, **347**
boreal 31
classification 30, *31*
climate influence on 150
cottongrass 345, **346**

definition 5, 15, 25, 368,
drainage 360
Estonian restoration 257–258, 329–333, **332**, **335**, 335
hydrochemistry 69
plant adaption 86, 139
pothole **345**
protection and 257
research **52**
storage capacity 58
Mise, Tatsuhei 205
Mississippi River
deglaciation, effects on 168
eustasy, isostasy in 142–143
fisheries 197
floods (1993) 66–68, **67**, **68**, (2011) 196
restoration projects 233
transportation on 207, **207**
zebra mussels in 132
Mississippi River delta **144**, **178**, 195, 279, **280**
case study 285–291, **286**, **287**, **288**, **289**, **290**, 298
Mississippi River pigtoe clam (*Pleurobema beadleianum*) 206
Missouri 9, *9*, Plate 4-16
see also Mississippi River delta; Missouri River
Missouri Coteau (North America) 299, 318–323, **319**, **320**, **321**, **322**, Plate 16-29
Missouri River **42**, **49**, 66, **67**, 168, **219**, 300
mitigation, wetland 234, **235**, 254, 255, 257, 269, 368, 232, *232*
molluscicides 132
Mobile Bay, Alabama 144, **291**, Plate 8-13
monkey puzzle (*Araucaria araucana*) 162
monsoon 270, 272, 346, 368
Montana **49**, 300
Monte Vista, Colorado 347, **349**
Monte Vista NWR, Colorado 347
moor 15, 368
moose (*Alces alces*) 125, **126**
Moose Jaw, Saskatchewan, Canada **319**, 323

mopane (mopani)
   (*Colophospermum
   mopane*) 275
moraine 181, 226, **337**
   Blue-Bear Lakes 342, 344, **347**
   Culebra Range 342, **345**, 357
   definition 368
   Trinchero Creek Valley **344**
   see also glaciation
moral suasion 227
Morganza, Louisiana 196
morning-glory (*Ipomoea* sp.) 291
Moscow 153–154
mose 15
mosquito fern (*Azolla cristata, A.*
   sp.) 89, 114, 312, Plate
   7-12
mosquitos 109, 110–111, **110**,
   130, 133, 159, 162, 223–
   224, Plate 7-6
moss (*Sphagnum cuspidatum, S.
   rubellum, S. warnstorfii, S.
   fuscum, S.* sp.) **80**, **100**,
   330, Plates 5-12, 6-30
mother-of-pearl 205, **206**, 206,
   Plate 11-11
Mt Maxwell, Colorado **343**
muck 19, 73, 76, 79, *79*, 84, 85,
   368
mud sedge (*Carex limosa*) 343,
   346
mugger crocodile (*Crocodylus
   palustris*) 273
Muir, John 218
multi-level approach (to wetland
   management) 30
Munsell Color 49, **50**, 76, 368,
   Plate 3-28
Murray-Darling Basin, Australia
   9, 263
muskeg 5, 15, 368
musk ox (*Ovibos moschatus*)
   352, **352**
muskrat (*Ondatra zibethicus*) 93,
   94, 97, 105, 125, 127, 128,
   130
mussels 108
   see also zebra mussel

nacre 205, 206, 368
   see also pearl farming
Namibia 265–266, 274, 276, 298
Närke, Sweden **164**

narrow-leaved cattail (*Typha
   angustifolia*) **22**, 86, **87**,
   94, 324
National Federal Conservation
   Act (Germany) 256–257
National Forest Service (U.S.)
   152
National Marine Fisheries Service
   56
National Oceanographic and
   Atmospheric
   Administration (NOAA)
   325
National Park Service (NPS)
   213
National Strategy on Biological
   Diversity (Germany)
   257
National Water Initiative,
   Australia 263
National Wildlife Refuge (NWR)
   system 10, 219, 253, 295
natural gas 181, 262
Natural Resources Conservation
   Service (NRCS) 11, 56,
   228, 233, 238
Nature Conservancy, The (U.S.)
   10–11, 107, 222, 233, 310
   see also Cheyenne Bottoms
Nature Conservancy of Canada
   323
navigation, effect on wetlands
   195, 198, 208, 223, 224,
   253, **253**
NAWMP
   see North American Waterfowl
   Management Plan
Nebraska Sand Hills **61**, **130**,
   170, 299, 300, 314–318,
   314, **317**, **317**, Plates 4-6,
   16-24
Nene goose (*Branta
   sandvicensis*) 122
Neoglaciation (climate phase)
   164, 166, 168, 171, *171*,
   173, 368
Neolithic 8, 166, **166**, **188**
Neosho Wildlife Area, Kansas
   130, **225**
Ness, Outer Hebrides 202
Netherlands 242
Nevada **25**, **87**, **89**, **95**, 308
New Brunswick, Canada 242

New England
   case study, coastal wetlands
      323–328, **325**, **327**, 333
   wetland vegetation 103, 191
   see also Maine;
      Massachusetts; New
      Hampshire; Plum Island
      Ecosystem; Vermont
New Hampshire 9, *9*
New Jersey *9*, 161
New Mexico 299, 300, **310**, 341,
   347
New Orleans, Louisiana **144**,
   178, **286**, Plates 8-13, 10-4
New South Wales, Australia 262
New York 9, 208, **208**, Plate
   11-13
New York State Barge Canal,
   United States **208**, 208
New Zealand
   amphibian extinction in 115
   Resource Management Act
      (1991) 263
   rope rush in 69
   Snares Island group **124**
   Subantarctic World Heritage
      Site **124**
   wetland losses 9, *9*
   wetlands governance 263
Nigeria 200, *270*
Niger River/Delta 1, 252
Nigula Bog, Estonia **17**, **27**, **32**,
   35, Plate 2-2A
Nile
   crocodile
      see African crocodile
   delta **3**, 193
   River 2, **3**, 193
Niobrara River, Nebraska **25**
Nishikawa, Tokichi 205
nitrate(s) (NO_3^-) 177, 196
   see also nitrogen
Nitrite (NO_2^-) 177
   see also nitrogen
Nitrobacter 177
nitrogen (N) xii, xiii, 175, 177,
   194
nitrogen cycle *xiii*, 186, 187, 189,
   194
nitrous oxide (NO_x) 177, **177**,
   187
   see also greenhouse gases
Nitrosomonas 177

No Net Loss policy 234
Norfolk Island pine (*Araucaria heterophylla*) 162
normalized difference vegetation index (NDVI) 36
North American Breeding Bird Survey (BBS) 54
North American Waterfowl Management Plan (NAWMP) 228, 321–333, **333**
North Carolina 144, **304**
North Dakota 318
see also Devils Lake
North Sea coast **46**, 141, 144, **145**, **163**, Plates 8-14, 9-9
North Slope, Alaska 336, 351, **351**
northern harrier (*Circus cyaneus*) 124
Northern Territories, Australia 262
northern water milfoil (*Myriophyllum sibiricum*) 96, **98**
northern white-cedar (*Thuja occidentalis*) 86
Norway
  Bolstadfjord **142**
  crustal uplift *143*
  Eidslandet **143**
  fjords 141–143
  Jostedalsbreen **361**
  Skansen Bog, Askøy Island **51**
  Nova Scotia, Canada 94, 168
nutria (*Myocastor coypus*) 131–132
Nymphaeaceae 95, **97**, **98**, 159

oak (*Quercus* sp.) 101, *101*, *102*
oasis, oases
obligate vegetation 100, *100*, **101**, 106, 159, 173, 368
see also wetland vegetation
Ohio
  River 168, 285
  wetland losses 9, *9*
oil-palm industry 203, 214
Okavango River/Delta 1, **2**, 11, **12**, **21**, **108**, 128, 191, 193, 210, **211**, 252, 264–265, **265**

Okefenokee Swamp, Georgia **90**, **97**, 113
Oklahoma **8**, **70**, 246–247, **246**, 247, 248, **249**, 300, Plates 13-9, 13-12
Old Wives Lake, Saskatchewan, Canada **99**, **320**, 320, **322**, **323**, Plates 6-29, 16-29, 16-33
Oldman River, Alberta, Canada **62**
oligotrophic wetlands 27, 79, **80**, 89, **180**, 368, Plates 5-12, 10-7
ombrotrophic wetlands 26, 29, 157, 330, 369
Ontario, Canada 168, **208**, Plate 11-13
organophosphate insecticide 111
osprey (*Pandion haliaetus*) 124
otter 125, 128, 134, 278
oxbow 369
ozone depletion *xiii*

Pacific Grove, California **82**
Padre Island National Seashore **2**, 291, **293**, **297**
Padre Island, Texas 1, **2**, 36, **119**, 279, 281
  case study 291–297, **292**, **293**, **294**, **296**, **297**
painted turtle (*Chrysemys picta*) 115–116, **116**, Plate 7-15
pakihi 369
Palau islands 113
paleosols 20, **75**, 369, Plate 5-3
Paleozoic coal 157–158
  see also coal
Palliser Triangle, Canada 320
palma pita (*Yucca treculeana*) **296**
palsa mires 15, 158, 369
paludification 136, 138, 155, 369
palustrine wetlands *23*, 24, 25, 29, 369
  see also bog; mire; peatland; playa; mudflat; marsh; salt pan
palynology 51, 369
Pamir Mountains 198
Panama, Farallon **323**
Pangaea 156, 157, *158*, 159, 173, 369

panne (pan) 324, **324**, **328**, 369
Pantanal caiman (*Caiman yacare*) 278
Pantanal jaguar (*Panthera onca palustris*) 278
Pantanal Regional Environmental Program 202
Pantanal wetlands, South America 63, 191, 210, 226, 252, 264, 265, 268, 276–280, **276**, **278**, 298, Plate 15-9
pântano 15, 369
  see also Pantanal wetlands
papyrus (*Cyperus papyrus*) 89, 275
Paraguay
  see Pantanal wetlands; Paraguay River
Paraguay River 63, 207, 265, 266, 277, 279, 298
páramo 3, 338, 339–340, **340**, 357, 369, Plate 17-4
Páramo de Piedras Blancas, Venezuela **340**
Parker River, Massachusetts 327, **327**, Plate 16-40
Parque Nacional Sierra Nevada, Venezuela 339, **339**
Pass A Loutre Wildlife Management Area 286
Pass á Loutre, Mississippi 285, 286, 287, 289
passage grave **166**
peanut 73
pearl farming 2, 191, 199, 204–207, **206**, *205*, 217, Plate 11-11
pearl mussel (*Margaritifera margaritifera*) 164, 204–205
  see also fresh-water pearl
pearl production
  see pearl farming
Pearl River, Mississippi **8**
peat moss (*Sphagnum* sp.) 92
peat **3**, **6**, 19, **20**
  mining **258**, 199, 202–203, **202**
  see also peatland
peatland 24–27, **26**
Peipsi, Lake
  see Peipsi–Pihkva

Peipsi-Pihkva lakes 328, **329**, 330, **330**
pelican 117, 122, 123–124, **124**, 133, 291
Pelican Island NWR, Florida 219
penguin **124**, 134, 171
pepperwort (*Marsilea vestita*) 99
Permanent Water Commission on the Okavango (OKACOM) 265
Permian 158, 173
pesticides
    pollution from 125, 134, 279
    taxes 228
Peter the Great 220
Philippines
    pearl production in 205, *205*
    Mt Pinatubo **148**
phosphorus 175, 177, 178, 186, 187, 189, 194, 225, 245
phosphorus cycle xii, *xiii*
phreatophytes 307, 369
Picher, Oklahoma **8**, **246**, 247, **247**, Plate 13-9
pickerel weed (*Pontederia cordata*) **22**, 94, **96**, Plate 6-21
Pico Águila, Venezuela **339**, **340**
Pico Bolívar, Venezuela 338
Pihkva
    see Peipsi-Pihkva lakes
Piirissaar (Estonia) 330, **330**
Pinchot, Gifford 219, 220
pine (*Pinus* sp.) 99, *102*, 161, 184, 330
Pinedale Glaciation 342
pink smartweed 349
piracema fish migration 278, 279
pirogue 17, **219**
pitcher plant **90**, 163
Pityoxylon 161
plains gartersnake (*Thamnophis radix*) **114**, Plate 7-13
plains zebra
    see zebra
plant hardiness zones 101–102, *102*, 106
    Culebra range 344
    definition 369
    Gulf of Mexico 279
    San Luis Valley 345
    USA **103**

plants, wetland
    see vegetation, wetland
Platte River 285
playas 15, 20, 21, 24,46, **47**, **229**, 300, 301, 333, 369
Pleistocene wetlands 163–174, 369
    glaciation and 181, 189, 329, 331, 340, 342, **342**
plover 118, 309
Plum Island Ecosystem (PIE) 326–328
Plum Island, Massachusetts **191**, 192, 335
pneumatophores 86, **87**, 92, 105, 272, **272**, 369
pocosin (Carolina Bay) 15, 301, 304, **304**, 369
Point Piedras, California **44**, 127
polar bear (*Ursus maritimus*) **127**, 352
pollen
    analysis 51, 164, 270
    diagrams **153**, **170**
    microfossils 25
    see also palynology
pollution
    abatement 225, 227–228, 361–362
    in amphibious extinctions 115
    atmospheric 184–185, 187, 257, 258
    chemical 12, *13*, 188, 204
    control (ecosystem service) 6, *191*
    see also fertilizer; heavy metals; herbicides; mining
Põltsamaa River (Estonia) 332, **334**
Polynesia 112
Pompano Beach, Florida **282**, Plate 15-13
pondweed (*Potamogeton* sp.) 95, **97**
pond skater
    see water strider
Port Bienville, Mississippi **8**
Port Isabella, Texas 291
Post, Lennart von 164
postage stamp(s) **5**
potassium 69, 80, **84**, 175, 178, 187, 189
potassium chloride 365

potassium-halide salts 305
pothole mire **345**
Prairie Pothole region, North America 9, **41**, 120, 255, 318, 323, 361
Presque Isle, Pennsylvania **5**, **83**, Plate 1-9
Presumpscot River, Maine 241–242
proglacial lakes 167–168, **169**, 173, 319–320, **320**
prop (stilt) roots 86, **87**, 92, 195, **195**, 272, **283**, 369
prop (stilt) roots 86, **87**, 92, 195, **195**, 272, **283**, 370
Prudhoe Bay, Alaska 352
Pueblo, Colorado **302**, **308**
puna peatland 370
pyrethroids 111
pyrites 178–179, **178**

Queen Charlotte Sound, Canada 141
Queensland, Australia 262
    see also Cairns
Quelccaya ice cap, Peru 171
Quivira NWR, Kansas **115**, **119**, **120**, 211, **224**, 233, 235, 312

raba 15
rabbitbrush **308**
Rachel Carson NWR, Maine xii, **78**, 325, **359**, Plate 1
radiolarians 82
raised bog
    see bog, raised
Ramsar Convention 10, 11, 14, 15, 192, 251, 370
    Site Information Service (RSIS) 11, 192
    wetland definition 5, 15, 28, 56
raptors 122–123, 124, 125, 133
rattlesnake 115
Raupo swamp 370
red admiral butterfly (*Vanessa atalanta*) **92**, Plate 6-14
red beds 20, **20**, 156, Plate 2-10
red-eared slider (*Trachemys scripta*) **115**, 116
redhead duck (*Aythya americana*) 121

Red Hills, Kansas **39**, Plate 3-11
red mangrove (*Rhizophora mangle, R. apiculata*) 86, **87**, 92, **195**, **283**
red maple (*Acer rubrum*) 101
Red River 151, 195, 300
red samphire (saltwort) (*Salacornia rubra*) **38**, 88, **89**, Plates 3-9, 6-8
Red Sea 112
redoximorphic
  definition 370
  features 73, 78, 82
  iron accumulation **80**, Plate 5-11
red-shouldered hawk (*Buteo lineatus*) 124
reedmace swamp 370
reedswamp 370
Regina, Saskatchewan, Canada **88**, **97**, **100**, **321**
reindeer
  see caribou
remote sensing (in wetland research) 31–45, 54, 57, 363
  see also aerial photography; satellite imagery
replacement cost 215, 216
  see also cost–benefit analysis
resin (in amber formation) 157, 160–161, 173
restoration 231–249
  definition 233
  ecology 232
Rex, North Carolina **304**
rhea (*Rhea americana*) 278–279
Rhine river **3**
rhinoceros, white
  see white rhino
rice cultivation 20, 73, **76**, 111, 172, 202, 204, 272
Riehl, Wilhelm Heinrich 219
rift system, East Africa
  see East African rift system
Riga, Gulf of 328
Rio Grande 307, 341, 345, 346–347, **349**
Rio Grande Canal, United States **349**
Ristinge Klint (Denmark) 71, Plate 4-23

riverine wetlands 23, 24, **25**, 29, 193
Rock Island, Illinois **207**
rock ptarmigan 352
Rocky Ford, Colorado **302**, Plate 16-4
Roosevelt, Theodore 219
roseate spoonbill (*Ajaia ajaja*) 284
Roskilde (Denmark) 167, **168**
rotifers **61**, 317, Plate 4-6
Rowley River, Massachusetts **17**, 60, 328, **328**, Plate 16-41A
Royal Gorge, Colorado **301**
royal palm (*Roystonea regia*) 284
Rudorff, Ernst 219–220
Russell Lakes State Wildlife Area, Colorado **19**, **45**, **63**, 348, **350**, Plate 17-26
Russia(n)
  Baikal Highlands 354
  Karelia, Russia **80**, **263**, Plate 5-12
  Kamchatka Peninsula, Russia **148**
  Mt Kliuchevskoi, Russia **148**
  Laptev Sea, Russia 354, 355
  Lena Delta Nature Reserve, Russia 355
  Lena River, Russia 336, 354–357, **356**, Plate 17-31
  Moscow 153–154
  Siberia 196, 261–262, **356**, Plate 17-31
  Tiksi, Russia 355
  Tula, Russia **263**
  Ural Mountains, Russia **158**
  Ust-Lensky Zapovednik, Russia 355
  Volga River, Russia 252, **252**, Plate 14-1
  Yakutsk, Sakha Republic, Russia 354
  Yantarny, Kaliningrad, Russia 163
  Yenisei River, Russia 261

Saami people 352
sabkha 15, 22, 72, 370
saddle-billed stork (*Ephippiorhynchus senegalensis*) **267**, Plate 14-20

Sadlerochit Mountains, Alaska **351**, Plate 17-28
sagebrush **342**
Sahara desert 6, 17
Saint Louis, Missouri 66
salada 305, 370
salamander 114, 158
  see also amphibians
salar 370
salina 305, 370
salmon
  see Atlantic salmon; chinook salmon; chum salmon; coho salmon
salt marsh **xii**, **17**, **23**, 59, 223, 335
  see also Everglades; Plum Island Ecosystem; Quivira NWR; Rachel Carson National Wildlife Refuge; Rowley River; Wells Reserve
salt production **68**, 199, **200**, **266**, 317–318, 320
salt sacaton grass **308**
saltcedar (*Tamarix* sp.) 88, **89**, **303**
  biocontrol of, upper Arkansas River valley, USA 305–309, **307**, **308**
Saltcedar Biological Control Consortium (SBCC) 307, 309
saltcedar leaf beetle (*Diorhabda "elongata," D. carinulata*) 307–309, **308**
saltgrass (*Distichlis spicata*) 324
saltmeadow cordgrass
  see cordgrass
saltmeadow rush (*Juncus gerardii*) 324
saltwort
  see red samphire
San Isabel National Forest, Colorado **129**
San Juan Mountains, Colorado **341**, **349**
San Luis, Colorado 213, **214**
San Luis Lake, Colorado 347, **348**, Plate 17-20

San Luis Valley, Colorado **19**, 336, 341, **342**
  case study 345–350, **348**, 349, 357, Plate 17-20
  see also Russell Lakes State Wildlife Area
sand seas 316, 370
sanderling (*Calidris alba*) 323, **323**
sandhill crane 347
sandpiper 117, 118
sandstone **20**, 81, 341, 342, **342**, Plate 2-10
Sangamon 140–141, **141**, 370
  see also Eemian
Sangre de Cristo Mountains (Colorado) **19**, **20**, **45**, **61**, **82**, **91**, **214**, **227**, **301**, 341, **341**, **348**, **349**, **350**
  see also Cucharas, Cuchara
Sarus crane (*Grus antigone*) 211
Saskatchewan Watershed Authority, Canada 323
satellite imagery
  Advanced Very High Resolution Radiometer (AVHRR) 47
  Earth Resources Technology Satellite 32
  see also Landsat
  Enhanced Thematic Mapper Plus (ETM+) *40*
  Ikonos 42, **42**, **234**
  Moderate Resolution Imaging Spectroradiometer (MODIS) 37, 40, 243, 277, **278**, Plate 15-9
  Multi-angle Imaging SpectroRadiometer (MISR) **329**
  Multi-spectral Scanner (MSS) 40, **338**
  Landsat 40, *40*, 42
  Landsat Data Continuity (LDC) Mission
  SeaWIFS **280**, Plate 15-10
  Thematic Mapper (TM) 40, *40*
Saville-Kent, William 205
scarlet ibis (*Eudocimus ruber*) **17**, **120**
Scheldt River, Netherlands 242
Scolopacidae 118
Scotland 184, *202*, 204, *210*

Scots pine (*Pinus sylvestris*) 184, 330
sea level change 136, 139–147, **140**, **141**, **142**, **143**, **145**, **146**, **147**, 155, 156, 163, 173, 286–287
  see also eustasy
sea mink (*Neovison macrodon*) 128–129
sea oats (*Uniola paniculata*) 292, 294
sea purslane (*Sesuvium portulacastrum*) 291
sedge (*Carex aquatilis*) 343, **345**
sedge meadow 244, 370
sediment
  bioclastic 81–82, **83**, 85
  burial 135, 173, 157, 159
  carbonate **19**, **83**, 140, 150, **150**, 179, 279, 282, Plate 2-5A
  clastic 81, 85, 279
  gravel 81, 144, 195
  grus 73
  loess 72, 301, Plate 5-3
  rock flour 17, **19**
  sand 59, **74**, 81, 83
  silt 59, 60, **74**, 81, 237
  till **6**, 72, **75**
  turbidity 17, 35, 60, 88
Senegal 92
serpulid worm 293–294
shadow price 215
  see also cost–benefit analysis
Shark Bay, Australia 156, **157**
Shark River, Florida 281
shoalgrass (*Halodule wrightii*) 293
shorebirds 4, 11, 117–118, **119**, 133, 370
shrimp 108, **201**, 204
shrimp aquaculture 204, 264, 270, 293
Siberia 196, 261–262, **356**, Plate 17-31
Siberian crane (*Grus leucogeranus*) 193
Siberian lemming (*Lemmus sibiricus*) 355
Sieperda Polder, Netherlands 242
Sierra de Perijá, Colombia **338**
Sierra Nevada de Santa Marta, Colombia **338**

Sierra Nevada Mountains **25**
Skagen Peninsula (Denmark) 144, **146**
Skaggerak Sea **146**
Skansen Bog, Askøy Island (Norway) 51
skunk-cabbage (*Symplocarpus foetidus*) **101**, *101*
slash pine (*Pinus elliottii*) 99
slate 81
slimstem reedgrass (*Calamagrostis stricta*) 343, 344
slough 15, 24, 370
Slovakia 257
  Hornád River **225**
  Lomnicky (mountain) **261**, **337**
  Stará Lesná **261**
  Strané pod Tatrami **337**
  Tatra Mountains 259, **261**, **337**
small-format aerial photography (SFAP) 42, **43**, 44
smooth cordgrass
  see cordgrass
snail 108, 125, 282, 293
snail kite (*Rostrhamus sociabilis*) 125, **125**, 133, 284
snake (Squamata) 116
Snares crested penguin (*Eudyptes robustus*) **124**
snipe 118
snowy egret (*Egretta thula*) **119**
snowy owl (*Nyctea scandiaca*) 353, 355
Society for Ecological Restoration International 231
Society of Wetland Scientists (SWS) 12
soda ash
  see sodium carbonate
sodium 69
  carbonate **266**
  chloride **84**
  sulfate **84**, 320
sodium-halide salts 305
soils, wetland
  characteristics 18–20
  classification 73, **74**, **75**, 76–78, 77, 79, 84
  acrotelm 180, 364

soils, wetland (*cont'd*)
  Aqualf 76, 84; Aquents 76;
    Aquepts 76; Aquolls 76,
    84; Aquox 76; Aquods,
    Aqualts
  hydric 19, 28, 72–73, 76–85,
    78, 79, 367, Plate 5-11
  see also acrotelm; catotelm
soil temperature logger **12**
Sonneratia 272
South Asia 111, 198
  see also India; Sundarbans;
    Thailand
South Carolina 129
South Dakota 168, 314
South Florida Water Management
    District 285
South Padre Island, Texas **36**,
    279, **296**
South Saskatchewan River,
    Canada 255
southwestern willow flycatcher
    (*Empidonax traillii
    extimus*) 309
Soviet Union 198, 257, 262
  see also Commonwealth of
    Independent States
soybean cultivation 73, 279, 287
Spanish dagger **296**
Spanish moss **283**
Spartina alterniflora, S. patens
    191, **191**, 324, **324**, **325**,
    326
spatterdock *(Nuphar luteum)* **22**
spike rush (*Eleocharis* sp.) 93,
    93, **235**
spiny softshell turtle (*Apalone
    spiniferus*)**115**
spoonbill 278, 284
spreading rope rush (*Empodisma
    minus*) 69
spruce (*Picea* sp.) *101*, 163, *170*,
    263, 343, **345**, **346**
Squaw Creek NWR, Missouri **94**,
    119
St. Lawrence, Canada 168
Stafford Act 281
stem hypertrophy (trunk
    buttress) 87
Steverson Lake, Nebraska **318**
Stevns Klint (Denmark) **147**,
    159
Stillwater NWR, Nevada **87**, **95**

stilt roots
  see prop roots
Stone Age
  amber jewelry 160, 173
  peat fossils **26**
storage capacity, wetland 58, 70,
    197–198
stork 267, 278, 284, Plate 14-20
Sub-Atlantic period *54*, 166
Sub-boreal period *54*, 166
succinite 162
  see also amber
Sudan 193
Sugar City, Colorado **303**
sugarcane 73
Sumerians 190
Sun Chariot (Solvognen) **167**
Sundarban Biosphere Reserve
  see Sundarbans
Sundarbans 4, 8, 195, 268, 269–
    274, *269*, *270*, **271**, *271*,
    272, 297, Plate 15-2
Sundari (*Heritiera fomes*) 270
suo 15
Suo journal 12
swamp cypress (*Taxodium
    distichum*) **87**, 101
Swamp Land Acts (U.S.) 253
swan 117, 121, 122, **123**
sweetgale (*Myrica gale*) 331
SWS
  see Society of Wetland
    Scientists
Syr Darya river 198, **199**, Plate
    11-6B
Syria 243

Tajikistan 198
takir 370
tamarack 163
Tanana River, Alaska 253
Tar Creek, Oklahoma **246**, Plate
    13-9
Tarli (*Sardinella melanur*) **201**
Tartu (Estonia) **137**, **140**, **202**,
    258, **259**, 334, Plate 8-1
Tatra Mountains 259, **261**, **337**
taxes (to encourage wetland
    preservation) 228–229
tea 204
teal 117, 121, **121**, Plate 7-25
Tectonic
Tennessee 206, 285

Tennessee River 285
  see also Mississippi Delta
Texas 279, 299, 333, Plate 15-10
  during Holocene thermal
    maximum 168
  hurricanes in 280, **280**, *280*
  plant hardiness 102
  playas in 300
  Rio Grande water compact
    347
  saltcedar biocontrol 307
  see also Laguna Atascosa
    NWR; Laguna Madre;
    Padre Island
Texas gopher tortoise (*Gopherus
    berlandieri*) **296**
Texas prickly pear (*Opuntia
    engelmannii*) **296**
terrestrialization 136, 155, 331,
    335, 370
Thailand
  aquaculture production 204
  Bight of Bangkok **76**
  mangrove losses 264
  pearl production *205*
  Songkhram River 202
Thalassia **83**
Thematic Mapper 40, **40**, 355
Thoreau, Henry David 218
Thiobacillus 178
Tibet Plateau 336
Tien Shan Mountains 198
Tierra del Fuego, Argentina 12,
    336
Tigris River 233, 243, **244**
Tigris–Euphrates valley 233, 243,
    244
Tiksi, Russia 355
tiling 223, *223*, 225
Tishomingo, Oklahoma **93**
toads 114
  see also amphibians
Tolland Man **167**
Tooma Experimental Station,
    Estonia 334
tourbière 15
transient electromagnetic (TEM)
    method 52
travertine 69
treatment wetland 234, 244–246,
    250, 370
tree snail (*Liguus fasciatus*)
    282

Treece, Kansas 247, **247**, **248**
Trinchera Peak, Colorado **342**, **343**, **344**
Trinchero Creek, Colorado **61**, **344**
Triploid white amur 105
tropical wetlands
  see wetlands, tropical
Trundholm Mose, Sjælland (Denmark) **167**
trunk buttress
  see stem hypertrophy
tsunamis 113
tufa 69, **70**
Tufted bulrush (*Trichophorum cespitosum*) 331
Tula, Russia **263**
Tule Lake, California 117
tundra 165, 191, 263, 336, 341, 351, 354–355, 357, 370
Turkey 243
Turkmenistan 198
turlough 371
Turner Falls, Oklahoma 22, **70**, **99**
Turtle (Testudines) 115–116
Tuttle Creek Lake, Kansas **68**

Uganda 193
Ukraine 162–163, 260
uniformitarianism 156, 371
Union Canal, Scotland **210**
United Kingdom
  Countryside Stewardship Scheme 229
  Water Act 257
  Wildlife and Countryside Act 257
  Wildflower & Wetlands Trust, UK 257
United Nations
  Millennium Ecosystem Assessment 358–359
  UNESCO Biosphere Reserve 266
  UNESCO World Heritage Site 273
United States Fish and Wildlife Service
  data dissemination 51
  duck stamp **253**
  management of Arctic wetlands 352

  role in wetland management 11
  waterfowl management 228
  wetland classification 22, *23*, 24
  wetland definition 15–18
  wetland plant categories 99, *100*
  wetland restoration definition 233
wet soil classification 76–77
Ural Mountains, Russia **158**
Uruguay 279
U.S. Army Corps of Engineers see Army Corps of Engineers
U.S. Bureau of Land Management (BLM) 56, 213, 352, 354
U.S. Department of Agriculture 56, 245
  plant hardiness zones *102*
U.S. Environmental Protection Agency (EPA) 11, 16, 56, 222, 233, 238, 245–246, 254–255
U.S. Fish and Wildlife Service see United States Fish and Wildlife Service
U.S. National Flood Insurance Program 291
U.S. Geological Survey 64
Ust-Lensky Zapovednik, Russia 355
Utah 308, *18*
Uzbekistan 198

Väinameri Sea (Estonia) **36**, **331**
Valentine National Wildlife Refuge (NWR), Nebraska **170**
Valgesoo bog (Estonia) **138**, **259**
Vancouver Island, Canada 141
vàrzea 200, 371
vegetation index 36, 371
vegetation, wetland 20–22, 86–106
  emergent plants **11**, 15–16, 20, 21, **21**, 29, 36, 87, **91**, **93**, **94**, **95**, 97, 98, 106, 364, 371, Plates 2-12, 17-26
  floating plants 20–21, **21**, **91**, 94–96, **95**, **97**, 106
  hardiness see plant hardiness zones

  shoreline plants 20, **21**, 29, **82**, 90–92, **91**, 97, 105–106
  submerged plants 21, **22**, 29, 87–88, **88**, **91**, 96–97, **98**, **99**, 196–197
  see also halophytes; hydrophytes
  vegetation zones 21, **22**, **23**, 27, 29, **38**, 90, **91**, 97–99, **99**, **100**, 105–106
Venezuela
  Cordillera de Mérida 336, **338**
  Mérida 338, **339**, **341**
  Mesa del Caballos 339
  Páramo de Piedras Blancas **340**
  Parque Nacional Sierra Nevada 339, **339**
  Pico Águila **339**, **340**
  Pico Bolívar 338
  Río Chama 338, 339
  see also Andes Mountains, páramo
Vensyssel, Jylland, Denmark **145**, Plate 8-14
Vermont 9, *9*
vernal pond 371
Victoria, Australia 262, *9*
Viking Ship Museum, Roskilde, Denmark **168**
Vikings 166–167, **149**, **168**
Vilayati babul (*Prosopis juliflora*) 213
Virginia 304
vlei 371
vloer 371
Volga River, Russia 252, **252**, Plate 14-1

wad 15, 371
Wager Bay, Canada 127
Wascana Creek, Regina, Canada **88**, **97**, **100**
Walnut Creek, Kansas **311**
water cycle **59**
  see also hydrologic cycle
water diversion
  in Aral Sea 198, *199*, **199**, 361, Plate 11-6B
  High Plains, USA **4**
  Mesopotamian Marshes **245**
  Ganges 273

watercress (*Rorippa nasturium-aquaticum*) **21**, Plate 2-12
water-hemlock (*Cicuta maculata*) 86
waterlily
  American white 95, **97**
  giant 95, **98**
  see also Nymphaeaceae
water smartweed (*Polygonum amphibium*) 99
water strider (Gerridae, *Gerris remigis*) 109, *109*
water table
  definition 371
  effect on carbon cycle 179, 180
  effect of ditching on 223–224
  effect of human extraction on 63, **64**
  effect on wetland storage capacity 197–198
  in hydroseral succession 136, 138
  raising, in wetland restoration 241, 242
  in wetland definition 16
water willow (*Justicia americana*) 93, **93**
waterfowl 117–125, **118**, **119**, **120**, **121**, *121*, **122**, **123**, **124**
  definition of 371
  impact of wetland conversion on 75
  importance of wetlands to 4, 11, 212
  incentives for habitat improvement 228
  preservation of habitat 253
  wetland mitigation and **235**
  see also "duck stamps"; North American Waterfowls Management Plan; Ramsar Convention;
Wegener, Alfred 157
Weir-Pittsburg coal bed, Kansas **158**
Welland Canal, Canada **208**, Plate 11-13
Wells National Estuarine Research Reserve (NERR), Maine 325, **325**, **326**, 335

Wembury Bay, Plymouth, England 141
West Antarctic Ice Sheet 140
West Indian mahogany (*Swietenia mahagoni*) 284
West Mineral, Kansas **183**
West Nile virus 110, **110**, **301**, **317**, Plate 7-6
West Virginia 9, *9*
western massasauga (*Sistrurus catenatus tergeminus*) 116
wet meadow
  Arkansas River valley 300, **302**
  Culebra Creek valley **214**
  Culebra Range 342, **344**
  definition 15–16, 371
  hay production 200
  Missouri Coteau 318, 319
  Nebraska Sand Hills 315, **316–317**
  Okavango **266**, Plate 14-20
Wet Mountains, Colorado **341**
wet prairie 371
wetland ecosystems, human use of 190
  see also ecosystem services
wetland loss 8–9, *9*
  Australia *9*
  Canada *9*
  China *9*
  Europe *9*
  Mississippi Delta 288, **288**
  Philippine mangrove swamps *9*
  USA, northeastern states *9*
wetland mitigation *see* mitigation, wetland
WetlandCare Australia 11
wetlands
  classification 22–24, *23*, **24**
  definition 15
  estuarine *see* estuarine wetlands
  lacustrine *see* lacustrine wetlands
  marine *see* marine wetlands
  palustrine *see* palustrine wetlands
  riverine *see* riverine wetlands

tropical 6, 9, 13, 17, 59, **80**, **98**, 171, 173, 191, 297
  see also coral reefs; mangroves; Mexico, Gulf of; Okavango; Pantanal; Sundarbans
Wetlands and Wildlife National Scenic Byway, Kansas **211**, 312
Wetlands International 11
Wetlands Reserve Program (WRP) 228, 254, 371
wheat **4**, **47**
whirligig beetles (Gyrinidae) 1–9
white ibis (*Eudocimus albus*) 284
white-faced ibis (*Plegadis chihi*) 347
white-fronted goose (*Anser albifrons*) 121–122
white mangrove (*Laguncularia racemosa*) 92
white rhino (*Ceratotherium simum simum*) 210
whooping crane (*Grus americana*) 4, **5**, 233, 309, **349**
Wildflower & Wetlands Trust, UK 257
Wildfowl & Wetland Trust (UK) 11
wild goose (*Anser anser*) **259**
willow (*Salix* sp.) **88**, *101*, 307, 308, 309, 343
Willow Bunch Lake, Saskatchewan, Canada 323
Wilson's phalarope 347
wind power 187
wind farms **188**, **229**
Wisconsin Glaciation **169**, **181**
wood duck (*Aix sponsa*) 121
wood stork (*Mycteria americana*) 284
worms 108
wrack 195
Wrangell-St. Elias National Park, Alaska 354
Wyoming 161, 182

Yakutsk, Sakha Republic, Russia 354

Yantarny, Kaliningrad, Russia 163
yellow water-crowfoot (*ranunculus flabellaris*) 98–99
Younger Dryas glaciation phase **143**, *171*
Yucatán Peninsula 279, **280**, Plate 15-10

Yukon Delta NWR, Alaska **353**, 354, Plate 17-30
Yukon River/Delta 353–354, **353**, 357, Plate 17-30
Yup'ik 352, 354

zakaznik 262
zapovednik 262
zebra (*Equus burchelli*) **128**

zebra longwing (*Heliconius charitonius*) 284
zebra mussel (*Dreissena polymorpha*) 132, **132**, 133, 208
Zimbabwe 200, 264
zinc 196
  mining **8**, 184, 246, **246**, 248

www.ingramcontent.com/pod-product-compliance
Lightning Source LLC
Jackson TN
JSHW062026230125
77636JS00003B/31